HARDY HEATHERS

FROM THE NORTHERN HEMISPHERE

They call my plant the Wandering Heath:
It wanders only in the West:
So flower the purple thoughts beneath
The sailor's, miner's, mother's breast.
O hearts of exile! still at home,
And ever turning while ye roam!

Q, *Wandering heath* (1895).

F. W. Burbidge's St Dabeoc's heath (*Daboecia cantabrica*)

There are divers sort of Heaths, some greater, some lesser,
some with broad leaves, and some narrow:
some bringing forth berries, and others nothing but floures.

John Gerard, *The Herball or generall Historie of Plantes* (1597).

Title page: John Ruskin's cross-leaved heath (*Erica tetralix*). "… I meant to have drawn every English and Scottish wild flower, like this cluster of bog heather … — back, and profile, and front." As a footnote, opposite the engraved drawing, Ruskin (1875: 11) wrote: "Admirably engraved by Mr. Burgess, from my pen drawing, now at Oxford. By comparing it with the plate of the same flower in Sowerby's work, the student will at once see the difference between attentive drawing, which gives the evidence of the cadence and relation of masses in a group, and the mere copying of each flower in an unconsidered huddle."

A BOTANICAL MAGAZINE MONOGRAPH

HARDY HEATHERS

FROM THE NORTHERN HEMISPHERE

CALLUNA • DABOECIA • ERICA

E. Charles Nelson

with watercolours by
Christabel King, Wendy Walsh,
Deborah Lambkin and Brita Johansson

and line drawings by
Joanna Langhorne and Stella Ross-Craig

Kew Publishing
Royal Botanic Gardens, Kew

First published in 2011 by
Royal Botanic Gardens, Kew,
Richmond, Surrey, TW9 3AB, UK
www.kew.org

ISBN 978-1-84246-170-9

British Library Cataloguing in Publication Data
A catalogue record for this book is available from the British Library

Production editor: Ruth Linklater
Design, typesetting and page layout: Christine Beard
New maps prepared by John Stone
Publishing, Design & Photography, Royal Botanic Gardens, Kew

Cover illustrations. Front, left to right: *Erica erigena* 'Irish Dusk', *E.* × *darleyensis* 'George Rendall', *Daboecia cantabrica*,
D. cantabrica f. *alba*, *D. cantabrica* 'Charles Nelson', *D. cantabrica* 'Atropurpurea', *D. cantabrica* 'Praegerae' and *D.* × *scotica*
'William Buchanan Gold'; back, left to right: *Calluna vulgaris* 'Beoley Crimson', *Erica* × *watsonii* 'Cherry Turpin',
E. carnea 'Springwood White'; front flap: *E. mackayana* 'Plena'. Painted by Christabel King.

Printed in Spain by Grafos S.A.

For information or to purchase all Kew titles please visit
www.kewbooks.com or email publishing@kew.org

Kew's mission is to inspire and deliver science-based plant conservation worldwide, enhancing the quality of life.

Kew receives half of its running costs from Government through the Department for Environment, Food and
Rural Affairs (Defra). All other funding needed to support Kew's vital work comes from members, foundations,
donors and commercial activities including book sales.

CONTENTS

LIST OF PAINTINGS

Hand-coloured, engraved cartouche from the ornamental title-page of *Centuria quarta* of *Flora graeca sibthorpiana*; the two heathers depicted are *Erica arborea* (white) and *E. manipuliflora*. The landscape that was depicted below the cartouche showed a view of Byzantium (modern Istanbul).

The other plants are, (left to right) *Frankenia pulverulenta* L., *Fritillaria sibthorpiana* (J. E. Sm.) Baker, *Tulipa gesneriana* L., *Daphne sericea* Vahl, *Ornithogalum sibthorpii* Greuter, *Gagea graeca* (L.) Terracc., *Aristolochia parvifolia* J. E. Sm. Identification of ancillary species follows Lack (1997b: 630–631).

PROLEGOMENON

The origins of this monograph need to be carefully sketched, as some readers may sense a disparity, an anachronism, between the main illustrations, the water-colours by Christabel King, and the present-day representation of heathers in gardens, garden centres and nurseries.

The paintings were planned, composed and completed under the supervision of the late David McClintock VMH, FLS (1912–2001) in the early and mid-1980s, with the intention that they would illustrate a work of this kind, a monograph linked to *Curtis's Botanical Magazine*, then, temporarily, called *The Kew Magazine*. There was an informal arrangement between us that when the monograph was complete, apart from the detailed botanical descriptions, I would write those descriptions. However, although David McClintock did produce a draft text, and did present this for consideration, the monograph he prepared was never accepted for publication. Therefore, the descriptions were never written, and, more importantly, the paintings remained unpublished.

Christabel King's paintings thus record heathers grown during the 1980s in the southeast of England, and they also reflect the views of that period. These plants were highly regarded then, perhaps holding the Royal Horticultural Society's now-abandoned Award of Merit, or perhaps they were unique within a particular species or hybrid. A good example is the white-flowered cultivar of *Erica mackayana* (Mackay's heath) called 'Dr Ronald Gray', which is far inferior to 'Shining Light', a clone introduced from Galicia in 1982 but not selected or named until 1989: in the early 1980s, 'Dr Ronald Gray' was unique, the only white *E. mackayana* in cultivation, and so merited inclusion. There was also a very personal element in the choice of the subjects for each painting; as far as surviving information indicates, many came from McClintock's own garden, called Bracken Hill, at Platt in Kent, and others from Cottswood, West Clandon, Surrey, the garden of Major-General and Mrs P. G. Turpin. Within those two gardens, both now in other hands and no longer "Meccas" for heather enthusiasts, numerous new heathers were cultivated and observed. Combined, Bracken Hill and Cottswood undoubtedly held one of the largest and most comprehensive collections of hardy heathers, both cultivars and species, then in existence.

No attempt was made by David McClintock to procure for the artist examples of each species from the wild, although wild-collected clones are included by chance. This ignoring of the wild plants, and the favouring of cultivars, always seemed to me to be ill-judged.

Concerning the composition of the plates, some apparently comprise a random selection of unrelated heathers. The juxtaposition of *Erica cinerea* (bell heather) and *E. ciliaris* (Dorset heath) on plate 13 does not necessarily reflect a close botanical relationship between those species, nor does the mixture on plate 17 of *E. australis* (Spanish heath), *E. sicula* (Sicilian heath) and *E. umbellata* (dwarf Spanish heath). Dismantling the plates by modern electronic means is entirely feasible, but the artist's skilful arrangement of the sprigs of the heathers has been respected.

There were also omissions, not from ignorance but for explicable, contemporary reasons. When the plates were being painted, several heathers from remote, little-known habitats were either not in cultivation in England, or were very recently introduced and had not matured or flowered properly. Indeed, some of these northern hemisphere heathers are still very uncommon, or very fleeting, in British and Irish gardens. *Erica andevalensis*, for example, was only introduced into cultivation from southern Spain in July 1982, while *E. sicula* subsp. *bocquetii*, from Turkey, did not reach England until July 1989, after the completion of these illustrations. The absence of *E. maderensis* (Madeira heath) is not so easy to explain, except by supposing that neither McClintock nor Turpin grew it well enough. Other omissions are several hybrids that were artificially created since the late 1980s: *E.* × *afroeuropaea*, *E.* × *arendsiana* (Arends's heath), *E.* × *gaudificans* (Edewecht heath), *E.* × *krameri* (Kramer's heath) and *E.* × *oldenburgensis* (Oldenburg heath) from Kurt Kramer in Germany, and *E.* × *garforthensis* (Garforth heath) raised by Dr John Griffiths in England, were not illustrated by Christabel King because they were in the future. By chance, *E.* × *griffithsii* (Griffiths's heath) was painted because a certain clone, subsequently (1990) named 'Heaven Scent', was selected for inclusion to represent *E. manipuliflora* (whorled heath). Study of this plant, and comparison with another series of hybrids produced by Dr Griffiths, caused its re-identification as *E.* × *griffithsii*. Thus, also by chance, *E. manipuliflora* was not represented in Christabel King's watercolours.

Considering the other genera, *Calluna* and *Daboecia*, there are more lacunae. *Daboecia azorica* was not painted — a space was left on the lower right of plate 6 but never filled. In the early 1980s, not only was *D. azorica* very rare in cultivation, and probably was not growing at either Bracken Hill or Cottswood, but also its status as a separate species was doubted, not least by David McClintock: in 1989, he reduced it to a subspecies of *D. cantabrica*. The hybrid between *D. azorica* and *D. cantabrica*, which McClintock named *D.* × *scotica*, is also meagrely represented by an atypical cultivar, the variegated 'William Buchanan Gold'.

In the case of *Calluna*, the two plates display a range of cultivars, including examples with "double" flowers and many with "coloured" foliage. Some of these are still widely cultivated and readily available; others are only to be obtained from a few specialist nurserymen. None of the spate of artificially produced clones that have so dramatically altered the economy of cultivated heathers in the last decade — the misnamed "bud-bloomers"— could have been illustrated, although this group of *Calluna* cultivars is represented by the old, wild-collected clone 'David Eason'.

The world of heathers epitomised by the plants given to Christabel King for painting in the mid-1980s has mostly passed into history. It was a staid, placid world of enthusiasts and small-scale nurseries, of island-beds thickly carpeted with heathers punctuated by "dwarf" conifers: evergreen, low-maintenance, all-year-round colour, longevity. "Bud-bloomers", many protected by plant breeders' rights, are nowadays produced on an industrial scale by the tens of millions each year and are sold as pot-plants, to be discarded when (or even before) the last vestiges of colour and life have seeped away. Some are even sprayed with dyes to produce "yellow" or "blue" heather: the ultimate indignity. Today, the vast majority of heathers are sold as neglectable, throwaway, buy-again-next-year merchandise, but this monograph is not about them.

1. INTRODUCTION

Erice.
Erice is called in greeke Ereice, it is named in english Heth, hather, or ling, in duch
Heyd, in french Bruyer, it groweth on frith
and wyld mores; some vse to make brusshes
of heath both in Englād and in Germany.

The names of herbes … Gathered by William Turner (1548).

Heather and heaths — interchangeable English names and imprecise as to genus or species[1] — in their various guises are familiar plants to the peoples of western Europe and the Mediterranean region. Vast tracts of north-western Europe have a natural or anthropogenic vegetation that is dominated by one or more heathers, especially prominent being *Calluna vulgaris*, ling. Heath or heathland are among the terms employed to describe and designate such "cultural landscapes" (Gimingham 1995; Thompson *et al.* 1995; Parry 2003) which have been characterised as "ecosystems … of particular interest because they are composed of naturally occurring, not cultivated, species and are systematically managed for herbivore production" (Gimingham *et al.* 1979: 365). The plants themselves, the heaths and heathers, all woody and evergreen, were of considerable economic importance in pre-industrial communities in many parts of Europe. Heathers and heathlands provide, or have provided, directly or indirectly: grazing or fodder for domesticated and wild animals and birds; fuel for cooking; wood for fencing and for carving into utensils and tobacco pipes; robust, long-lasting materials for creating shelters, partitions and roofs; bedding for animals and humans alike; flexible shoots that can be twisted into ropes or woven into baskets; nectar for honey; beeswax for candles and hence illumination; dyes; poultices to stop bleeding; palliatives; tisanes; even beer which some claim to be the fabled Viking ale, "the most powerfullest drink ever known"; and, as their names indicate, brushes, besoms and brooms.

Erica scoparia was deliberately named because the shrub's slender, flexible shoots were used for making besoms, *scopae*: "… *& ex hac scopæ apud Monspelienses conficiuntur*", wrote Caspar Bauhin in his magisterial work, ΠΙΝΑΞ [*Pinax*] *theatri botanici* (1623: 486), "… and from this besoms are produced at Montpellier". In the broad margin of his own copy of the 1671 reprint of *Pinax*, the great Swedish botanist Carl Linnaeus (Carl von Linné) inscribed "Erica scoparia"[2] which became, and remains, that particular heather's Latin name. *Calluna*, from καλλύνω (kallyno), has a similar connotation. Richard Salisbury, who coined that name, echoed Bauhin when he wrote in 1802: "… and so I have named it *Calluna*, on account of its very frequent use in the making of brooms." As the Revd William Turner (1548), the "Father of English botany" (Jackson 1877), explained, *Erica* is an archaic plant name known to the naturalists and poets of Ancient Greece, and in modern Greece the same word is still used for heathers. The third genus in this monograph got its name from an Irish saint!

The genera included here are *Calluna*, *Daboecia* and *Erica*: these form the tribe Ericeae within the subfamily Ericoideae of the family Ericaceae (Stevens *et al.* 2004).[3] To this trio might have been added *Andromeda* (bog rosemary) because it is among the genera for which The Heather Society is the duly appointed International Cultivar Registration Authority (ICRA). Despite intertwining nomenclature going back almost three centuries, *Andromeda* is not related closely to the others and belongs to a different subfamily, Vaccinioideae (Stevens *et al.* 2004: 184), within the Ericaceae, a position justified on morphologic and genetic grounds (see Kron *et al.* 2002; Stevens *et al.* 2004). Although originally included in the monograph that David McClintock planned, *Andromeda* has been excluded from this one. Formerly *Bruckenthalia* was also in this list of names, making a fifth genus within the ICRA's domain, but it is no longer regarded as a separate genus and the single species, originally described as a species of *Erica*, has been re-absorbed into the mega-genus *Erica* (see Oliver 1989, 1996, 2000), as was *Pentapera* several decades ago (Webb & Rix 1972; Stevens 1978).

Calluna is widespread throughout western Europe, dominating moors and heaths especially in western parts. This is the heather that turns the seemingly barren mountains purple when it is in full bloom (Fig. 1): to quote Professor Charles Gimingham (1972),

> At times, it is true, heathland gives an impression of barren waste and windswept monotony,
> but when in late summer the rich purple of heath plants in flower spreads across the country,
> native and visitor alike respond to the beauty of the landscape.

There is just the one species of *Calluna*, *C. vulgaris*, called ling in several languages. It is endemic to the northern hemisphere, and to the "Old World": Europe, western Asia and north-western Africa. Indubitably *Calluna* is an alien in North America (Nelson 2010a), and also in the southern hemisphere.

Daboecia has a much more restricted range and, like *Calluna*, is confined to the northern hemisphere, to the western fringe of Europe. One species, *D. azorica*, is found on the Azores, the volcanic archipelago that straddles the Mid-Atlantic Ridge — *Calluna* also occurs there, as well as one species of *Erica*. The second, the principal species of *Daboecia*, *D. cantabrica*, grows in western Ireland, and on the periphery of the Bay of Biscay from southern Brittany into northern Portugal.

Erica is different. To start with, there are at least 810 species — no exact figure can be given. The vast majority of these (around 690, it is estimated[4]) grow only in a relatively restricted area of South Africa, once called Cape Province (or the Cape of Good Hope): this area is mainly within the region designated now as Western Cape Province. Hence, for a long time, these species, many of which have large, attractive flowers and are highly ornamental, have been called the "Cape heaths" (in South Africa they are known as "Cape Ericas"); they are not included in this monograph although occasional reference is made to them. There are *Erica* species elsewhere in Africa, as explained later, but the deserts of northern Africa, between the Equator and the Mediterranean Sea, are a formidable barrier. Thus, those *Erica* species that inhabit sub-Saharan Africa do not, with a single exception, occur in north Africa or Europe. The exception, the only heather to grow both north and south of the Tropic of Cancer (23°26'22" N), is *E. arborea* (tree heath) which ranges from the south of France to Mount Kilimanjaro (3°S, 37°E) in northern Tanzania (Beentje 2006).

The species of *Erica* treated here — the hardy heathers — are those recorded from the northern hemisphere, or, more precisely, those that occur north of the Tropic of Cancer: as just noted *Erica arborea* grows south of the Equator and so is the only "hardy", northern species inhabiting the southern hemisphere. Twenty-two Erica species are recognised in this monograph (see Table 1). Also included are three naturally-occurring hybrids between these species, and seven other hybrids that have formed accidentally in cultivation or been deliberately created between the hardy heaths (Table 2). To these I

Fig. 1 (opposite). The landscape of south-western Scotland turned purple with ling in full bloom; near Wanlockhead, in the Lowther Hills, Dumfries and Galloway, September 1994. © P. & A. Macdonald, & Scottish Natural Heritage.

Table 1

List of hardy heaths and heathers, including natural hybrids, native north of the Tropic of Cancer.

Calluna vulgaris	*Erica cinerea*	*Erica sicula*
Daboecia azorica	*Erica erigena*	*Erica spiculifolia*
Daboecia cantabrica	*Erica lusitanica*	*Erica × stuartii = E. mackayana × tetralix*
Erica andevalensis	*Erica mackayana*	*Erica terminalis*
Erica arborea	*Erica maderensis*	*Erica tetralix*
Erica australis	*Erica manipuliflora*	*Erica umbellata*
Erica azorica	*Erica multiflora*	*Erica vagans*
Erica carnea	*Erica platycodon*	*Erica × watsonii = E. ciliaris × tetralix*
Erica ciliaris	*Erica scoparia*	*Erica × williamsii = E. tetralix × vagans*

Table 2

Horticultural hybrids involving hardy, northern heaths and heathers; only those that have been given a Latin name are listed below (the parental species are accounted for in alphabetical order).

Daboecia × scotica = D. azorica × cantabrica	*Erica × gaudificans = E. bergiana × spiculifolia*
Erica × afroeuropaea = E. arborea × baccans	*Erica × griffithsii = E. manipuliflora × vagans*
Erica × arendsiana = E. cinerea × terminalis	*Erica × krameri = E. carnea × spiculifolia*
Erica × darleyensis = E. carnea × erigena	*Erica × oldenburgensis = E. arborea × carnea*
Erica × garforthensis = E. manipuliflora × tetralix	*Erica × veitchii = E. arborea × lusitanica*

have also added two deliberate hybrids that have a Cape and a hardy heath as parents: *E. × afroeuropaea* and *E. × gaudificans*. A few other, as yet unverified and so unnamed, artificial hybrids exist and are noted where relevant.

Hardiness is a relative and nebulous attribute. In the context of wild and garden plants, it generally signifies a capacity to survive undamaged through the winter outdoors without protection in a particular region. Many factors determine hardiness, or provide the environment that allows a species to inhabit a place, whether a moorland or a garden. Studies have shown that some "hardy" heather species, including *Erica ciliaris* (Dorset heath), *E. erigena* (Irish heath) and *E. mackayana* (Mackay's heath), are more frost resistant than, for example, *E. cinerea* (bell heather) (Bannister & Polwart 2001; Bannister 2007), yet the first three do not range as far north nor reach as high an altitude as bell heather and all three are extremely restricted in their occurrences in Ireland and Britain. Similarly, a heather can thrive in a garden on the south-facing side of a peninsula while a few kilometres away, a little inland but at the same latitude and with apparently the same climate, that particular heather needs glasshouse protection in the winter.

The hardy heathers are especially fascinating because many have disjunct, or discontinuous, patterns of distribution. Indeed the same can be said of the genus *Erica* as a whole, with clusters of species in Europe, west tropical and east Africa, Madagascar and some Indian Ocean islands, and South Africa, separated from each other by large tracts of territory in which no heaths are known to grow (see Fig. 62, p. 101). Disjunct distribution patterns in flowering plants have been a research interest of mine for almost four decades — and the first to enthral me was *E. vagans* (Cornish heath; see p. 132). Abandoning *Erica* for a few years, I then worked on *Adenanthos*, an Australian genus, also with a disjunct distribution across the southern part of the continent (Nelson 1981d), a member of that great, largely southern-hemisphere family, the Proteaceae. In South Africa, Proteaceae and *Erica* dominate the fynbos vegetation.

Returning to Ireland having completed my doctoral research on *Adenanthos*, I renewed my interest in *Erica*, mapping the distribution of *E. mackayana* in Connemara and, with Peter Foss, *E. erigena* in Connemara and Mayo. Those three Irish heathers — *E. vagans*, *E. mackayana*, *E. erigena* — share the same enigma: why are there large disjunctions in their distribution patterns? There is one small colony of *E. vagans* in County Fermanagh, Northern Ireland (Figs 2–4), about 500 kilometres from the nearest population in Britain on the Lizard Peninsula in Cornwall. *E. mackayana* and *E. erigena* do not occur naturally in Britain; the gap between the Irish populations and the nearest ones on the European mainland exceeds 900 kilometres for *E. mackayana* and 1,000 kilometres for *E. erigena*. As in the case of *E. erigena*, there is a discontinuity of around 1,000 kilometres between the Connemara colonies of *Daboecia cantabrica* (St Dabeoc's heath) and those in north-western Spain and north-western France.

Fig. 2. *Erica vagans* (Cornish heath) in County Fermanagh (E. C. Nelson).

Fig. 3. The habitat of the Cornish heath at Carrickbrawn, County Fermanagh (E. C. Nelson).

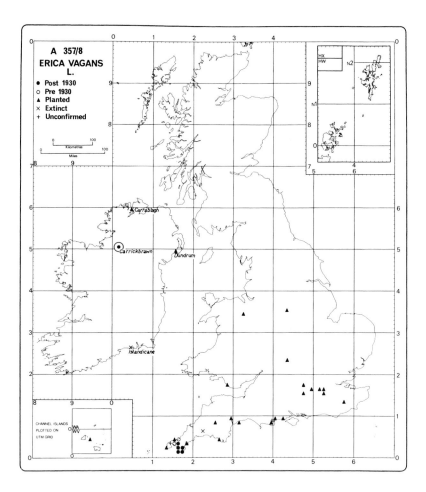

Fig. 4. Distribution map of *Erica vagans* in Ireland and Britain (from Nelson & Coker 1974); for an updated map see Preston *et al.* (2002).

While the indigenous heathers of Ireland remain my main and special interest, similar disrupted patterns of distribution are known for other northern-hemisphere *Erica* species. There is a bewildering colony of *E. cinerea* in Algeria, about 700 kilometres from the southernmost population of bell heather in the Languedoc region of France. Why does *E. multiflora* (many-flowered heath) inhabit Cyrenaica in northern Libya, around 700 kilometres east of the nearest plants on Malta or at Levkas in western Greece? The same region also is one of the habitats of *E. sicula* (Sicilian heath); its fragmented distribution pattern is most extraordinary with populations scattered between north-western Sicily, Cyrenaica, Cyprus, Lebanon and southern Turkey. *E. terminalis* (Corsican heath) also has an interrupted range, occurring in south-eastern Spain and northern Morocco, Sardinia, Corsica and southern Italy. The easternmost populations of *E. spiculifolia* (formerly *Bruckenthalia spiculifolia*; Balkan heath) in the extreme north-east of Turkey are widely separated, and there is another gap of around 350 kilometres between the species' type locality in western Turkey on Uludăgh (the mountain in Asia Minor known as Bythinian (or Mysian) Olympus) and the nearest colony to the west in the Balkans. The distribution of *E. arborea* is perhaps the most remarkable, with isolated colonies on Emi Koussi, a dormant stratovolcano in the middle of the Sahara, and on the Arabian Peninsula and, as mentioned already, in eastern Africa.

Add to these the heathers of Macaronesia, the collective toponym used for the Atlantic archipelagos, the Canary Islands, Madeira and the Azores (plus the Cape Verde Islands which do not harbour any *Erica* species), and there are more riddles that one day might be answered, but only, I suggest, by postulating, as always in plant geography, the least improbable improbabilities. How did *Daboecia*, with *Calluna* and a single species of *Erica* reach the Azores, islands that did not exist until they erupted close to the Mid-Atlantic Ridge sometime during the last five million years? The Canary Islands are closer to Africa than the Azores are to Europe, yet they are also young volcanic islands that would have been utterly sterile, like Iceland's Surtsey was in the 1960s, when they emerged. How did *E. arborea* and *E. platycodon*, a close ally of *E. scoparia* (besom heath), reach them?

What these fractured patterns suggest is that in the past most, if not all, of the northern hemisphere's *Erica* species had more southerly ranges, some extending deep into northern Africa. When the hemisphere's climate was cool during the glacial maxima (ice ages) of the Quaternary, these heathers retreated southwards. When the hemisphere warmed, and deserts expanded, the plants migrated north again, and the populations on the desert fringes died out except in a few favoured places such as Cyrenaica and Emi Koussi. On the Atlantic margins, similar migrations occurred. In addition, as the ocean levels fell, vast new areas of land were exposed on the continental shelf within the Bay of Biscay and to the south-west of Ireland, and heathers were surely among the plants that colonised this area. When the climate ameliorated and the ice caps melted, ocean levels rose again. For a while, migration routes along the continent shelf allowed some heathers to reach (or return to) Ireland, but not Britain, but soon these routes were submerged and severed.

These expansions and contractions of ranges, and the advances and retreats of species, were not confined to the most recent glacial and post-glacial periods. Some ancient layers of peat found near Gort in County Galway on the eastern fringe of The Burren in western Ireland contained evidence from much earlier epochs — preserved in the peat were the flowers of a heather that grew thereabouts several hundred thousand years ago. That heather can be identified — it was *Erica scoparia* (Jessen *et al.* 1959; Godwin 1975; Nelson 2009a). The enigmas multiply. The tiny, distinctive pollen grains of *E. spiculifolia* (formerly *Bruckenthalia spiculifolia*) have been extracted from peat in eastern England, north-eastern Scotland and the Shetland Isles (Whitington 1994; Nelson 2009a). If the identification is correct, these ancient grains indicate a northward expansion, and a subsequent retreat, by the Balkan heath of more than 13° latitude, or at least 2,000 kilometres, from its present northern limits in the Carpathian Mountains (cf. Niedermaier 1985).

The hardy, northern heathers have been cultivated for several centuries. At first only the "typical" wild plants were available, but gardeners have an ineluctable fascination with the anomalous, even the monstrous and bizarre, variations that occasionally occur both in the wild and in gardens. For example, from the mid-seventeenth century, white-flowered ling (*Calluna vulgaris*) has been grown in English gardens — but not because of its repute as a lucky plant which is a fantasy that originated within Queen Victoria's family! Nowadays such variants, termed cultivars because they are "varieties" deliberately maintained in cultivation by gardeners, are still cherished and, as in the 1600s, the best, or most distinctive, may be given names. Since the mid-1600s gardeners have propagated all sorts of oddities from sports, "reversions" and "witches brooms" (McClintock 1984c) — heathers with "double" flowers (McClintock 1977), or buds that never open; heaths that are prostrate or grow into tight, non-flowering hummocks; heathers with golden foliage, or variegated leaves, or brightly discoloured young shoots. Hundreds of cultivars of *Calluna*, *Daboecia* and the hardy species and hybrids of *Erica* are available today, and new ones are introduced every year.

Until the mid-twentieth century, every hardy heather cultivar was a "natural" production — either a chance seedling or a sport (a mutation). The range of habits and foliage colours that they display is astonishing. There are prostrate, lawn-making heathers as well as upright ones, and species that can form veritable trees. To the innumerable shades of green of the foliage, add yellows, golds, oranges, reds, bronze and silver, and these tints can change with aspect, weather and season. The flowers also vary through myriad hues from deep purple-maroon, recalling "black" tulips, to pure white. Green is rarely mentioned as a flower colour, yet it is also in the northern heathers' palette for the flowers of some of the besom heaths are closest to green.

In this monograph, the cultivars described in Appendix I (pp. 351–385) are those that currently (2007) hold the Royal Horticultural Society's Award of Garden merit (indicated by ♛) or are recommended

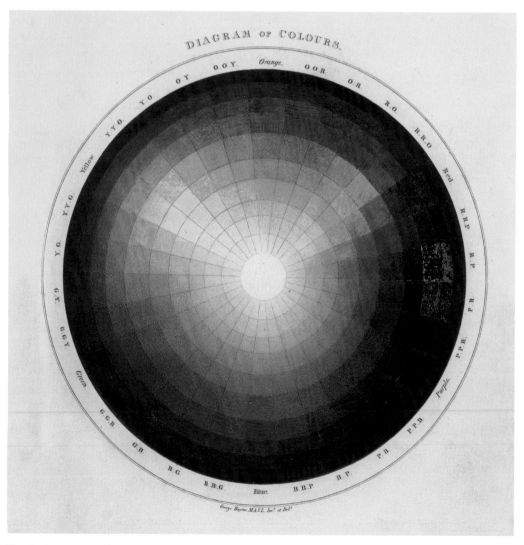

Fig. 5. George Haytor's "Diagram of colours" produced to enable the more exact description of the colours of the flowers of heathers, especially the Cape heaths, growing at Woburn Abbey in 1824 (from *Hortus ericaeus woburnensis*).

by The Heather Society (indicated by 🌸) (Small & Cleevely 1999; see also Davis 2006; Everett 2006). The AGM heathers should be readily available in Great Britain, at least. Because Christabel King's watercolours are the core of this monograph, descriptions have also been included for those cultivars that she painted but which are not among either the RHS's AGM plants or The Heather Society's recommendations. Many other named cultivars are mentioned, and some descriptions are provided for newly named clones of the artificial hybrids. Cultivar names that have been formally registered with The Heather Society are indicated by the symbol ® followed by the registration number (see Nelson & Small 2000, and subsequent supplements to the *International register of heather names* published annually in the *Yearbook of The Heather Society* (2001–2003) and *Heathers* 1–8 (2004–2011)).

Work for this monograph started in the autumn of 2005 and writing was completed in May 2008 (the text was updated in places in June 2010, July 2011 and October 2011). The descriptions of species and hybrids are all new, prepared using fresh specimens provided principally by fellow heather enthusiasts in England, Ireland and the USA. A few were compiled using the heathers growing in the garden of Tippitiwitchet Cottage, in the East Anglian fens. I also have examined specimens in herbaria in the United Kingdom and Ireland. Some data — for example, the dimensions of seeds — have been extracted from published descriptions which were consulted also to ascertain, or check, overall dimensions of mature plants; the sources of these extraneous data are cited.

The descriptions of the AGM-holding cultivars and those recommended by The Heather Society are derived from the data held on The Heather Society's database, which are, in turn, often taken from G. P. Vickers' *A report on a trial to determine the characteristics of cultivated heathers … at Harlow Car Harrogate* (1976; see also Anonymous 1983) or, for the newest, from the most recent edition of Anne and David Small's *The Heather Society's handy guide to heathers* (2001). For a small number of cultivars, new descriptions have been written, or the existing information amended or augmented, following examination of specimens and correspondence with colleagues. On the whole, the cultivar descriptions are brief, and the dimensions given will be for plants that were five years old — these data are often derived from The Heather Society's trials held in the 1970s at the Northern Horticultural Society's garden at Harlow Carr, Yorkshire (this is now one of the gardens owned by the Royal Horticultural Society).

Describing flower colours exactly has long exercised the heather fraternity. Indeed, thanks to one of our number, a pioneering "wheel of colours" (Fig. 5) (cf. Tucker *et al.* 1991), composed by the artist George Haytor, was published in 1825 within the catalogue of the heaths grown at Woburn Abbey (Sinclair 1825). The difficulties were clearly expressed in a letter (Sinclair 1825: 39–42) from Haytor to his patron, John Russell (1766–1829), the heather-obsessed sixth Duke of Bedford:

> At your Grace's desire I write the following few lines of explanation, to accompany my Diagram of Colours; … During a visit to your Grace, at Woburn Abbey, Mr. Sinclair, who superintends the Garden department, was one morning showing me the new Heathery, when, upon expressing my admiration of the great beauty and variety of the forms and colours of the blossoms of the various plants, he lamented much that no work existed to render clear and definite all communications on the subject of the precise colour of any new or rare blossom.

> In attempting at present to describe the colour of any given flower, the term red, he remarked, was applied to so many blossoms of various hues under that denomination, that it was impossible to define precisely what was meant by it; and, to give me an evidence of the want he complained of, he took me to several flowers generally called red, but varying so much, that, to my eye as a painter, they each demanded a different appellation for their colour. …

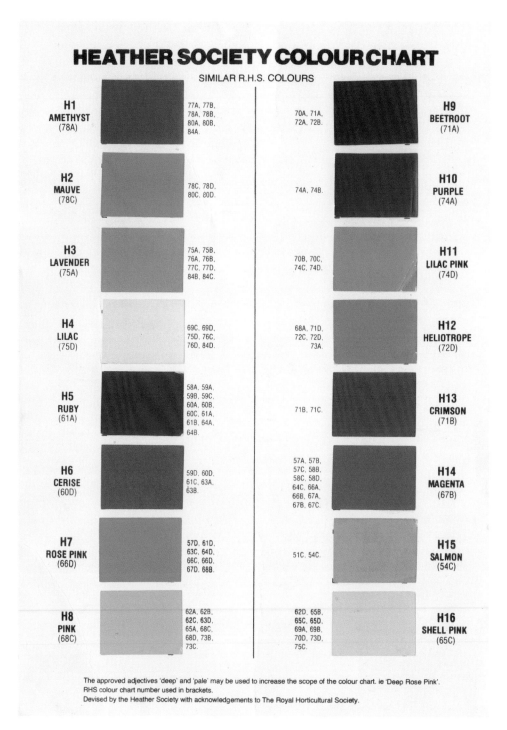

HEATHER SOCIETY COLOUR CHART

SIMILAR R.H.S. COLOURS

H1 AMETHYST (78A)	77A, 77B, 78A, 78B, 80A, 80B, 84A.	70A, 71A, 72A, 72B.	**H9** BEETROOT (71A)
H2 MAUVE (78C)	78C, 78D, 80C, 80D.	74A, 74B.	**H10** PURPLE (74A)
H3 LAVENDER (75A)	75A, 75B, 76A, 76B, 77C, 77D, 84B, 84C.	70B, 70C, 74C, 74D.	**H11** LILAC PINK (74D)
H4 LILAC (75D)	69C, 69D, 75D, 76C, 76D, 84D.	68A, 71D, 72C, 72D, 73A.	**H12** HELIOTROPE (72D)
H5 RUBY (61A)	58A, 59A, 59B, 59C, 60A, 60B, 60C, 61A, 61B, 64A, 64B.	71B, 71C.	**H13** CRIMSON (71B)
H6 CERISE (60D)	59D, 60D, 61C, 63A, 63B.	57A, 57B, 57C, 58B, 58C, 58D, 64C, 66A, 66B, 67A, 67B, 67C.	**H14** MAGENTA (67B)
H7 ROSE PINK (66D)	57D, 61D, 63C, 64D, 66C, 66D, 67D, 68B.	51C, 54C.	**H15** SALMON (54C)
H8 PINK (68C)	62A, 62B, 62C, 63D, 65A, 68C, 68D, 73B, 73C.	62D, 65B, 65C, 65D, 69A, 69B, 70D, 73D, 75C.	**H16** SHELL PINK (65C)

The approved adjectives 'deep' and 'pale' may be used to increase the scope of the colour chart. ie 'Deep Rose Pink'.
RHS colour chart number used in brackets.
Devised by the Heather Society with acknowledgements to The Royal Horticultural Society.

Fig. 6. The Heather Society's colour chart produced to enable the more exact description of the colours of the flowers of heathers, but it was limited to the colours of the hardy species of *Erica*, as well as *Calluna vulgaris* (ling) and *Daboecia* species.

While various modern horticultural colour charts of greater precision than Haytor's exist, The Heather Society compiled its own version with 16 colours (Fig. 6) — from amethyst (H1) to ruby (H5), purple (H10) and shell-pink (H16), with names and codes (prefixed by H, as in the four examples just given). These can be qualified by "pale" or "dark" to give 48 possible shades of "purple heather"; when white is included there are 49. In the cultivar descriptions, colours are given by referring of The Heather Society's colour chart (using H1–16); sometimes the equivalent colour from the Royal Horticultural Society's colour chart is provided (using RHSCC) instead.

There is no equivalent Heather Society chart, or codes, for shades of green or other foliage colours because these are far too complicated to be condensed into a single, handy card.

———————————

I have not written a manual on heather gardening, on pests, diseases, mulching and hoeing, although their cultivation and propagation are explained briefly. These topics are well treated by others in the various books about heathers that have been published during the last four decades — I commend these older books, especially Terry Underhill's *Heaths and heathers* (1971, 1990), as invaluable sources of information about gardening with heathers. The Heather Society has published a series of booklets entitled "The world of heathers": *Everyone can grow heathers!* by Daphne Everett (2000b) is brief yet comprehensive, while *Recommended heathers* by David Small and Ron Cleevely (1999) provided a quick guide to those cultivars that are considered to be outstanding. The most recent, authoritative book about heathers, and thoroughly recommended, is David Small and Ella May Wulff's *Gardening with hardy heathers* (2008); they cover many more cultivars than are noticed here.

My interest in heather is botanical and historical and that informs this book. For several decades I have worked closely with a number of botanical artists, especially with Wendy Walsh, and so the portrayal of heathers is another and a very appropriate theme of this monograph. My earliest independent research on the ecology of *Erica vagans*, guided by Dr Paddy Coker, remains fundamental to my fascination for these plants. In working with David Small (1939–2010), and David McClintock (1913–2001), both former presidents of The Heather Society, on the *International register of heather names*, published in two volumes and eight parts between 2000 and 2005, I gained a better understanding of the nomenclature and history of *Erica*, *Daboecia*, *Calluna* and *Andromeda*. Indeed working on this monograph has enlightened me further, and shown me that there is still a great deal of work needed on all aspects of heathers, from their nomenclature to the dispersal of their seeds, from their external morphology to their "genetic fingerprints".

Notes

[1] The *Oxford English dictionary* (Second edition. 1989) entry for heather includes this somewhat out-of-date, preliminary explanation: "Some recent botanical writers have essayed to limit the originally local names heath, ling, heather, to different species; but each of these names is, in its own locality, applied to all the species there found, and pre-eminently to that locally most abundant. On the Yorkshire and Scottish moors, the most abundant is *E. vulgaris* [*Calluna vulgaris*], which is therefore the 'common ling' of the one, the 'common heather' of the other. But in other localities, esp[ecially] in the south-west, *E. cinerea* is the prevalent species, and is there the 'common heath' ..." (*OED Online*. URL http://dictionary.oed.com/cgi/entry/50103837 accessed 21 July 2007).

McClintock (1980) wrote, in *A guide to the naming of plants*, that "there is no fixed distinction between a Heath and a Heather. Heath tends to be used for the Ericas [sic] ...".

In 1955, in his seminal book *The Englishman's flora*, Geoffrey Grigson wrote:

> This [*Calluna vulgaris*], and the commoner species of *Erica*, are heather in Cheshire, Lincolnshire,
> Yorkshire, Westmorland, Cumberland, Northumberland, Scotland, Ireland; hadder or hedder
> in East Anglia, Yorkshire, Westmorland, Cumberland; heath in Devon, Somerset, Hampshire,
> Lancashire; and ling from East Anglia and Shropshire to Scotland, and in Ireland.

I must also note that whereas many of the Latin names affixed to these plants have been in use for two and a half centuries, and their application has been constant, the vernacular names in English have not been remained unchanged. For example, *Erica tetralix*, a name published in 1753 by Carl Linnaeus, is now generally known as cross-leaved heath, yet was named "bog heather" in Grigson's book and six decades earlier had been called "bell heather" by William Robinson (1898). Robinson (1898) employed "Scotch heather" for *E. cinerea*, another of Linnaeus's names. Unlike Latin names, the usage of vernacular names is not regulated by an internationally accepted code and this allows for different applications. The capricious state of vernacular nomenclature is well illustrated by the name Irish heath (Nelson 2004b) which in Ireland and Britain usually is applied to *E. erigena*, but in North America and in German-speaking nations (as Irische heide) is usually attached to *Daboecia cantabrica*.

[2] Copy in The Linnean Society of London; inscription is on p. 485.

[3] In earlier classification schemes, *Daboecia* was consistently placed a subfamily of the Ericaceae quite separate from *Calluna* and *Erica*. For example, Stevens (1971) designated the tribe Daboecieae, within the subfamily Rhododendroideae, and *Daboecia* was the tribe's only member.

[4] Oliver (1995) estimated there were 650 species in southern Africa (south of the Limpopo River), to which must now be added the 95 species with indehiscent fruits which subsequently were subsumed into *Erica* (Oliver 2000), giving around 750.

However, the most up-to-date estimate of species numbers in the mega-genus, following revision of those in tropical Africa, posits 692 species in the Cape Floristic Region, with a grand total of 816 species throughout Africa and Eurasia (M. D. Pirie, E. G. H. Oliver & D. U. Bellstedt, pers. comm. October 2011).

2. CULTIVATION: PAST AND PRESENT

Notwithstanding the commonness of our British heaths,
they deserve a place in small quarters of humble flowering shrubs, where by the beauty and long continuance
of their flowers, together with the diversity of their leaves, they make an agreeable variety.

T. Martyn, *The gardener's and botanist's dictionary* (c. 1795).

... these hardy species when cultivated, without admixture of other plants ...
form an interesting feature in the flower-garden ... .

G. Sinclair, Preface, *Hortus ericaeus woburnensis* (1825).

A SHORT HISTORY OF HARDY HEATHERS IN GARDENS

It is unlikely that the European hardy heathers were deliberately cultivated before the end of the sixteenth century. None of the early herbals, including that of the Englishman John Gerard (1597), contains references, direct or indirect, to these plants in gardens. They were known in the wild and were plentiful therein, so if they were required as simples for physic they could easily be gathered. There was no need to cultivate them for any other purpose — they were commonplace and ubiquitous in those regions where they were regarded as useful.

From seventeenth-century garden catalogues, we can be certain that hardy heathers were being grown in botanical collections by the mid-1600s at least. However, published dates of introduction of the non-indigenous heather species into British gardens (see, for example, Harvey 1988) are unreliable partly because of nomenclatural difficulties, as I will explain. Both *Erica arborea* (tree heath) and *E. erigena* (Irish heath) are attributed to the mid-seventeenth century, the latter said to have been grown as early as 1648, and the former, at Oxford, in 1658, yet neither of these dates is likely to be correct.

The University of Oxford's *Hortus botanicus* contained seven different heathers when the Revd William Browne and Dr Philip Stephens compiled a new catalogue late in the 1650s. Two of these were ling (*Calluna vulgaris*), the "ordinary" purple-blossomed form and the form with white flowers (Stephens & Browne 1658: 59); McClintock (1984a) erred in giving 1685 as the earliest date for the latter in Oxford Botanic Garden. What the other heathers were must remain open to doubt given that the phrase-names employed by the cataloguers cannot be equated unequivocally with the names that we employ now. A good example of the possibility of serious confusion is the phrase-name "*Erica vulgaris hirsuta*"; if long-accepted synonymy is applied, a plant labelled "*Erica vulgaris*" should be renamed *Calluna vulgaris*. However, an herbarium specimen[1] in a collection associated with the Bobarts, both named Jacob, who were successive curators (*Hortus Praefectus*) at Oxford in the late 1600s, demonstrates that "*Erica vulgaris hirsutior*" (which was a contemporary synonym, according to Ray (1670), of "*Erica vulgaris hirsuta*") was one of the names applied to cross-leaved heath, called *E. tetralix* nowadays, and so was not a hairy-leaved variety of ling (as stated by many authors including McClintock 1966: 41–42; cf. Nelson 2008a). Yet, in Thomas Johnson's (1633) revised edition of John Gerard's herbal, this phrase name is attached to a wood-cut (Fig. 7) that shows *E. lusitanica* (cf. Ramón-Laca & Morales 2000) — there is no specimen of this exotic heath in the Bobarts' herbarium.

The question must be asked: did the collections at Oxford in 1658 include any *Erica* species from southern Europe as well as native English heaths? Four other heathers were catalogued, all identified by reference to Thomas Johnson's 1633 edition of Gerard's *Herball*. These were "Erica major flore purpurea", "Erica tenuifolia", "Erica major flore albo" and "Erica pumila Belgarum Lobelii". Going by the illustrations in Johnson's version of Gerard's *Herball* (Figs 8, 9 & 10), the first three were

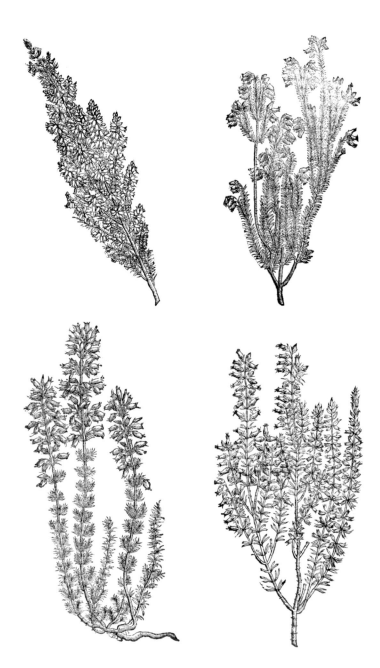

The four heathers, illustrated in Thomas Johnson's revised edition of John Gerard's *Herball* that were supposed to have been growing in the Oxford *Hortus Botanicus* during the late 1650s. The wood-cuts are all based on water-colour paintings, dating from the mid-1560s, preserved in *Libri picturati*, now in Jagellonian Library, Kraków, Poland.

Fig. 7 (top left). "*Erica vulgaris hirsuta*. Rough leaved Heath."; the illustration is of *Erica lusitanica* (Portuguese heath).

Fig. 8 (top right). "*Erica maior flore purpureo*. Great Heath with purple floures."; the wood-cut portrays *Erica australis* (southern heath).

Fig. 9 (bottom left). "*Erica tenuifolia*. Small leafed Heath."; this is a fine representation of *Erica cinerea* (bell heather).

Fig. 10 (bottom right). "*Erica maior flore albo Clusii*. The great Heath with white floures."; this is *Erica erigena* (Irish heath).

respectively *E. australis*, *E. cinerea* and *E. erigena* (cf. Ramón-Laca & Morales 2000). The fourth name is yet another synonym of *E. tetralix*. However, without voucher specimens from the Oxford *Hortus botanicus* to verify the existence of *E. lusitanica* (Portuguese heath), *E. australis* (southern heath) and *E. erigena* (Irish heath), there must remain considerable doubt about their introduction into English gardens by the middle of the seventeenth century.

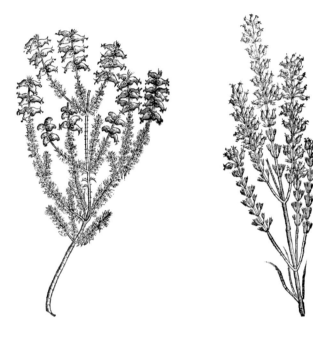

The other heathers illustrated in Johnson's 1633 edition of Gerard's *Herball* were

Fig. 11 (top left). *Erica ciliaris* (Dorset heath) which Gerard and Johnson called "*Erica tenuifolia caliculata*. Challice Heath."

Fig. 12 (top right). *Erica umbellata* (dwarf Spanish heath) is the species represented in this woodcut: "*Erica ternis per intervalla ramis*. Heath with three branches at a ioint."

Figs. 13 & **14** (bottom left and right). *Erica carnea* (winter heath); no. 13 "*Erica pumila, 3. Dod.* Dodonæus his Dwarfe Heath.", and no. 14 "*Erica Coris folio* VII. *Clusij*. Creeping Dutch Heath." (although it does not grow in Holland). The engraving of *Erica Coris folio* VII is the same as *Erica altera* of Charles l'Écluse's *Rariorum aliquot stirpium per Pannoniam* ... (1583).

At the end of that century, plants and seeds of the heathers indigenous to the Canary Islands were received at Chelsea Physic Garden (see p. 206) (Francisco-Ortega & Santos-Guerra 1999). We have little idea how these heathers were cultivated, but the chances are that the Canarian plants were destined for a greenhouse; they are likely to have been grown indoors in containers, rather than in the open air, planted out in beds.

Identifying the heathers cultivated in the early 1700s is neither easier nor more certain. *Erica multiflora* (whorled heath) is said to have arrived in English gardens in 1731, but whether it did is questionable because, again, there is no certainty that that name was then applied to the species we know by the same pair of Latin words (see, for example, McClintock 1984a).[2] English nurserymen did not stock the exotic, hardy heathers until the last half of the eighteenth century and selling the familiar native species would have been a pointless waste of time then. While white-flowered heathers had been on record for more than a century, as just noticed, heathers were not yet appreciated as garden-worthy by those who had the wealth to create ornamental gardens, although they may have been collected by the more curious gardeners. Indisputable evidence that non-indigenous *Erica* species were in cultivation in north-western Europe, at least in specialised collections, by the mid-1730s is contained in the herbarium of George Clifford's garden at Hartecamp in The Netherlands, and in the associated book by Carl Linnaeus (1738). According to Linnaeus (1738: 147–148), Clifford grew perhaps five different African heaths, including *E. curviflora*, and the herbarium specimens indicate that two forms of *E. tetralix* (Fig. 15) and perhaps also *E. arborea* (tree heath) were in the Hartecamp collection.[3]

Philip Miller, Curator of the Chelsea Physic Garden and the leading British horticulturist of the mid-eighteenth century, listed a few hardy heathers in the several editions of his famous *Gardener's dictionary*. He dismissed heathers generally as "seldom grown in gardens" and stated that they "may not be had from nurseries". Instead Miller advised taking plants from the wild "with a ball of earth to their roots" during autumn and transplanting them into the garden. Around a dozen heaths were specified by Richard Weston, a Leicester-based plantsman and thread-hosier, within his *Botanicus universalis et hortulanus* of 1770, suggesting that an increasing number were being propagated. Among the small number of English nurserymen to offer hardy heathers for sale during the later decades of the eighteenth century was William Malcolm of Kennington Turnpike, outside London, whose *Catalogue of hot-house and green-house ... plants* (1771) contained four heaths that were considered suitable for outdoor cultivation, although exactly what these species were is not easy to determine — under "Hardy trees and shrubs" (Malcolm 1771: 24–39) he offered "Erica Triflora African Tree heath" (but also listed as a shrub for the greenhouse), *Daboecia cantabrica* (as "Erica Daboecii Cantabrian heath"), "Erica Lusitanica Portugal upright red heath" which, as discussed later (p. 213), may have been *Erica australis*, and "Erica Paniculata Ethiopian Heath" (possibly *E. arborea*).

Paintings and other works of art made from living plants, and herbarium specimens provide excellent evidence that the hardy heathers were being grown by discerning plantsman in the last quarter of the 1700s, and we also have first-hand accounts of the origin and cultivation of these plants. Peter Collinson, a Quaker plantsman with an important garden at Mill Hill in Hendon, north of London, had a flowering plant of St Dabeoc's heath, raised from Spanish seed, in 1765.

Fig. 15 (opposite). *Erica tetralix* (cross-leaved heath), "Erica myrica folio hirsuta Boerhavis ... ex rubro nigricans scoparia", was probably growing in George Clifford's garden at Hartecamp in the mid-1730s. The heath also was abundant in the wild, in bogs in Gelderland where Carl Linnaeus botanised in June 1735 while he was briefly in that state (one of the constituent states of the United Netherlands) to obtain his medical degree at Harderwijck University (Stafleu 1971: 10–11). This handsomely mounted, pressed specimen would have been examined by Linnaeus (© Natural History Museum, London).

Georg Ehret painted this handsome heather in 1766 (see Knapp 2003: 260), probably at Bulstrode in Buckinghamshire, seat of the dukes of Portland (see p. 69). Not long afterwards, *Daboecia cantabrica* had also been imported from the west of Ireland (see p. 71).

Margaret Bentinck (1715–1785), Duchess of Portland, who was applauded as being "well acquainted with English plants", had a famous garden at Bulstrode which attracted botanists and artists as well as her friends, one of whom was Mrs Mary Delany (1700–1788). Mrs Delany's fame rests largely on her *hortus siccus* of "paper mosaicks", collages of cut and shaped tissue paper: "I have invented a new way of imitating flowers … ." Among the hundreds of plants "imitated" in this unusual way were about a dozen heaths (Hayden 1976; see Laird & Weisberg-Roberts 2009) including St Dabeoc's heath ("Andromeda Daboecia") dated 1774 and annotated "Bulstrode". Mrs Delany also fashioned the late-summer flowering heathers, *Erica cinerea* (Bulstrode, 25 September 1776), *E. ciliaris* (no details recorded) and *E. vagans* (as "Erica multiflora"; Bulstrode, 3 September 1776), and the spring-blooming *E. arborea* (28 April 1777, Kew) and *E. erigena* (7 May, St James's Place, which was her London home).[4]

In 1789 William Aiton accounted for more than half a dozen exotic, hardy heathers in the Royal Gardens at Kew, as well as white-flowered variants of the native heaths, *Calluna vulgaris*, *Erica cinerea* and *E. tetralix*. Franz Bauer (1759–1840), an Austrian botanical artist who came to England in 1788 and was employed by Sir Joseph Banks to paint the plants cultivated at Kew, recorded some of these heathers, notably *E. umbellata* (dwarf Spanish heath) (see Plate 12). Bauer made 30 portraits of *Erica* species which were published in *Delineations of exotick plants*, an extraordinary work composed only of hand-coloured engravings. Richard Anthony Salisbury (né Markham; 1761–1829), who was to produce a remarkable monograph on *Erica* at the beginning of the next century, assembled a collection of heaths (among many other plants) in his garden at Chapel Allerton, a village situated between Leeds and Harrogate, and issued a catalogue of his entire collection in 1796. He had 62 *Erica* species in cultivation, including the native ling (then still called *Erica vulgaris*) but he had not planted it — "In horto sponte", he wrote (Salisbury 1796), "in the garden of its own accord". As for other heathers, he had white-flowered *E. cinerea*, and *E. scoparia* (besom heath) (Salisbury 1796).

Undoubtedly the fashion for Cape heaths which raged for around half a century from the 1780s stimulated the cultivation of hardy species too. This phenomenal horticultural craze (beyond the scope of this work[5]) resulted in the production of innumerable artificial hybrids between the Cape species (Nelson & Oliver 2005a), the first attempt to cross-pollinate a European and an African species (see p. 328), and also a procession of books about heaths and heathers of which Henry Andrews's two multi-volume series, *Coloured engravings of heaths* (4 volumes; *CE*) and *The heathery* (6 volumes), published between 1794 and 1830 (Cleevely & Oliver 2002; Cleevely *et al*. 2003), were undoubtedly the most remarkable. Andrews depicted four European heathers among 284 Cape heaths (and their hybrids) and these were most probably grown under glass at that time: *Erica australis* (*CE* **3**: tab. 150; plate dated July 1805) (see Plate 16), *E. multiflora* (*CE* **2**: tab. 111; plate dated 1 December 1801), *E. terminalis* (Corsican heath) (as *E. stricta*: *CE* **2**: tab. 134; plate dated 1 November 1801) and *E. umbellata* (*CE* **2**: tab. 143; plate dated 1 July 1804).

Given the limited number of hardy species of heathers available to gardeners, attention naturally focused on plants with unusual habits, variously coloured foliage and different floral shades (remembering that the hardy heathers do not have a medley of floral colours, just variations on "purple" and "pink", and white). Gradually their versatility was recognised — no longer were the heathers only grown in pots, as single specimens. At Woburn Abbey by 1825 an entire parterre, a curved wedge around 100 feet long and 50 feet broad at its widest, was devoted solely to hardy heaths:

... these hardy species, when cultivated, without admixture of other plants, in a parterre, form an interesting feature of the flower-garden. ... There are ... thirty-five species and varieties of this beautiful family capable of being cultivated in the open air all the year ... . (Sinclair 1825).

The Woburn Abbey parterre (Figs 16 & 17; Sinclair 1825), probably designed by the Duke of Bedford's head-gardener, George Sinclair (Padley 2006), was formal, like the gaudy parterres stocked with half-hardy plants that were to be so familiar later in the nineteenth century. It surely counts as the first heather garden but it does not seem to have been influential for very few gardeners are known to have copied or even emulated it (see below). Formal plantings of hardy heaths have rarely been an enthusiasm of gardeners.

WEST VIEW OF THE HEATH HOUSE.

Fig. 16. "West view of the heath house" at Woburn Abbey in 1824, with the parterre of hardy heathers (see plan, Fig. 18) (from Sinclair 1825).

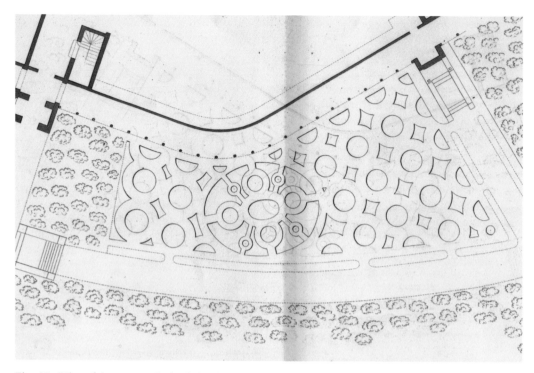

Fig. 17. "Plan of the parterre for hardy heaths" (detail), from George Sinclair's (1825) catalogue of the Woburn Abbey heathers.

Instead, in most nineteenth-century gardens, hardy heathers were relegated to edging and an occasional flower-bed. They were not usually displayed *en masse*. Professor Thomas Martyn, Philip Miller's editor, signalled the style in the late 1700s: "Notwithstanding the commonness of our British heaths, they deserve a place in small quarters of humble flowering shrubs, where by the beauty and long continuance of their flowers, together with the diversity of their leaves, they make an agreeable variety" (Martyn c. 1795). The same method of planting was exemplified half a century later in plans published by James Lothian who was gardener during the early 1840s at Ormsary House on the Kintyre Peninsula in western Scotland (Nelson & King 2007). In his pioneering book about "alpine and rock plants", Lothian (1845) portrayed a pond (Fig. 18) surrounded by "a border ... filled with dwarf American shrubs; and, separating the walk from the border, [was] an edging of Calluna vulgaris, or common ling, mixed with the white variety." That such schemes were common during the mid-nineteenth century is obvious from other reports. J. L. Snow (1847), writing about "heath edgings", suggested that "no one can form an idea of [their] pleasing effect". In 1848 at Trentham Park, Staffordshire, a bed containing "American plants" was "edged with common Heath, thrusting itself luxuriantly upon the lawn." The effect was "charming" ("G." 1848: 703), and the common heath (presumably ling) was "enhanced considerably by an irregular patch of wild Heath, which is planted in the lawn, a little in advance of the edging." At Trentham, native heathers and other "dwarf evergreens" kept "up during winter both character and interest" ("G." 1848: 687). The idea of beds edged by heather was also promoted by the Irish horticulturist and journalist, William Robinson, in the first edition of his iconoclastic book *The wild garden* published in 1870: "These tiny shrubs and their allies in size might form a sort of edging or marginal line round a bed of choice shrubs planted in peat, as they frequently are and must be in gardens." Robinson (1870) also noted the increasing number of cultivars that were being propagated: "... from time to time sports have appeared ... which nurserymen have preserved; and thus, where you see a good collection of these, the variety of gay colour is quite surprising" (see also p. 45). He continued to applaud heathers in his other famous book *The English flower garden*, which went through 15 editions between 1883 and 1933, yet Robinson was soon to abandon skimpy edgings in favour of massed heathers in large groups.

At Alton Towers, also in Staffordshire, there was once a heath parterre containing a collection of all the hardy heaths then available in the trade. Alexander Forsyth, the Earl of Shrewsbury's head-gardener at Alton Towers, thought the effect of these heathers in the flowering season was excellent, but, because they were "not handsome when not in flowers" he was ordered to remove them by Lady Shrewsbury "very much against her loving lordship's wishes" (Forsyth 1873). This garden, overflowing with statues, bridges, temples, seats and conservatories, was also a "landscape filled with trees, rhododendrons and other shrubs, heather and rockwork absorbing the architectural features into an increasingly picturesque jungle" (Bisgrove 1990: 172). In Scotland, at Drummond Castle near Crieff in the late 1870s, a parterre of elaborate design was formed, including circular beds containing five different species of hardy heaths planted in a flower-like pattern (Anonymous 1877: 688–689). The Drummond Castle parterre, unlike the one at Woburn Abbey half a century earlier, contained other plants and was rigidly formal, being based on the saltire of St Andrew (see Coats 1970; Anonymous 1877).

At this period, heathers acquired a new use — as temporary bedding plants. They were arranged to "fill in" when the more tender exotics had passed their best in the labour-intensive gardens of the late Victorian era. This was the practice at Heckfield, Hampshire, during the early 1880s. Heathers and

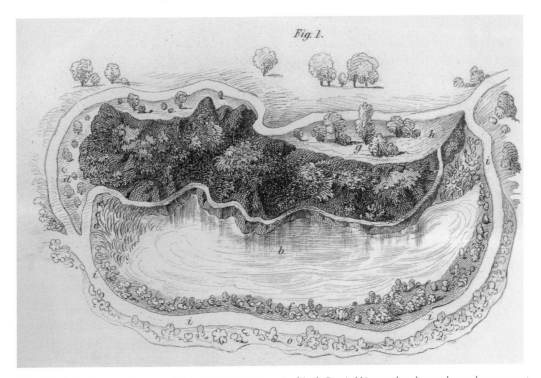

Fig. 18. A plan for a rockery, as published in 1845 in James Lothian's *Practical hints on the culture and general management of alpine or rock plants*; it is possible that this was made at Ormsary House, Argyllshire.

other ornamental shrubs were planted in the vacant spaces, and within a few days "the gay pompadour dress of the garden" gave place to a "sombre but more seasonable velvet attire" allowing the garden to stand "arrayed in fitness but still in exceeding beauty for the winter months" ("X" 1884: 371).

Towards the turn of the nineteenth century, heaths and heathers began to be used on a more extensive scale and references to heather or heath gardens become more frequent. In the issue of William Robinson's periodical *The garden* on 15 December 1888, there is an extract from *The field* (most probably penned by Robinson himself) extolling hardy heaths (Field 1888). The writer recalled that "We saw last year some very large and effective beds of heathers in Sir W[illiam] Bowman's[6] garden in Surrey. It is the bold artistic arrangement that is called for and not the niggling, dotted botanical way." Heather gardens progressed, and in 1911, William Dallimore of the Royal Botanic Gardens, Kew, remarked that "Heaths are now included amongst the plants of which a special feature is made in gardens, and the "Heath Garden" is becoming as welcome as the "Bamboo Garden," "Wall Garden," or "Water Garden" (Dallimore 1910: 201). William Robinson (1897) was among the writers who used the term, reporting that a "singularly pretty Heath garden" had been made by Sir Philip Currie at Hawley in Hampshire. "In front of his house he has kept a piece of the Heath land of the district almost in its natural state, save for a little levelling of old pits." Exactly what Currie created at Minley Manor, the name of his family's estate, is not clear, but "garden" implies some management if not also deliberate planting. (Harry Veitch, a prominent nurseryman and expert plantsman, is reputed to have laid out a garden at Minley Manor in the last decade of the nineteenth century.) Immediately after noting Sir Philip Currie's heath garden, Robinson (1897: 496) commented that where

> ... Heaths abound, there is no occasion to cultivate them, although we cultivate nothing prettier; but certain varieties of these Heaths are charming and deserve a place in the garden or wild garden. In places large enough for bold Heath gardens it would be charming to plant them, but a small place is often large enough for a few beds of Hardy heaths. Once established, they need very little attention.

Admitting that he had begun "rather slowly and doubtfully" with heathers, William Robinson created several separate heather gardens in the grounds of his Sussex home, Gravetye Manor, describing them individually in his chronicle of the making of the manor's garden (Robinson 1911). He was "charmed" first by "the little Alpine Forest Heath" (*Erica carnea*) although he noted that it was "generally so ill-grown that we seldom see the fine effects it may give". Next, he was "charmed" by some of the "bright forms of the Grey Heather" (the name he used for bell heather, *E. cinerea*);

> ... and so I went on until one cold winter's day, when searching in a nursery for wholly different plants, I saw a beautiful little bush of rosy-tinted flowers, and this was the Portuguese Heath (*E. lusitanica*). That settled me as a Heath-lover, and I have planted as much as I could get of it since. (Robinson 1911: 130).

> [1890.] In taking away some poor trees on the bank west of house, seized the opportunity to make a bold bed of Heaths in place of an awkward bank. Mostly the finer, not British, Heaths such as *E. Codonodes* [*sic* = *E. lusitanica*], Mediterranean Heath [*E. erigena*], and some of our rarer native kinds, were planted. (Robinson 1911: 48).

> [1892.] Planted nearly all obtainable kinds of British Heaths and other hardy Heaths in large beds in ground north of the house, intending to let them, once established, form wild masses, i. e. not cultivated much after first planting, and planting hardy bulbs, and things for the wild garden, between them. (Robinson 1911: 71–72).

For heather gardens, Robinson laid down "rules" which are still followed: "The larger Heaths ... should be massed in visible groups, and the dwarf ones seen in dwarf masses only, and not treated as mere specks on 'rockeries' or used as edging plants only" (Robinson 1911: 131).

Other gardeners, it seems, followed the model of Gravetye Manor when making heather gardens. At Nymans, Ludwig Messel (1847–1915) designed and planted a heath garden which is sometimes claimed to be "the first ever to be planted in England" (Coats 1970: 120; Thomas 1979; Bisgrove 1990), although it can hardly pre-date that at Gravetye Manor given that Messel only acquired this property in Sussex in 1890. Messel was certainly influenced by William Robinson who visited Nymans frequently and wrote a foreword for a catalogue of the plants grown at Nymans (L. Messel 1918; Allan 1982). In that foreword, Robinson lauded Messel as "one of the first to enjoy the charms of a garden of all the heaths hardy in our country", and he also praised the garden as "a very good one, varied here and there with groups of cyclamens and many of the choicer rock plants." The heathers thrived at Nymans, "especially *Erica australis* which forms huge banks of pink bells" (M. Messel 1918: viii), and there was a 20-year old *Daboecia cantabrica* ten feet in diameter (L. Messel 1918: 57).

About 1905, Mark Fenwick (1860–1945)[7] began to make a heather garden covering around three-quarters of an acre at Abbotswood in Gloucestershire (Everett 2007; Coats 1970: 163) as an adjunct to a large ten-acre garden designed and laid out with the assistance of Sir Edwin Lutyens. A photograph of Abbotswood and its heather garden was published in 1913 within Gertrude Jekyll's book *Wood and water gardens* (1913: opposite p. 199; see below) (Fig. 19). Graham Stuart Thomas (2003: 110) recalled the Abbotswood heather garden as "the biggest surprise of all" within a "superlative" garden that was "large and lovely". The heather garden still flourishes at Abbotswood, more than a century after its creation (Everett 2007).

Fig. 19. An early twentieth-century heather garden (in the foreground) at Abbotswood in Gloucestershire; this photograph was published in 1913 in Miss Gertrude Jekyll's book *Wall and water gardens with chapters on the rock-garden and the heath-garden*.

Notable heather gardens were formed also at the Royal Horticultural Society's gardens, Wisley, in Surrey (before 1934), and at Tyninghame, East Lothian, in Scotland where the Earl of Haddington's "special penchant for heaths and heathers of all kinds, and these with rhododendrons and azaleas filled" part of the garden (Thomas 2003: 122). Coats (1970) asserted that the Tyninghame heath garden was "as well planted as any in Scotland."

> Heath gardens have not only great appeal, but sterling qualities too. For if thoughtfully planted, they can provide flowers and colour almost all the year round. They look particularly well in Scotland, and the view of Tyninghame's pepper pot turrets seen over the rose, grey and sea-green colouring of Lord Haddington's heath garden is a telling one. (Coats 1970: 105).

These were all essentially informal gardens with the heathers planted in "broad swathes" usually composed of single clones, and the shrubs were allowed to grow unchecked, giving a semi-natural appearance. They could lead out into the wild, to the surrounding heathland, or they could allow the natural heathland to be annexed to the more formal garden as appears to have been the approach of Mackay Hugh Baillie Scott (1865–1945) in Berkshire where a house, appropriately named Heather Cottage, and garden, both to his designs, were under construction in the early 1900s (Scott 1906; Ottewill 1989: 137) (Fig. 20). "On sunny hills, where purple heather grows, purple heather shall be the dominant note in his [the small householder's] garden scheme ...", wrote Baillie Scott (1906):

> Of natural surroundings ... there seems nothing so good as a setting for a house as a dark moorland covered with the purple of the heather, and that lesser kind of gorse which seems to have been clipped into neat round bushes by Nature's invisible shears.

Robinson's friend and fellow-journalist Miss Gertrude Jekyll also promoted the heath-garden as a separate horticultural entity in the fifth edition of her book *Wall and water gardens* (1913).

Fig. 20. Water-colour by M. H. Baillie Scott, showing the setting, including heathers (foreground), for the house named Heather Cottage that was being built in Berkshire; from his book *Houses and gardens* (1906).

> One of the wilder paths from the rocky places may well lead to the Heath-garden. ... The
> main intention of the Heath-garden is the conversion of a piece of profitless sandy waste to
> a place of beauty and delight. The fact of the very poor soil restricting the number of plants
> that can be used may be regarded as an advantage, because it tends to simplicity and breadth,
> those precious qualities that so often are lacking in ordinary aspects of horticultural practice.

Miss Jekyll's heath-garden was not a monoculture of heathers but could accommodate other members
of the Ericaceae, and a range of other shrubs, such as brooms (*Cytisus*, *Genista*), rock-roses and sun-
roses (*Helianthemum*, *Cistus*), harebells (*Campanula rotundifolia*: "nothing is easier to grow in quantity
from seed") or the blue-flowered sheep's-bit (*Jasione montana*: "so welcome ... an annual, but once in
place will reproduce itself from seed."). While the choice of plants was restricted, she cautioned that
there were too many heathers ("some fifty varieties") available from nurseries:

> ... far too many for use in any one Heath-garden; for a good stretch of one kind at a time
> may make a delightful picture, while a dozen of each of the fifty kinds will only show as a
> collection of samples; though clever combination, and above all plenty of space, will allow
> a fair number of varieties.

Jekyll (1913: 193) suggested making "long-shaped wedges or drifts of a few of the best varieties.
These range from an unusually tall white to a dwarf form looking almost like a closely tufted moss".
Paths within the heather garden could be "natural sandy earth" or a short turf of grasses with
"an abundance of the sweet wild Thymes, while the paler-leaved *Thymus lanuginosus* may well
carpet some of the smaller grey-leaved Heaths" (Jekyll 1913: 197). Or, if heathers and whortleberry
(*Vaccinium myrtillus*) are already growing there and the plants are not too old, they can be "cut down
close to form a delightfully springy tuft ... which will only want mowing one, or at most twice a
year" (Jekyll 1913: 199).

The most diligent exponent of heaths during the first half of the twentieth century, apart from a
few nurserymen with vested interests, was Arthur Tysilio Johnson at Bulkeley Mill near Conwy in
North Wales. Johnson and his wife Nora formed a garden that "was perhaps the most important and
original of what are called "ecological" gardens today, where each plant's preferences were studied,
bringing into things a wholesomeness and a gentleness new to the [twentieth] century" (Thomas
2003: 146). In 1928, A. T. Johnson wrote that:

> ... in these days of rock-garden and the wild and woodland garden, when most garden-
> lovers are striving to get away from the formal, to foster a more artistic spirit and create
> a more natural effect, the Heaths must come into their own, since for all these phases of
> garden work they are ideal subjects.

The subheadings within the fourth chapter of Johnson's book on *The hardy heaths* (1928) provide
a convenient outline of the potential uses of "Heaths as garden subjects": "Heaths in the rock-
garden"; "As edging and hedge plants"; "For carpeting"; "In the mixed border"; and "As bedding
shrubs". The final section was headed "Heath walks", and concerned the making of heathery paths,
as previously advocated by Gertrude Jekyll:

> There is yet another opportunity for the use of Heaths, and that is the covering of walks
> as a substitute for turf. Every one who has an observant eye must have noticed how
> moorland tracks are often closely carpeted by heaths. They are pleasantly green at all
> seasons, delightfully springy to walk upon, and wear extraordinarily well. There is no
> reason whatsoever why we should not again take a leaf out of Nature's book and introduce
> that feature to our gardens. Heath walks or drives are most appropriate in wild or open

woodland gardens, in the Pinetum, among Rhododendrons and other shrubs. For the Heath Garden itself there could be nothing better, and in many rock-gardens they would be most distinctive and harmonious.

The most suitable Heath for this purpose is *E. cinerea*, but *Calluna vulgaris* is also good, and these do quite well together. The best way to lay down a Heath walk is by sowing the seed broadcast about the end of May. If the surface is hard, stony, and well drained, as a path should be, so much the better will the seedlings grow. The seeds only require to be raked in, this being followed by a good rolling, and the latter should be repeated once or twice during the first summer. For one year the seedlings may be allowed to grow at will, but when they are about a year old they may be mown, the machine being set rather high, especially to begin with. It is a good plan to top-dress lightly with sand and sifted loam about a month before mowing commences, and this dressing may be worked-in with a besom and rolled. (Johnson 1928: 39–40).

NURSERIES, NURSERYMEN AND PUBLICATIONS

The creation of heather gardens would not have been possible without the ready availability of cultivars, and the less common species. While nurseries had been propagating hardy heathers on a small scale since the late 1700s, there were very few named "varieties" (cultivars) until the late 1800s. "Varieties" of *Calluna vulgaris* available before the 1850s included plants with yellow foliage, silvery foliage, "double" flowers, crimson flowers as well as several with white blossoms; these were usually dubbed with Latin names, adjectives signifying a distinctive characteristic (*lutea*, *argentea*, *plena*, *rubra*, *alba*, for example). The earliest to be given more distinctive names seem to have been: 'Alportii' (upright, dark green foliage, crimson flowers) and 'Hammondii' (vigorous, light green foliage, white flowers) in the 1850s; and 'Foxii' (dwarf, mid-green foliage, purple flowers) and 'Serlei' (upright, dense green foliage, white flowers) in the 1860s.[8] The history of double-flowered *C. vulgaris* is very complicated (McClintock 1966c). William Curtis (1746–1799), botanist, nurseryman and founder of the *Botanical magazine*, apparently had a plant as early as 1789, presumably in his nursery at Queen's Elm, Brompton. Both white and pink double-flowered clones of *Calluna* were known in the early 1800s to the German plantsman and author of a monograph on *Erica*, Carl Friedrich Waitz (1805).

The earliest distinct "varieties" of St Dabeoc's heath were the white-flowered, which was collected as early as 1813, and a dwarf one which was grown in Britain and on the continent in the 1820s. These were soon followed by other colour variants: present-day clones named 'Atropurpurea' (deep purple) and 'Coccinea' (ruby-crimson) probably do not represent the plants grown in 1830s even though they retain names first published then. The extraordinary clone with striped flowers called 'Bicolor' (see p. 360) made its entrance in the 1870s, while William Robinson (1911: 132) praised an anonymous "very bright and finely-coloured red form which is to be found in some nurseries and should be sought out and increased ...". It may have been 'Coccinea'.

Turning to the hardy *Erica*, as with *Calluna* "varieties" any that were regarded as garden-worthy were given Latin names indicating their distinguishing characteristics. Atypical plants with white flowers, as in *Calluna* and *Daboecia*, were among the earliest to be propagated and named (generally as *alba*) (McClintock 1984a). Occasional wild-collected oddities, usually named after their finders, were propagated and distributed: *E. lawsoniana* (dubiously represented today by *E. mackayana* 'Lawsoniana'; see Nelson (2000d)), *E. maweana* (correctly *E. ciliaris* 'Mawiana'; see Nelson (2003b)), and *E. stuartii* (now *E.* × *stuartii* 'Stuart's Original': ® E.2007.03; see p. 230) were all introduced to gardeners in the late 1800s. *E. crawfordii* (indistinguishable, probably, from *E. mackayana* 'Plena'; see Nelson (1995b); p. 230) was found in 1901. The move away from Latin epithets to "fancy" names — in other words, names not in Latin form — occurred in 1911 (McClintock 1987) when James Backhouse & Son Ltd, The Nurseries, York, issued a special catalogue of *Erica carnea* cultivars. This chamois-leather encased publication promoted a dozen clones of *E. carnea* that had been selected by Richard Potter (1844–1922) (Backhouse 1911); each clone was illustrated by two coloured photographs. Evidently Potter had an "eye" for a good plant, and while on plant-hunting holidays in Switzerland had collected cuttings of the best heathers he saw. The names chosen for the resulting introductions were apposite. Members of the British royal family were prominently represented: 'King George V' (p. 366), 'Queen Mary', 'Prince of Wales' and 'Queen of Spain'; followed by the Backhouses and their friends, 'James Backhouse', 'C. J. Backhouse', 'Mrs Sam Doncaster'[9] and the mysterious 'Thomas Kingscote'. The other clones received undistinguished names and Potter himself was never accorded the accolade of having a heather named after him!

The first British nursery to specialise in hardy heathers was Maxwell & Beale of Broadstone in Dorset. In 1921, this firm introduced four new cultivars: *Erica cinerea* 'Frances's Variety' and 'Pallida' (although this epithet had been used before), and *E. tetralix* 'Ruby's Variety' (named after Miss Ruby Beale) and 'Rufus' (see Jones 1978). The driving force behind this nursery was Douglas Fyfe Maxwell (1891–1963) who was indefatigable in spotting variations among wild heathers, propagating these and introducing them. By 1926 Maxwell & Beale listed 63 cultivars, and in Maxwell's book *The low road: hardy heathers and the heather garden*, published the following year, 106 cultivars are mentioned.

Some other nurserymen also took a keen interest in hardy heaths, but not to the exclusion of shrubs from many different families. James Smith of Darley Dale in Derbyshire had been marketing newly selected heaths since the 1850s. The nursery is notable as the place where *Erica* × *darleyensis* was first noticed and propagated in the 1890s. Veitch of Exeter, Devon, was the source of *E.* × *veitchii* in the early 1900s. The Daisy Hill Nursery in Newry, County Down, was another that produced novel cultivars in the first decades of the twentieth century (Nelson & Grills 1998).

D. F. Maxwell's oddly titled book *The low road: hardy heathers and the heather garden* was a landmark in the progress of heathers as hardy, outdoor plants for British gardens. It was the first book devoted to the hardy species and their cultivation. The next year A. T. Johnson's *The hardy heaths and some of their nearer allies*, pocket-sized and obviously about heathers (67 cultivars were mentioned), was issued by The Gardeners' Chronicle Company. Whereas Maxwell's book was never reissued, Johnson's was updated and a second edition (including more than 140 cultivars) was released in 1956. Johnson, meanwhile, had published numerous articles about hardy heaths, and heathers are mentioned in his other books including *The mill garden* (1950). Frederick J. Chapple's *The heather garden* first appeared in 1952, with a second edition in 1960; this book was revised again and reissued in 1964 by which time The Heather Society had been established with Chapple as its first President.

The Heather Society was formed following a meeting, chaired by Sir John Charrington (see p. 357), held in the headquarters of the Royal Horticultural Society in London on 20 February 1963. The Society, which had been Charrington's idea (Charrington 1962; MacLeod 1963; Patrick 1963), was undoubtedly influential in promoting the use of heathers in gardens during the next three decades. It was appointed the International Registration Authority (now International Cultivar Registration Authority) for heathers at the International Horticultural Congress held in Tel Aviv, Israel, during March 1970. Between 1971 and 1981 the Society organised a series of trials of cultivars of hardy heaths at the Northern Horticultural Society's Harlow Carr Gardens, Harrogate, the invaluable results of which were published (Vickers 1976; Anonymous 1983; Julian & Newton 2004). Similar trials took place subsequently under the auspices of Royal Boskoop Horticultural Society in The Netherlands (see, for example, van de Laar 1998).

During the 1960s and 1970s there were several important specialist heather nurseries in Britain, of which Beechwood Nurseries at Beoley, Worcestershire, run by Joseph Weetmen Sparkes (1894–1981) (Jackson 2009) — "he was always "Sparky" to me" (Nicholson 1982) — was the earliest and one of the most productive of outstanding new cultivars (Everett 1983). Of the 110 heathers currently holding the Royal Horticultural Society's Award of Garden Merit, ten came from Sparkes's nursery. John Letts had a garden and nursery at Windlesham in Surrey, and in 1966 issued his catalogue in book form: *Handbook of hardy heaths and heathers*. His garden, Foxhollow, demonstrated how heathers could be used to fine effect. At the same period, Brian Proudley (d. 2003) and his wife Valerie established their nursery, first in the village of St Briavels and later at Blakeney, both places within the Forest of Dean, Gloucestershire. They published *Heathers in colour* in 1974, and it remained in print for more than 15 years. Sparkes, Letts and the Proudleys were early supporters of The Heather Society. Other publications promoting hardy heathers as garden plants appeared in the last three decades of the twentieth century, of which probably the most notable was Terry Underhill's *Heaths and heathers* (1971), greatly enlarged, updated and republished in 1990 as *Heaths and heathers the grower's encyclopaedia*.

From its start The Heather Society attracted members from other parts of Europe and elsewhere in the temperate zone. Its membership peaked at around 1,500 between 1982 and 1986 (McClintock 1987). A sister society, Ericultura, was formed in The Netherlands in 1971, and at one time it had almost one thousand members (van de Laar 1978: 11). Gesellschaft der Heidefreunde was formed in Germany in 1977. All three European societies published journals: *Heathers* (formerly *Yearbook of The Heather Society*) has been issued every year since 1963 (except 1968) while its companion *Bulletin* comes out three times each year and is now in its seventh volume; *Ericultura* was a quarterly; *Der Heidegarten* is issued twice a year. Both Dutch and German heather enthusiasts have had the benefit of books: Harry van de Laar's *Het Heidetuinboek* published in 1974 was translated into German and English (van de Laar 1978), but *Heidegarten* by Lothar Denkewitz (1987) was available only in German. There has never been a society devoted to heathers based in France but *La bruyère* by Baron Bernard de la Rochefoucauld (1979, 1997) championed these shrubs to Francophone gardeners.

The Heather Society's influence also permeated North America. Following a visit in May 1977 with the Northern Group of The Heather Society to the garden at Harlow Carr, Ken Wilson (a Yorkshire-born horticulturist; see p. 385) returned to Vancouver, Canada, determined to help establish a "Pacific North West" group (cf. Nelson 2008c). Formally inaugurated as the Pacific Northwest Heather Society in August 1978 with members in British Columbia and the states of Washington, Oregon and California, this group flourished and eventually transformed into the North American Heather Society, a federal organisation now comprising six societies, five in the USA and one in Canada. The new society assisted in the introduction of cultivars and in establishing trials to investigate, for example, the performance of heathers in Californian gardens (Eighme 1982). One of the founding members was Mrs Dorothy Metheny who wrote and illustrated *Hardy heather species and some related plants* (1991).

In the future, the heyday of gardening with heathers in western Europe, and particularly in Britain and Ireland, may be determined as the 1970s and 1980s, even though there are probably more named cultivars extant today than there have ever been. In those decades, heathers were promoted as labour-saving, ground-covering shrubs, eminently suitable for the small, suburban garden. They

Fig. 21. Informal heather garden in the University of Liverpool's Ness Botanic Garden, Wirral, Cheshire (E. C. Nelson).

Fig. 22. The informal garden of heathers created by the late Colonel Jim Thomson at Manchester in northern California (E. C. Nelson).

Fig. 23. A formal "Irish knot garden" made with cultivars of the Irish heath (*Erica erigena*) by Daphne Everett at The Bannut near Bromyard in Herefordshire, photographed in 2007. The principal heathers used are 'W. T. Rackliff' (dark green) and 'Golden Lady' (yellow) with 'Irish Dusk' as a cone-shaped "dot plant".

Fig. 24. The harp, Ireland's national emblem, on the Ennis Road roundabout on the outskirts of Limerick, composed from two cultivars of *Erica* × *darleyensis* (Darley Dale heath), 'Darley Dale' and 'Silberschmelze', with the yellow-foliaged *E. vagans* 'Valerie Proudley' (Cornish heath). (© John Herriott 2007.)

were planted in island beds mingled with slow-growing conifers and other dwarf evergreen shrubs, and there left to their own devices. The best known exponent of island beds densely planted with heathers and conifers is Adrian Bloom of Bressingham in Norfolk (see, for example, Mikolajski 1997: 11). The same scheme worked in larger gardens, as at Great Comp, in Kent, where a spacious vista owed "its low maintenance and serenity to a concentration of conifers and heathers" (Thomas 1984). In his books *Colour in the winter garden* (1957) and especially *Plants for ground-cover* (1970), Graham Stuart Thomas was among the authors who during the late twentieth century extolled the virtues of hardy heathers although they were simply echoing, or "interpreting in modern prosaic terms" (Bisgrove 1990: 287), the ideas and practices of Gertrude Jekyll and William Robinson. In the small garden of a tidy-minded late twentieth-century gardener "... multi-coloured heathers and dwarf conifers provide[d] that firmness and delight so admired in the seventeenth century, while offering a very distant reminder of our heather-clad hills and ... more rugged wilderness ..." (Bisgrove 1990: 291). The heaths and heathers, "social shrubs of dwarf bushy habit" to use Thomas's terminology (1970), furnished "interest of as varied a nature and colouring as any." Yet, achieving the perfect heather garden, or the perfect combination of heathers with other plants, even slow-growing conifers, is not as simple as it may seem. Thomas asserted in one of his later books that:

> It is no use being small-minded about planting these little shrubs. How often does one see a collection of a dozen or so different sorts planted in a prescribed area, all distinct in colour and growth. This is the worst sort of planting of them, brought about by our acquisitive tendencies and a desire for variety coming to the fore in our minds after a visit to a garden centre. (Thomas 1997).

None of the virtues of heaths and heathers has dissipated. They are among the most useful and versatile plants, for informal plantings (Figs 21 & 22) and formal ones whether on a small (Fig. 23), or a grand scale (Fig. 24). They are as attractive as ever, when well grown and well maintained.

> Heathers spell the open air, the moorland, the wide sweeps of hill and dale, and the companionship of grasses and small-leafed shrubs and little trees. They bring to the garden, if rightly used, a smell of the wild open country. (Thomas 1996).

CULTIVATION AND CARE IN GARDENS

The hardy heathers are easy, rewarding and durable plants, although they may not suit every garden or every gardener. While the vast majority of heathers are planted directly into the ground, there is an increasing trend towards growing them in containers such as troughs and window-boxes. The most important thing to remember about growing heathers in containers is that the watering regime needs to be carefully controlled; the soil should never be allowed to dry out nor should it ever become waterlogged.

As with any group of plants as diverse as the northern, hardy heaths, there can be no generalised rules of cultivation, maintenance and propagation. Even within a species or hybrid, individual cultivars may require different treatments to achieve their best. Moreover, every garden is different, with a unique set of characteristics — soil, climate, aspect, shade, will vary from garden to garden, imperceptibly perhaps, yet certainly. Even within a small garden, there are innumerable nuances that are detectable only by experience — the simplest relate to aspect, whereby one part may get no sun at certain seasons, while other parts perhaps have either morning or afternoon sun. And, for some

heathers aspect is very important. Other well-established plants, especially trees and large evergreen shrubs, will affect conditions too, through shading and, particularly, their extensive root systems.

There is a general fallacy that all heathers need lime-free, peaty soil, even pure peat, to flourish: this is not a new prescription but goes back into the early nineteenth century, if not earlier. While heathers will all grow in neutral to mildly acidic (with pH7 or less), lime-free soil, there are quite a few that will flourish in calcareous soils, with a pH higher than 8, including *Erica carnea* (winter heath), *E. erigena* (Irish heath) and their hybrid *E. × darleyensis* (Darley Dale heath). Others, for example, *E. manipuliflora* (whorled heath), *E. terminalis* (Corsican heath), *E. vagans* (Cornish heath) and *E. arborea* (tree heather) have some tolerance of lime and will thrive in soils with pH7–8, and we are learning that this is also the case for some of the recently created hybrids such as E. × *gaudificans* (Edewecht heath) and *E. × oldenburgensis* (Oldenburg heath) (see, for example, Canovan 2005).

Surprisingly, perhaps, some cultivars of *Daboecia cantabrica* (St Dabeoc's heath) and *D. × scotica* display tolerance to lime in the soil as long as this is well-drained.

Drainage is a factor that is rarely mentioned when the growing of heathers is discussed, but several of the northern heaths that are exceptionally rare in cultivation, being only in specialist collections and unavailable from commercial nurseries, undoubtedly require very free-draining soil that is also moisture-retentive (a extremely difficult balance to achieve). They are best grown in voluminous containers, such as troughs or old ceramic sinks, and would probably benefit from the addition of limestone chippings to the soil. These include *Erica sicula* (Sicilian heath) from its various habitats and *E. sicula* subsp. *bocquetii* (Bocquet's heath).

Although few people attempt to grow hardy heathers as pond-side or bog-garden shrubs, there are some which might succeed, given that their natural habitats include lake-side and riverine habitats, where they have their roots in water. *Erica erigena*, which is often half-submerged in the winter around the loughs and in the bogs of western Ireland, and *E. andevalensis*, which grows close to rivers that must, when they are in spate, overflow the plants, are perhaps suitable for wet, but not perpetually waterlogged, places. Some heathers will rapidly succumb if their roots become waterlogged even briefly, including *E. mackayana* and *E. multiflora*, as I can vouch.

Heathers are plants of open habitats although some will tolerate shade and even grow in open woodland. In gardens, too, heathers thrive only in situations where there is no shade from surrounding plants and buildings, or very little. If there is too much shade, the plants will not be floriferous and will become scruffy with unattractive, long, leafy shoots. "Dry shade", when both water and light are severely restricted, as under wide-spreading trees, is inimical to heathers.

Otherwise, hardy heathers are undemanding plants. In the early years after planting, an annual mulch with an appropriate mixture such as very well rotted leaf mould, shredded bark or moss peat may be beneficial especially on dry soils to help to retain moisture. A mulch will also help prevent annual weeds from becoming a nuisance. Any weeds that do appear should be removed by hand; hoeing around heathers is not good practice because they tend to be surface-rooting. Slow-release fertilisers like bone-meal or blood-and-bone can be applied, but in moderation — on the whole heathers do not need to be fed — but even this is not necessary if the heathers are mulched. Once established heathers should not need extra watering.

Fig. 25 (opposite). "Heathers spell the open air ..." (G. S. Thomas): this 1923 poster for London's Underground, simply titled "Heather", is a stylised portrait of bell heather (*Erica cinerea*) by Irene Fawkes (1886–1977) (Nelson 2006d). This was one of a series of posters promoting the use of the Underground as a means of getting into "the open air", and to places such as the Royal Botanic Gardens, Kew (Riddell & Stearn 1994). (©TFL from the London Transport Museum collection).

In cultivation heathers do not, in general, suffer from pests or diseases, although some soil-borne fungi can affect them. Plants that show signs of distress should be removed and burned to prevent the spread of the disease. The heather beetle, *Lochmaea suturalis* (Thomson, 1866), which does attack plants of *Calluna vulgaris* in the wild, is unlikely to be a nuisance in gardens except those close to severely infected heathland. The beetle will sometimes also feed on *Erica tetralix* and *E. cinerea* (Pakeman *et al.* 2002). The imago and larva of the beetle graze on the foliage of ling, and it is likely that in the wild the beetle's populations are controlled by natural predation by insectivorous birds and other natural enemies including parasites (Pakeman *et al.* 2002). Gall-midges can also infect plants causing not loss of foliage but the formation of malformed shoot tips that look a little like miniature pineapples (hence the name pineapple galls). These can occur sometimes on cultivated heathers; *E. carnea* (Harris 1966; Shaw & Sutherland 1966; Jones & Jones 1981) and *E. vagans* (Maginess 1981) seem to be the only ones affected in Britain. Cutting off the galls and burning the infested shoots will control, but probably not eliminate, the insect. On the other hand, Daphne Maginess (1981) reported that after a patch of *E. carnea* 'Ruby Glow' was badly infested, she cut it hard back and burnt the prunings, and there was no reappearance of the galls in subsequent years on the regenerating plant. (For reports of galls on heathers in the wild, see p. 217, note 7.)

PRUNING

Pruning of cultivated heathers is a vexed and disputed topic. Some enthusiasts, for example the late David McClintock, advocated that heathers should never be pruned; he "deplore[d] the fetish that all heathers must be cut, clipped, shaved", resorting to the pun that clipping was a "shear waste of time" (McClintock 1974a). That is perhaps the ideal situation, and certainly saves time and energy, but only given unlimited space and providing a ragged, "wild" appearance is not unpleasing.

In most situations, however, especially in small gardens, some control needs to be exercised and annual pruning is desirable. There are several well-established rules for pruning heathers. Winter-flowering heathers (*Erica carnea*, *E. erigena* and *E.* × *darleyensis*) should be trimmed only in spring or early summer, immediately after their flowers fade, because later pruning will remove the flower-initials for the following season which form immediately the current blooms fade. Summer-flowering heathers may be pruned in late winter or early spring; this applies to *Calluna* and *Daboecia* as well as *Erica* cultivars. However, cultivars that have colourful young shoots in springtime — these are almost all hybrids of *Erica*, apart from certain selections of *Calluna* — should be trimmed immediately after the flowers have faded in late summer or autumn.

Graham Stuart Thomas (1997), after long experience, made the following recommendations about clipping heathers.

> I think this is most easily done with long-handled shears with horizontal blades. One can walk over the dwarf shrubs with impunity and, if it is done every year, the clippings will be short, need not be picked up and will soon be obscured by the new growths. ... The clipping serves several purposes: it keeps plants compact and thus more resilient to autumn's fallen leaves, and it tidies away the dead flower spikes; furthermore, by circumspect use of the clippers an undulating natural effect can be achieved, some demanding closer clipping than others.

There are no fixed dimensions for heathers — some are taller, some broader — and any pruning regime needs to strike the balance between space occupied and vigour of growth. The basic intention of pruning should be to produce a neat plant and remove unsightly or dead material, particularly the withered flowers. *Daboecia* plants are easiest as they only need to have the old flowering spikes removed. In *Calluna* the aim should be to trim back to just below the withered flowers.

Bear in mind that some summer-blooming heathers remain attractive throughout the autumn and winter by virtue of the russet or even silvery tones of the old flowers, or of the fruits in the case of *Daboecia* ("one of the best of the winter colours", according to McClintock (1974a)). Heathers that provide this subtle extension of interest in the garden need to be treated separately, and pruning should be done only after the plants cease to be attractive in late winter.

It is also important to remember that severe clipping not only results in ugly plants but also will retard and reduce flowering. Although heathers that are used in formal gardens will need to be clipped regularly so that they retain their shape, continuous heavy clipping will reduce the life of these plants. Excessive pruning every year will also weaken heathers, especially those that will not resprout from old wood.

On the other hand, old heathers can often be rejuvenated by a very drastic pruning, by cutting them back to short stumps from which new shoots will sprout (as they would when the heathers are burnt in their natural habitats). However, caution is required as not every species will respond by resprouting. Those known to have a capacity to regenerate in this way include *Erica arborea*, *E. australis*, *E. lusitanica* and *E. vagans*. Others such as *E. ciliaris*, *E. mackayana*, *E. × stuartii*, *E. multiflora* and *E. manipuliflora* should do likewise. *E. carnea*, *E. erigena* and their hybrid, *E. × darleyensis*, as well as *E. × griffithsii*, will respond to hard-pruning by resprouting from stumps that can be as short as 7 cm in the case of *E. carnea*, but should be twice as long as that for the others (Small 2008).

Experimental work on *Calluna* (summarised by Gimingham 1972) indicates that stems more than about 15 years old have a very greatly reduced capacity to resprout after clipping or burning. Extrapolating from this, it is most unlikely that ancient bushes of *Calluna* will react to drastic pruning by producing a new crop of shoots.

Old heather plants can be "rejuvenated" by layering — in effect, propagating new plants from the decrepit one (see below).

VEGETATIVE PROPAGATION

Named cultivars *must* be propagated by vegetative methods (using cuttings, or by layering), and the same methods need to be applied for any unnamed plants that are particularly good and need to be multiplied so that the progeny are identical to the parent. Seed collected from cultivars — indeed, from the vast majority of plants grown in gardens — will not "come true": the seedlings will usually be a ragbag of colours and habits. However, growing heathers from seeds need not be discouraged (see below) if you want novelty and are prepared to accept that many seedlings will be rather "ordinary". In particular, the vast majority of seedlings raised from white-flowered variants of heathers which normally have "pink" flowers, will not be white; this applies even when both the seed-parent and the pollen-parent are white-blossomed.

Vegetative propagation in hardy heathers is usually easy, whether by cuttings or layers, although a few species and cultivars may be slow to root and the percentage take may be small. Layering is simple and usually sure because the "parent" plant remains in the place where it is growing and continues to nurture the shoots while they are layered. Old plants can "rejuvenated" by layering. Layering is achieved by pinning shoots down so that they are at least touching the soil or, better still, are partly buried. This can be done with a peg made from a loop of strong wire, or a small stone. Cover the part that is pinned down with a small amount of peat/sand/soil mixture, but make sure the shoot tip is showing. Better rooting may be achieved if the shoot being layered is bent sharply upward at the point of pinning (see Underhill 1971: 70; 1990: 91–92). Layering can be done at any time of year. When a layer has rooted, it should be carefully severed from the parent plant but left in place for about a month. After that period the new plantlet can be lifted and transplanted, and if the parent is an old, moribund heather, it can be removed and destroyed.

When propagating by cuttings, there is a choice of season (see Lamb (1994) or Small (1995) for more details). In the late summer (July–August, in the northern hemisphere), small (± 3 cm long), semi-mature shoots can be selected and inserted in a cold frame or in pots or "seed trays". Larger, hardwood cuttings (± 4 cm) can also be used, taken in mid-autumn (October, in the northern hemisphere); these can also be placed in a cold frame or pots, and should be potted on in April. By the late summer the young plants should be 10–15 cm across and in bloom.

In the case of *Calluna* cultivars, the young, leafy growth that appears in spring above the previous year's flowers can be used for cuttings — this is handy as the cuttings can be taken when the shrubs are pruned. The new growth should be at least 1 cm long, and the cuttings should include about 1 cm of the older growth. The dead flowers on these spring cuttings can be removed simply by rubbing your finger and thumb along the stem (Small 1995).

In all cases, any flower buds or open flowers should be removed. For cuttings taken from summer-blooming heathers, the flowering portion should be cut off with a sharp knife. It may be difficult to find suitable non-flowering shoots on spring-blooming heathers during the summer, but flower buds can be removed by gently rubbing your thumb and finger *upwards* along the stem (Small 1995). It is not, as often said, essential to remove the leaves from the lower portion of the stems.

The rooting medium recommended by Lamb (1994) is moss peat. Small (1995) suggested three parts moss peat and one part Perlite (or a gritty, lime-free sand). If using a cold frame, after watering-in the cuttings cover them with transparent plastic sheet which needs to be inspected and turned every ten days or so. If using pots or trays, water-in the cuttings and allow them to drain for about 20 minutes. Then enclose the pots or seed trays in clear polythene bags, without holes, and put in a shady place, such as against a north-facing wall (in the southern hemisphere, a south-facing wall), so that no direct sunlight can fall on the containers. Leave these for several months, checking occasionally that the polythene is fogged — if not, water again, allow to drain, and replace in the bag (Small 1995). At the same time any cuttings that have died should be removed to prevent disease.

For commercial propagation, if bottom-heat is available, rooting time is greatly reduced, and hardwood cuttings can be taken in late winter (February); these should root within seven to nine weeks (Lamb 1994).

It is possible to store heather cuttings — this is a valuable tip if you chance on an interesting heather in the wild. Small (3–5 cm long) cuttings of non-flowering shoots, if possible, carefully removed from the parent plant should be promptly enclosed in a sealable plastic bag into which a tiny amount of water (a few drops at most) is introduced (if the cuttings are already damp with dew or rain, no additional water is needed). These can be stored in a cool place, away from direct sunlight, for several weeks. *Calluna* cuttings kept in store rooted best when planted in a propagation unit to commence rooting after only two weeks' storage, but nearly as well after four and seven weeks' storage (Small 1973). In my experience, hardy *Erica* species can be treated in the same way. It is essential that the cuttings are not allowed to dry out: they have to be inserted into a plastic bag almost instantly, and the bag must be sealed immediately. In the wild, a small piece of *Sphagnum*, if it is available, can be used in place of water.

While some people may deplore any interference with wild plants, the careful removal of a small number of cuttings (as suggested above) from an interesting heather is not likely to cause any harm. These cuttings are much more likely to result in successful propagation than the drastic, and reprehensible, method of digging up a whole plant with a ball of soil. Moreover, transplanting mature heathers, even well-grown ones within a garden, is a perilous procedure that usually results in the death of the plant.

Testing whether cuttings have rooted is simple. Having left the cuttings for about eight to ten weeks, gently pull, and if there is resistance, roots will have been produced. If a cutting has not rooted, it will come out of the soil easily, and can be replaced.

RAISING SEEDLINGS

Heathers can be grown from seed, but, to repeat, it is essential to remember that seedlings will not necessarily resemble the parent plant, so this method of propagation is not appropriate for cultivars.

It is important to use fresh seed as the viability of seeds and their capacity to germinate does decrease with age, although there is evidence that for at least one species, *Erica australis* (Valbuena & Vera 2002), stored seed will germinate better under certain circumstances. Seed can usually be collected from the hardy species about two months (eight to twelve weeks) after the conclusion of flowering (after pollination). Individual species differ in the time taken for their fruits to ripen and dehisce, and the swelling capsules within the withered flowers need to be examined regularly to determine when they are ready for harvesting — they will go brown and start to split open (Griffiths 1985). In *Daboecia*, the corolla falls away after fertilisation, and thus the capsules can be observed easily; they can be harvested when they begin to show signs of dehiscing and releasing their seeds. The simplest way to collect seed of *Calluna* and *Erica* is to gather the withered flowers, spread these on a sheet of white paper and leave in a dry, airy place. When thoroughly dry, these old flowers should be gently rolled between thumb and finger to shatter them. A fine sieve can be used to separate the seeds from the chaff.

Seeds should be sown as soon as possible after collection in the spring or in the autumn onto the surface of the soil; they must not be buried because light seems to be a requirement for germination in some species at least (*Erica arborea, E. tetralix* and *E. vagans*; cf. Valbuena & Vera 2002). Sellers (1999) recommends a mixture of a proprietary (lime-free) compost suitable for Ericaceae, mixed with an equal proportion of lime-free sand and grit (or Perlite). Using a seed tray 7.5–10 cm deep, fill this with a layer of fine gravel (about ± 0.5 cm), a layer of grit (± 1 cm), and then fill with the prepared mixture, gently firming it down to give a flat sowing surface. Stand the tray in a shallow container of rainwater until the surface of the soil is moist, and then allow to drain.

The easiest way to sow the seeds evenly is to mix them with some *dry* silver sand and sprinkle this mixture evenly over the surface. To inhibit moss, and to help retain moisture, a *very* shallow layer of finely sieved silver sand can be added. Place the tray on some capillary matting or in a shallow tray containing a layer of fine gravel, so that water can be absorbed by capillary action. The compost should not be allowed to dry out, but the tray should not stand in water. Mist-spraying the surface twice a day even after the seed has germinated will help to keep the soil moist (Sellers 1999).

Germination will take several weeks, maybe as long as three months, or even one year. Allow the seedlings to remain in the seed trays for at least a further three months until they are large enough to handle before attempting to transplant them into individual pots or compartmented trays.

While field research does indicate that some of the European heaths germinate most readily after fire (Valbuena *et al.* 2000; Valbuena & Vera 2002), the seeds of most hardy species will germinate readily without any pre-treatment as long as they are fresh. On the other hand, the dormancy exhibited by the seeds of some Cape heaths has been shown to be broken principally by the smoke from fires. Thus when germinating seeds of the African *Erica* species it may be necessary to use "dehydrated smoke" to break dormancy and promote or improve germination. "Instant SmokePlus seed primer", developed at the National Botanical Garden, Kirstenbosch, Cape Town, is a simple, proven method that obviates the dangers of using real fire to "smoke" the seeds (see Kotze 1996).

NOTES

[1] The Bobarts' *Hortus siccus* (herbarium) contains two *Erica* specimens: *E. tetralix* (labelled "Erica vulgaris hirsutior Common woolly Heath", and *E. cinerea* (labelled "Erica tenuifolia Thinn leafed Heath"). *Calluna vulgaris* is also represented by a single specimen (labelled "Erica vulgaris Comon Heath"). (http://herbaria.plants.ox.ac.uk/BOL/home/default.aspx accessed 26 July 2007.)

The specimens in this collection are labelled only with Latin polynomials and common names. There are no indications of provenance. Vines & Druce (1914) "considered that the specimens were collected entirely by [the younger] Bobart and that the majority were from the Botanic Gardens and Oxfordshire" As just shown, this *Hortus siccus* "provides a means of verifying the identities of names used in the two 'Catalogues' published for the Oxford Botanic Gardens in 1648 and 1658 and illustrates the range of plants that were being grown during the period" (http://herbaria.plants.ox.ac.uk/BOL/home/default.aspx).

[2] More probably, this was Cornish heath (*Erica vagans*), which by then was known to grow in Cornwall, and had been listed in 1724 as "Erica foliis Corios multiflora ... *Fir-leaved Heath with many Flowers*" by Dillenius (1724: 471).

[3] *Erica erigena* (as *E. hibernica*, syn. *E. viridipurpurea*) is also listed on the George Clifford Herbarium database (http://internt.nhm.ac.uk/jdsml/research-curation/projects/clifford-herbarium/list.dsml accessed 21 July 2007), but the specimen certainly does not represent that species (cf. Ross 1967: 69, 70; Nelson & Cafferty 2005). Only one of the three Clifford specimens listed as *E. arborea* appears to be correctly identified.

[4] Mrs Delany's "mosaick" of "Scarlet-flower'd heath" (see plate 15, p. 25 in Laird & Weisenberg-Roberts 2009) is one that incorporates a portion of the plant itself, in this case one flower and fragment of a leafy shoot. Hayden (1976: 13) stated that this showed "*E. coccinea* ... possibly *E. pillansii*": later, Hayden (1980: 143) identified this "mosaick" as possibly portraying *E. cruenta* which is also the most recent opinion (det. Dr E. G. H. Oliver) (see Edmondson & Nelson, in Laird & Weisenberg-Roberts 2009: 268–269).

[5] The craze was "fuelled" initially by the Cape heaths that were raised from seed collected by Francis Masson (1741–1805) during his first visit to the Cape of Good Hope, which had lasted two and a half years between October 1772 and March or April 1775. Aiton (1789) attributed to Masson the collection or introduction of 18 Cape species during 1774; two of his collections are dated 1773 and three are dated 1775. In all, more than half of the 41 *Erica* species listed at Kew in 1789 had been raised from seed obtained by Masson in southern Africa (Nelson & Oliver 2005a). Masson returned to England in 1775, and visited the Azores in 1777 (see Bradlow 1994). In 1786, he went back to the Cape where he remained for more than nine further years. By 1792, the specially built greenhouse in the Royal Garden at Kew contained more than 80 different *Erica* species. Aiton (1811) attributed an additional 63 Cape *Erica* species to Masson, collected and introduced between 1787 and 1795, some through James Lee's Hammersmith nursery.

Other collectors followed, including the Scot James Niven (cf. Nelson & Rourke 1993), and the enigmatic Jacob (or James) Mulder who apparently collected only for Richard Salisbury (cf. Salisbury 1802).

[6] This was undoubtedly Joldwynds, between Dorking and Guildford, the property of the ophthalmic surgeon and anatomist, Sir William Bowman (1816–1892). An earlier account (Alemon 1881) described Joldwynds as "a collector's garden in the best sense ... well ordered and well managed." While there was no description of a heather garden, the writer noted that "Conifers, herbaceous plants, Heaths, alpine plants, bog plants, rock plants, Lilies, all have their appropriate nooks, while bedding plants are conspicuous by their absence." Robinson (1897: 496) alluded to "some of the best and most successful" beds of heathers "on the level ground ... in the late Sir William Beaumont's garden in Surrey." But that also surely refers to Bowman's Joldwynds, and the surname is a mistake. I can trace no Sir William Beaumont known to have gardened in Surrey.

[7] Fenwick may have been influenced by an unidentified heather garden in Northumberland which he had recalled in 1932 as "the oldest heather garden I know" (Everett 2007), rather than by Gravetye Manor.

[8] As noted by Nelson & Small (2000), the origins of these epithets are not known, and even the spelling of the names is uncertain. The individuals after whom the heathers were named — Alport (or Allport), Fox, Hammond and Serle (or Searle or Serl) — cannot be identified.

[9] I am aware of two copies of this catalogue. One is in the Lindley Library of the Royal Horticultural Society, London. The second is in a private collection and bears the signature "Sam[l]. Doncaster", but no other annotations.

3. *CALLUNA* SALISB.: LING

LING. *n. s.* [*ling*, Islandick.]
I. Heath. This sense is retained in the northern counties; yet *Bacon* seems to distinguish them.
Heath, *li* and *ng*, and sedges. *Bacon's Natural History.*
Samuel Johnson, *A dictionary of the English language.* 1755.

This shrub constitutes a very distinct genus, which from its utility in making broom,
I have called Calluna.
R. A. Salisbury, *The generic characters in The English botany ...* 1806.

The genus *Calluna* contains a single species, *C. vulgaris*, which ranges from Flores, the westernmost island of the Azores (31°W) in the mid-Atlantic Ocean (Masson 1779; Tutin 1953; Schäfer 2002), eastwards through almost all of western and northern Europe into west-central Siberia (to approximately 95°E) (cf. Beijerinck 1936, 1940; Malyschev 1997: 231), and from Skipsfjorden (71°1'N), close to Nordkapp, the northern extremity of the Norwegian mainland (Alm 1999), southwards into the north of Morocco (Fennane & Ibn Tattou 2005) (Fig. 26). With the aid of humans, *Calluna* has also invaded and colonised substantial areas in the temperate zones of the western and southern hemispheres and, consequently, is sometimes treated there as a pest, a weed to be eradicated ruthlessly (for example, in Australia (see Sheppard *et al.* 2003) and New Zealand (see Syrett *et al.* 2000)). The worldwide export of peat for horticultural use is aiding the dispersal of *Calluna*; seedlings have been found in nurseries in South Africa which used imported Irish peat in potting mixes (E. G. H. Oliver, pers. comm.).

The success of a plant like *Calluna* in occupying new territories has much to do with its capacity to inhabit inhospitable habitats ranging from the edges of coastal cliffs where it can be seared by salt-laden winds and has adopted a completely prostrate habit, to the margins of active fumaroles where it will be heated by sulphur-laden steam. Ling tolerates high concentrations of some toxic elements (Marrs & Bannister 1978; Oxbrow & Moffat 1979), so it will colonise the spoil heaps of abandoned lead mines, as at Parys Mountain, Anglesey, north Wales, or abandoned copper mines in Avoca, County Wexford, Ireland. Research on the plants from those two localities indicates that the heather's capacity to tolerate high concentration of copper, zinc and lead, for example, is linked to the presence of symbiotic mycorrhizal fungi in their roots (Bradley *et al.* 1981; Gibson & Mitchell 2006).

Calluna differs from *Erica* and *Daboecia* in several significant characters. Although this characteristic is hardly ever mentioned, its leaves are sessile and have pronounced basal auricles, downward-pointing triangular lobes (Fig. 27D). In northern species of *Erica*, the leaves are always stalked and never have auricles. The leaves in *Daboecia* have prominent stalks, and they, too, lack auricles. In *Calluna*, the leaves are arrayed along the stems in alternating pairs, and so form into four rows; each leafy shoot, because of the tiny, tightly packed leaves, becomes four-sided. In *Erica*, the leaves are in whorls of three or (usually) four and are never so tightly packed as to form a four-sided shoot. The leaves of *Daboecia* are arranged loosely and alternately. In terms of the flowers, in northern-hemisphere *Erica* species and in *Daboecia*, the calyx lobes (sepals) are relatively small, usually less than half the length of

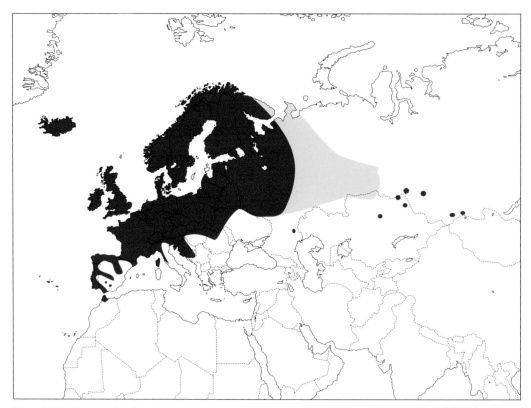

Fig. 26. Distribution of *Calluna vulgaris* in Eurasia and North Africa.

the corolla, so the two perianth whorls are clearly different in appearance. Contrariwise, in *Calluna*, the calyx lobes (sepals), forming the outer perianth whorl, are always slightly longer than the inner, corolla lobes, and are the same colour and texture as the corolla lobes. Those are the obvious differences. Others include the way the seed capsules open — septicidal in *Calluna* (and *Daboecia*) but loculicidal in *Erica* — and the number of ovules — 6–8 per locule in *Calluna*, more than 50 per locule in *Daboecia*, and 10 or more (to ±60) in *Erica*. For detailed discussion of the distinctions between the three genera see Oliver (2000).

There is a vast literature in a uncounted number of languages about *Calluna vulgaris*, its ecology and the ecology of the heathland and moorland that it dominates throughout north-western Europe. There is an entire monograph detailing the diversity in gross morphology that *Calluna* displays in one small country, Holland, an intense and, at times, hard-to-understand tome of 180 pages by the Dutch botanist Dr Willem Beijerinck (1891–1960). This includes an hierarchical classification of the species subdivided into varieties, forms and subforms (Beijerinck 1940 — I happen to possess the copy that Dr Beijerinck sent to Professor William T. Stearn). On a lesser scale, but no less carefully researched, is Professor C. H. Gimingham's biological flora, published in the *Journal of Ecology* (Gimingham 1960), in which many aspects of the wild plant are succinctly summarised. There are also numerous works that describe, or refer to, the (former) uses of *Calluna* in the economy of places where it grows naturally.

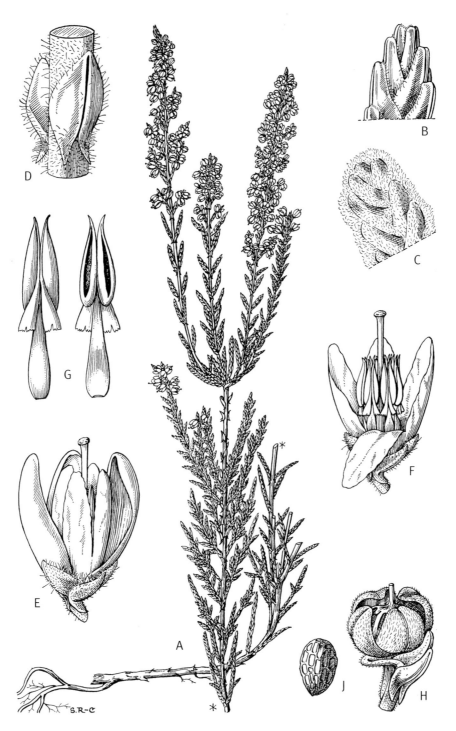

Fig. 27. *Calluna vulgaris.* **A** part of a plant × 1; **B** upper part of one of the short barren branches × 12; **C** part of barren branch of a hirsute variant × 12; **D** leaves on flowering stem × 12; **E** flower, a bract and one sepal removed × 12; **F** flower, sepals removed × 12; **G** stamens in back and front views × 24; **H** fruit, the persistent calyx, corolla and stamens removed × 8; **J** seed × 30. Drawn by Stella Ross-Craig.

CALLUNA AND LING

The Latin name for the genus was provided by Richard Salisbury in the preamble to the monograph on *Erica* that he read to the members of the Linnean Society of London on 6 October 1801, and which was published the following year in the Society's *Transactions*. Writing in Latin, Salisbury (1802: 317) declaimed:

> Mirum fortasse nonnullis videatur, Ericam *vulgarem* desiderari in sequentibus paginis: sciant autem velim, hancce stirpem, si quæ alia in toto Ordine, proprium constituere genus: jure antiquiore profecto suum nomen retinuisset, sed cum tot aliæ stirpes, apud omnes Botanicos jam eodem cognomine gaudeant, satius duxi hanc unam novo insignire titulo: itaque *Callunam* appelavi, ob usum ejus frequentissimum in scopis conficiendis: essentia generis, quâ differt ab *Ericá*, est in paricarpii valvis ad latera loculorum dehiscentibus, septis axi relictis: habitus, absque ullo rudimento petioli pedunculive, omnino sessilis.
>
> … in the works of all botanists may they [heaths] now enjoy the same generic name, [but] I have rather taken the lead to distinguish this one with a new title: and so I have called [it] *Calluna*, on account of its very frequent use in the construction of brooms: …

Liddell & Scott (1890) in their *A Greek–English lexicon* quoted Classical authors to explain καλλύνω, (*kallyno*), derived from καλός (*kalos*, beautiful). Sophocles used the word to mean "to beautify" and Aristotle and Polybius to mean "to sweep clean". Κάλλυντρον (*kallyntron*), according to Plutarch and Clement of Alexandria, is any implement for cleansing, a broom or a brush. Whether Salisbury intended the double meaning, to sweep clean and to beautify, we cannot tell.[1]

Salisbury (1802, 1806) did not formally make the combination *Calluna vulgaris* (as required under the present rules of nomenclature). The first person to do this was a somewhat obscure Manchester-based physician, Dr John Hull (1761–1843), a graduate of the University of Leiden, in the second edition of his *British flora* (1808: 114). Three years earlier, in the third edition of *Flore française* (Lamarck & Candolle 1805), the Swiss botanist Augustin-Pyramus de Candolle (1778–1841) had published the binomen *Calluna erica*, but the specific epithet *erica* is not the earliest available, being preceded by Linnaeus's (1753) *vulgaris*. Although Salisbury's and Hull's names, as a combination, have been available for two centuries, some nurserymen were still using *Erica vulgaris* for ling in the 1990s!

In fact, Richard Salisbury should not have renamed *Erica vulgaris* because that seems to have been the species on which Carl Linnaeus (1754b) based the generic description of *Erica*. In other words, *Calluna* should have stayed as *Erica*, and the multitude of other heather species should have been placed in a genus with another name. This was pointed out by, among others, Sir James Edward Smith, President of The Linnean Society of London, and one-time friend of Salisbury (see also Oliver 2000: 5). In the second edition of *The English flora* Smith (1824: 224) commented, with regard to Salisbury's, and Hull's *Calluna*:

> Although there is but one known species of this genus, the most common, if not perhaps the original, *Erica* of Dioscorides, Tournefort, or Linnaeus, its generic distinctions are so very important, that I gladly concur with Mr. Salisbury, who first pointed out those distinctions. To avoid the inconvenience of giving a new generic appellation to the hundreds of plants, familiar to every body as *Ericæ*, or Heaths, he has judiciously called our common ling, *Calluna*… .

Ling also has interesting etymological roots — Icelandic, according to the classic *A dictionary of the English language* compiled by Dr Samuel Johnson in the early 1750s. In Denmark, there are many toponyms incorporating "Lyng" and "Lung"; in the past, many of the places so named were covered with peat bogs, where turf was cut for fuel. The Danish words "ling" and "lung" are the same word, and a derivative "lunge" is the name for the organs called in English the lungs (Lange 1999). The derivation of the English word "lung" is clear: according to the *Oxford English dictionary*, "The lungs were so called because of their lightness", the word having the same etymology as "lights", a collective name for the lungs of such beasts as sheep, pigs and bullocks when used as food, especially for cats and dogs. Extrapolating from Danish words for types of soil, Lange (1999) suggested that "there can be no doubt that [in Danish] *lyng* or *lung* originally alluded to the fact that heather turf is light (in weight) when dry." Lange (1999) concluded:

> Thus it is not difficult to discover a credible explanation of the word lyng, a turf, and extend this is lyng, the area, and from that to a place name like Højby Lyng. ... Finally, when we attempt to find an explanation for the plant name lyng, it is in fact quite simple. ... We all know the concept of contact association. ... When the "heath blooms", it is the plants of the heath that blossom and especially the heather. It is a simple process whereby the word lyng has gone from meaning the area — originally the type of soil in that area — to the meaning of the plants, or the single plant, growing there.

CALLUNA: BOTANICAL ILLUSTRATIONS

William Turner's *A new herball* (1551), published in London, was the first British work in which ling was illustrated, and the first to contain a record of a heather growing wild in Britain: "Plini ... wryteth that erica groweth in woddes[2] which I could yit neuer se. For our heth groweth in playnes and in wylde groundes, and in moyste places, and upon sum wodles hylles. ... The hyest hethe that euer I saw groweth in northumberland which is so hyghe that a man may hyde hymself in." Turner's printer, like so many during the succeeding decades, was content to recycle an illustration, and so he copied (or used a existing copy of) an earlier wood-cut of ling. There was not much to choose from — so William Turner's *A new herball* included a reduced copy of the splendid engraving of *Calluna vulgaris* from Leonhart Fuchs's *De historia stirpium commentarii insignes*, published in Basle in 1542, one of the finest of the sixteenth-century herbals.

Leonhart Fuchs (1501–1566) had employed Albrecht Meyer to paint the plants he wanted to illustrate, Heinrich Füllmaurer to transfer Meyer's images to woodblocks, and Veit Rudolf Speckle to cut the images into blocks of wood for printing. The results are some of the best images of plants printed in botanical books before modern times. Some copies of *De historia stirpium* ..., Fuchs's "botanical masterpiece" to quote Agnes Arber (1986), are hand-coloured in watercolours, although the book would not originally have been coloured. One such copy is in the library of the Linnean Society of London (Plate 3).

Careful study of Meyer's portrait of *Calluna vulgaris* (Plate 3) shows that he used for his model a sprig on which the flowers had faded and fruits had formed. At this stage, the outer segments, which form the calyx, tend to curl inwards at the tips and the persistent style becomes more prominent due to the enlargement of the fruit. He depicted these characteristics faithfully.

John Gerard (1546–1611/12) and his publisher, John Norton, also made use of existing wood-blocks to illustrate *The Herball* (1597), so the earliest illustrations of *Calluna vulgaris* printed in Britain were all second-hand. Norton had acquired these wood-blocks from Nicolaus Bassaeus of Frankfurt am Main. Those portraying ling had been used before in *Eicones plantarum* (1590), the illustrated companion for *Neuw Kreuterbuch* (1588) by Johann Dietrich of Bergzabern (better known by the Latin version of his name, Jacobus Theodorus Tabernaemontanus (c. 1520–1590)).

Gerard (1597) has two entries for *Calluna*, the first being his "*Erica vulgaris, sive Pumila*. Common or dwarfe Heath" which had "small, soft" flowers "of a light red colour tending to purple". The accompanying woodcut is rather stiff and formalised, and shows a fruiting sprig of ling. In *Eicones plantarum* (Tabernaemontanus 1590) the same image was labelled "Weiss hend", white heath! Gerard's second entry for *Calluna* was headed "*Erica pumila alba*. Dwarf heath with white flowers", and is simply the white-blossomed ling. Again the wood-cut is stylised, stiff, and shows a piece of *Calluna* in fruit. In Tabernaemontanus's *Eicones plantarum* (1590) the same image was labelled "Hend" (Nelson 1998).

For the revised edition of Gerard's great herbal, edited and enlarged by Thomas Johnson (1633), the Bassaeus wood-blocks were abandoned — maybe they were not available any more — and Johnson's printer used a new set, this lot obtained from the Amsterdam printer Christophe Plantin. The Plantin wood-blocks had also been used before, in publications by the Flemish botanist Charles de l'Écluse (Carolus Clusius), whose name will often be repeated here, and most of the images of heathers can be traced back to l'Écluse's visit to Spain and Portugal in 1564–1565 (Ramón-Laca & Morales 2000; Ramón-Laca *et al.* 2005). However, the woodcut of *Calluna vulgaris* was not derived from an Iberian plant, but was a redrawn and greatly reduced version of the woodcut from Fuchs's *De historia stirpium* (1542).

Between Fuchs's *De historia stirpium* of the early sixteenth century and today there was really little room for improvement, except in printing techniques and the use of colour. However, two artists may be singled out because of their use of enlargement: John Ruskin (1819–1900) and Dr Arthur Henry Church FRS (1865–1938). Ruskin, Slade Professor of Art at the University of Oxford, might be described as an amateur botanist, being best known as a social critic and an art critic. Some of Ruskin's original drawings and paintings survive in the Ashmolean Museum, including a splendid study of "lateral sprigs" and flowers of *Calluna vulgaris* (Fig. 28). As far as I am aware, this has never before been reproduced, and Ruskin's purpose in painting it is not recorded. However, he did publish a study of two sprigs of ling — "Blossoming — and stricken in days" (Ruskin 1875: frontipiece) — to illuminate a comment in the fourth chapter of his strange book, *Proserpina. Studies of wayside flowers while the air was yet pure among the Alps and in the Scotland*

Fig. 28. John Ruskin's original watercolour illustrations of ling (*Calluna vulgaris*).

and England which my father knew. "Take a spray of ling, and you will find that the richest piece of Gothic spire-architecture would be dull and graceless beside the grouping of the floral masses in their various life."[3] Church's (1908) dramatic image of a section through a single flower of *Calluna* (Fig. 29) was prepared to illustrate his lectures on floral biology. He also made diagrams of the flowers of *Erica cinerea* (Fig. 167, p. 262) (see Nelson 2000) and *E. carnea* (Fig. 66, p. 116) in similar style. Seen together, Harry Church's enlarged drawings of heathers demonstrate not just the differences between *Calluna* and *Erica* in terms of the structure of the flowers, but also show the difference between those *Erica* species possessing spurs, the appendages on the stamens at the point where the filament and anther fuse, and those which lack these enigmatic structures (see p. 104). Ruskin's simpler watercolour (Fig. 28) illustrates the alternating arrangement of outer segments of the calyx with the shorter inner segments of the corolla in the flower of *Calluna*.

Fig. 29. Arthur Church's stylised *Calluna* flower.

VARIATIONS IN THE WILD AND IN CULTIVATION

Given its distribution, ranging for about 125° longitude west to east and 36° latitude north to south, *Calluna* is remarkably uniform. No one has succeeded in splitting the genus into well-marked species nor describing distinct varieties or subspecies that occupy separate areas: all the subforms, forms, varieties and subspecies that have been described by Jansen (1935), Beijerinck (1940) and others, merge into each other almost imperceptibly. Only the plants with pure white flowers and golden anthers, *C. vulgaris* f. *alba*, are signally distinct.

In cultivation, however, given that clones are selected for marked characteristics or abnormalities, these imperceptible gradations do become more clear-cut, so a garden of *Calluna vulgaris* clones becomes a patchwork quilt of colour, form and texture. William Robinson (1838–1936), the Irish horticulturist best remembered for his disapproval of the bedding-out of subtropical perennials, appreciated the diversity of heathers. In his pioneering book, *The wild garden* (1870; see Nelson 2003a: 171; 2010: 171), among the native plants that he praised was *Calluna*.

> That the Heath family is likely to afford much interest I need hardly remind any person who has seen the wide spread of beauty on our heaths and mountains in summer or autumn. But of the variety of loveliness which exists among our native heaths few people have any idea: not even the sportsman or botanist, who continually wanders over their native wilds, or the plant collector, with a quick eye for everything attractive or noble in the way of a plant. The species themselves are of course very beautiful; but from time to time sports

have appeared amongst them which nurserymen have preserved; and thus, where you see a good collection of these, the variety of gay colour is quite surprising. Though I knew all the species and admired them, I had no idea of the beauty of colour afforded by the varieties till I visited the Comelybank nurseries at Edinburgh a few years ago, and there found a large piece of ground covered with their exquisite tints, and looking like a most refined flower-garden. But if all this beauty did not exist, the charms of the usual form of the species, as spread out on our sunny heaths, should suffice to warrant their culture on the rockwork or among dwarf shrubs.

As for the Ericas, all are worthy of a place, beginning with the varieties of the common ling (Calluna vulgaris) — the commonest of all heaths. It has "sported" into a great number of varieties, many of which are preserved in nurseries, and these are the kinds we should cultivate. Some of them are better, brighter, and different in colour; others differ remarkably in habit, some sitting close to the ground in dense, green, tiny bushes; others forming fairy shrubs of a more pyramidal character, and all most interesting and pretty.

Some *Calluna* clones change foliage colours between the later part of spring, when they produce new shoots and leaves, and winter when the cold affects the mature foliage — good examples are 'Firefly' (Plate 2), 'Robert Chapman' (Plate 1) and 'Sir John Charrington' (but there are many others). Moreover, the colour of the foliage varies subtly with aspect — for example, view a plant of 'Beoley Gold' from the south, and it will be a different colour than when viewed from the north (Rogers 2006).

The most notable recent introductions into horticulture, the cultivars that are propagated in tens of millions annually, are commonly known in English as "bud-bloomers" (a misnomer now, alas, entrenched; see Jones 1997). In 2003, approximately 65 million were sold in Germany alone (Schröder 2005). In these plants, the flower is malformed in such a way that the bud never opens. Several different type of "bud-flowers", the original term coined by Dr Willem Beijerinck (1940; Turpin 1979), were recorded by Jansen (1935) and Beijerinck (1940). In *Calluna vulgaris* f. *diplocalyx* Jansen (1935), the perianth is composed only of a duplicated calyx, four whorls of two sepals, without any corolla or stamens (Beijerinck 1940: 134). 'David Eason' (Plate 2, p. 54) belongs in this *forma*. In *C. vulgaris* f. *clistanthes* Jansen, all the floral parts are present, including corolla and stamens, but some parts may be deformed and the bud never opens: "Flores semper clausi", the flowers always closed (Jansen 1935). Beijerinck (1940: 134) wrote of *C. vulgaris* f. *clistanthes*: "Flowering-time as a rule [is] somewhat later than with the ordinary Scotch heather, so that ... this form is conspicuous by the fresh colour of its flowers" when most of the other plants have finished blooming.

This capacity to retain colourful buds long after the flowers on "normal" plants have faded and turned whitish is exploited in the "bud-bloomers". Because pollination and fertilisation cannot occur, the physiological changes associated with seeds and fruit formation, including the withering of the corolla and the breakdown of floral pigments, are retarded. These cultivars are selected for retention of colourful buds into late autumn and early winter, especially so that they appear fresh around the festivals of All Saints (1 November) and All Souls (2 November). In some parts of Europe, especially in German-speaking regions, it is traditional to decorate family graves with wreaths of evergreens and flowers on All Saints' Day (*Allerheiligen*). Pot-grown *Calluna*, being hardy and resilient, makes an ideal dwarf plant for use as a grave decoration, and the recent explosion in the production of "bud-bloomers" has very profitably capitalised on this repetitive, annual market.

ETHNOBOTANY

Ling has many other uses, and there is a substantial literature about these (including Howkins 1997; Milliken & Bridgewater 2004). Today the most important are undoubtedly as a decorative ornamental garden plant, as a source of pollen and nectar for honey bees, and, in the wild, as the plant that dominates moorland. Heather honey (produced from *Calluna vulgaris*) is amber-coloured, aromatic, can have a slightly bitter taste, and is thixotropic, becoming liquid when stirred or shaken.

Former uses of ling, or present-day small-scale uses, include thatching (M'Nab 1832; Evans 1977) (Fig. 30), fodder for cattle, kindling, and the making of brushes and besoms (Fig. 31). Ropes were made from plaited heather (Figs 32 & 33) in Scotland[4] (Milliken & Bridgewater 2004) and the Isle of Man (Mabey 1996). Dried heather was used to provide bedding for humans and beasts. According to the sixteenth-century Scots historian George Buchanan (quoted by Beith 2004: 222), a heather bed

> ... may vie in softness with the finest down, while in salubrity it far exceeds it; for heath, naturally possessing the powers of absorption, drinks up the superfluous moisture, and restores strength to the fatigued nerves, so that those who lie down languid and weary in the evening, arise in the morning vigorous and sprightly.

William McNab, Superintendent of the Royal Botanic Garden, Edinburgh, wrote an encomium to a heather bed in his booklet about Cape heaths (1832): "... I can state, from my own experience, that a dry bed of native heather is, to the weary traveller in many parts of the Highlands, a real

Fig. 30. A reconstruction of a "heather house" at Florencecourt, County Fermanagh, Northern Ireland, a property owned by the National Trust (E. C. Nelson).

luxury."[5] Samuel Gray (1821: 399) noted that to make a bed, ling was "confined in a frame with the flowering tops uppermost." Yet, it was surely not always so neat and well-arranged, as when a party of students participating in a botany field-class led by William Hooker slept "elbow to elbow on a large bed of heather spread out over the floor of a cottage" in the Highlands (Allen 1978). Indeed, heather had a reputation in the Scottish Highlands for aiding sleep; a poultice of heather was used to relieve insomnia while, even in the late twentieth century, some of "the older people still put sprigs of [heather] under their pillows" (Beith 2004). Mary Beith (2004) also asserts that "an occasional cup of heather tea does wonders for the 'nerves'."

Ling can be used to dye wool and cloth (Mahon 1982; Milliken & Bridgewater 2004), giving yellow (when used with alum; Gray 1821) or green. It was also used to tan leather (Gray 1821). A novel modern application is the use of a mixture of peat with heather in filters to reduce odours and "other aerial emissions" from buildings housing livestock (Phillips *et al.* 1995). Underhill (1990: 32) reported the use of heather (presumably *Calluna*) by a Swedish steel manufacturer for cleaning the scale off rolled steel plate.

Gray (1821) stated that the "young tops brewed along with half as much malt" made a "good beer", and in recent decades in Scotland, the brewing of a beer flavoured with ling has been revived (Milliken & Bridgewater 2004). Heather ale was drunk as a tonic (Beith 2004). Whether this even approximates to the legendary Viking ale is impossible to tell (Nelson 2000b), but at least one Scottish writer, Katherine White Grant (quoted by Beith 2004: 116), recorded that whisky was "often known euphemistically as 'heather ale' ...", casting doubt on the beverage's true formulation. And, whisky was drunk from cuachs or quaichs, rare examples of which were made from slivers of ling stems (Fig. 34).

Fig. 31 (above right). "Renges or pot scrubbers of ling ... Sold at 1½d. each. Rannock, Perthshire."; acquired in 1897 by Sir William Thiselton-Dyer; Economic Museum, Royal Botanic Gardens, Kew [Cat. No.: 51213].

Fig. 32 (right). Rope made from heather stems twisted together as "used by people in the Isle of Skye"; Economic Museum, Royal Botanic Gardens, Kew [Cat. No.: 51218].

Fig. 33. Iain Clachair (Ian Campbell) plaits a rope from heather stalks on South Uist, Outer Hebrides, Scotland, in 1930; a photograph by Margaret Fay Shaw (olim Mrs J. L. Campbell; 1903–2004), originally published in her book *Folksongs and folklore of South Uist* (1955) (reproduced by courtesy of The Canna House Archives, National Trust for Scotland, Isle of Canna).

Iain Clachair (Iain Mac Dhomhnaill 'ic Iain Bhain) (c. 1858–1932), crofter and mason of South Lochboisdale, was also, like his brother, a poet and he gave to Margaret Fay Shaw songs and cures that were published in *Folksongs and folklore of South Uist* (1955).

Fig. 34. A cuach or quaich, a drinking cup, made from small fragments of heather (*Calluna vulgaris*). The wooden staves, as in a barrel, are "feathered" into each other and then bound together with cane. This example was presented to the Royal Botanic Gardens, Kew, by Archibald Campbell (Glenlyon, Ballanaenge [?Balnahanaid]); Thiselton-Dyer; Economic Museum, Royal Botanic Gardens, Kew [Cat. No.: 51214].

Examples of cuach made in this manner are apparently rare. One, composed of slivers of oak and bound with iron rings, which was once owned by Nance Tinnock, keeper of the ale-house in Mauchline which was occasionally visited by Robert Burns, is preserved in Burns' Cottage Museum.

WHITE HEATHER

As remarked above, John Gerard (1597) noticed that some plants of ling were white-, not pink-flowered. He knew, probably having found some himself, that white-blossomed *Calluna* grew wild on "the downes neere unto Gravesend" in Kent. While Gerard's report is the earliest record for white heather in Britain, Tabernaemontanus (1590) should be credited with the first record of it anywhere in Europe. Gerard did not impute any special significance to the white variant, nor did white heather have anything but a curiosity value for botanists, like any anomalous plant, until the second half of nineteenth century. There is nothing in botanical literature, nor in any other publications, even to hint that white heather was anything but a freak of Nature: a plant incapable of synthesising the anthocyanins that turn its calyx and corolla from colourless — white — to pink (Fig. 35).

The dramatic alteration in white heather's allure resulted from an event on a heather-clad Scottish hillside in 1855. Even then, a further 13 years elapsed before the news emerged into the public domain and white heather acquired its fabulous reputation. On Saturday 29 September 1855, when riding on Creag nam Ban, a hill not far from Balmoral, Crown Prince Friedrich Wilhelm of Prussia asked the 14-year old Princess Royal, Princess Victoria, the eldest daughter of Queen Victoria and Prince Albert, to be his future wife. The story goes that "Fritz" (the Crown Prince's nickname) picked a piece of white heather and gave it to the princess, because, so "Fritz" said, white heather was an the emblem of "good luck".

The original source for this tale is Queen Victoria's book *Leaves from the journal of our life in the Highlands from 1848 to 1861*. When published in 1868, the book was a sensational best-seller, selling 100,000 copies within three months. Thus, the earliest reference to *lucky* white heather really should be dated 1868.[6] The tradition attributing good luck to white heather achieved popularity only afterwards, during the second half of Victoria's reign (McClintock 1970b); evidence from other publications supports this. In books dealing with the language of flowers published before about 1870, there are no entries for white heather, only for "heath" and its meaning is always "solitude". However, in the Revd Robert Tyas's *Speaking flowers* (1875: 98–99) there is an entry for white heather and its meaning is given as "good luck": "it is so regarded in Scotland, as we read in our beloved Queen's "Journal in the Highlands"...", Tyas explained. Under "Good luck", Tyas (1875: 145) noted that it was signified by "white Heather (Calluna vulgaris, β alba)".

Fig. 35. *Calluna vulgaris* in bloom, with "normal" pale mauve flowers (left) and with white flowers (right) (E. C. Nelson).

It is probably not a coincidence that bouquets of pink roses and white heather adorned the dresses of the bridesmaids who attended the Princess Royal at her wedding to the Crown Prince on Monday 25 January 1858, but that heather cannot have been white-flowered ling (Nelson 2006c, 2009b).

White heather was frequently among the flowers adorning other princesses, especially at their weddings. For hers, in March 1872, Princess Louise's[7] wedding dress of white satin, covered with a deep flounce of Honiton point lace, was trimmed with cordons of orange blossoms, white heathers, and myrtle. On the same occasion the bride's sister, Princess Beatrice, wore a wreath of white heather, and the bridesmaids' silk dresses were trimmed with cerise roses, white heather and ivy. The bridesmaids at the wedding on 27 April 1882 of Prince Leopold, Duke of Albany, to Princess Helena of Waldeck-Pyrmont, wore floral head-dresses of an unusual composition: white heather, violets and primroses. Queen Victoria's youngest daughter, Princess Beatrice, was married in July 1885, and her dress was trimmed with orange blossom, white heather and myrtle, while for her departure from Osborne House, the bride "wore a white bonnet entirely covered with white heather, in which was placed a branch of orange-blossoms." Orange blossom, white heather and myrtle also were selected for the bridal flowers at the wedding on 6 July 1893 of the Duke of York to Princess Mary ("May") of Teck; her wedding dress had elaborate "trails" of orange-blossom gathered into clusters with sprigs of white heather, and she carried a bouquet of white flowers (Nelson 2006c, 2009b).

White heather is a token between couples, boding good fortune or good luck. There is no element of chance — of chancing upon a plant of white-blossomed heather in the wild — in the same way as a lucky four-leaved clover. In any case, the element of chance, if it ever was a part of the lore, has been eliminated by the ready availability at every season of a white-blossomed heather, whether *Calluna* or *Erica* (for example, *E. lusitanica*; see p. 216), or even dried inflorescences of *Limonium sinuatum* (sea lavender) (Richards 1990).

1. CALLUNA VULGARIS. Ling, heather

Calluna vulgaris (L.) Hull, *British flora* **1**: 114 (1808). Type: [Lapland], *C. Linnaeus*. Lecto.: Institut de France, Paris: sheet 141 (Jarvis & McClintock 1990; Jarvis 2007: 501).
Basionym: *Erica vulgaris* L. *Species plantarum*: 352 (1753).
C. erica DC., *Flore française*: 680 (1805).
C. sagittifolia (J. Stokes) Gray, *Natural arrangement* **2**: 399 (1821).
C. atlantica Seemann, *J. Bot.* **4**: 305–306 (1866) (see Nelson 2010a).

ILLUSTRATIONS. J. Curtis, British entomology 3: tab. 145 (1826); W. Fitch, J. Bot. 4: tab. 53 (1866); S. J. Roles, Flora of the British Isles: illustrations pt 2: tab. 932 (1960); S. Ross-Craig, Drawings of British plants, part 19: tab. 21 (1963); R. W. Butcher, A new illustrated British flora: 104 (1961).
BIOLOGICAL FLORA OF THE BRITISH ISLES. *J. Ecol.* **45**: 455–483 (1960).
DESCRIPTION. *Evergreen, much-branched, hemispherical, bushy shrub,* capable of regenerating after fire from stem bases; stems erect or spreading, or rarely completely prostrate, to 0.25–0.5 m tall — in exceptional circumstances old, well-protected shrubs can exceed 2 m (Praeger 1934; Beijerinck 1940) — or (when stems prostrate) a dense hummock or spreading mat. *Primary root* can extend 2.5 m deep (Beijerinck 1940). *Shoots* hirsute when young with simple hairs of varying length, gradually becoming glabrous; upright shoots often rooting at the base; prostrate shoots also often with adventitious roots (which, in cuttings, commence to emerge above the leaf nodes (Small 1974); bearing numerous, short, non-flowering, densely

Calluna vulgaris   Painted by Christabel King

PLATE 1

LEFT: **'E. F. Brown'**. Upright, neat, open plant (±0.25(–0.4) m tall × ±0.45–0.6 m spread); foliage light green; flowers lavender (H3); September–December. (Painted 18 October 1985.)

Described as a good late-flowering cultivar with very pretty flowers, this cultivar was found by E. F. Brown, an American airman and friend of John Letts, in Germany, before 1966.

UPPER CENTRE: **'Beoley Crimson'**. Open, erect shrub (±0.3(–0.45) m tall × ±0.4 m spread); foliage dark green; flowers bright crimson (H13); September–October. (Painted 23 September 1985.)

This ling was outstanding in bloom during The Heather Society's trials (Vickers 1976). 'Beoley Crimson' came from Beechwood Nursery, Beoley, Redditch, Worcestershire, England, and was introduced by J. W. Sparkes and P. W. Sparkes in conjunction with Tabramhill Gardens, Nottingham, about 1970. The plant responds to pruning by producing longer flower-spikes.

LOWER CENTRE LEFT: **'Golden Feather'**. Low, spreading shrub with a feathery appearance due to the long spreading shoots (±0.25(–0.35) m tall × ±0.8 m spread); foliage rich orange-bronze in spring, turning golden in summer, and acquiring tints of red in autumn; in winter the foliage is reddish orange; flowers mauve (H2), sparse; August–October. (Painted 25 March 1986.)

This is another of the introductions of J. W. Sparkes, Beechwood Nursery, Beoley, Redditch, Worcestershire, England, dating from the early 1960s. It is very susceptible to wind-burn (Small & Small 2001), and the foliage will go brown in damp conditions (Underhill 1990). Slinger (2007) described 'Golden Feather' as an "attractive and most reliable plant ... . A small branch with its branchlets does indeed resemble a feather to some degree but what I like particularly ... is the fact that the golden hue never varies throughout the year. I have never known it to produce a flower."

LOWER CENTRE MIDDLE: **'Robert Chapman'** 🏅 🏆 Compact, floriferous ling, producing a shapely, rounded shrub (±0.25 (–0.5) m tall × ±0.6(–0.8) m spread), with shoots of varying heights; foliage changes with the seasons: yellow-green appearing "gold" (summer), to orange with scarlet (autumn), then red (winter and spring); flowers lavender (H3); August–October. (Painted 26 March 1986.)

This "very popular" ling was found and introduced by J. W. Sparkes, Beechwood Nursery, Beoley, Redditch, Worcestershire, England, before 1962. Writing in the late 1960s, Slinger (2007) commented that 'Robert Chapman' "is a plant that has come to stay. ... [It] has young foliage of gold and this ripens and matures to flame and red and this colour is held throughout the winter." The aspect from which a plant is viewed is significant: colours are more pronounced on the sun-facing side (south in the northern hemisphere, north in the southern) (see Rogers 2006).

Robert Chapman, the eldest son of Charles Chapman, a director of J. V. White of Birmingham, was a "little hard-working lad that pushed a barrow in his parents' wholesale market in Birmingham" (Underhill 1990). AGM 1993.

LOWER CENTRE RIGHT: **'Silver Knight'** 🏅 Vigorous, upright heather, yet retaining quite a neat habit because there are relatively few side-shoots (±0.4 m tall × ±0.5 m spread); foliage downy, appearing grey and in winter turning to deeper purple-grey; flowers single, lavender (H3); August–September. (Painted 22 September 1985; from Daphne Everett.)

This silvery clone was named 'Silver Knight' because "it stands straight up like a knight in armour". Found by J. W. Sparkes, Beechwood Nursery, Beoley, Redditch, Worcestershire, England, 'Silver Knight' was introduced either by J. W. Sparkes or J. F. Letts about 1966. One of its several sports is 'Velvet Fascination' (see p. 358).

LOWER RIGHT: **'Cottswood Gold'** (GELBE MAASSEN) (*C. vulgaris* f. *alba*). Spreading heath with upright shoots (±0.3 m tall × ±0.45 m spread); foliage bright yellow throughout the year; flowers white; August–September. (Not dated.)

A spontaneous seedling, found by Major-General Patrick G. Turpin in his garden at West Clandon, Surrey, England, in 1974. The heather was named after the Turpins' house, Cottswood (® no. 16). This is similar to 'Beoley Gold' (see p. 353) and 'Gold Haze' (see p. 354) but, according to Underhill (1990), 'Cottswood Gold' is better at retaining its foliage colour.

In certain nurseries in Germany, particularly those associated with Maassen of Straelen, a combination of 'Cottswood Gold' with two other *Calluna* cultivars is marketed as "Beauty Ladies" (a registered trade-mark). The established cultivar name 'Cottswood Gold' has been replaced, improperly, by GELBE MAASSEN which is deemed to be a trade designation.

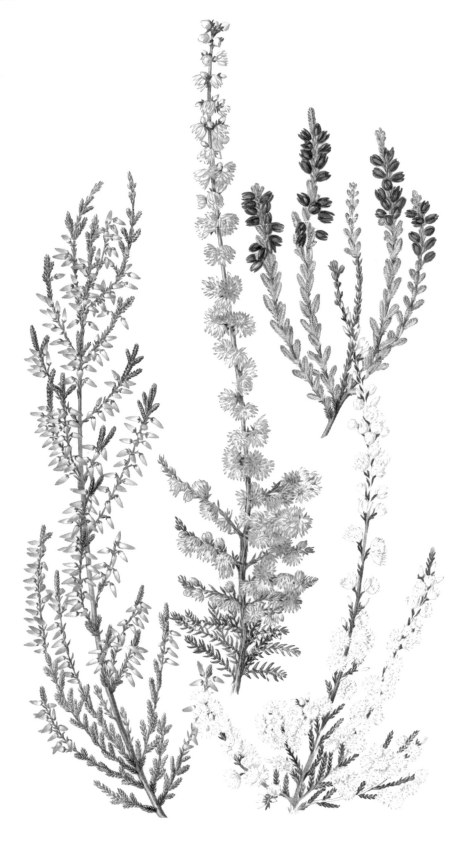

leafy, axillary shoots on which the leaves are imbricate and arranged strictly in 4 rows; leaves on main shoots larger and more widely spaced. *Leaves* very small, ±1–2 (–3) mm long, ±0.5–1.5 mm broad, in opposite pairs, usually dark green, sometimes tinged with red or yellow (depending on season or stress), glabrous to very densely hirsute (when the foliage will appear silver or grey); leaf boat-shaped, sessile, flattened above, edges slightly curved and prolonged into two triangular, short (less than half length of lamina), downwards-pointing, basal lobes (auricles); margins strongly revolute and parallel on underside completely concealing lower surface of leaf; sulcus narrow, linear. *Inflorescence* one-flowered; numerous inflorescences clustered into raceme-like or panicle-like "inflorescence". *Flower* weakly zygomorphic, pale mauve (but varying greatly in cultivated clones) or very rarely colourless (white), terminal on a greatly condensed shoot with 1 leaf at base and 2–3 leaves or prophylls above closely appressed to involucre; *peduncle* absent; 4 translucent, scariose bracts with densely ciliate margins closely appressed to calyx, forming involucre. *Calyx* comprising 2 pairs of free (not fused), oblanceolate, concave sepals, to ±4 mm long, to ±2 mm across, petaloid, the same colour as the corolla, glabrous, scariose, persistent in fruit and tending to curl inwards when dry. *Corolla* ±2–2.5 mm long, campanulate, 4-lobed, lobes erect, glabrous. *Stamens* 8, ±2 mm long, slightly declinate, included; *filament* ±1 mm long, ±0.2 mm broad towards base, narrowing to ±0.1 mm at apex, flattened, glabrous, hooked at based, geniculate towards apex; *anther* ± 1 mm long; *thecae* fused except toward "horn-like" apices, brown, glabrous; pore lateral, longitudinal, extending almost from tip to base; two cream-white *spurs*, ±0.5 mm long, sometimes with forking apices, densely covered with minute hairs. *Ovary* squat-globose, c. 1.5–2(–2.5) mm long, variously hirsute at least in upper portion, emarginate; 6–8 ovules per locule; *style* ±3 mm long, ±0.5 mm in diameter at base, erect, curving slightly, glabrous, tapering towards apex; *stigma* 4-lobed. *Seed* ±0.6 × 0.35 mm. n = 8.

Calluna vulgaris   **Painted by Christabel King**   PLATE 2

LEFT: 'David Eason'. Spreading, straggling heath (±0.2(–0.4) m tall × ±0.45 m spread); foliage mid-green; flower buds mauve (H2), never opening; August–November.

This was one of the first of the so-called "bud-bloomers" and was found on the moors beside Broadstone Golf Links, Dorset, England, by C. D. Eason of Maxwell & Beale Ltd. Named after Charles Eason's son, the clone was introduced by Maxwell & Beale in 1935.

CENTRE: 'Annemarie' 🎖 🏆 Tall, spreading heath (±0.5 m tall × ±0.6 m spread); foliage dark green; flowers "double", rose pink (H7), in long spikes; base of the flower is not white; September–October.

This "double"-flowered ling arose as a sport on 'Peter Sparkes' (see below). It was found in 1973 by Kurt Kramer, Edewecht-Süddorf, Germany, and introduced by Kramer in 1977. In the opinion of Small & Small (2001) this is one of the "more outstanding double" lings (see also Platt 1982). AGM 1993.

UPPER RIGHT: 'Firefly' 🎖 🏆 Stiffly erect heath (±0.45 m tall × ±0.5 m spread); foliage terracotta in summer, changing to brick-red in winter; flowers dark mauve (H2); August–September. (Painted 14 September 1987.)

Found and introduced before 1977 by Clive Benson, Farington, Preston, Lancashire, England. According to Small & Cleevely (1999), when in flower the colour is "peach". Rogers (2006) showed that when viewed from different aspects, the colour of the foliage was remarkably different. In other words, considerable care is needed when planting 'Firefly' to ensure that the plants can be seen from the best standpoint. AGM 1993.

LOWER RIGHT: 'Kinlochruel' (*C. vulgaris* f. *alba*) 🎖 🏆 Compact, low, spreading heath (±0.25 m tall × ±0.4 m spread); foliage bright deep green acquiring bronze tints in winter; flowers "double", pure white, in long curving spikes; August–September.

Brigadier and Mrs E. J. Montgomery found this spectacular ling as a sport on 'County Wicklow' during 1969 in their garden, Kinlochruel, Colintraive, Argyll, Scotland. Mrs Montgomery registered the name (® no. 4) in 1977. AGM 1993.

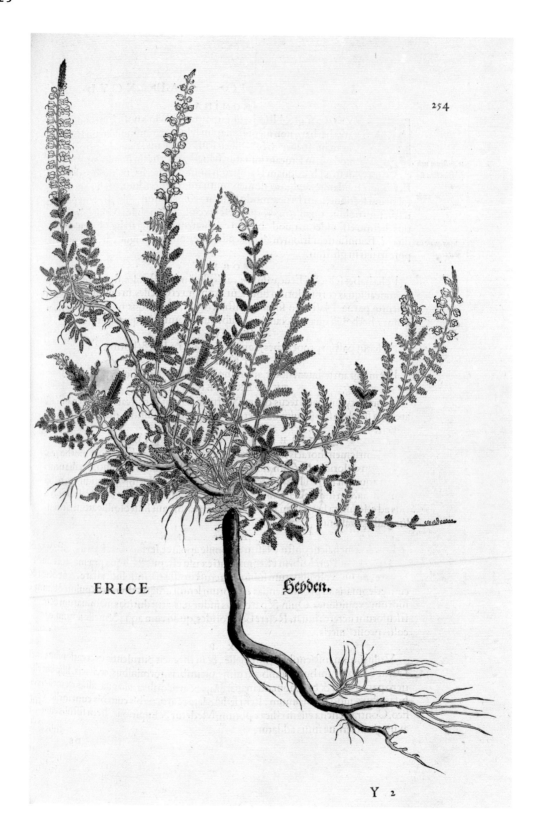

254

ERICE   Heyden.

Y 2

FLOWERING PERIOD. (July to) September–October (to November); in cultivation, a few cultivars can continue in flower into January.

DISTRIBUTION & ECOLOGY. Western Europe including Azores, Iceland and Faeroe Islands, eastwards into Siberia. North Africa: northern Morocco (Ball 1877; Fennane & Ibn Tattou 2005; Villar (1993) states that it extends south into Mauritania, but this seems an error). In Spain it ascends to 2,000 m altitude (Villar 1993), and *Calluna* is reported reaching about 2,700 m on La Fibbia in Canton Ticino (Tessin), southern Switzerland (Hegi 1927: 5 (3): 1650). **Maps:** Beijerinck (1936, 1940), Gimingham (1960), Bolòs & Vigo (1995: 32), Malyschev (1997: 231).

In Ireland, *Calluna vulgaris* occurs from sea-level to 1,040 m in Macgillycuddy's Reeks, County Kerry, while the highest altitude attained in Britain is 1,095 m on Beinn a'Bhuird in Aberdeenshire, Scotland (*fide* McClintock 1989e; Preston *et al.* 2000). *C. vulgaris* is a member of the European Boreo-temperate element of the flora of Ireland and Britain (Preston & Hull 1997).

Calluna was as much affected by glacial advances and retreats as the species of *Erica* and *Daboecia*. Evidence from studies of DNA suggests that ling survived in refuges in south-western Europe including the north of Spain, the Pyrenees and the Massif Central (Rendell & Ennos 2002). In the context of Britain, *Calluna* will have disappeared from much of the region at the height of the last glaciation simply because the land was completely submerged under an ice sheet; it may have survived in the extreme south of what is now England. Ling recolonised northern parts soon after the ice retreated, and its subsequent expansion was probably aided by deforestation (Godwin 1975: 294–296). Genetic evidence indicates that the present populations of *Calluna* in Scotland are more similar to those in Belgium than they are to populations in northern Spain, the Pyrenees and southern France (Mahy *et al.* 1998), and this seems to lend support to the idea that the species recolonised northern and eastern parts of its present range in an east to west direction from the region of southern Britain when Britain was still joined to continental Europe (Mahy *et al.* 1998: 659).

How *Calluna* reached the Azores (Fig. 37) is a mystery that may only be explained by long-range dispersal by wind or birds. On a very local scale, Bullock & Clarke (2000) have shown that wind will disperse *Calluna* seeds from the parent bush for distances up to 80 metres, although the vast majority (more than 90%) of the seeds will fall to the ground within a metre of the parent plant (Bullock & Moy 2004). Given that a single bush of *Calluna* is capable of producing up to one third of a million seeds (Bullock & Moy 2004), at least some seeds will travel more than a metre. Yet this cannot account for dispersal from Europe across the Atlantic Ocean to the Azores.

Wind is also capable of blowing withered heather flowers, which can contain some trapped, viable seeds (see Gimingham 1972: 88), substantial distances (Legg *et al.* 1992), and so the old, dead flowers could act as long-distance disseminules. In February 1881, "large quantities of plant débris", to quote H. P. Guppy's classic book on *Plants, seeds and currents in the West Indies and the Azores* (1917: 425), "mostly fruits of *Calluna vulgaris* mixed with blossoms of *Erica tetralix*, were, during a

Calluna vulgaris   **Painted by Albrecht Meyer**   PLATE 3

Published in Leonhart Fuchs, *De historia stirpium commentarii insignes* (1542).

Fig. 36 (left). *Calluna vulgaris* in bloom in mid-September in coastal heathland in Brittany, France (E. C. Nelson).
Fig. 37 (right). *Calluna vulgaris* growing near fumaroles at Furnas, São Miguel, Azores (E. C. Nelson).

gale ... blown across the Cattegat from the Swedish coast to the eastern shores of Jutland, a distance of 110–120 kilométres." Wind-dispersal of withered flowers containing viable seeds is also recorded in *E. australis* (cf. Valbuena *et al.* 2001: Table 1).

As for dispersal by flying animals, this cannot be ruled out. Birds may have carried seeds from continental Europe to the Azores. The main birds that are reported feeding on *Calluna* in Britain are red grouse (*Lagopus lagopus*) and ptarmigan (*Lagopus mutus*). In the case of the red grouse, flowers and fruits are consumed in the autumn and winter (Gimingham 1972; Savory 1978), and *Calluna* seeds have been found in their droppings (see, for example, Welch 1985; Legg *et al.* 1992). Many other birds also feed on *Calluna* including migratory species such as snow bunting (*Plectrophenax nivalis*) (Sandeman 1957).

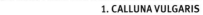

Fig. 38. *Calluna vulgaris* established in acid, peaty soil on the surface of limestone pavement on the coast of The Burrren, County Clare (E. C. Nelson).

Fig. 39. *Calluna vulgaris* with shrubby cinquefoil (*Potentilla fruticosa* L.) and devil's-bit scabious (*Succisa pratensis* Moench) near Mullach Mór in The Burren National Park, County Clare. (E. C. Nelson).

NATURALISED POPULATIONS. *Calluna vulgaris* is now naturalised in many remote places, to which it has been transported, deliberately or unwittingly, by human beings! There are populations in eastern (Lawson 1866; Seemann 1866; Boivin 1966: 417; Barclay-Estrup 1988, 1991; Nelson 2010) and western North America (Barclay-Estrup 1974; Small 2004a) (see below), New Zealand (Sykes 1981) especially in Tongariro National Park, Tasmania where it was planted originally for erosion control (Rozefelds *et al.* 1999; Sheppard *et al.* 2003), Tristan da Cunha (Wace & Holdgate 1958), the Falkland Islands (Broughton & McAdam 2002), and, in the sub-Antarctic, on Crozet Island (Frenot *et al.* 2001, 2005).

In Australia, ling was placed on the Federal Government's "Alert list for environmental weeds" (plants that are targeted for eradication) in 2001; although not controlled by legislation, the sale and importation of *Calluna* are discouraged and gardeners are asked to select alternative species for their gardens (Sheppard *et al.* 2003).

In New Zealand, *Calluna vulgaris* is listed under the National Pest Plant Accord (NPPA)[8] as an "unwanted organism" and plants (except double-flowered cultivars) cannot be sold, propagated or distributed. As ling threatened to overwhelm the natural vegetation in the Tongariro National Park, one of its natural pests, the heather beetle (*Lochmaea suturalis*), was selected as a "promising biological control agent". After tests showed that this beetle would not be a threat to indigenous plants, the insect was released in an attempt to eliminate *Calluna* from the park (Syrett *et al.* 2000).

That *Calluna* can still extend its range has recently been demonstrated in Georgia (Transcaucasus), where the species was not hitherto recorded. In 2000, ling was found on a peat-bog complex at Ispani near Kobuleti, close to the Black Sea coast, more than 150 kilometres from its nearest known locality in north-eastern Turkey, and more than 1,000 kilometres from its principal habitats in Russia (Kaffke *et al.* 2002). Studies of the pollen profile obtained from a peat core taken in the centre of the colony showed no traces of *Calluna* pollen in the layers below 70 cm of the surface, indicating that ling was a modern arrival on the bog, not a long-established, indigenous plant. Kaffke *et al.* (2002) date these earliest *Calluna* grains to about 1920. They have suggested plausibly that seeds of *Calluna* were carried to the Ispani II bog, a unique "percolation mire" within the Kolhketi wetlands, on the "boots, clothes or coring devices" of visiting prospectors or scientists, among whom was the Russian ecologist Vladimir Semenovich Dokturovsky (1884–1935), a specialist in mires, who visited the Ispani site in the early 1900s; he was present in 1930 when delegates to the Second International Congress of Soil Scientists were photographed taking a core on the Ispani I bog (reproduced in Kaffke *et al.* 2002).

***CALLUNA* IN NORTH AMERICA.** *Calluna* was reported from North America, particularly Newfoundland, by a number of reputable, early nineteenth-century botanists, including David Don (1834), George Bentham (1839: 613) and Sir William Hooker (1840), without provoking much interest and no excitement. Their information derived from William Eppes Cormack (1796–1868)[9], a Newfoundland-born entrepreneur and explorer, who had studied at the universities of Glasgow and Edinburgh and consequently, under the influence of Professor Robert Jameson, developed a keen interest in natural history (Story 2000). David Don, in a footnote to his "new arrangement of the Ericaceae", stated that specimens of *Calluna* "were contained in a collection of dried plants from Newfoundland, given me by Mr Cormack, who assured me they had been collected in that country" (Don 1834: 150; see also Watson 1865; Wallace 1903). *Calluna* still grows in a scattering of about half a dozen localities, all on the Avalon Peninsula in the extreme south-east of Newfoundland (Barclay-Estrup 1988; 1991: 52–53; Rouleau & Lamoureux 1992: map 313; Day 1995; Meades *et al.* 2000).

However, after a small colony of apparently indigenous *Calluna* was found "growing within 20 miles of Boston" at Tewksbury, Massachusetts, a kerfuffle arose in American horticultural and botanical circles. The "discoverer" was later revealed as Miss Margaret Strachan (Wallace 1903). She

Fig. 40 (left). The colours in The Bannut, Bringsty, Herefordshire, in early September derive mainly from cultivars of *Calluna vulgaris* (E. C. Nelson).

Fig. 41 (right) *Calluna vulgaris* cultivars in The Bannut, Bringsty, Herefordshire, in early September (E. C. Nelson).

showed the heather to her father, a native of Kincardineshire, who confirmed its identification with Alexander Skene, another Scot (Wallace 1903; Wiksten 2000). A pot-grown plant of *C. vulgaris*, derived from the Tewkesbury site, was exhibited at the Massachusetts Horticultural Society on 13 July 1861 (Wiksten 2000) by a "perfectly upright and straightforward" young gardener called Jackson Dawson. This precipitated a famous cartoon lampooning the Flower Committee of the Massachusetts Horticultural Society published in *Vanity fair* on 22 February 1862 (see Wiksten 2000).

The argument about whether *Calluna vulgaris* was indigenous to North America was soon joined by several distinguished botanists including Hewett Cottrell Watson (1865) and Berthold Seemann (1866). The latter wrote: "Dr. [David] Moore has kindly supplied me with fresh specimens of the Heather which he received some years ago from Newfoundland, and which

Fig. 42. Walter Fitch's hand-coloured engraving prepared using a plant grown at the Royal Dublin Society's Botanic Gardens (now the National Botanic Gardens), Glasnevin, Dublin. The seed was sent from Newfoundland, and Berthold Seeman called this plant *Calluna atlantica*.

has been growing since then side by side with the common European Heather in the Glasnevin Gardens." Seemann (1866), whose paper was illustrated by a fine portrait of *Calluna* by Walter Fitch (Fig. 42), regarded the Newfoundland plant as a separate species which he named *C. atlantica*: "I fully believe the Newfoundland plant is a distinct species ... which I have also seen from Iceland and the higher Alps" (see Nelson 2010a).

Today, few disagree with the proposition that *Calluna vulgaris* is an introduced alien plant in North America, although the date of introduction remains uncertain (cf. Barclay-Estrup 1991).

ETYMOLOGY. Latin: *vulgaris* = common.

VERNACULAR NAMES. Ling, heather, Scotch heather, cat heather (English); vanligur heiðalyngur (Faeroes); Besenheide, Heidekraut, Herbstheide, Sandheide (German); lyng (Danish; see Lange 1999); lyng, røsslyng (and numerous variants) (Norwegian; see Alm 1999); bruyère callune, brande, callune vulgaire, callune, béruée, bruyère commune, bucane, fausse bruyère, grosse brande, péterolle (French); brezo (Spanish: for the numerous names used in Spain, see Villar (1993), Pardo de Santayana *et al.* (2002)); queirò, urze (Portuguese). Willem Beijerinck (1940: 148–149) listed names used for *Calluna* from many parts of Europe.

For discussion of the enigmatic name "cat heather" see under *Erica cinerea* (p. 264).

CULTIVARS. See pp. 53, 55 and Appendix I (pp. 351–359).

NOTES

[1] Smith (1824) alluded to this dual meaning: "... *Calluna*, from χαλλυνω; ... is doubly suitable, whether, with Mr. Salisbury and Dr. Hull, we take it to express a *cleansing* property, brooms being made from Ling; or whether we adopt the more common sense of the word, to *ornament* or *adorn*, which is very applicable to the flowers."

[2] Pliny the Elder (Gaius Plinius Secundus, AD 23/24–79), the Roman naturalist who was killed by the eruption of Vesuvius in AD 79, would have known several different heaths. As a much-travelled soldier, he almost certainly saw *Calluna vulgaris* (Turner's Northumbrian "heth") when he was in the north-western region of the Roman Empire. In his *Natural history*, Pliny mentioned *Erica* in the context of honey and most probably was referring to one of the southern European heathers, maybe *E. arborea* or *E. manipuliflora* — "*Tertium genus mellis minime probatum silvestre, quod ericaceum vocant. convehitur post primos autumni imbres, cum erice sola floret in silvis, ob id harenoso simile ...*", which Bostock & Reilly (1855) translated as follows: "The third kind of honey, which is the least esteemed of all, is the wild honey, known by the name of ericaceum. It is collected by the bees after the first showers of autumn, when the heather alone is blooming in the woods, from which circumstance it derives its sandy appearance." (Philomen Holland's 1601 translation reads: "There is a third sort of wild honie, which the Greekes call Ericæum, and is of least reckoning. It is gathered after the first raine in Autumne, when the heath and lings only bloum in the woods, whereupon it seemeth as if it were sandie.") The autumnal flowering is wrong for tree heather, *E. arborea*, but is correct for *E. manipuliflora* — these are the heathers that would have been best known to the Greeks.

[3] I am grateful to Professor Stephen Wildman (Director, Ruskin Library and Research Centre, Lancaster University) for confirmation of the correct text. Rendall (1934: 253) erred in citing the source of this sentence as Ruskin's *Fors clavigera*, and also seems to have been the source of an erroneous transcription in which "spray" has been replaced by "sprig".

[4] It is widely reported that a fragment of rope, composed of twisted woody stems, excavated at Skara Brae, Orkney, was made from heather (*Calluna vulgaris*) — for example, Lambie (1994: 14) stated that among "primitive tools and animal bones they discovered Seomain fraoich, Heather rope" (see also Darwin 1996). This is *incorrect*. I am grateful to Dr David Clarke (Keeper of Archaeology, National Museums of Scotland, Edinburgh), who was in charge of the excavation when the rope was discovered, for correcting the identification (e-mail, 25 September 2007). According to Dr Clarke, the rope fragment was "originally identified as heather — more because recent ropes of that type in the Northern Isles are of heather than anything else". However, when the Skara Brae fragment was studied by the late Dr Camilla Dickson, she determined that the plant used was *Empetrum nigrum* (crowberry) (see Dickson & Dickson 2000). The re-identification was noted by Milliken & Bridgewater (2004).

[5] As a footnote McNab (1832) quoted Sir Walter Scott (Canto I, from "The Lady of the Lake"):

> ... the stranger's bed
> Was there of mountain heather spread,
> Where oft a hundred guests had lain,
> And dreamed their forests' sport again;
> Nor vainly did the heath flower shed
> Its moorland fragrance round his head.

[6] There is an instance of Queen Victoria endorsing white heather as a lucky emblem before the publication of her book. On 3 September 1862, the recently widowed Queen was at Laeken in Belgium to meet Prince and Princess Christian of Denmark and their daughter Alexandra, the fiancée of the Prince of Wales, "Bertie". "Her whole appearance", wrote Queen Victoria in her private diary, "was one of the greatest charm, combined with simplicity and perfect dignity. I gave her a little piece of white heather, which Bertie gave me at Balmoral, and I told her I hoped it would bring her luck" (Hibbert 1984: 167).

[7] In an obituary of William Page Wood, Baron Hatherley (1801–1881), one-time Lord Chancellor, it was revealed that white heather also was involved in the bethrothal of Princess Louise (see Nelson 2006c, 2009b):

> Another example of the friendly interest with Lord Hatherley's manly and simple rectitude of nature inspired in members of the Royal House was the gift to him by the Princess Louise of a spray of white heather, broken by her from the bush at which the Princess and the Marquis of Lorne plighted their troth. It was carefully preserved by him to the last in his drawing-room at Great George-street. (*The Times* 11 July 1881: 4).

[8] The National Pest Plant Accord (NPPA) is a co-operative non-statutory agreement between the Nursery and Garden Industry Association, regional councils and New Zealand government departments with biosecurity responsibilities. (URL http://www.biosecurity.govt.nz/pestsdiseases/plants/accord.htm accessed 30 July 2007).

[9] In Dr Paul Barclay-Estrup's account of *Calluna* in eastern Canada, the earliest record is attributed to "the collections and list of Bachelot de la Pylaie" (Barclay-Estrup 1988, 1991; for discussion see Nelson 2010a). Auguste Jean Marie Bachelot de la Pylaie (1786–1856), French explorer, botanist and archaeologist, visited Newfoundland in 1816 and 1819–1820 when he observed plants and collected specimens. No confirmation that Bachelot de la Pylaie collected or observed *Calluna* in Newfoundland has been traced, and Hooker's account (1840) is much more likely to have been derived from Don's (1834) footnote.

4. *DABOECIA* D. DON: SAINT DABEOC'S HEATHS

The plant you intitle *Erica S. Dabeoci* I am in some doubt whether it be a genuine species of *Erica*,
the flower falling away, & ye fruit seeming to be different.

Revd John Ray to Edward Lhuyd, 11 June 1701.

Connemara Heath (E. Dabeoci). — A beautiful dwarf bush, the name of which has been so often changed by
botanists that it is difficult to find it by name in books, and I give it by the Linnean name here.

William Robinson, *Gravetye Manor* (1911: 132).

Dabeoc (pronounced dah-vock, and so sometimes nowadays spelled Davoc), whose name is perpetuated in these heaths, was an early saint of the Irish church who is otherwise almost unremembered: he was, perhaps, like Saint Patrick, of Welsh origin. Saint Dabeoc's festival days are 1 January, 24 July and 16 December. The reason for his connection with this shrub is quite forgotten and irretrievable (Nelson 2009d: 52–55). To add confusion to the unknown, the Latin name is a misspelling[1], with the central vowels of Dabeoc transposed as if the saint's name contains the ligature œ, pronounced as a simple sound, a monophthong or "improper diphthong". Sometimes the name is even printed *Daboëcia*, without any justification.

Irish hagiographies report that Dabeoc (or Beoc, or Mobeoc, or Bean; see Farmer 1997) flourished from the latter part of the fifth century into the early sixth century (Nelson 1984b). The annals of the Irish church report his death in the year 516 and record that he founded a monastery at Lough Derg in County Donegal in north-western Ireland. An island in that lough became revered throughout Christendom during the Middle Ages because of a cavern known as Saint Patrick's Purgatory where, according to legend, Patrick had a vision of the Otherworld. The island continues to serve as a place of penitential pilgrimage.

Long ago, a tradition arose in western Ireland that this purple-flowered shrub was associated with Saint Dabeoc, but the heather does not grow naturally around Lough Derg nowadays. In fact, Saint Dabeoc's heath inhabits a relatively small area in counties Mayo and Galway: the nearest locality to Lough Derg is around 140 kilometres distant to the south-west. This geographical disparity between the saint and his heather has led to suggestions that the saint commemorated was not the Donegal monk but another early saint from a place nearer the plant's area of occurrence (see Praeger 1925; Nelson 1984b).

Daboecia is a genus of only two species, one of which, *D. azorica*, is restricted to the Azores in the central North Atlantic Ocean. The principal species, *D. cantabrica*, the type of the genus, is a sprawling, evergreen subshrub, a native of the north-western corner of Portugal, northern Spain from Coruña to Nararra, and the south-western extremity of France as well as parts of western Ireland.

Daboecia is distinguished from *Erica* by several characters, including a deciduous corolla. The Revd John Ray was perhaps the first to recognise the difference between *Erica* and *Daboecia* as long ago as June 1701 (Gunther 1928: 280–281): "The plant you intitle *Erica S. Dabeoci* I am in some doubt whether it be a genuine species of *Erica*, the flower falling away, & ye fruit seeming to be different."

The generic name *Daboecia* is generally attributed to David Don (1834), but it was certainly in print many decades earlier in Thomas Martyn's edition of Philip Miller's *The gardener's and botanist's dictionary*. The exact publication date of the *Dictionary* fascicle containing *Daboecia* has not been determined, but it was certainly before 1797 and probably as early as 1794 (see Nelson 2004c, 2011b). We also know from correspondence between Richard Duppa and Sir James Edward Smith that the name *Daboecia* was familiar to English plantsmen in 1827 (Smith 1832: 125; Nelson 2000a).

2. DABOECIA CANTABRICA. Saint Dabeoc's heath

The first published mention of a heather from Cantabria, in other words northern Spain, with large flowers and myrtle-like leaves that were white underneath — *Erica Cantabrica flore maximo, foliis Myrti subtus incanis* — is found in Joseph Pitton de Tournefort's *Elemens de botanique* (1694: 475). The source of the French botanist's information is not exactly recorded; perhaps he himself gathered specimens during his exploration of the Pyrenees in 1681, but this seems unlikely given his use of the geographical epithet *cantabrica*. Tournefort (1656–1708) distributed specimens of this Cantabrian plant to fellow botanists including his pupil, the Leicestershire-born lawyer William Sherard (1658/9–1728).

In the spring and summer of 1700, the Welsh naturalist, antiquarian, lexicographer and Keeper of the Ashmolean Museum in Oxford, Edward Lhuyd (*olim* Llwyd or Lhwyd; 1659/60–1709), travelled to Ireland, where he collected several hitherto unknown plants including a heath with large purple flowers (Gunther 1945; Mitchell 1975; Nelson & Walsh 1995). Lhuyd, "a brilliant and many-sided genius", worked along the north and west coasts of Ireland, reaching the southern part of County Mayo, around Westport, when this heather was in blossom. This must have been towards the end of May 1700[2] — Saint Dabeoc's heath can be in blossom by this time, although the main flush of flowers is usually not until July, by which month Lhuyd was far from Connemara and Mayo. On his return to Oxford, he favoured several friends with dried, pressed specimens of the heather; one example of these is preserved in the Revd Adam Buddle's herbarium (now incorporated into Sir Hans Sloane's herbarium in the Natural History Museum, London, volume 126, folio 41; see Dandy 1958a).

In letters written during the summer of 1700, Lhuyd described the Irish heather to others (see Mitchell 1975; Nelson 1979a), and speculated about its relationships. Writing to Dr Tancred Robinson about various aspects of Ireland's natural history, Lhuyd reported (Lhwyd 1712: 525) that:

> In most of the Mountains of *Galloway* and *Mayo* grows an elegant sort of Heath, bearing large Thyme-leaves, a Spike of fair purple Flowers like some *Campanula*, and viscous Stalks. I know not whether it be any thing related to the *Cisti Ladaniferæ*.

To Dr Richard Richardson of North Brierly, near Bradford, Yorkshire, he wrote in similar vein (see Owen 1922; Dandy 1958a: 156):

> In the moors of yᵉ County of Mayô & Galloway grows a very elegant sort of Heath so common that yᵉ people have given it yᵉ name of Frŷch Dabeôg I. Erica Dabeoci & sometimes yᵉ women carry sprigs of it about them as a Preservative against Incontinency. I calld it Erica maxima viscosa, rubra; Rosmarini foliis brevioribus flosculis Campanula minoris ... .

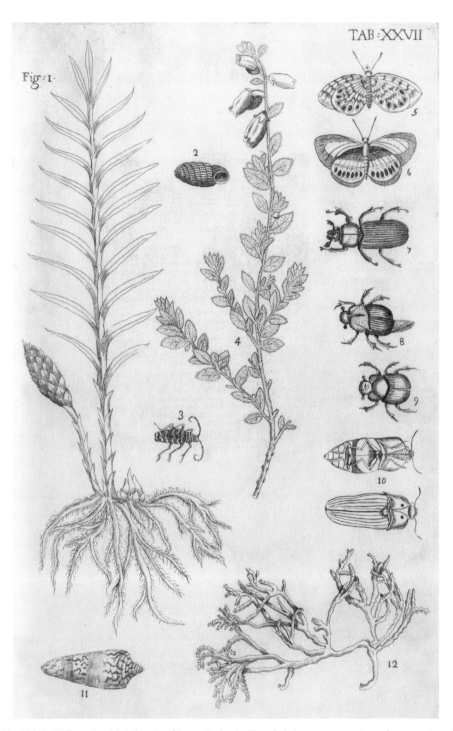

Fig. 43. *Tabula* 27 from the third decade of James Petiver's *Gazophylacium natura et artis* — the engraving of a sprig of St Dabeoc's heath (numbered 4) depicts one of Edward Lhuyd's specimens. This engraved image is the lectotype of *Daboecia cantabrica* (Huds.) K. Koch.

Lhuyd's specimens furnished several British authors with material for their publications. The first to notice it was Ray in *Historia plantarum* ... (1704: Book XXX: 98): the same heather is in a list of plants that Ray had received from James Petiver (Ray 1704: Appendix, 244). Ray reported that the heath was frequent only in the rough and spongy mountains of Mayo and through the whole of Iar-Connacht in county Galway — "In montibus *Mayo* squalido & spongioso solo frequens est, ut & per totum *Hiar-Connacht in Gallovidia*." The most remarkable thing about Ray's contribution is that he correctly equated Tournefort's Cantabrian heather with Lhuyd's Irish one. This accurate identification is most probably attributable to Sherard, who had written on 29 May 1701 from "Badmington" (Gunther 1945) to Lhuyd, stating:

> As to yᶜ curious plants you have innrich'd me wᵗʰ I can say little, however shall venture to make some conjectures or queries abᵗ them, since yᶜ are pleas'd to comand it of me. I find (as Mr. Ray complains to me abᵗ my own) 'tis hard to judg of plants by dry'd specimens, especially where they are not perfect & well preserv'd, wᶜʰ is not always practicable, especially in travelling.

In an accompanying, annotated list of plants, which clearly included Lhuyd's Irish specimens, Sherard had noted (under number 4) that "I had this (or some very like it, wᶜʰ upon comparing yᶜ specimens I shall easily judg) from Dr. Tournefort, by yᶜ name of, Erica Cantabrica, fl. max. fol. myrti subtus incanis. Inst. r. herb." Lhuyd then seems to have labelled his specimens with Tournefort's name, including whichever one he sent to Ray, although from the letter (already quoted) written by Ray on 11 June 1701 (Gunther 1928: 280–281), it is evident that Lhuyd also used the name *Erica Sancti Dabeoci*.

Shortly after Ray's *Historia plantarum* was published in the summer of 1704, James Petiver (1663/4–1718), fellow botanist and author, who had also been favoured by Lhuyd with a specimen of the heather, used that pressed sprig as the template for an engraving published in the third decade of *Gazophylacium natura et artis*, an illustrated part-work begun in 1702 — the heather appeared in the third decade: figure 4 on *tabula* 27 (Fig. 43). Petiver's illustration was the first, by almost 90 years, of the many pictures of Saint Dabeoc's heath that have been published in botanical books (Fig. 44).

Johann Jacob Dillenius (1684–1747), editor of the third edition of Ray's *Synopsis stirpium britannicarum*, included Saint Dabeoc's heath

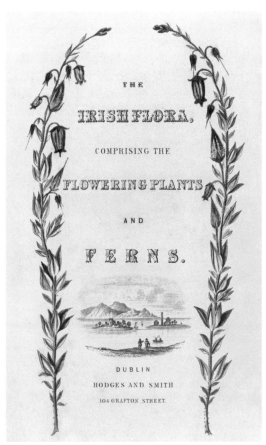

Fig. 44. St Dabeoc's heath depicted on the ornamental title-page found in some copies of the 1845 re-issue of Katherine Sophia Baily's anonymously published *The Irish flora*. The hand-coloured engraving is not signed.

as the sixth member of the genus *Erica*, employing the long, elegantly descriptive phrase-name *Erica Cantabrica flore maximo, foliis Myrti subtus incanis* of Tournefort, and repeating information first published by Ray (1704; Dillenius 1724).

The Revd Dr Caleb Threlkeld (1676–1728) listed this western Irish species also using Tournefort's phrase-name *Erica Cantabrica flore maximo, foliis Myrti subtus incanis* but, significantly, added *Erica Hibernica folijs myrti pilosis subtus incanis, Erica Sancti Dabæoci* (Threlkeld 1726): the addition may be translated as "The heather from Ireland with leaves of a myrtle, hairy, white beneath, The heather of Saint Dabæoc".

Threlkeld probably never saw this plant — he is not known to have travelled into western Ireland. The next Irish botanical author, the Revd John Keogh (c. 1681–1754), made no mention of Saint Dabeoc's heath in his herbal *Botanalogia universalis hiberniae* (1735). Dr Patrick Browne (c. 1720–1790) could not have failed to notice this spectacular species in his native county, Mayo. Indeed he collected it on The Reek, Croagh Patrick, a place as revered as Lough Derg. Browne's flora of Mayo (Nelson & Walsh 1995), a work not published until more than two centuries after his death, contained Saint Dabeoc's heath as *Andromeda daboeci*, an illegitimate name which Carl Linnaeus had coined for it in 1767 (cf. Jarvis 2007: 289). Browne also noted, as a synonym, *Erica daboeci*, another illegitimate name first used by Linnaeus in *Flora anglica* (1754a; cf. Jarvis 2007: 499).

HISTORY IN CULTIVATION: EVIDENCE FROM PAINTINGS

While native in western Ireland, it is unlikely that *Daboecia cantabrica* was cultivated in Irish gardens before the early nineteenth century, when there was a resurgence of interest in matters botanical.

Daboecia cantabrica is not native in Great Britain but was cultivated near London by the Quaker plantsman Peter Collinson (1694–1768), an avid collector of unusual plants, by 1765. In his garden at Hendon, north of London, Collinson had raised plants from seeds sent from Spain by Dr William Bowles (1705–1780), a native of County Cork who had risen to become the Director-General of the Spanish mines. In 1763, Bowles had gathered seeds of Tournefort's *Erica cantabrica flore maximo* when he was exploring north-eastern Spain. He sent some to Collinson, and the resulting seedlings blossomed in August 1765 (O'Neill & Nelson 1995), causing Collinson to exclaim, in a letter to Carl Linnaeus written on 17 September 1765, that "It is an elegant plant, and makes a pretty show." Thus an Irishman introduced into cultivation an Irish heather from the north of Spain.

One year later, almost to the day, Georg Dionysius Ehret (1708–1770), one of the finest botanical artists of the eighteenth century, saw and drew the heather and dissected some flowers; his watercolour and pencil sketch "Drawn at B. Sep[br] 10. 1766"[3] survives in the collection of the Natural History Museum, London (Knapp 2003: 260).

Another painting of Saint Dabeoc's heath executed during the late 1700s, although not dated, by Simon Taylor (1742–?1796), is among the collections of the Earl of Derby.[4] Taylor was described by John Ellis, in a letter to Carl Linnaeus written on 28 December 1770 as "a young man ... who draws all the rare plants of Kew garden ...; he does it tolerably well" (Smith 1821: **1**: 255; Desmond (1994) attributed this remark to Ehret). The painting, on vellum, presumably depicts a plant that was growing in the Royal Gardens, Kew.

There is also evidence from paintings that plants were introduced into cultivation from Ireland about this time. An undated watercolour of Saint Dabeoc's heath ("*ERICA dabeocia*") by the Yorkshire artist James Bolton (d. 1799), now also in the Earl of Derby's collection, portrays a flowering sprig that Bolton "... took ... from an healthy Plant, in the Garden of John Blackburn Esq.[r]

Daboecia cantabrica **Painted by Wendy F. Walsh**   PLATE 4

Published in E. C. Nelson & W. F. Walsh, *The flowers of Mayo* (1995).

of Orford, near Warrington Lancashire." "The Root was sent from Ireland" to Blackburn, whose garden was described by Sir James Edward Smith as "one of the oldest and richest botanic gardens in England" (see Henrey 1975: **II**: 640). *Daboecia cantabrica* became commercially available soon after its introduction. William Malcolm of Kennington Turnpike offered the "Cantabrian heath" for sale in his 1770 catalogue, and by the early 1790s, Smith (1791) was able to write that "Erica Dabeoci ... is frequently cultivated with us in gardens."

Daboecia cantabrica '**Charles Nelson**'   Painted by Wendy F. Walsh   PLATE 5

Published in E. C. Nelson & W. F. Walsh, *An Irish flower garden replanted* (1997).

VARIANTS AND MALFORMATIONS

While the majority of wild plants have purple flowers, white-flowered plants (Fig. 47) occur sporadically as isolated individuals throughout the species's range. Any plant, wild or cultivated, with white blossoms can be named *Daboecia cantabrica* f. *alba* (Sweet) D. C. McClint. (McClintock 1984a: 191). Employing Sweet's epithet *alba* as a cultivar name ('Alba') is common practice, but several different clones are being propagated under this "umbrella" name and so its use is to be firmly discouraged. This "albino" form generally lacks any trace of red pigment in the flowers and was collected in Ireland as early as 1813 by Dr George Williams, Professor of Botany at the University of Oxford (Nelson 1982a; McClintock 1984a), but apparently it was not until July 1831 that it was brought into cultivation, when a single, weak plant, found at Clifden, Connemara, County Galway, was grown in the Trinity College Botanic Garden, Dublin (Webb 1970; McClintock 1970a; Nelson 1982a).[5] James Townsend Mackay, Curator of the College Botanic Garden, sent specimens to Professor William Hooker in Glasgow on 9 September 1831, saying "I have as yet but a single plant, which I suppose is the only one in cultivation" (Nelson 1982a). By December, he had managed to propagate the white-blossomed *D. cantabrica*, for he sent the Revd Henry Ellacombe (1790–1885) a "... rooted cutting from the only flowering plant in cultivation", adding that the variety "is an interesting shewy [*sic*] plant which I have only very lately sent to a few friends" (Nelson 1982a).

Two years later, the young Connemara-resident botanist, William McCalla (c. 1814–1849), described an "old plant" of an unusual variant which he had seen " ... on the side of a mountain, remote from any of its own species ... the blossom perfectly white, the calyx red ..." (McCalla 1834; Nelson 1982a, 1983).

Fig. 45. St Dabeoc's heath with western gorse (*Ulex gallii* Planch.) and ling (*Calluna vulgaris*), in bud, in Connemara, County Galway (E. C. Nelson).

Fig. 46. St Dabeoc's heath with tormentil (*Potentilla erecta* (L.) Räuscel) near Pontoon, County Mayo (E. C. Nelson).

Fig. 47. White-flowered St Dabeoc's heath, *Daboecia cantabrica* f. *alba* growing wild at Ballyconneely, Connemara, County Galway (E. C. Nelson).

Another wild-occurring variant has almost pure red flowers and was first recorded in 1938 by Mrs Praeger, wife of Dr Robert Lloyd Praeger, near Errisbeg Mountain at Roundstone in Connemara (see below). Although it has been reported on a few occasions since the 1940s, the red-flowered Saint Dabeoc's heath is the rarest of the wild colour variants.

As well as colour forms, various peculiar malformations have been noticed in the wild in Ireland, and some have been propagated, giving rise to odd, yet attractive cultivars. 'Celtic Star', for example, has greatly enlarged, brightly coloured red sepals which contrast with the lavender-coloured corolla (Nelson 1989b, 1994a) (Fig. 48).[6] 'Charles Nelson' (Plate 5; Fig. 206, p. 361) has "double" flowers (McClintock 1982a). 'Doris Findlater' has erect, not nodding, crimson flowers (Nelson 1986a).

This latter variant also occurred in cultivation; seedlings with erect blooms of various shades — white ('White Blum'), pink ('Pink Blum') and purple ('Purple Blum'; ® no. 178) — were noticed by the Dutch plantsman Herman Blum (1934–2003) (Dahm 2004; Small 2004b) in the heather garden at De Voorzienigheid, Steenwijkerwold, Holland, and McClintock (1984b) assigned these to *Daboecia cantabrica* f. *blumii*. Normally, in both *Daboecia* species the pedicel elongates and curves downwards as the flower matures so that at anthesis the corolla is inverted and the mouth faces down. After pollination the corolla falls off, and the pedicel becomes erect again so that the capsule is upright. There is also one instance known of *D. cantabrica* with a corolla split into irregular segments: the cultivar 'Covadonga', which has no Irish parallel, came from Cordillera Cantabrica in northern Spain.

CULTIVATION & PROPAGATION

Daboecia cantabrica can be grown easily in lime-free soil, yet it is also remarkably tolerant of slightly alkaline soil containing lime (see, for example, Jones 1977; Stow 2004). It is hardy and may be raised from seed or propagated by cuttings. Named cultivars, being clones, have to be propagated vegetatively. Gardeners are usually recommended to plant heathers in groups of about half a dozen plants of the same cultivar creating bold blocks and a "chess-board" effect. However, St Dabeoc's heath is amenable to mixed planting with cultivars, colours and shades mingled randomly (Fig. 49).

Daboecia cantabrica (Huds.) K. Koch, *Dendrologie* **2** (1): 132 (1872). Type: icon, fig. 4, tab. 27 in J. Petiver, *Gazophylacii naturæ et artis decas tertia* (1704), lecto. (Nelson 2000a).
Basionym: *Vaccinium cantabricum* (L.) Huds., *Flora anglica*: 143 (1762).
Erica daboecii L., *Species plantarum*: 509 (1762; second edition); Huds., *Flora anglica*: 166 (1778; second edition).
Menziesia polyfolia Juss., *Ann. Mus. Natl. Hist. Nat.* **1**: 52–56 (1802).

ILLUSTRATIONS. J. Sowerby, *English botany* **1**: tab. 35 (1791); J. Curtis, *British entomology* **12**: tab. 574 (1835); S. J. Roles, *Flora of the British Isles: illustrations* pt 2: tab. 925 (1960); S. Ross-Craig, *Drawings of British plants*, part 19: pl. 29 (1963); R. W. Butcher, *A new illustrated British flora*: 99 (1961); W. Walsh, *An Irish florilegium*, pl. 17 (Walsh *et al.* 1983); Anonymous (1985: 19); W. Walsh, *Flowers of Mayo*, pl. VI (Nelson & Walsh 1995: 101).
BIOLOGICAL FLORA OF THE BRITISH ISLES. *J. Ecol.* **46**: 205–216 (1958).
DESCRIPTION. *Evergreen shrub* to ±0.6(–1) m tall, spreading and often entwined with other shrubs in natural habitats; all young aerial parts covered with sticky, gland-tipped hairs; branches and

Fig. 48. An unusual variant of St Dabeoc's heath with enlarged, coloured sepals, growing wild on Errislannan Peninsula, Connemara, County Galway. This clone is in cultivation as 'Celtic Star' (see p. 74) (E. C. Nelson).

Fig. 49. The best way to grow St Dabeoc's heath: a mixed planting of clones with different flower colours; in a Californian garden (E. C. Nelson).

branchlets woody, with brown flaking bark, becoming glabrous, decumbent, then ascendent; flowering shoots erect; root system extensive, creeping. *Leaves* arranged alternately, ±1(–1.5) cm long, ±0.5 cm broad, oval to elliptical with pointed apex, dark glossy green above, with dense felt of very closely appressed white hairs underneath; margins curled downwards and inwards, entire; *petiole* covered with sticky, gland-tipped hairs. *Inflorescence* a raceme with ±6–12(–15) flowers; peduncle erect. *Flower* solitary, nodding at anthesis, on short, curving pedicel. *Calyx* of 4 free, green, lanceolate segments, ±4 mm long. *Corolla* purple (rarely red or white), lower side paler, urn-shaped, ±1–1.5 cm long, with 4 recurved, obtuse lobes at mouth, with gland-tipped or non-glandular hairs of unequal lengths on outside. *Stamens* 8, ±8.5 mm long, included; *filament* ±4.5 mm long; *anther* very dark, thecae ±4.5–5 mm long. *Style* straight, erect, persistent in fruit; *stigma* obtuse; *ovary* ±2 mm long, pyramidal to conical, tapering into base of style, with stout, gland-tipped hairs and also very short non-glandular hairs at base; ±40 ovules per locule; *nectary* glabrous. *Fruit* a capsule, opening septicidally, covered with gland-tipped hairs, *pedicel* recurving so that the capsule is almost erect. *Seed* red-brown, ovoid to ellipsoid, 0.65–0.75 × 0.5–0.6 mm, papillate (Fagúndez & Izco 2004b). n = 12 (Woodell 1958).

FLOWERING PERIOD. (May–) June–September (–November).

DISTRIBUTION & ECOLOGY. Europe: western Ireland, confined to the region between Clew Bay and Lough Corrib; extreme south-western France and a few widely scattered localities elsewhere in the south and west; northern Spain from the Pyrenees westwards and mainly on northern side of Cordillera Cantabrica; north-western Portugal only. **Maps:** Fig. 51; Dupont (1962: 118, carte 10); Webb (1983: 157); Preston *et al*. (2002).

In Ireland Saint Dabeoc's heath ascends from sea-level to around 580 m altitude on the mountains of Connemara (Webb & Scannell 1983; Preston *et al*. 2002). *Daboecia cantabrica* is a member of the Oceanic Southern-temperate element of the Irish flora. *Erica erigena* (see p. 119) is also included in this element (Preston & Hill 1997), as are *E. ciliaris* (see p. 162) and *E. vagans* (see p. 132), although neither of these heathers may be indigenous to Ireland.

Daboecia cantabrica is a naturalised alien in New Zealand (Sykes 1981), and has also been reported as an exotic from sites in Great Britain (in 13 separate 10 km squares) ranging from Jersey in the Channel Islands to Yorkshire, with a concentration in the Dorset/Hampshire region where it was reported as naturalised as early as 1871 (cf. Simpson 1960; Preston *et al*. 2002). Whether the heather persists unaided by cultivation in any of these localities is not known (Nelson 2004a) and it is not listed by Clement & Foster (1994).

Studies by Valbuena & Luz (2002) showed that high temperatures have a dramatic effect on seed germination in *Daboecia cantabrica*. Temperatures of 100°C or higher, applied for between two and four minutes, increased germination significantly. Older, stored seeds needed to be heated to a higher temperature than freshly collected seeds to reach similar percentage germination in the laboratory. On the other hand, experiments conducted by González-Rabanal & Casal (2004) led them to conclude that seed of *D. cantabrica* showed no significant response to the shock of high temperatures.

ETYMOLOGY. Latin toponym: *cantabricus* = from Cantabria, northern Spain.

VERNACULAR NAMES. Robert Lloyd Praeger (1925) wrote: "Dabœcia must stand as the botanical designation. But it we cannot now make the Latin name right, we certainly have no justification for making the English name wrong, and the plant remains St. *Dabeoc's* heath."

Fraoch Dabeoc, fraoch na h-aon choise, fraoch gallda (Hogan *et al*. 1900) (Irish); Saint Dabeoc's heath (English); Connemara heath(er), Irish heath (mainly North America); Gemeine Kriechheide

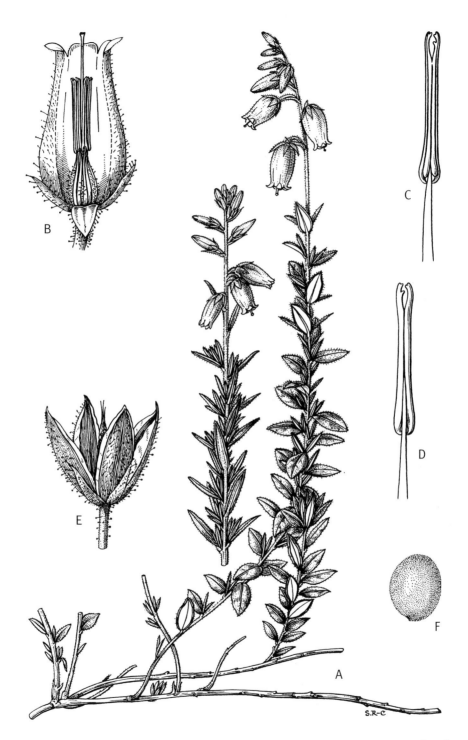

Fig. 50. *Daboecia cantabrica.* **A** part of a plant and a younger flowering stem × 1; **B** flower — part of corolla cut away × 4; **C** anther and part of filament in front view × 8; **D** anther and part of filament in back view × 8; **E** fruit × 4; **F** seed × 30. Drawn by Stella Ross-Craig.

(Dippel 1887), Irische Glockenheide, Irische Heide, Glanzheide (German); bruyère de Saint Daboec [*sic*], Daboecie des Monts Cantabres (French: *fide* http://www.jpdugene.com/; accessed 3 February 2007), bruyère des monts Cantabriques; ainarra, txilarra, ainarra kantauriarra (Basque: *fide* http://www.jpdugene.com/; accessed 3 February 2007)); brezo cantábrico (Spanish; for the numerous other names used in Spain, see Pardo de Santayana *et al.* 2002); irländsk ljung (Swedish).

Daboecia cantabrica and *D.* × *scotica*   Painted by Christabel King   PLATE 6

The plants represented are (in order by stems on bottom margin, FROM LEFT TO RIGHT):

D. cantabrica unnamed ("mauve"). (Painted 11 August 1985.)

D. cantabrica f. **alba.** (Painted 11 August 1985.)

D. cantabrica '**Charles Nelson**' 🌟 Sprawling, low-growing heath (±0.3 m tall × ±0.45 m spread); foliage green; flowers mauve (H2), the first flush being normal with properly formed stamens, and ovary, style and stigma present and not malformed; subsequent flowers squat and malformed, full of petaloid staminodes and extra corollas, and so "double", and usually lacking stamens, ovary, style and stigma; June–October. (Painted 6 August 1985.)

Unlike normal-flowered plants, in 'Charles Nelson' the withered flowers do not drop off. It is desirable to trim off the old racemes in late autumn, because the plump, dead flowers can turn mouldy and look unsightly. The fact that the first flowers have normal stamens and gynoecium, and a deciduous corolla, is inexplicable.

I found this intriguing monstrosity near Carna in Connemara, County Galway, Ireland, during the summer of 1978, and the original plant was still growing at the locality in 1999, but by then it was greatly overshadowed by bracken and was struggling to survive. This oddity was named by David McClintock (1982a; see also Nelson 1994b; Nelson & Walsh 1984, 1997).

Recently, Kurt Kramer, Edewecht-Süddorf, Germany, and Jens Kjærbøl, Bryrup, Denmark, have raised seedlings with similar "double" flowers in various shades of lavender, white and "red", and also with upright flowers in the manner of *D. cantabrica* f. *blumii*.

D. cantabrica '**Atropurpurea**'. Vigorous, tall heath (±0.4 m tall × ±0.7 m spread); glossy foliage deep holly-green; flowers small, purple (H10); June–October. (Painted 6 August 1985.)

This name has been in use since the 1830s: the Mackies, nurserymen of Norwich, employed it. It is questionable whether any extant clone bearing this name is the same as Mackies' (assuming theirs was a single clone). Moreover, at present more than one clone is being propagated under the name 'Atropurpurea'. The exact source of the plant used as a model by Christabel King is not recorded, so the identity of the illustrated plant must remain doubtful.

D. cantabrica '**Praegerae**' 🌟 Vigorous, spreading heather (±0.4 m tall × ±0.7 m spread); foliage green; flowers cerise (H6); June–October. (Painted 6 August 1985.)

This was the first of the so-called "red" cultivars, and it still has its merits. The flowers have no trace of blue in them.

Mrs Hedwig ("Hedi") Praeger (née Magnussen; d. 1952), wife of the eminent Irish naturalist Dr Robert Lloyd Praeger (1865–1953), found this plant, which is named after her, in Connemara, County Galway, Ireland, in the late summer of 1938. That September, propagation material was brought to the National Botanic Gardens, Glasnevin, whence it was distributed during 1941 (under various names) to Irish nurseries. Desmond Shaw-Smith, Ballawley Park Nursery, Dundrum, County Dublin, was the first to list this clone, and make it available commercially, and to establish the epithet 'Praegerae' (see Nelson 2000a; Nelson & Walsh 1984, 1997; Walsh *et al.* 1983; for another illustration, see Walsh *et al.* 1983: plate 17 (Wendy Walsh)).

D. × *scotica* '**William Buchanan Gold**'. Compact, spreading shrub (±0.25 m tall × ±0.45 m spread) with yellow shoots; foliage dark green flecked with red, gold and yellow throughout the year; flowers dark crimson (H13); May–September. (Painted 11 August 1985.)

Reputed to have been a sport on 'William Buchanan'; found about 1970 (McClintock 1978) by Clive Baulu, Hardwicks Nursery, Newick, Sussex, England, and introduced about 1975 by Robinson's Nursery, Knockholt, Kent.

On a purely historical note, when this plant was removed from *Erica* into *Menziesia* by the French botanist Antoine Laurent de Jussieu (1748–1836), one of a highly distinguished dynasty of botanists (Stafleu 1971), he named it "*Menziezia polyfolia*"[7] (Jussieu 1802), giving the French equivalent as "Menziezie (à feuilles de polium)", the Menziesia with leaves like "polium": "Nous proposerons de la nommer *menziezia polyfolia*, parce que ses feuilles ressemblent en effet beaucoup à celles du *Teucrium polium*, L. et de l'*andromeda polyfolia*, L."

ETHNOBOTANY. Perhaps the heather's inexplicable association with Saint Dabeoc caused the other, equally irrecoverable tradition. When the Irish habitat of this plant was discovered in the last year of the seventeenth century by Edward Lhuyd, he recorded that this heather was "so common that ... the women carry sprigs of it about them as a Preservative against Incontinency." In the particular context, "Incontinency" should be read as unchastity. Dr (later Sir) James Smith (1791) noticed that this superstition was mentioned by the Revd John Ray — "*Mulierculae superstitionae surculos ejus serum circumferunt adversus incontinentiam*" — adding, with a flash of wit rare in botanical publications: "Irish girls gird themselves with its long trailing branches as a protection to their chastity — With what success he unluckily has omitted to inform us."

The account makes clear that the plant was worn: it was not made into a medicine and applied directly to the body nor was any part or extract ingested. Given this, Allen & Hatfield's (2004: 121) suggestion that Lhuyd's account indicates that *Daboecia* may have had a medicinal role in western Ireland against urinary incontinence seems dubious. There is no report from other parts of its natural range of any similar medicinal use for Saint Dabeoc's heath.

CULTIVARS. See p. 79 and Appendix I (pp. 360–362).

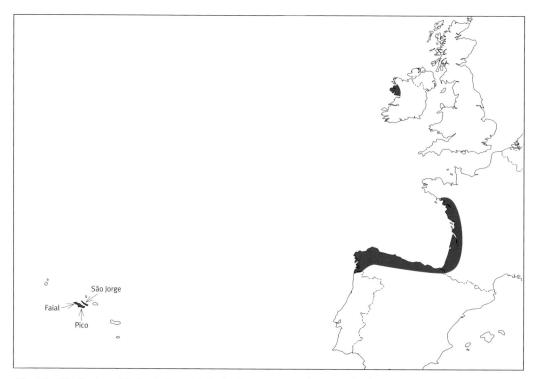

Fig. 51. Distribution of *Daboecia*; *D. azorica* in the Azores; *D. cantabrica* in Ireland and south-western Europe.

3. DABOECIA AZORICA. Queiró, Azores heath.

Daboecia azorica is a low-growing, neat, evergreen shrub with brilliantly coloured flowers, the vast majority of plants producing blossom that is almost pure red in colour (Figs 52–55), quite unlike the generally purple flowers of *D. cantabrica* from Ireland and the European mainland. For many decades it was regarded merely as *D. cantabrica*, but following a visit to the Azores in 1929, Tutin & Warburg (1932) decided it represented a separate species, distinguishable from the other by "its habit, smaller, proportionately broader leaves, and the shape, size, and colour of the flowers." In their protologue they omit any mention of the absence of hairs on the corolla, although they do state that the peduncle, pedicel, sepals and ovary were more glandulose ("*majus glandulosi quam in ...* ") than in *D. cantabrica*.

Does *Daboecia azorica* represent a separate species, or merely a subspecies? I am not aware of any DNA studies on *D. azorica*, and whether such studies would illuminate the problem is perhaps doubtful. In the end, it is a matter of opinion, and the judgements of individual botanists will inevitably differ. The Azorean populations of Saint Dabeoc's heath are distinct and, crucially, they must have a peculiar history. The archipelago is entirely volcanic in origin, the oldest rocks being about 5 million years old (Scarth & Tanguy 2001), but none of the lava on the islands where *Daboecia* grows today is more than 1 million years old. The oldest lava exposed on Faial is around 730,000 years old, while the oldest surface rocks on São Jorge have been dated at 550,000 years. In contrast, the most ancient lava on Pico is less than 37,000 years old. The spectacular stratovolcano of Pico, which has *Daboecia* growing close to the summit, is a young volcano, not an old one covered with new lava: "Pico is the newest addition to the Azores" (Scarth & Tanguy 2001). As none of the plants and animals that now inhabit the Azores can have been there before the islands emerged from the Atlantic, the present flora and fauna are all relatively recent immigrants, or are directly derived from such immigrants. There has been time for speciation, as evidenced by the number of endemic plants that occur on the Azores, including several taxa in the Ericaceae.

In *Flora europaea*, Webb (1972a) retained *Daboecia azorica* as a distinct species, but McClintock (1989a) relegated it to the status of a subspecies of *D. cantabrica*, advancing as reasons the fact that the Azorean plant "differs morphologically only in the smaller size of its parts and the absence of hairs on the corolla". McClintock also noted that "pressing personal wishes to the contrary by both the authors of [*D. azorica*] previously inhibited me from altering its status." From unpublished correspondence, it is evident that he and Webb debated the matter: rejecting McClintock's view that "free interbreeding with fertile offspring" is an argument for relegating it to subspecific rank, Webb (*in litt.* to D. McClintock, 9 June 1983[8]) stated that his "<u>acid</u> test of a distinct species is: can you distinguish from the other species over 99% of well-grown plants", and that in the case of *D. azorica* his confident answer was "Yes." "Finally", wrote Webb, "there is the argument: <u>quieta non movere</u>. If you were able to produce evidence from wild populations that Tutin's and my differentia were wrong, then you would be justified."

Certainly, *Daboecia azorica* in its native habitat is very distinct: a diminutive, bushy shrub, unlike the lax, lanky plant of Ireland and northern Iberia. The two species are separated in the wild, and there can be no naturally-occurring intermediates, although the plethora of garden-raised seedlings has substantially blurred the differences in cultivation. So I side with Webb, accepting that *D. azorica* in the wild is a distinct species.

SAINT DABEOC'S HEATH IN THE AZORES

Daboecia azorica is designated (from March 2002) as a "strictly protected species" under Appendix 1 of the Convention on the Conservation of European Wildlife and Natural Habitats ("Berne Convention"). The species is also protected by the European Union's "Habitats Directive" (Council Directive 92/43/EEC on the Conservation of natural habitats and of wild fauna and flora).

This delightful plant occurs only on three of the islands in the Azorean archipelago: Faial, Pico (Godman 1870: 193; Trelease 1897: 130; Tutin & Warburg 1932), and São Jorge (Sealy 1949; Silva *et al.* 2005). I have seen it myself on Faial and Pico, and there are numerous herbarium specimens to vouch for the species' presence on these islands since the late 1770s. On the majestic cone of Pico (the summit being 2,351 m altitude), *Daboecia* is recorded at altitudes up to 2,200 m (Tutin 1953, Sjögren 2001). Reports that *D. azorica* grows on Terceira and Flores (for both see Godman 1870: 193) were queried recently (see Silva *et al.* 2005: 141). Tutin & Warburg (1932) had written that the plant was "probably also [on] Flores and Terceira, but we have seen no specimens from these islands", whereas Sealy (1949) stated that it had been reported from four islands, including Terceira. São Miguel was included by Schäfer (2002) as a habitat, but there is no verification for this (it is queried by Silva *et al.* 2005). There have never been records of *D. azorica* on Corvo, Graciosa or Santa Maria (Sjögren 2001; Schäfer 2002; Silva *et al.* 2005).

The earliest known herbarium collections of the Azorean *Daboecia* were made by the Kew plant-collector Francis Masson (1741–1805), who visited the archipelago in the summer and autumn of 1777 (Bradlow 1994). He gathered Saint Dabeoc's heath on "Fayal & Pico, in summis montibus" (**BM**!). One of the earliest published accounts was by John and Henry Bullar in their quaint book *A winter in the Azores: and a summer at the baths of the Furnas*. They described making a trip "with a party, some on foot and some on asses, to see the Caldeira" on Faial:

> ... we began the gradual and easy ascent ... at first through fields, and then over heath and grass, interspersed with a few wild flowers; among which, a bright-belled heath, of deep scarlet, was scattered among the herbage, that resembled in colour the fruit of the wild strawberry, and was entirely different from the common heath so abundant in these islands; the unobtrusive blossoms of which seldom attract the eye. (Bullar & Bullar 1841: **2**: 2–3; see Watson 1843: 7.)

A few years afterwards, the English botanist Hewett Cottrell Watson (1804–1881) travelled to the archipelago aboard HMS *Styx* on a survey voyage conducted by the Royal Navy. Describing the hinterland of Flamengos (the same place that the Bullars had visited) on Faial, he (1843: 7) noted that *Daboecia azorica* was "extremely abundant on the hillsides" between the village and the Caldeira. On 1 July 1842, having crossed from Faial to Pico, Watson set off early to climb "the Peak" (Watson 1843: 401): "... as we ascended above the clouds, and attained an elevation that gave us a full view of the upper part of the Peak, now seen rising into a clear blue sky ... . *Calluna vulgaris*, and [*Daboecia azorica*] ... were interspersed in a few places between the larger shrubs, over spaces from which the latter had probably been burnt or cut and carried away."

On Faial, *Daboecia azorica* is easy to see on the upper slopes of Caldeira, both outside the crater where the roadway ends, and inside through the tunnel. There is noticeable variation in the colour of the flowers, with some plants having pale red corollas (Fig. 54). On Pico, the heather is abundant on the lower slopes, where the metalled road ends, growing with *Calluna vulgaris* and *Erica azorica*. Again there is variation in the colour of the flowers, and on a visit in 2003, I also noticed that some plants had distinctly narrow, almost conical corollas (Fig. 55).

White-blossomed plants are well-recorded, the earliest being those collected on Faial by Hewett Watson in June 1842. They are assigned to *Daboecia azorica* f. *albiflora* D. C. McClint. However, as far as is known, no white-flowered clone of this species has been introduced into cultivation.

In the Azores, the endemic queiró (Saint Dabeoc's heath — it seems illogical to use the name of an Irish saint to denote it!) grows in a variety of associations, often with *Calluna vulgaris* and *Erica azorica*. Most remarkable, however, is the startling combination of heather and lichen (Fig. 53): the colours are so dramatic. Against a background of black volcanic gravel, silver-grey lichens meld with the rich green foliage of the heather while above dangle the red flowers. Nowhere else, in my experience of heathers, is there such a starkly vivid combination of colours

Daboecia azorica   **Painted by Stella Ross-Craig**   PLATE 7

Original watercolour for tab. 45, *Curtis's Botanical Magazine* **166** (1949).

Fig. 52. *Daboecia azorica* on the upper slopes of Pico, Azores, with the paler pink flowers of the native thyme (*Thymus caespititius* Brot.) (© Christopher Massy-Beresford).

— black, palest grey to almost white, glossy green and brilliant glowing ruby. This association of lichens and heather is found on Pico wherever volcanic gravel has been excavated. In the resulting "scrapes", *Daboecia* evidently finds a congenial seed-bed among the burgeoning lichens that flourish in the moist, mild environment. To recreate this black-silver-green-red combination in a garden seems impossible!

CULTIVATION

Daboecia azorica was apparently not introduced into cultivation by any of the earlier collectors, even by the indefatigable Francis Masson. Although Conrad Loddiges of Hackney had a dwarf variant of Saint Dabeoc's heath (named "*nana*"), its portrait (Loddiges 1833: tab. 1907) does not match the Azorean species, and so it is likely to have been a seedling from *D. cantabrica*. Credit for introducing *D. azorica* is generally given to Dr Edmund Warburg in 1929 (Bean 1933; Sealy 1949), and the species received an Award of Merit when shown at the Royal Horticultural Society in June 1932. Warburg's introduction survived in British gardens until at least the early 1950s: Colonel (later knighted) Frederick Claude Stern (1884–1967) grew it at Goring-by-Sea and provided the plant that was painted by Stella Ross-Craig for *Curtis's botanical magazine* (Sealy 1949) (reproduced in Plate 7, p. 83).

Daboecia azorica was growing in North American gardens by the early 1940s, perhaps first in British Columbia, Canada, and thence in the Seattle area. Metheny (1991, 2007) grew the Azorean species for at least two decades, and while it was often affected in very cold weather, the well-established plant was not killed.

It is generally agreed that *Daboecia azorica* requires lime-free soil for successful cultivation. Seeing the plant in the wild, growing on volcanic gravel, also suggests that it needs a very free-draining soil.

Fig. 53. *Daboecia azorica* growing in the wild, Pico, Azores, with a seedling of the endemic *Euphorbia stygiana* H. C. Watson, and silver-grey lichen (E. C. Nelson).

Fig. 54. Pink-blossomed *Daboecia azorica* growing in the wild, Faial, Azores (E. C. Nelson).

Fig. 55. *Daboecia azorica* with narrow, more tubular flowers, growing in the wild, Pico, Azores (E. C. Nelson).

Stoker (quoted by Sealy 1949) cultivated *D. azorica* in a rock-garden; it grew "slowly and solidly". Sealy (1949) also reported that *D. azorica* grew for some years on the Rock Garden at the Royal Botanic Gardens, Kew, until killed by the severe winter of 1939–1940. Stern (Sealy 1949) recorded that plants were "hard hit" during the winter of 1946–1947, disliking especially snow. On the other hand, Metheny (1991) implied that a covering of snow served as protection against very low temperatures, and she commented elsewhere (Metheny 2007) that "the considerable hardiness ... is attested by their having survived unharmed the cold snaps that seem to be our lot every five years."

As well as Warburg's 1929 introduction, Don Richards and David McClintock brought material into cultivation in 1974 (see Griffiths 1999). And there probably were other, unreported introductions.

It is possible that plants selected from the highest altitudes on Pico might be hardier. However, Barry Starling (1998; Nelson 1999d) noted that this "is one of the few ericaceous plants which thrived for me in full sun ... . The secret is to plant it where the wood of the branches is fully ripened by the sun before winter. Soft tips may be damaged by frost but in all but the most severe winters the little shrub sprouts and grows away in the spring."

A. T. Johnson (see below) grew *Daboecia azorica* in the 1930s and 1940s — "since its introduction" — in his garden in North Wales. They were planted on a woodland slope in "drier, more open, gravelly places" as "fill-ups": "... we have found [it] quite reasonably hardy, and its ruby-crimson bells on spikes of some six inches, have a distinction of their own" (Johnson 1950). Johnson raised seedlings from *D. azorica*, if the following comments are any indication: "I have observed ... considerable variability in seed-raised plants, both in the colour of the flowers and hairiness of the foliage, this quite apart from the results of hybridization with *D. cantabrica* which is said to have occurred ..." — I will revert to his hybrids later.

Metheny (1991) noted spontaneous self-sown seedlings in "the peat-sand soil around the perimeter of the parents", and suggested that propagation was not easy. With bottom-heat, she rooted two cuttings in June. Richards (1976) also noted it was very difficult to root cuttings. In this context, it is important to recall that Mrs Metheny, in some undated notes (see Metheny 2007), observed in her garden that the prostrate shoots, close to the ground, did not produce adventitious roots: *Daboecia azorica* seems to reproduce by seed, not by adventitious rooting, a thesis supported by observations in the wild.

However, Allen Hall reports (pers. comm., 2007) that he has not had any difficulty propagating the species. Of the six cuttings he took in 2005, five rooted, and by the spring of 2007 they were all good plants.

Daboecia azorica Tutin & E. F. Warb., *J. Bot.* **70**: 12 (1932). Type: 'Pico', *T. G. Tutin & E. F. Warburg*, holo. **CGE**.

D. polifolia var. *nana sensu* Siebert (1844: 289).

D. cantabrica (Huds.) C. Koch subsp. *azorica* (Tutin & E. F. Warb.) D. C. McClint., *Bot. J. Linn. Soc.* **101** (3): 280 (1989).

ILLUSTRATION. S. Ross-Craig, *Curtis's Bot. Mag.* **166**: tab. 46 (1949) (including analysis) (Plate 7 and Fig. 56).

DESCRIPTION. *Dwarf, evergreen shrub* ±0.1–0.15 m tall when in flower, stems creeping, crowded and forming a clump or mat; shoots with very short, simple, non-glandular hairs mixed with longer, stouter, gland-tipped hairs (<0.6 mm long). *Leaves* oval to elliptical, with very short non-glandular hairs and longer, stouter, gland-tipped hairs on upper surface; when young, rolled into ± a cylinder; when mature, margins slightly inrolled; sulcus becoming open, undersurface thickly

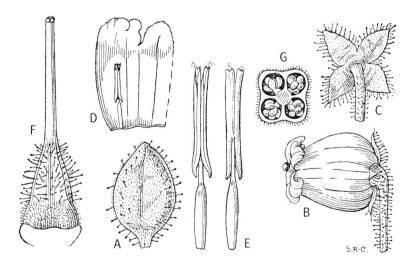

Fig. 56. *Daboecia azorica*. **A** leaf, upper surface, × 4; **B** flower in side view, × 3; **C** part of pedicel and calyx, × 4; **D** part of corolla, inside, showing one stamen, × 3; **E** stamen in face and back views, × 8; **F** gynoecium, × 8; **G** transverse section of ovary, × 8. Drawn by Stella Ross-Craig.

covered with white felt. *Inflorescence* a raceme with ±4–7 flowers; peduncle erect, red to brown. *Flower* solitary, nodding at anthesis; on curving *pedicel* of variable length, to 2 cm long, to 0.5 mm diameter, broadening towards apex, covered with gland-tipped hairs of various lengths (non-glandular hairs absent). *Calyx* 4-lobed, lobes triangular, fused at base, ±4 mm long, hirsute on outside with simple, very short non-glandular hairs covering whole surface, plus stout, gland-tipped hairs of various lengths on margins and a few on base, glabrous inside; erect in bud and fruit, bases swelling in fruit so that the corolla is forced to fall away. *Corolla* crimson to red or rarely pink or white, glabrous, <1 cm long, 4-lobed; lobes rounded, ±2 mm long, recurving. *Stamens* 8, ±6 mm long, included; *filament* ±3.5 mm long, ±0.3 mm across at base, straight, erect, glabrous, tapering upwards; *anther* erect, ±0.5 mm across in centre, thecae very dark, ± 3 mm long, fused together, lower lobes diverging away from filament, papillate; upper lobes each with an irregular slit, the margins slightly recurving. *Style* ±5 mm long, ±0.5 mm in diameter, terete, erect, glabrous, continuous with apex of ovary; *stigma* not broader than style; *ovary* ± pyramidal-conical, ±2.5 mm tall, ±1.5 mm across at base, covered with gland-tipped hairs of various lengths, and some very short, simple, non-glandular hairs about the base, ±36 ovules per locule; *nectary* encircling base, not prominent. *Seed* red-brown, ovoid to ellipsoid, 0.65–0.75 × 0.5–0.6 mm, papillate (Fagúndez & Izco 2004b).

FLOWERING PERIOD. April–June(–July) in wild.

DISTRIBUTION. Azores only; reported reliably on Faial, Pico and São Jorge. Fig. 51.

ETYMOLOGY. Latin toponym: *azoricus* = Azorean

VERNACULAR NAMES. queiró (Portuguese); Azores heath (English; McClintock 1980b); Azoren Glockenheide (German).

CULTIVAR. See Appendix I (p. 359).

HYBRIDS

In the early 1910s, Robert Vernon Giffard Woolley, Lapworth, Warwickshire, England, used a brush to cross two cultivars of *Daboecia cantabrica* — the pollen-parent was 'Atropurpurea' (see p. 79) — and one of the seedlings, a prostrate plant with globular lavender-coloured flowers, was named 'Sweet Lavender' (Woolley 1939). This is the earliest example of artificial pollination and hybridisation recorded in *Daboecia*.

There can be no natural hybrids in *Daboecia* because the two species grow in separate localities. However the fact that they have been cultivated side by side for around seven decades does allow for cross-pollination. This happened famously in William Cullen Buchanan's garden in Bearsden, one of the suburbs of Glasgow, Scotland, before 1960, and so the nothospecies has been named *D.* × *scotica* (Figs 57 & 58). However, perhaps as much as two decades earlier, spontaneous hybrids were reported in Wales by Arthur Tysilio Johnson (1873–1956). I am not aware of any unequivocal records of deliberate cross-pollination; all the clones of *D.* × *scotica* that have been selected and named to date are either chance seedlings or seedlings from open-pollinated plants. However, David Wilson (British Columbia, Canada) has deliberately back-crossed *D.* × *scotica* 'William Buchanan' with *D. cantabrica* and obtained seedlings (McClintock 1992b).

4. DABOECIA × SCOTICA. Hybrid Saint Dabeoc's heath. [*D. cantabrica* × *D. azorica*].

There are several curious references in A. T. Johnson's book *The mill garden*, published in 1950, to hybrids between the two species of *Daboecia*: for example

> *Daboëcia azorica* we have had in an open part of the woodland slope since it was first issued, but here there is what are purported to be some hybrids between that excellent heath and *D. cantabrica* ... . Quite hardy, a deeper green in foliage with a marked absence of the hairiness of the Azorean, these little bushes flower luxuriantly. The long spikes hung with bright blood-red bells are yielded so freely they almost conceal the leafage and the plants may be increased easily by cuttings.

He also wrote about plants that differed from "typical *D. azorica* ... in their smoother leaves and deeper green", which were "none the less cheerful in summer. We suspect *D. cantabrica* to have had a hand in their making and they do possess a hardiness and profligacy which suggests a hybrid origin." Johnson (see pp. 25–26) was a thoroughly competent plantsman, and his reference to hybrids between *Daboecia azorica* and *D. cantabrica* is entirely credible. He had mentioned the hybrid as early as 1942 in an article in the *Journal of the Royal Horticultural Society* (Johnson 1942; McClintock 1969). Thus it is probably to this brilliant Welsh gardener and horticultural author, and not to the Scot William Buchanan, that credit should go for first producing the hybrid now called *D.* × *scotica*. It should be stressed that there

Fig. 57 (left). *Daboecia* × *scotica*; cultivated at Beech Park, Clonsilla, County Dublin (E. C. Nelson).
Fig. 58 (right). *Daboecia* × *scotica*; cultivated at Woodfield, Clara, County Offaly (E. C. Nelson).

is no evidence that Johnson set out deliberately to cross these two species, no more than there is for Buchanan. Moreover, there is no evidence that any of Johnson's hybrids were distributed, or that they have survived. Thus the first clones to be recognised were from Buchanan's Bearsden garden.

Quite a few others also raised the hybrid unwittingly. An "eighty-one year old reader" (Chapple 1964: 107) — subsequently identified as Hugh B. Logan "of Inverness, now dead" (McClintock 1969: 451) — sent "a few plants of an interesting cross" from California to Fred Chapple sometime before 1964. These had "large bells" varying in colour from "rosy-purple" to "rosy-crimson". These Californian plants, proving robust and hardy in Britain (Chapple 1964), were sold on to James Smith of Darley Dale, Derbyshire, but were lost, perhaps in the winter of 1962–1963. D. F. Maxwell (see p. 27) is also stated to have "pollinated some Daboecias" and had a miniature that was "about 6 inches high with tiny leaves and crimson flowers", but these, too, were lost (McClintock 1969). Elsewhere, one of the Misses Logan Home of Edrom Nurseries in Berwickshire, Scotland, obtained seed from *Daboecia azorica* which was growing near a plant of *D. cantabrica* 'Praegerae', and "various *D. azorica* came as well as an apparent hybrid" (McClintock 1969: 452): this was named 'Bearsden' because it first bloomed "at the time of Mr Buchanan's death" (McClintock 1969). Although Miss Logan Home lost the seed-parent, 'Bearsden' survived.

William Buchanan (1887–1964), a keen plantsman with particular interests in growing ferns and alpines, "dug up a handful of seedlings" and gave these to Major J. ("Jack") H. Drake (1909–1997), the owner of Inshriach Nursery at Aviemore, Inverness-shire, Scotland. Three seedlings were selected, and they were marketed as *Daboecia azorica* × *polifolia* seedlings nos 1, 2 and 3. In 1966, Drake sent two of the clones (nos 1 and 3) to the Royal Horticultural Society's Gardens, Wisley, for trial, and on 11 July 1968, no. 1 was given an Award of Merit and then named 'William Buchanan'. No. 3, which Drake had thought the better plant, was subsequently named 'Jack Drake'. No. 2, which had "large, rose-crimson" flowers, was discarded and never given a cultivar name.

Since then, several more clones have been selected and named from different sources (see Appendix I, p. 362).

CULTIVATION & VARIATION

Daboecia × *scotica* is a relatively easy plant to accommodate; chose a site that is sunny, with neutral to acid soil. The cultivars must be propagated vegetatively by cuttings or layering.

The flowers can be bright purple through shades of red and crimson to pure white (*Daboecia* × *scotica* f. *albiflora* D. C. McClint.; McClintock 1984a: 191).

Daboecia × scotica D. C. McClint., *The Garden* **103** (3): 116 (1978) (Type: 'William Buchanan', "in horto D. McClintock, Bracken Hill, Platt, Kent, culta", 11 [*recte* 16] September 1968[9], *D. McClintock* **E**! (holo.) (see Miller & Nelson 2008).

When McClintock's (1989a) decision to make *Daboecia azorica* a subspecies of *D. cantabrica* is deemed to be correct and appropriate, *D.* × *scotica* is treated as a nothosubspecies, *D. cantabrica* nothosubsp. *scotica* (D. C. McClint.) E. C. Nelson (Nelson 1996b).

DESCRIPTION. *Variable, evergreen shrub* intermediate between its parents in habit and size, there being no reliable characters by which it may be distinguished from both; forming a low, dense, bushy plant; *shoots* erect, usually much less than 0.5 m long, but they have been recorded to 1 m (McClintock 1978); *leaves* to 9 mm long, usually smaller, glossy, dark green above, with white undersides; *flowers*

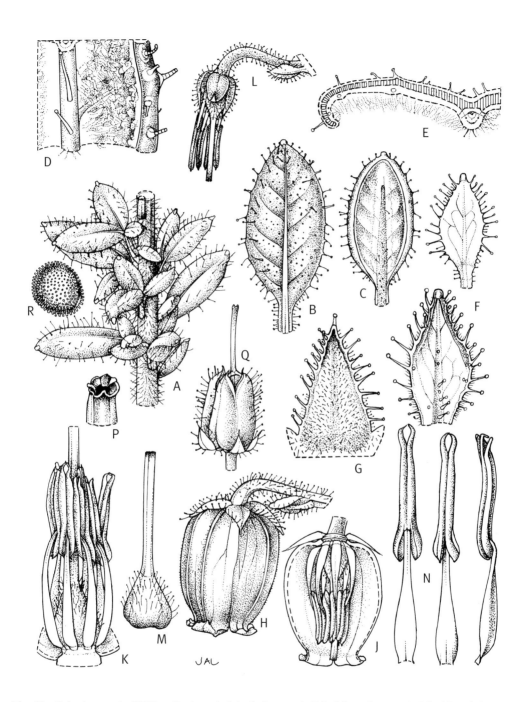

Fig. 59. *Daboecia × scotica* 'William Buchanan'. **A** leafy shoot × 4; **B** leaf from above × 6; **C** leaf from below × 5; **D** leaf indumentum × 18; **E** section through leaf × 6; **F** bracteole from above and below × 13; **G** calyx lobe × 6; **H** flower × 5; **J** section through flower × 4; **K** gynoecium and stamens × 3.5; **L** sepals, stamens and ovary × 3.5; **M** gynoecium × 5.5; **N** stamens from back, front and side × 4.5; **P** style apex (stigma) × 21; **Q** capsule × 3; **R** seed × 8; specimens supplied by Allen Hall and E. C. Nelson. Drawn by Joanna Langhorne.

bright purple to pure white. *Corolla* less than 10(–11) mm long, not glabrous (as in *Daboecia azorica*) but with sparse, scattered hairs, some gland-tipped (not as numerous as in *D. cantabrica*). Fertile, capable of producing seeds when open-pollinated.

FLOWERING PERIOD. May–June but recurrent and can be in bloom August–October(–November).

ETYMOLOGY. Toponym: *scoticus*, Scottish, alluding to the origin of the clones known to McClintock (1978). In fact, as just explained, there were reports of the hybrid from other parts of Great Britain, particularly north Wales, before any from Scotland.

VERNACULAR NAMES. Hybrid Saint Dabeoc's heath (English; McClintock 1980b); Schottische Glockenheide (German).

CULTIVARS. See p. 79 and Appendix I (p. 362).

NOTES

[1] On this apparently irremediable misspelling, see, for example, Praeger (1925), Chittenden (1931), Nelson (2000a).

[2] On 15 May 1700, Lhuyd was in Sligo (see Dandy 1958a: 156).

[3] "B" may stand for Bulstrode, home of the Duchess of Portland, one of his patrons. Ehret was there often in the 1760s teaching the Duchess's daughters to paint (Calmann 1977). This suggests that Collinson had distributed seedlings to other keen plantsmen, although there could have been other introductions.

[4] "Erica Daboecii Sp. Pl. 509 now Andromeda": collection of the Earl of Derby, Knowsley Hall (DL0000698, folio no. 305 in vol. VII of the Ehret/Taylor volumes). I have not seen this work. (J. Edmondson, pers. comm., 14 December 2006.)

[5] McClintock (1984a) reported plants in cultivation in Warsaw "by 1822 and 1824", but in his earlier paper (McClintock 1970a) he stated that "the earliest record I know is at Warsaw in 1832, where the plant is specifically said to come "from Ireland". He gave no sources for these statements.

[6] A similar aberrant plant, with enlarged, upturned sepals was seen by members of The Heather Society, led by Teresa Farino, on 12 July 2007, two kilometres west of Puerto de Piedraluengas (1,440 m asl), Cantabria.

[7] Jussieu should have used the orthography *polifolia*: on the epithet *polifolia*, see Nelson & Oswald (2005). *Menziesia* (not *Menziezia*) was named after the Scottish botanist, Archibald Menzies (1754–1842), surgeon and naturalist on HMS *Discovery*, Captain George Vancouver.

[8] A copy of this letter in among Major-General P. G. Turpin's papers, which were given to The Heather Society by Mrs Cherry Turpin.

[9] The date on the label is 16 September 1968, not 11 September 1968 as stated in the protologue (McClintock 1978).

5. *ERICA* L.: HEATHS AND HEATHERS

Genus vastum pulcherrimum, polymorphum,
sed characteribus certis nequaquam in genera pauca dividendum.
A vast, very beautiful and polymorphic genus,
but in no wise divisible into a few genera by any reliable characters.

G. Bentham, *Prodromus systematis naturalis regni vegetabilis* (1839).

George Bentham (1800–1884) (Fig. 60), Fellow of the Royal Society (elected 1862) and President of the Linnean Society of London (1861–1874), a man of private means with an aptitude for logic and taxonomic botany and a prodigious capacity for work (Stevens 2003), was not understating facts. *Erica* is a very beautiful but also an enormous and unwieldy genus. He undertook to study it in the late 1830s so that he could write the treatment of *Erica* which was published at the end of December 1839 (Nelson 2005e) by his friend and mentor, the Genevese botanist, Augustin-Pyramus de Candolle (1778–1841) in his "great *Prodromus*", a work that became "an institution" during Candolle's lifetime (Stafleu 1967: 70).

Since 1839, other botanists have tried to impose "order" on *Erica* and its occasional satellite genera. The genus has been shrunk or enlarged as each new generation of botanists has attempted its own solution to the enigma of its subdivision (for an excellent review of this, see Oliver 2000: 3–18). More than that, *Erica* was expanding as botanical exploration of Africa progressed and piles of specimens, almost inevitably including undescribed species, were amassed — not to mention the often secret

contribution of European nurserymen who created countless undocumented hybrids[1] (Nelson & Oliver 2005a). Bentham (1839) accounted for 429 species, of which just 14 were from the northern hemisphere. Guthrie & Bolus (1905) treated 469 species from South Africa, and there existed at least 157 others in the so-called "minor" African genera (Brown 1905–1906; Oliver 2000: 12). In the second decade of the twenty-first century, *Erica* is reckoned to contain about 820 species, 85% of these occurring only in the Cape Floristic Region of southern Africa (Pirie *et al.* 2011).

Yet, as far as the northern species were concerned, there were very few "new" species to discover after Bentham's time — Bentham knew almost every one, even if he did not regard some as worthy of a name or the rank of species.

Fig. 60. George Bentham (1800–1888); undated portrait by Daniel Macnee (© Royal Botanic Gardens, Kew).

CLASSIFICATION & ARRANGEMENT

Dr E. G. H. Oliver (2000), working principally with the species indigenous in southern Africa, reviewed Bentham's vast and beautiful genus and concluded that some small genera should no longer be regarded as distinct, because there were overlapping characters within them, and between them and *Erica* as hitherto defined, and recent nuclear and rDNA studies have confirmed that these genera are "nested" within *Erica* (Pirie *et al.* 2011). Among the genera that disappeared following Ted Oliver's work (1996, 2000) was the monospecific Eurasian *Bruckenthalia* which had been excluded from *Erica* in *Flora europaea* (Webb 1972a; Webb & Rix 1972). The other Eurasian segregate *Pentapera*, containing perhaps two species, was partly reabsorbed into *Erica* in *Flora europaea* (Webb & Rix 1972) and fully so in *Flora of Turkey and the east Aegean islands* (Stevens 1978).

The earliest attempt at arranging *Erica* species into convenient, smaller groupings was, not surprisingly, made by Carl Linnaeus in *Species plantarum* (1753). He noticed differences in the stamens of the 23 species that he knew. (Around half of the known heathers[2] have small spurs (also called, variously, appendages, awns or horns; see below, pp. 104–105), usually visible to the naked eye, where each anther joins the filament.) Thus Linnaeus divided *Erica* in two: there were ten species with two-horned anthers — *Antheris bicornibus* (Linnaeus 1753: 352) — and thirteen with simple, obtuse, emarginate anthers — *Ant[h]eris simplicibus obtusis emarginatis* (Linnaeus 1753: 354). In the second edition of *Species plantarum*, the terminology was a little different. Linnaeus (1762) replaced *bicornibus* by *biaristatis*, and *simplicibus* by *muticis*. That clarifies what he intended: the species with spurs were two-horned or two-bristled whereas those which lacked spurs were "*simplicibus/muticis*". However, Linnaeus seems to have been rather careless, or did not look closely enough inside the heather flowers, because he made a number of incomprehensible mistakes in his arrangement.[3] For example, he placed *Erica herbacea* (synonymous with *E. carnea*), *E. scoparia* and *E. umbellata* in the group which were stated to possess spurs (*bicornibus/biaristatis*): their stamens do not possess spurs. Contrariwise several Cape species of *Erica* with very prominent spurs ended up in his *simplicibus/muticis* division.

The morphology of the stamens, and their number in each flower, remained important when considering how to subdivide *Erica*. Richard Salisbury (Fig. 61) cultivated 62 species of *Erica* (including *Calluna*) during the 1790s in his Yorkshire garden, prepared a new edition of Carl Thunberg's dissertation on *Erica* (Salisbury 1800), and also studied hundreds of specimens before reading a definitive paper about the genus to the Linnean Society of London on 6 October 1801. In his monograph, Salisbury (1802) indicated that it was difficult to define the generic limits. He went on to prepare a classification scheme that was remarkably prescient, as we will see, but it remained essentially unpublished. Had Salisbury published his opinions, he

Fig. 61. Richard Anthony Salisbury (1761–1829), né Markham; portrait by W. J. Burchell (1817) (© Royal Botanic Gardens, Kew).

would have treated *Erica*, as he knew it, not as a genus but as an order comprising about 60 genera, one third of these containing a single species, and another third no more than four species! Salisbury's attempt to partition the genus employed the stamens, as well as the arrangement of the leaves and the flowers on the shoots. Using a commendably simple system of abbreviations to denote the states of critical characters, Salisbury (1802) coded the 246 species that he knew. For example, *Erica ciliaris* (Dorset heath) was succinctly analysed as **3. *m. i. ax.*** which means:

3 — 3 leaves in a single whorl

m — anthers muticous: lacking spurs

i — stamens included: not exceeding the length of the corolla and thus not visible

ax — the "inflorescence" axillary.

Using only these four characters, Salisbury (1802) arranged the species he had studied into sequence (Table 3), and then subdivided this into clusters, but he did not take the scheme any further in the Linnean Society paper — he did not give names to the clusters.

Table 3

Northern hemisphere *Erica* species according to Richard Salisbury's "Synopis Specierum ..." (1802).

Salisbury's epithet	Salisbury's code	Current name
Scoparia	3. m. i. p:ax.	*E. scoparia*
Spiculifolia	3. m. i. ax.	*E. spiculifolia*
Procera	3:5. c. i. t.	*E. arborea*
Polytrichifolia	3:5. c. i. t.	*E. lusitanica*
Saxatilis	3–5. m. e. p:ax.	*E. carnea*
Lugubris	4. m. e. p:ax.	*E. erigena*
Multiflora ★	5–6. m. c. p:ax.	*E. multiflora*
Manipuliflora ★	3. m. c. p:ax.	*E. manipuliflora*
Vaga ★	5–6. m. c. p:ax.	*E. vagans*
Lentiformis	3. m. e. t.	*E. umbellata*
Pistillaris	4. c. i. t.	*E. australis*
Ciliaris	3. m. i. ax.	*E. ciliaris*
Botuliformis	4. c. i. t.	*E. tetralix*
Multicaulis	6. c. i. t.	*E. terminalis*
Mutabilis	3. c. i. lat.	*E. cinerea*

Numerals indicate number of leaves in a single whorl.

Letters c / m indicate presence (c) or absence (m) of spurs on filaments

Letters e / i indicate state of anthers: exserted (e) or included (i). Note that in the published paper the letter c is used in three species (marked ★), but these seems to be typographic errors for e.

Letters ax / p:ax / t / lat describe the inflorescence: axillary, pseudo-axillary, terminal or lateral

There are various fragments of evidence suggesting that Richard Salisbury continued to work on *Erica* after the publication of his monograph, although his botanical interests were many and various. In 1806, he commented that "I have at present under consideration how far [the inflorescences] may enable me conjointly with other differences, to separate the vast genus of Erica into more." For good measure, he added: "This cannot be done in a hurry, nor till every known species has been examined" (Salisbury 1806: 2). He seems to have separated *Erica* into more genera in a manuscript that is now lost, although an extant letter that he wrote to a fellow botanist, as well as a neglected flora by a London pharmacist and his son, and George Bentham's work, all provide clues to Salisbury's thoughts. In a remarkable way Salisbury pre-empted what is now being suggested by DNA studies. He intended that a primary division should be between the hemispheres, and so he placed all the northern species that he knew in a division which he intended to name "Boreales". The southern species went into a different one to be called "Australes". As noted, this scheme was never published and Salisbury's successors, including Bentham, shied away from any geographical division, including both northern and southern species within certain of their genera, subgenera or sections. While a geographical subdivision may be derided as artificial, not natural, the clear indications from research on DNA sequences suggest the contrary (see Pirie *et al*. 2011). That Salisbury should have suggested this two centuries ago was probably mere chance; in the extant letter he gave no reasons for his proposed subdivision into "Boreales" and "Australes", nor did he give any defining characters that could be used to separate the species. As far as we can tell, there is no consistent set of morphological or anatomical differences between the heathers of the two hemispheres; one cannot allocate a species unerringly by its outward appearance.

Other botanists, in the interval between Salisbury's 1802 paper and Bentham's 1839 treatment, tried to divide *Erica* into small genera. The brothers David and George Don made a stab at this task. David's 1834 paper, "An attempt at a new arrangement of the Ericaceæ", was explicitly prepared "with the view of assisting my brother" in compiling his three-volume work, *A general system of gardening and botany*. David noted that

> Europe and Africa alone contain the normal *Ericeæ*, well characterized by their persistent corolla, the maximum of which is at the Cape of Good Hope, a spot where so many families of plants are found huddled together in strange confusion, as if Nature had at length deprived herself of sufficient space for their more equal distribution. (D. Don 1834: 150–151.)

He continued:

> As happens in other very natural families, the characters of the generic groups in the *Ericaceæ* are not so strongly marked as in those that are less so; but we are not on that account to give up the idea of dividing them, and to retain three or four hundred species in one genus, as has been done in the case of *Erica*, which I have attempted to subdivide into a number of minor groups; and, whatever opinion may be formed of their title to rank as separate genera, the arrangement of the species will, I trust, be found more natural than any hitherto proposed. (D. Don 1834: 151.)

Don proceeded to fragment *Erica* into 21 genera, most of which comprised only Cape species. The northern hemisphere species were assigned as follows: four species remained in *Erica*; *Gypsocallis* received seven; *Calluna* retained its solitary species. David Don (1843: 160) also recognised *Daboecia* — in fact he was the first properly to publish the generic name (see p. 66). Somehow he did not account for *Erica* (formerly *Bruckenthalia*) *spiculifolia*. Not unexpectedly, George Don (1834) dutifully accepted his brother's assistance and in the third volume of his gardening encyclopaedia followed suit, maintaining *Erica* (into which he placed both *E. sicula* and *E. spiculifolia*, misspelled *spiculiflora*), as well as *Gypsocallis* (which he enlarged to 49 species of which just six were of northern origin), *Calluna* and his brother's *Daboecia*.

Other botanists were at work in the mid-1830s on the problem of *Erica*. One of them, the German pharmacist Johann Friedrich Klotzsch (1806–1860), whose principal interest was fungi, had access to recent collections from southern Africa as well as Eurasian material. Using material available in Berlin, he described 26 new segregate genera, including *Pentapera* (see p. 290) for the Sicilian heath, *E. sicula*, yet he also reduced all of the segregate genera designated by David and George Don to synonymy.

George Bentham's solution was also to re-absorb some of the separate genera that Klotzsch and the Dons had split from *Erica*, particularly, in the context of the northern hemisphere species, *Gypsocallis*, and to apportion the species into sections and these, in turn, he assigned to four subgenera (Table 4). Bentham, like Salisbury and the Don brothers, accepted that *Calluna* was a distinct and separate genus containing one species. Like the Dons, too, he acknowledged that *Daboecia* was

Table 4

Northern hemisphere *Erica* species in George Bentham's (1839) proposed classification scheme.

Subgenus Ectasis
    Section Callicodon
        E. carnea
        E. carnea β *occidentalis* [= *E. erigena*]

Subgenus EuErica [= Erica]
    Section Loxomeria Salisb.
        E. ciliaris
    Section Eremocallis Salisb.
        E. tetralix
        E. mackayi [= *E. mackayana*]
        E. cinerea
        E. cinerea β *maderensis* [= *E. maderensis*]
        E. stricta [= *E. terminalis*]
        E. australis
    Section Pyronium Salisb.
        E. umbellata
    Section Gypsocallis Salisb.
        E. multiflora
        E. vagans (including *E. manipuliflora*)
    Section Arsace Salisb.
        E. polytrichifolia [= *E. lusitanica*]
        E. arborea (including *E. acrophya*)
    Section Chlorocodon
        E. scoparia (including *E. scoparia* β *parviflora*)

different and deserved to remain apart; Salisbury (1802: 323) also held this opinion but in the early 1800s Saint Dabeoc's heath was treated as a species of *Menziesia* (cf. Jussieu 1802; Gray 1821). Two other genera, both containing a single species, Carl Ludwig von Reichenbach's *Bruckenthalia* and Klotzsch's *Pentapera*, and both designated after the publication of Salisbury's monograph, were also maintained by Bentham.

THE GRAY CONNECTION

A puzzling aspect of Bentham's account of *Erica* in Candolle's *Prodromus* is his use of a suite of 22 names for the sections[4] that he defined, which he explicitly attributed to Richard Salisbury. Bentham (1839: 614) acknowledged that most of the sections had already been indicated by Salisbury ("… *plurima jam à Salisburio indicata ...*"), or defined by Klotzsch, and he also noted at least one section that he decided not to publish — Salisbury's *Tylospora*[5] comprising only *E. australis* (see p. 222). Salisbury had died in 1829, a decade before Bentham was working on *Erica*, and given that there is no evidence that George Bentham had any long-standing interest, reaching back to the 1820s, in heathers (see Oliver 2000: 9) it is rather unlikely that the two botanists ever discussed *Erica*, let alone the subdivision of the genus. Where, or how, did Bentham get the names he used?

There is a clue in the earlier publication of several of Richard Salisbury's names within a little-known, indeed "deliberately and effectively ignored" (Gunther 1975), two-volume work entitled *A natural arrangement of British plants … including those cultivated for use*, published in London late in 1821. *Gypsocallis* and *Eremocallis* made their appearance in this book, and it is probably more than significant that they are among the genera named and briefly defined in the enigmatic manuscript letter written by Richard Salisbury which is preserved in the Botany Library, Natural History Museum, London.

A natural arrangement of British plants has Samuel Frederick Gray's (1766–1828) name on the title-page. He was a Londoner, "Lecturer on Botany, the Materia Medica, and Pharmaceutic Chemistry." Gray's authorship of *A natural arrangement* has been disputed (Browne 2004; Cleevely 2004) — at least, in some autobiographical notes, his younger son, John Edward Gray (1800–1875), claimed that he had suggested the project to his father and also done much of the work:

> As my botanical teachers always dilated on the advantages of the Natural System of Jussieu and Mirbel, it occurred to me that a work on British Plants on the system would be useful. I consulted my father [Samuel], he who was a botanist of the Ray School who had never been induced to adopt the Linnean System. He entered readily into the plan but suggested its being on Ray's System. At length he was convinced that Jussieu was more concerned with the state of the science and he said he regarded it as the natural results of Ray's plan of studying plants and he found a bookseller who was ready to undertake the work.
>
> The permission which Sir Joseph Banks had accorded to me of attending his Library became of great advantage. After arranging the plants according to Jussieu's genera I went there to consult all the more modern works on European plants; and there met Dr. Decandole (the elder), and Dunal who were here to consult the Banksian and Linnean Herbarium and they gave me their most hearty assistance and sympathy and I also here made the acquaintance with Richard Anthony Salisbury, who helped me in every way and afforded me the use of all the MS, which he had composed with the intention of publishing a Genera Plantarum. (Quoted from Gunther 1974: 60.)

Thus, the younger Gray had access to all of Salisbury's manuscripts including, evidently, material on *Erica*. It is quite possible that the extant letter (see Appendix II, pp. 387–388), from which both the name of the recipient and the date have been deliberately removed, was addressed to him. He could well have asked Salisbury for additional information about his proposed segregate genera: "I cannot rummage in my papers to send you my full generic characters now," Salisbury replied.

The connection goes deeper, for, as Gunther (1975) relates, Salisbury offered to leave his library "and his fortune" to John Gray on condition that young Gray would pursue a career in botany. Gray declined. After Salisbury's death, his papers and herbarium passed to the botanical explorer William Burchell "on whose decease in 1863 the [manuscripts] and drawings passed into the hands of Dr. J. E. Gray, by whom they were given to the [Natural History] Museum in 1865" (Sawyer 1971). Given this, it is very likely that it was Burchell who made Salisbury's notes on *Erica* available to George Bentham.[6] According to Sawyer (1971) "Salisbury's notes and drawings of Ericaceous Plants were transferred [to the Natural History Museum] from the Royal Gardens at Kew in 1881", suggesting that Bentham had retained these notes after completing the monograph for Candolle. There is a further twist. As Gunther (1974: 60 n110; 1975: 42) pointed out, "In memory of his friend, Gray in 1866 edited from Salisbury's [manuscripts] his *Genera of plants* ...".

Stepping back to the Grays' book, it is remarkable that they abandoned Linnaeus's generic name *Erica* (except in synonymies), clearly indicating that they accepted Salisbury's precept, announced in the letter, that "The old Genus Erica containing 400 species ... I now regard as an Order."[7] Thus, in *A natural arrangement of British plants* (Gray 1821: 398) cross-leaved heath (called "headed heath") and bell heather ("grey heath") were consigned to Salisbury's *Eremocallis* as *E. glomerata* and *E. cinerea* respectively — *Eremocallis glomerata* was a new name derived from Salisbury's (1796) *Erica glomerata*. Cornish heath was transferred to Salisbury's *Gypsocallis* as *G. vagans* and was also dubbed "loose gypsum-heath" (Gray 1821: 398–399). Ling stayed in Salisbury's *Calluna*, but with the binomial *C. sagittaefolia* (from Jonathan Stokes's (1812) *Erica sagittaefolia*) and the moniker "arrowleaved ling" (Gray 1821: 399). Saint Dabeoc's heath lingered at little longer in Smith's *Menziesia* (Gray 1821: 399).

THE NORTHERN HEMISPHERE ERICA SPECIES

When George Bentham set out his classification of *Erica*, which explicitly relied on what Salisbury had written in a now-lost manuscript, he did not take into account Salisbury's notion that northern and southern (Cape) *Erica* species were not closely linked. Some of Bentham's sections, like the Dons' genera, contained species from both hemispheres. For example, in Bentham's scheme (1839), section *Gypsocallis* comprised seven species, two from the north (*E. multiflora* and *E. vagans*) and five from the Cape. In *Flora capensis* (Guthrie & Bolus 1905) the section was amended again; nine Cape species were included. Hansen (1950) included three northern species (*E. manipuliflora* (as *E. verticillata* Forssk.), *E. multiflora* and *E. vagans*). In the most recent "complete list" of South African *Erica* species (in Schumann & Kirsten 1992: 260–262[8]), ten Cape species were assigned to section *Gypsocallis*. None of this makes sense now, in the light of DNA evidence that the European and African clades are separate (McGuire & Kron 2005; Pirie *et al.* 2011), although we must wait for a lot more DNA-sequence data before making a final pronouncement about the north/south partition. In their gross morphology these species may be quite similar, and so may *seem* to be allied one to another, but they cannot be closely related if the African and European species really occupy different clades; in other words, occur on separate branches of the "family tree". They cannot have evolved from the same ancestral heather, a proto-*Erica* "gypsocallis", as implied by the designated sections.

Herein I have deliberately abandoned Bentham's sections, despite the fact that some European floras still employ them (see, for example, *Flora iberica*). Continuing to use them, and even adding to their number (as Hansen (1950) did), is not rational. The subgeneric classification of *Erica* is in turmoil for the time being, a plaything for botanists who are enthralled by molecular taxonomy. Bentham's vast and very beautiful genus remains bewildering, seemingly not divisible by reliable morphological characters.

Despite this, I have grouped the northern hemisphere species of *Erica* together into what may be entirely artificial "chapters", on the simple basis of similarities in overall appearance, by their gross morphology. Avoiding Latin names, I have preferred to use English headings — The cross-leaved heaths; The Pentaperan heaths — because these are less liable to be misconstrued as a "new" classification replete with formal, valid nomenclature. This may not please readers who are systematic botanists, but it is hoped it will be less intimidating to the others, the keen plantsmen and gardeners who will, I hope, read and enjoy this monograph.

ERICA L., *Species plantarum* **1**: 352 (1753).

(For more comprehensive synonymy see Oliver 2000: 98–100).

Type species *E. cinerea* L.; lecto: selected by Jarvis & McClintock (1990) (see also D. Don 1834; Jarvis 1992, 2007: 498; Barrie 2006).

Generic description (from Oliver 2000).

Prostrate shrublets, erect shrubs, or trees to 15–20 m tall. *Leaves* 3- or 4-nate in whorls, occasionally 6-nate, rarely opposite or spirally arranged, mostly linear-oblong to ovate, trigonous, with revolute margins almost touching on underside (ericoid), less commonly broad, flat and open-backed but then less than 10 × 5 mm, sometimes very reduced and scale-like or enlarged coloured and sepaloid below the inflorescences. *Flowers* always axillary, in 1–12 whorls variously arranged in elongate racemose to umbellate florescences, either at ends of most leafy lateral branches or very reduced lateral branchlets or only the main branches, sometimes forming dense synflorescences; pedicel usually present, sometimes very reduced, very rarely non-existent; *bract* always present, on the main axis or partially recaulescent to fully recaulescent forming the abaxial lobe/segment of calyx; *bracteoles* 1 or 2, or absent, sometimes included in the calyx as the lateral lobes/segments. *Calyx* hypogynous, (1–)4–5-partite or -lobed, small and leaf-like or large and showy, rarely longer than corolla, sometimes enlarging and thickening considerably in fruiting stage. *Corolla* persistent, hypogynous, (2–)4–5-lobed, tubular, ampullaceous, urceolate, globose, ovoid, campanulate, cyathiform, obconic or funnel-shaped, small to large and brightly coloured, dull coloured in most wind-pollinated species. *Stamens* (3–)8–10, free or completely fused, included or exserted; *anthers* dorsifixed or basifixed, with or without simple or elaborate dorsal appendages, thecae partly united or sometimes free, with small subterminal pores or pores sometimes as long as thecae; *pollen* shed as tetrads or sometimes as monads. *Ovary* (1–)4(–8)-locular, with 1–180 ovules per locule, sessile or stalked, nectaries usually prominent around base, sometimes absent in wind-pollinated species; *style* filiform to cylindrical, rarely non-existent, exserted or included; *stigma/style* complex simple-truncate, capitate, peltate, cyathiform or funnel-shaped and considerably enlarged in wind-pollinated species, rarely distinctly 4-lobed. *Fruit* mostly a dehiscent loculicidal capsule, sometimes a dry berry or drupe, pericarp hard and woody to very thin and papery; *seeds* with thick testa, mostly alveolate, some smooth or papillate, in indehiscent fruits the testa thick to very thin, transparent or almost non-existent; pits in inner periclinal and anticlinal walls present or absent. 2n = 24, 36.

GEOGRAPHICAL DISTRIBUTION

Europe (18 spp.): north-western Norway to southern Spain, Ireland, Portugal eastwards to Finland and Romania; Mediterranean islands including Balearics, Corsica, Sardinia, Sicily, Malta, Crete. Macaronesia; Azores (1 sp.); Madeira (3 spp.); Canary Islands (2 spp.). North Africa (north of Tropic of Cancer) (9 or 10 spp.): Morocco eastwards to Libya; Chad, Ethiopia, Somalia and southwards into east tropical Africa. Asia (4–5 spp.): Turkey to Lebanon, Cyprus, Saudi Arabia, Yemen. Fig. 62.

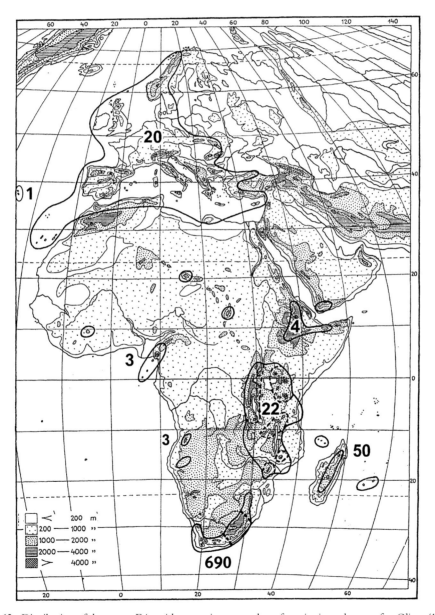

Fig. 62. Distribution of the genus *Erica* with approximate number of species in each area; after Oliver (1994a, 2000) with corrections.

NOTES ON MORPHOLOGY

Oliver (2000: "Character assessment", 22–60) has discussed the morphology and anatomy of *Erica*, *Calluna* and *Daboecia* in detail; the following notes on characters, in the same order, relate to the northern hemisphere species of *Erica* only; see also Table 5.

HABIT: Northern hemisphere *Erica* species range in habit from trees to low-growing but erect, bushy shrubs. None of the northern species form prostrate shrublets. Most of the species are capable of resprouting from the root-stock or stem bases after fire or other damage to the aerial parts. None of the northern species can be described as a "single-stemmed reseeder", a habit common among the southern African species. *Erica mackayana* and its hybrid, *E.* × *stuartii*, are capable of regeneration from the damaged ends of their slender roots, as shown by tufts of shoots on the face of turbary banks in western Ireland.

Oliver (2000; see also Kubitzki 2004: 173) gave 30 metres as the height attained by tree-growing species of *Erica*. This figure derives from an account of *E. arborea* (tree heather) growing in the Canary Islands published by Herman Knoche (1924; cf. Page 1995). Modern accounts of the islands' flora have the maximum height attained by this species as around half that, to 15 metres (Bramwell & Bramwell 1974: 171; Hohenester & Welss 1993: 180), with Kunkel (1977) stating that on La Gomera *E. arborea* does attain 20 metres in height.[9] Annotations on herbarium specimen suggest that *E. arborea* may reach 18 metres in height on the Arabian Peninsula, but the accuracy of these measurements is quite unknown.

BRANCHES: The presence or absence of an indumentum can be useful in separating species, but there is a wide range of variation, and "glabrous" plants can be found among species that have hirsute or pubescent stems. Plumose hairs are present only in *Erica arborea* and *E. australis*; all the other species have simple hairs.

LEAVES: The leaves in northern hemisphere species are typical of the Ericeae; they are narrow, needle-like with a sulcate abaxial surface. However the amount of inrolling of the margins does vary and in some species, the leaves can be flat with an exposed abaxial surface. The presence or absence of prominent gland-tipped hairs on the leaves and petioles can be useful in separating species but, as with branch indumentum, there is variation within species.

Glandular and eglandular plants can be found in populations of some species (for example, *Erica ciliaris*, *E. mackayana*, *E. tetralix*).

LEAF WHORLS: In northern hemisphere *Erica* species, leaves are arranged in whorls of either 3 (for example, *E. ciliaris*) or 4 (for example, *E. tetralix*, *E. mackayana*). Atypical plants, with fewer or more leaves in each whorl, can occur within populations (Brough & McClintock 1978) and such plants have sometimes been given names (for example, *E. tetralix* var. *quinaria*).

INFLORESCENCE: As stressed by Oliver (2000: 29–32) the structure of the inflorescence in *Erica* is polymorphic, and much misunderstood.

BRACT AND BRACTEOLES: A bract is always present in the Ericeae but its position relative to the calyx can vary; in some species the bract is at the base of the pedicel, but in others it may be inserted around the mid-point of the pedicel. Two bracteoles are also present and their position between the bract and calyx can also vary; sometimes they are close to the apex and so appear to be part of the calyx. In *Erica spiculifolia* the two bracteoles are fully recaulescent and having suppressed the two lateral sepals, form part of the calyx so that they appear to be absent (Oliver 2000: 37).

Table 5

Erica species of the northern hemisphere: a summary of general distribution and characters.

Erica species	Eur	NAf	Mac	WAs		Stm	Spur	Exs	Pol	Cal	Ov	Cor	Sti	2n	Lfwh
E. spiculifolia	■		■	■		8			○	lo	±12	4	±eq	36	2–6
E. andevalensis	■					8	■		❖	sp	10–30	4	cap	24	4–5
E. mackayana	■					8	■		❖	sp	∝	4	cap	24	4
E. tetralix	■					8	■		❖	sp	40–50	4	cap	24	4
E. arborea	■	■	■	■		8	■		❖	lo	25–30	4	cap	24	3
E. lusitanica	■					8	■		❖	lo	25–25	4	cap		3–4
E. australis	■	■				8	■		❖	sp	±20	4	cap	24	4
E. cinerea	■	■				8	■		❖	sp		4	cap	24	3
E. maderensis			■			8	■		❖	sp	±30	4	cap		3
E. terminalis	■	■				8	■		○	sp	±25	4	cap		4–5
E. ciliaris	■	■				8			❖	sp	20–30	4	cap	24	3
E. carnea	■					8		■	❖	lo	20–25	4	eq	24	4
E. erigena	■	?				8		■	❖	lo	10–12	4	eq	24	4
E. manipuliflora	■			■		8		■	❖	lo/sp	±21	4	±eq		3–4
E. multiflora	■	■				8		■	❖	lo		4	eq	24	4–5
E. umbellata	■	■				8		■	❖	sp/lo	±12	4	eq	24	3
E. vagans	■					8		■	❖	lo	±8	4	eq	24	4–5
E. azorica			■			8			❖	lo		4	pel		3–4
E. scoparia	■	■				8			❖	lo		4	pel	24	3–4
E. platycodon			■			8			❖	lo		4	pel		3–4
E. sicula subsp. bocquetii				■		10			❖	sp/lo	>50	5	cap		3–4
E. sicula subsp. sicula	■	■		■		10	±		❖	sp		5	cap		4–5

Key to columns (left to right)

Eur	native in Europe
NAf	native in north Africa
Mac	native in Macaronesia
WAs	native in western Asia
Stm	number of stamens per flower
Spur	■ spur present on stamen at base of anther
Exs	■ stamens exserted at anthesis
Pol	❖ in tetrads; ○ in monads (single grains)
Cal	lo = calyx segments fused at base with free lobes; sp = calyx segments (sepals) not fused
Ov	number of ovules per locule
Cor	number of corolla lobes
Sti	style apex expanded (peltate: pel), slightly expanded (capitate: cap), or equal in diameter to distal part of style (eq) (± indicates more or less).
2n	diploid chromosome number
Lfwh	number of leaves in a whorl

CALYX: The number of calyx segments or lobes always equals the number of corolla lobes in the northern species of *Erica*. The majority have four sepals or are four-lobed, but in the species formerly in *Pentapera* (*E. sicula*) the calyx is usually, but not always, formed from five sepals.

In *Erica spiculifolia* there are only two sepals; as noted above, these are fused with two bracteoles to form the four-lobed calyx.

Oliver (2000: 38–39) stated that for the whole genus *Erica* most species have free sepals. However in the northern species, despite descriptions that suggest otherwise (for example, Webb & Rix 1972), more than half the species have sepals that are fused for about one third of their length into a cup-shaped or conical calyx.

There is not as much variation in texture and relative size of the calyx in the northern species as there is in southern African *Erica* species. The calyx never exceeds the length of the corolla and is usually less than half that length. In some northern species the calyx lobes are coloured and petal-like although not greatly enlarged.

Other calyx characters — glands and sulcus — do not vary significantly in the northern species, and are not diagnostic.

COROLLA: As noted, the number of corolla lobes equals the number of calyx segments or lobes. In northern species, with the exception of those formerly in *Pentapera*, there are four lobes, the predominant state in the genus (Oliver 2000: 40). Some individual plants may have atypical numbers of lobes but this is not of taxonomic importance although the anomaly may serve to characterise cultivars (as, for example, in *Erica mackayana* 'Lawsoniana'; see Nelson 2000d).

The corolla lobes are never longer than the corolla tube. There is some variation in the surface characteristics of the corolla with a few species having some indumentum especially on the outer surface of the lobes.

STAMENS: The number of stamens always is twice that of the corolla lobes. Thus in the majority of species there are eight stamens, with the exception of the Pentaperan species which may have ten. Anomalous individual plants can occur within wild populations and in cultivation, but their atypical stamen complements are not of taxonomic significance.

In hybrids, and also in Irish populations of *Erica mackayana* (Nelson 1995b), the stamens can be grossly malformed or transformed into petaloid structures; the cultivar 'Plena' exhibits this state.

Stamen position at anthesis: In the majority of the northern species, the stamens are fully included in the corolla at, and after, anthesis. In others, the filaments continue to elongate as the flowers mature, and so the anthers emerge, wholly or partly, from the corolla.

Filaments are never fused in northern *Erica* species. Only in *E. lusitanica* and its hybrid, *E.* × *veitchii*, do the filaments have any indumentum.

Spurs are present in ten of the northern species (and also in *Calluna*). As defined by Oliver (2000: 44), the spurs are "dorsal appendages arising from the point of attachment of the filament to the anther" (see also Hermann & Palser 2000). They vary in shape and size. In northern species the spurs are either always present and obvious to the naked eye, or always absent, with the exception of *Erica sicula* in which vestigal spurs are described (Stevens 1978) although in the majority of specimens that I have examined I been unable to detect any vestiges of spurs.

Most species which lack spurs have exposed, exserted anthers at anthesis. *Erica ciliaris* is a notable exception. In the northern *Erica* species which possess spurs, all anthers remain included even after anthesis. The function of the spurs has not been adequately explained, but it is likely that these prominent structures have an important role in pollination (cf. Church 1908, quoted by Nelson 2000c). Their form and position suggest that they may transmit vibrations from the corolla to the stamens, so that the pollen grains are shaken out of the anthers (like pepper from a pepper-pot), and perhaps they are an indication of species that are predominantly buzz-pollinated (cf. Hermann & Palser 2000). Sonication (buzz pollination) is known to occur in some species of *Vaccinium* and *Arctostaphylos*, but has not been reported from *Erica* as far as I can ascertain.

The purpose and function of the spurs remains one aspect of *Erica*, especially the northern species, that require detailed investigation.

Anthers are composed of two thecae, which are usually erect, distinct and separate. In a few species the thecae are divergent. In a few taxa each pair of thecae are fused together, wholly or partly, by the "inner" walls.

Before anthesis, all the anthers within a flower adhere to form a "collar" around the style, the point of adhesion of adjacent thecae being their pores. This "collar" of anthers has to be fractured before the pollen grains can escape from the thecae.

Pollen: Except for *Erica terminalis* and *E. spiculifolia*, which have pollen in monads, all the northern *Erica* species have pollen in tetrads. The size of pollen tetrads has been employed to distinguish pollen of *Erica* taxa preserved in peat profiles.

Pollen may be produced in enormous quantities, especially in the wind-pollinated besom heaths (Herrera 1987: Table 1).

GYNOECIUM: Oliver (2000) noted that the number of ovules, as well as their position and the nature of the placenta, was important in delimiting the African species.

Ovules: The number of ovules per locule is given when this could be determined or when published figures could be traced. In northern species of *Erica* the number of ovules per locule ranges from about 10 to about 60.

Nectary: In *Erica* the nectary usually is a darker, lobed ring encircling the base of the ovary. In wind-pollinated species no nectary is present.

Style and stigma: All the northern *Erica* species have erect, relatively straight styles, which may taper upwards and are usually much longer than the ovary. The stigma, strictly, is the set of four, minute lobes situated at the apex of the style (Oliver 2000: 48), these lobes being the terminal portions of the carpels. In some northern species the apex of the style around these lobes is not markedly expanded, but in others the style tip is greatly enlarged and the four separate lobes cannot be distinguished. In the wind-pollinated species, the style apex is very greatly enlarged to form a conspicuous, peltate structure. The expanded style apex is generally referred to as the "stigma" (cf. Oliver 2000: 48). Following Oliver (2000) I also retain stigma as the term for the apex of the style.

Fruit: The capsules of all the northern *Erica* species dehisce loculicidally. The wall of the capsule when ripe is usually relatively thin and papery, so that it readily disintegrates when the ripe capsule is gently rubbed. Only in one species, *E. maderensis*, does the fruit have substantial, thickened, woody valves, a character that clearly separates it from *E. cinerea*.

Seed: There is substantial variation in the size of the seeds of northern *Erica* species, and in the ornamentation of the testa. One species, *E. australis*, is unique among the northern species in the possession of a prominent elaiosome (see p. 221). Such structures are usually associated with seed collection and dispersal by ants, but ants will collect seeds that do not have an elaiosome (for example of *Calluna vulgaris* and *Erica tetralix*; see Brian 1977: 55, Mallik *et al*. 1984)).

ETYMOLOGY

Ερείκη (*ereikê*) was a name of a plant known to the Ancient Greeks; it was used for making fires. The name occurs in the works of two poets Aeschylus and Theocritus.

Aeschylus was born in 525BC in Attica. As well as being a poet he was a soldier, a combatant at the battles of Marathon (490BC) and Salamis (480BC). He was also a frequent visitor to Sicily. In 456BC he was killed, according to legend, when an eagle dropped a tortoise on his head — a lammergeier might have been the culprit as this bird is known to drop bones from great heights so that they are smashed open although whether lammergeiers also do this with tortoises is not recorded. Within his trilogy Oresteia, in *Agamemnon*, Aeschylus alludes to the use of ερείκη to form one of the chain of blazing beacons that transmitted the message about the fall of Troy from that tragic city to Mycenae. Clytemnestra addresses the Chorus of Argive Elders (Johnston 2007: 17):

> Hephaestos, god of fire, sent his bright blaze
> speeding here from Ida, his messenger,
> flames racing from one beacon to the next—
> from Ida to Hermes' rock in Lemnos.
> From that island the great flames sped
> to the third fire, on the crest of Athos,
> sacred to Zeus, and then, arcing high,
> the beacon light sprang across the sea,
> exulting in its golden fiery power,
> rushing on, like another sun, passing
> the message to the look-out towers
> at Macistus. The man there was not sleeping,
> like some fool. Without a moment's pause,
> he relayed the message, so the blazing news
> sped on, leaping across Euripus' stream,
> to pass the signal to the next watchmen,
> at Messapion. Those men, in their turn,
> torched a pile of dried-out heather[10], firing
> the message onward. The flaming light
> was not diminished—its strength kept growing.
> Like a glowing moon, it jumped across
> the plain of Asopus, up to the ridges
> on mount Cithaeron, where it set alight
> the next stage of the relay race of fire. ...

To quote Alice Lindsell, writing in 1937, "This is interesting because, on the mountains about Mycenae, *Erica arborea* still grows magnificently, and it is still the common or only fuel of the Argolis, and it is still called ρείκι [pronounced ríki]."

Theocritus lived in the third century BC, and may have been born at Syracuse in Sicily. He is also associated with the Aegean island of Cos, and with Alexandria in Egypt. In his famous series of bucolic verses, the *Idylls*, the word appeared again in its Doric form, ἐρείκα (*ereikā*) (*Idyll* V, line 64).

Lacon.

Thence, where thou art, I pray thee, begin the match, and there sing thy country song, tread thine own ground and keep thine oaks to thyself. But who, who shall judge between us? Would that Lycopas, the neatherd, might chance to come this way!

Comatas.

I want nothing with him, but that man, if thou wilt, that woodcutter we will call, who is gathering those tufts of heather[11] near thee. It is Morson. ...

Alice Lindsell (1937) pointed out that *Idyll* V was "staged on the Sila Mountains of S. Italy, which are not limestone like Syracuse, but gneiss like ... much of Greece", so *Erica arborea* is the most likely plant. Theocritus does not refer again to the heather nor to the purpose the woodcutter had in gathering tufts.

Greeks today, as in the past, can only know from local experience a few heathers, and the spring-blooming *Erica arborea* is certainly the most abundant, widespread and prominent. A second species, the autumn-flowering *E. manipuliflora*, grows throughout Hellas but is a lesser shrub, and so probably not so conspicuous a plant. Other heathers exist within the Hellenic world but are so uncommon that it is highly unlikely they would feature as subjects in literature, even in writings that have an explicit botanical character. These other species include *E. multiflora*, *E. spiculifolia* and *E. sicula*, as well as *E. carnea* and *Calluna vulgaris*.

Many eighteenth- and nineteenth-century authors (for example, Hereman 1868: 217), as well as some more recent ones (especially on the internet), include an interpretation of *Erica* that derives it from ἐρίκω, *erico*, I break. Thomas Martyn's edition of Miller's *The gardener's and botanist's dictionary* (*c.* 1795) has "... from ἐρίκω, or ἐρείκω, frango, to break; from its supposed quality of breaking the stone in the bladder."

Randal H. Alcock, in his *Botanical names for English readers* (1876), objected.

ERI´CA, Dioscorides. G., *ereice*. The derivation is said to be from *ereico*, I break; because it was formerly supposed to have the power of destroying calculi of the bladder. Yet the old botanists give but little prominence to this supposed property. Matthiolus mentions it, but from Dioscorides downwards the chief virtue ascribed to the genus is as an application to the bites of snakes. It puzzles me where they got so many snakes in Europe, especially in England, for we have only one venomous kind, and its bite is seldom a very serious affair.

This seems borne out by the entry in John Gerard's *Herball* (Johnson 1633: 1386):

¶ *The Temperature.*

Heath hath, as *Galen* saith, a digesting facultie, consuming by vapors : the floures and leaues are to be vsed.

¶ *The Vertues.*

A The tender tops and floures, saith *Dioscorides*, are good to be laid vpon the bitings and stingings of any venemous beast : of these floures the Bees gather bad hony.

B The bark and leaues of Heath may be vsed for, and in the same causes that Tamariske is vsed.

The "vertues" of tamarisk, according to Gerard, did not include breaking stones in the bladder.

Notes

[1] In the introductory remarks to their treatment of *Erica* in *Flora capensis*, Francis Guthrie & Harry Bolus (1905: 5) wrote:

> This genus is remarkable for an unusual degree of variability in the form of almost all its organs. It is therefore one difficult of definition as to its species, and of arrangement into satisfactory natural groups. Many of the species are obviously allied to others in very different sections; and in most of the sections and subgenera it is necessary to note exceptions to the general technical characters. Many authors have treated of them with great divergency of views; and the earlier botanists unduly multiplied the species as they arrived from the Cape. At the end of the 18th, and the early part of the 19th centuries, the heaths became fashionable in European gardens, were much hybridized and copiouslyfigured. This has added to the difficulty of definition, and still more to the confusion of the synonymy [*sic*]. In respect of the latter, we have had largely to rely upon others, and can but hope at most to have cleared up some few of the obscurities by which the genus has been surrounded, leaving many in which, owing to absence of types and imperfect descriptions, it may continue to be involved.

[2] According to Volk *et al.* (2005), 478 taxa possess anthers with "appendages" and 406 lack "appendages", while 60 are scored for both states. The database contains 914 taxa (including subspecies).

[3] Linnaeus's errors undoubtedly confused his students and successors who, relying on *Species plantarum*, sometimes misnamed *Erica* specimens. A good example is Pehr Forsskål who collected *E. arborea* in Arabia but named it *E. scoparia*. Various authors have pointed out that Linnaeus erred, including Tausch (1834), Richter (1840) and, more recently, the Dutch-born *Erica* scholar Dr Hans Dulfer (1963).

[4] The names of sections designated and described by Bentham (1839) within the genus *Erica* that are attributed to Richard Salisbury were: *Amphodea* (Sect. vi; p. 618), *Arsace* (Sect. xlviii; p. 689), *Bactridium* (Sect. xviii; p. 637), *Callibotrys* (Sect. xii; p. 624), *Ceramus* (Sect. xx; p. 641), *Diphilus* (Sect. xxxiii; p. 662), *Ephebus* (Sect. xxxix; p. 670), *Eremocallis* (Sect. xxxv; p. 665), *Evanthe* (Sect. xiv; p. 628), *Gigandra* (Sect. viii; p. 621), *Gypsocallis* (Sect. xxxvii; p. 667), *Heliophanes* (Sect. xlii; p. 682), *Hermes* (Sect. xxxiii; p. 662), *Loxomeria* (Sect. xxxiv; p. 665), *Melastemon* (Sect. xliv; p. 683), *Myra* (Sect. xix; p. 640), *Orophanes* (Sect. xl; p. 673), *Pilostoma* (Sect. ix; p. 622), *Platyspora* (Sect. xxv; p. 649), *Pleurocallis* (Sect. xiii; p. 625), *Pyronium* (Sect. xxxvi; p. 666) and *Trigemma* (Sect. xxviii; p. 655).

[5] "Hæc species Salisburio sect. propriam Tylosporam constituit" (Bentham 1839: 666). The section was eventually designated and named by Dr Irmgard Hansen (1950), using Salisbury's name (see Nelson 2007a).

[6] It is a plain coincidence that when Bentham was working on *Erica* he was the Secretary of the Royal Horticultural Society of London, a post earlier occupied by Richard Salisbury.

[7] There is a remarkable letter among Sir James Edward Smith's papers in the archives of the Linnean Society of London, from that excellent plantsman and botanist (and Vice-President of the Linnean Society) Aylmer Bourke Lambert (1761–1842). Dated 30 November 1811, Lambert remarked wryly that

> Our old friend is working on <u>Erica</u> making ∞ Genera out of them & a Natural Order.
> Oh Lack Oh Lack
> & a well a day,
> the Genus again by Salisbury.

Lambert was alluding either to *The beggar's opera*, by John Gay, which also contains the line "A lack, and well-a-day!" in Air LIII "I am a poor Shepherd undone" (from Act 3 scene 4), or possibly to a once-performed (on Saturday 2 April 1796) play, allegedly by William Shakespeare, *Vortigern*, wherein (Act 2 scene 4) the Fool sings a song with the refrain "Lack, lack, and a well a day!"

[Original ms, in Smith correspondence vol. 6: 144; see Dawson 1934.]

[8] The list was developed by Inge and Ted Oliver for the herbarium arrangement based on the *Flora capensis* sections and numbering of species (Guthrie & Bolus 1905), with additions following work by Dulfer (1965), and then the Olivers' reshuffling of both with the further additions of all the species named after 1965, with renumbering (E. G. H. Oliver, pers. comm., 2008).

[9] In a survey of "records" for heaths and heathers, McClintock (1989e) also noted heights exceeding 20 metres including trees at Agua de los Llamos, La Gomera (*fide* Kunkel 1977: 76). A note reading "40m branches on fallen trunk, La Gomera ..." has to be an error (McClintock also attributed this to Kunkel, but dated the source as 1987).

[10] In the translation by Edmond Doidge Anderson Morshead, the fuel became furze: "... And from the high-piled heap of withered furze / Lit the new sign and bade the message on. ...". Robert Browing has "... Kindling with flame a heap of grey old heather ...", and Gilbert Murray "... They beaconed back, then onward / with a high Heap of dead heather flaming to the sky. ..."

[11] As Lindsell (1937) pointed out, Charles Stuart Calverley translated ἐρείκα as bracken.

WINTER- AND SPRING-FLOWERING HEATHERS

Erica carnea, E. erigena

Two species, *Erica carnea* (winter heath, mountain heath) and *E. erigena* (Irish heath, Mediterranean heath), comprise an apparently natural group within the European *Erica*. Their geographical ranges are widely separated and do not overlap. These heaths are remarkably similar in morphology, being only clearly distinguishable by their habits: *E. carnea* is a low-growing, spreading subshrub, whereas *E. erigena* forms a upright, bushy shrub that can be as much as 3 metres tall and tree-like (with a single, main trunk) — in fact, large bushes of Irish heath ('Superba' and 'W. T. Rackliff') in David McClintock's garden in Kent were sufficiently substantial and protective for birds, including linnets, to chose to nest in them (McClintock 1974b). The close similarity between the two heathers means that dried, pressed specimens (as housed in herbaria) are so alike that it is almost impossible to determine which species is involved without having information about habit or locality of collection. Ross (1967: 62; plate 1), in collaboration with Dr C. R. Metcalfe, showed by sectioning stems that the distinctive habits are reflected in anatomical differences; but, alas, these anatomical features do not serve as clear characters for separating herbarium or even fresh specimens. The prostrate habit of *E. carnea* is surely an adaptation to its habitats on the mountains of central and south-eastern Europe, where plants are buried under deep snow each winter; it is significant that heavy snow will easily shatter the branches of *E. erigena*. When these features are carefully considered, it might be argued that the Irish heath is merely an erect, geographic variant of the winter heath, adapted for milder, snowfree habitats, and that it should be designated as a subspecies of the winter heath. This has, in fact, been done, and the name *E. carnea* subsp. *occidentalis* (Benth.) M. Laínz may be used for *E. erigena* by anyone so minded to treat these heathers as belonging to the one species. It is noteworthy that George Bentham (1839) was so uncertain about these plants representing different species that he treated the Irish heath as only a variety of *E. carnea*.[1] Whether he had seen either of them growing in natural habitats is uncertain, so he is likely to have relied on pressed and dried herbarium specimens which, as just noted, are notoriously difficult to identify (cf. Ross 1967; McClintock 1989e).

In cultivation, the two species will hybridise readily, and the habit of the progeny is intermediate between the parent species: *Erica* × *darleyensis* forms a more upright, taller plant than *E. carnea* yet never attains the stature nor the upright, almost tree-like habit of *E. erigena*. Hybrids are never produced in the wild, as these plants do not grow in the same or even closely adjacent habitats.

A complication arose, however, during the past decade when the results of some pioneering DNA analyses were published (see e.g. Small & Kron 2001; McGuire & Kron 2005) as these appeared to indicate that *Erica carnea* and *E. erigena* are not closely allied, despite superficial appearances. More recent studies sampling hundreds of *Erica* species (Pirie *et al.* 2011), each sample supported by a verified, herbarium voucher specimen, does not rule out this surprising possibility. Pirie *et al.* (2011) have found that in southern Africa, *Erica* species which have similar floral morphology, often the principal basis of our classification systems at species level, do not necessarily have a common ancestry, and that the classification scheme for *Erica*, initiated by Richard Salisbury and George Bentham (see pp. 94–98) does not enshrine actual genetic relationships: it is entirely artificial. More data about the DNA profiles of European species of *Erica* are needed to resolve this conundrum but we may have to face several inconvenient facts: that appearances are deceptive; and that *Erica* species which look alike may be the chance product of confusing patterns of convergent evolution.

5. ERICA CARNEA. Winter heath, mountain heath.

Erica carnea, it is generally stated, has been cultivated in Britain since 1763 when it was introduced by the Earl of Coventry (who is also credited with introducing *E. australis*, see p. 221) (for example, Aiton 1789, 1811). It is now one of the most widely grown heathers, valued for its winter blossom; some cultivars can begin to show colour in the autumn (for example 'King George' which came from Switzerland, and 'Eileen Porter', an enigmatic plant raised before 1936 by James Walker Porter (see p. 329)), while others carry the flowering season almost into early summer. Winter heath is tolerant of lime in the soil so it is can be accommodated in most gardens, and it is frost-hardy: temperatures down to -23°C are tolerated (Bannister 1996). With a low, spreading habit, *E. carnea* makes an excellent ground-covering evergreen that needs little attention once established.

This species inhabits open alpine and subalpine moors and rocky places, on calcareous or neutral soils, and also open conifer woodland. *Erica carnea* is recorded as reaching an altitude of about 2,400 metres.

Being distributed in central Europe, the winter heath was known to botanists by at least the mid-1500s. Charles de l'Écluse (Carolus Clusius) and others described the species and illustrated it in some of the earliest local floras. No fewer than three woodcuts were made for l'Écluse, to illustrate *Rariorum aliquot stirpium per Pannoniam, Austriam, & vicinas quasdam prouincias observatarum historia* (l'Écluse 1583). The engraved woodblocks that were used to print these illustrations were re-used several times, so the images appear in several different works, including John Gerard's famous herbal, and, re-numbered, in l'Écluse's own *Rariorum plantarum historia* (1601). *Erica Coris folio* VII (of *Rariorum plantarum historia*), flowering in June, had been collected between Prague and Nurnberg in 1579, and also close to Lunz am See, in Austria, on the ascent of Herrenalpe ("*Herren alben*"), above the second lake, Mitter-See, blooming in May (this was *Erica altera* of *Rariorum aliquot stirpium per Pannoniam ....*) (see Fig. 14, p. 15). *Erica Coris folio* VIII also came from mountains in the vicinity of Lunz but was collected in August, and

Fig. 63. *Erica carnea* in black pine (*Pinus nigra*) forest at Kraubath, Austria (© Dr Helmut Pirc). The soil here is thin and poor in nutrients; the base rock is serpentine.

Fig. 64 (right). The lectotype of *Erica carnea*: *Erica Coris folio* IX from Charles de l'Écluse's *Rariorum plantarum historia* (1601).

Erica carnea, *E. erigena* and *E.* × *darleyensis*   Painted by Christabel King   PLATE 8

LOWER LEFT: *E. carnea* 'Myretoun Ruby' 🌿 ♀ Moderately vigorous heather (±0.25 m tall × 0.5 m spread); foliage dark green; flowers richly coloured, opening heliotrope (H12) and gradually turning magenta (H14) to crimson (H13) or purple (H10); January–May. A garden seedling raised by Allan Porteous, Myretoun, Menstrie, Clackmannanshire, which was introduced by Delaney & Lyle, Alloa, Clackmannanshire, by 1965. This outstanding cultivar is frequently praised: Yates (1978) declared it was "one of the finest garden plants". AGM 1993.

LOWER MIDDLE: *E. carnea* 'Ann Sparkes' 🌿 ♀ (*E. carnea* f. *aureifolia*). Spreading, low-growing subshrub (±0.15 m tall × 0.25 m spread); foliage golden orange "with a spattering of bronze" (Chapple 1964), in very cold periods turning crimson; pale flowers opening rose-pink (H7) but deepening to heliotrope (H12); February–May. 'Ann Sparkes' was a sport on 'Vivellii' noticed and propagated by J. W. Sparkes, Beechwood Nursery, Beoley, Redditch, Worcestershire, about 1955. Named after his two sisters-in-law, Sparkes introduced this before 1964. AGM 1993.

LOWER RIGHT: *E. carnea* 'Westwood Yellow' 🌿 ♀ (*E. carnea* f. *aureifolia*). Compact heather (±0.15 m tall × 0.3 m spread); foliage yellow all year; flowers shell-pink (H16) aging to heliotrope (H12); February–April. A seedling found in John Letts's garden, Foxhollow, Westwood Road, Windlesham, Surrey, in 1973; he gave it to John Hall, Windlesham Court Nursery, who named (® no. 15) and introduced it in 1977. AGM 1993.

MIDDLE LEFT: *E. carnea* 'Springwood White' 🌿 ♀ (*E. carnea* f. *alba*). Vigorous, creeping sub-shrub, capable of trailing down slopes and forming a mat to 4 m across (±0.15 m tall); foliage bright green; flowers white with prominent brown anthers protruding; December–May. Mrs Anna Walker (1866–1948), Springwood, Stirling, found this heather on a mountain side in northern Italy during a thunderstorm, before 1925 (Walker 1934, 2007). The clone was originally named 'Springwood' (Johnson 1935) by F. J. Chittenden, Director, RHS Gardens, Wisley, but "White" was tagged on later, when it received an Award of Merit, to distinguish it from 'Springwood Pink'. 'Springwood White' was originally propagated by Mrs Walker's gardener, Robert Howieson Syme (1873–1946). An early description read: "excels not only in the purity of its white, but in the individual blossoms which are very large, and flowering spikes of 6 to 7 inches in length are not uncommon" (Johnson 1932). In the early 1950s, A. T. Johnson (1951, 1956) recorded that the plant he had received from Mrs Walker then covered an area 15 ft (4.6 m) across. AGM 1993.

UPPER LEFT: *E. carnea* 'Vivellii' ♀ Floriferous, neat, spreading heather (±0.15 m tall × 0.35 m spread); young foliage green soon changing to dark bronze-green, especially intense in very cold periods; flowers heliotrope (H12) aging to magenta (H14).

Although 'Vivellii' is still widely available, this clone is no longer recommended by The Heather Society, which prefers 'Adrienne Duncan' (see p. 364). Paul Theoboldt, Aulendorf, Württemberg, Germany, collected this clone in the Engadine Alps, Switzerland, during 1906, and he named it after his former employer, Adolf Vivell (1878–1959), landscape architect of Olten, Solothurn, Switzerland. The cultivar was introduced by Theoboldt in 1909. "In its lowly stature and the colour of its foliage and flowers 'Vivellii' is quite unique" (Johnson 1928: 66). Leslie Slinger (2007) applauded it as "very distinct ... its unusual bronze foliage distinguishes it from all others and its red flowers come quite late in the winter." He suggested that 'Vivellii' "must be found a place" in a heather garden. AGM 1993.

CENTRE: *E. erigena* 'Irish Dusk' 🌿 ♀ Neat, upright, compact bush (±0.6(–1.2) m tall × 0.45 m spread); foliage dark green flushed with silver-grey; flower buds salmon (H15), opening to rose pink (H7); November–May.

This splendid plant was collected on 7 April 1966 by David McClintock from a swarm of seedlings that had sprouted on the north-western shore of Lough Carrowmore, County Mayo, Ireland; it "stood out" because of the "good pink flowers in contrast to the usual paler, and bluer, colour" (McClintock 1982a). 'Irish Dusk', so named because of "its dusky colour and the state of my ancestors' island" (McClintock 1982a), was introduced in 1972 and has also been sold (incorrectly) under the name 'Irish Salmon'. AGM 1993.

UPPER CENTRE RIGHT: *E.* × *darleyensis* 'George Rendall'. Tidy, spreading, bushy shrub (±0.35 m tall × 0.75 m spread); young shoots with red tips turning to pink and cream as the flowers fade; flowers opening pink (H8) darkening to heliotrope (H12), calyx lobes pale mauve tipped creamy white; November–May.

Charles Douglas Eason (c. 1890–1972), chief propagator in Maxwell & Beale's nursery, Broadstone, Dorset, found this chance seedling. 'George Rendall' was introduced by Maxwell & Beale Ltd in 1935. The heather was named by Eason after his younger son, William George Rendall Eason: George Rendall was also the name of the boy's maternal grandfather (Eason 1993). AGM 1969.

UPPER RIGHT: *E.* × *darleyensis* 'White Perfection' 🌿 ♀ (*E.* × *darleyensis* f. *albiflora*). Vigorous, with upright branches (±0.4 m tall × 0.7 m spread); young shoots yellow, mature foliage bright green; flowers white (sometimes described as creamy white) produced on long spikes, calyx lobes at least ¾ length of corolla, anthers partly emergent, tan; December–April. Described as "outstanding" (Small & Small 2001), with excellent contrast between the flowers and foliage, this was a sport on 'Silberschmelze', found by H. Knol, Gorssel, Netherlands, who propagated and introduced it about 1972. AGM 1993.

certainly the sparse flowers look well past their prime (this equates with *Erica tertia* of *Rariorum aliquot stirpium per Pannoniam ...*). The ninth *Erica* — *Erica tertia* of the earlier book — came from mountain woods in Pannonia, a district on the middle Danube, and bloomed in April and May: this illustration (Fig. 64), as printed in l'Écluse's *Rariorum plantarum historia* (1601), has been selected as the lectotype for the species (Ross 1967; Jarvis & McClintock 1990) and is a reasonable representation of the winter heath. There was also a fine, albeit rather stylised, woodcut of *E. carnea* in Rembert Dodoens' *Stirpium historiae pemptades sex, sive libri XXX* (1583); the dwarf heath with reddish flowers came from localities that were arid and uncultivated, and from woods, in Germany, particularly near Nurnberg in Bohemia: in April, it may be found with elegant little flowers, according to Dodoens (1583). This was also reprinted in John Gerard's herbal (see Fig. 13, p. 15).

One consequence of the earlier authors' descriptions and illustrations was that Linnaeus published at least four different names for this single species, including *Erica carnea* and *E. herbacea* (Ross 1967). For a while in the late twentieth century, the latter name held temporary sway because it was the name used in *Flora europaea* (Webb & Rix 1972), but first McClintock & Wijnands (1982) argued for the reinstatement of *E. carnea*, and then Brickell & McClintock (1987) successfully proposed the conservation of that more familiar name over *E. herbacea*.

VARIATION IN THE WILD AND IN CULTIVATION

There is not a vast range of variation in the wild, at least judging by published descriptions and accounts, and most concern flower size and colour (cf. Strid 1986).

However, being such a popular species with gardeners, there is substantial variation among the cultivated clones. Selection will inevitably favour certain fashionable, or distinctive, variants, including plants with yellow/golden foliage (*Erica carnea* f. *aureifolia* D. C. McClint.; McClintock 1988a) or with dark purple or bronze foliage (for example, 'Vivellii'). Flower colour in cultivars varies from rich dark pink to pure white (*E. carnea* f. *alba* (Dippel) Braun-Blanq.; McClintock 1984a: 191) (Fig. 64). As noted, the flowering period of clones also varies from early autumn to early summer.

Breeding and selection programmes have been undertaken to obtain better cultivars with large, brightly coloured flowers that remain fresh for a much longer period, and some of these new cultivars are "promising", as we say about such unfamiliar and untried plants. In 1986, encouraged by Major-General Pat Turpin, a small-scale programme of deliberate cross-pollination of *Erica carnea* was undertaken by John Hall senior, of Windlesham Court Nursery in Windlesham, Surrey. Hall hoped to produce a more compact plant than 'Springwood White' and also a variation of 'Myretoun Ruby', so these two cultivars were used as parents. The results included several good cultivars. 'Whitehall' (® no. 83) is a fine white-flowered plant with bright green foliage and a compact, upright habit. 'Pink Mist' (® no. 84) was a seedling from 'Springwood White' crossed with 'Myretoun Ruby', while 'Jean' (named after John Hall's wife; ® no. 80) came from 'Myretoun Ruby' × 'Heathwood'.

CULTIVATION

As noted, *Erica carnea* can be grown in any good garden soil, and being tolerant of lime, it does not need special lime-free soil. Pruning is rarely needed, but when it is, this must be done immediately the old flowers fade, because the flowering initials for the next year's blooms are formed soon afterwards. Cultivars should only be propagated vegetatively, by cuttings or by layering.

Fig. 65. *Erica carnea* in cultivation (E. C. Nelson).

HYBRIDS

Erica carnea is not known to have formed any natural hybrids in the wild where it inhabits the same localities as other *Erica* species. In cultivation, *E. carnea* has yielded one chance hybrid with *E. erigena*, namely *E.* × *darleyensis* (see p. 316); the hybrid first occurred in a Derbyshire nursery in the late nineteenth century and has since been produced artificially by both amateur and professional plant-breeders in Great Britain, Germany and Canada.

Artificial pollination using *Erica carnea* as a seed- or pollen-parent with other *Erica* species has also yielded now-established hybrids, or at least viable seeds. In 1986, from *E. arborea* pollinated with *E. carnea*, Kurt Kramer (Edewecht-Süddorf, Germany) produced *E.* × *oldenburgensis* (see p. 346); the reciprocal cross was a failure. In 1987, Kramer obtained viable seeds from *E. spiculifolia* pollinated with *E. carnea*, and the progeny was named *E.* × *krameri* after him (see p. 343); again the reciprocal cross was not successful. Dr John Griffiths (Leeds, Yorkshire) got viable seed from *E. carnea* × *lusitanica*, but the resulting plants were lost before any decision could be reached about their status, while the reciprocal cross was a failure for him. Kramer has raised seedlings from *E. lusitanica* × *carnea*, five of which flowered for the first time in the Spring of 2008 (pers. comm., 5 May 2008). *E. tetralix* × *carnea* yielded viable seeds for Griffiths, and the seedlings appeared to have hybrid characteristics, but they did not thrive and were lost before flowering. The same cross failed for Kramer.

Unsuccessful crossings using *Erica carnea* as seed- and/or pollen-parent recorded in recent decades include those involving *E. australis* (Griffiths, Kramer), *E. cinerea* (Griffiths, Kramer), *E. terminalis* (Griffiths), *E. umbellata* (Griffiths, Kramer) and *E. vagans* (Griffiths).

Erica carnea L., *Species plantarum*: 355 (1753) *nom. conserv.* Type: *icon.* "Erica Coris folio IX": C. de l'Écluse, *Rariorum plantarum historiae* I: 44 (1601): lecto.: selected by Ross (1967) (see Jarvis 2007: 498).

E. herbacea L., *Species plantarum*: 352 (1753) nom. rej. (Jarvis 2007: 499).

E. pallidipurpurea L., *Species plantarum*: 354 (1753) nom. rej. (for typification see Nelson 2002; Jarvis 2007: 500).

E. purpurascens L., *Species plantarum*, ed. 2: 503 (1762) nom illeg. (Jarvis 2007: 500–501).

E. mediterra [*sic*] L., *Diss. de Erica* 10 (1770) (Jarvis 2007: 500).

E. mediterranea L., *Mantissa pl. alt.*: 299 (1771).

E. saxatilis Salisb., *Prod. Chap. All.*, 295 (1796).

E. carnea subsp. *mediterranea* (L.) Borja, *Anales Inst. Bot. Cavanilles* **12**: 155 (1954).

ILLUSTRATION. L. Denkewitz, *Heidegarten*: 170 (as *Erica herbacea*) (1987).

DESCRIPTION. *Evergreen shrub* to 0.2(–0.3) m in height, branches decumbent with flowering portion ascending; shoots glabrous; infrafoliar ridges extending to below the next whorl of leaves; adventitious roots produced from shoots (in cuttings, rootlets emerge above the leaf nodes (Small 1974)). *Leaves* in whorls of 4, ascending, ±5–8 mm long, ±0.8 mm across, linear, glabrous, upper surface flat; edge with inconspicuous forward-pointing spicules; underside convex, margins recurved and concealing lower surface, contiguous; sulcus linear, almost closed, margins with minute, simple hairs; petiole glabrous. *Flowers* (Fig. 66) lavender-mauve to pink or rarely white, (1–)2–3 in axillary clusters on very reduced short side-shoots, sometimes a terminal umbel present; *pedicel* usually markedly curved, glabrous, green or red, to 3–5 mm long, ±0.3 mm diameter, broadening markedly just below calyx (to ±0.5 mm diameter); bract below or at middle, with paired bracteoles above or about middle; bract triangular, ±1.5 mm long, margins glabrous below but with some simple hairs towards and at apex, sulcus minute, obscure; bracteoles similar, shorter but often broader. *Calyx* 4-lobed, ±3–4 mm long; lobes ±1.2 mm across, oblanceolate-triangular, fused at the base for 0.5 mm, waxy, thicker in texture than corolla, coloured (usually similar to that of corolla), glabrous except for a very few spicules at apex, apex slightly thickened, sulcus present, obscure, ±0.6 mm long. *Corolla* ±9 mm long, ±3 mm diameter, ovate, lobes usually erect to 1 mm long, ±1.5 mm wide at base, overlapping alternately, rounded, margins smooth, glabrous. *Stamens* 8, longer than corolla, ±9 mm long; *filament* ±7 mm long, straight, hooked at base and attached under the nectary, 0.5 m broad, strap-shaped with slight waist below anther, white or pink with brown tinge towards anther, glabrous; *anther* erect, entirely free when mature, *thecae* ±1.5 mm long, ± 0.5 mm across, pore narrow-elliptical, ±0.8 mm long, surface with microscopic spicules. *Style* ±7 mm long, slender, 0.3 mm diameter, tapering only slightly to apex; *stigma* not enlarged, ±0.25 mm diameter; *ovary* cylindrical, ±2 mm tall, ±1 mm in diameter across centre, "stepped" with narrower upper part (±0.7 mm across), prominently ribbed, glabrous, with ±20–25 ovules per

Fig. 66. Arthur Church's diagram of a flower of *Erica carnea*.

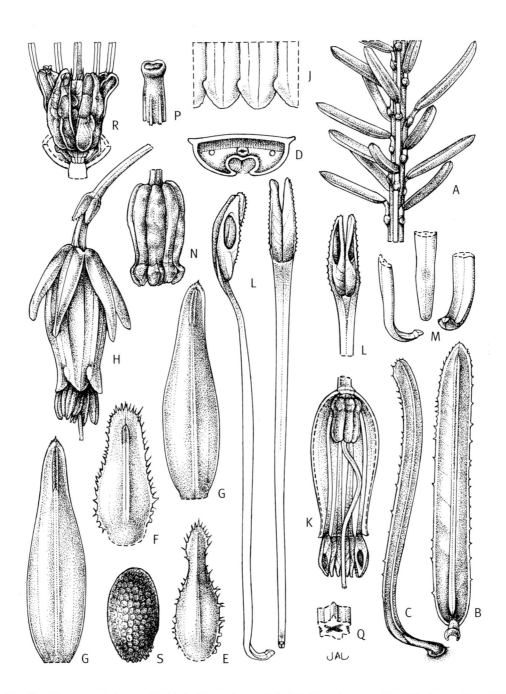

Fig. 67. *Erica carnea* 'Springwood White'. **A** leafy shoot × 3.5; **B** leaf from below × 10.5; **C** leaf from side × 10.5; **D** section through leaf × 26; **E** bracteole from back × 13.5; **F** bract from back × 22.5; **G** sepal from back × 14; **H** flower from side × 6.5; **J** lobes of corolla × 7; **K** section through flower × 6.5; **L** stamens from front, side and back × 23; **M** detail of filament bases × 24; **N** ovary and nectary × 16; **P** style apex (stigma) from side × 24; **Q** section through style × 38; **R** capsule × 5.5; **S** seed × 23; from specimens grown in RHS Garden Wisley, no. W 885501A. Drawn by Joanna Langhorne.

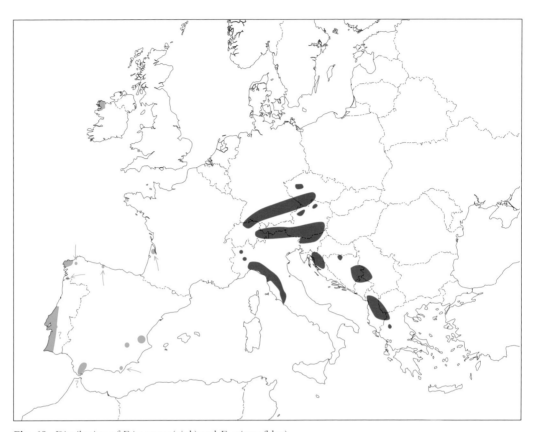

Fig. 68. Distribution of *Erica carnea* (pink) and *E. erigena* (blue).

locule; *nectary* lobes prominent, nectar profuse. *Seed* 1.0–1.2 × 0.6–0.65 mm (Nelson 2006a). n = 12 (McClintock 1979b; see Nelson & Oliver 2005b).

FLOWERING PERIOD. March–June in the wild, but several months earlier in cultivation in milder regions, and sometimes even as early as October.

DISTRIBUTION & ECOLOGY. Central Europe (Czech Republic, southern Germany) southwards to south-eastern France and northern Italy, and into the Balkans as far south as the north of Greece (Northern Pindhos) (Strid 1986: 738). **Map:** Meusel *et al.* (1978: 331 as *Erica herbacea*). Fig. 68.

On the calcareous rocks of the northern Alps in central Europe, *Erica carnea* can be an understorey shrub in mixed oak (*Quercus petraea* or *Q. robur*)/Scots pine (*Pinus sylvestris*) woodlands at low altitudes, and it also occurs in subalpine regions in open, mixed shrub communities (Dunk *et al.* 1987). In Greece and Albania, *E. carnea* is reported to occur only on serpentine soils (Greuter 1975).

ETYMOLOGY. Latin: *carneus*, flesh-coloured.

VERNACULAR NAMES. Winter heath, mountain heath, snow heath (English); Schneeheide, Frühlingsheide, Fleischrotblütige Heide (Dippel 1887) (German); bruyère carnée, bruyère couleur de chair, bruyère des neiges, bruyère blanc rosé (French); vårljung (Swedish).

CULTIVARS. See p. 113 and Appendix I (pp. 364–368).

6. ERICA ERIGENA. Irish heath, Mediterranean heath.

Erica erigena is unique among the heathers growing wild in Ireland, as well as among those that inhabit other parts of Britain and Ireland, because it blooms in the late winter and spring (Figs 68–70); all the other species are summer-flowering shrubs. Indeed, it is one of a small group of plants that blossom early in the year; the most conspicuous shrub in flower with the Irish heath is gorse (*Ulex europaeus*). Irish heath has a markedly discontinuous distribution pattern in western Europe and, moreover, appears to be a relatively recent colonist in western Ireland. The heather occurs in several distinct and widely scattered habitats within Spain, Portugal, France and Ireland, separated from the nearest colonies by hundreds of kilometres. A report that *E. erigena* is also native in north Africa, in northern Morocco ("ruisseaux à l'oeust de Sebta": Stefanesco & Vilhena 1966; Fennane & Ibn Tattou 2005) may have to be discounted as a misidentification of *E. multiflora* — I have not seen any specimens of *E. erigena* from Morocco.

Confusing these two species is not a new pastime: it has been going on since the days of Linnaeus. The thorough muddle of names for this species and the very similar *Erica carnea*, involving also *E. multiflora* — "*Similis* E. multiflorae, *sed Corolla omnino ovato*" (Linnaeus 1771: 229) — was only sorted out in 1967 by Robert Ross (1912–2005) in a closely argued paper that established, alas incorrectly, the binomial *E. hibernica* (meaning Irish) for this heather. Ross (1967) was able to demonstrate that the long-used name *E. mediterranea* was applied by Linnaeus to the species from the Alps, because the type specimen in the herbarium of Joachim Burser (1583–1649) (see p. 134) had been gathered in the Austrian Alps. Ross had to retract *E. hibernica* after it was pointed out that "*hibernica*" (meaning "of winter") had been used by gardeners for a Cape heath — "[une] jolie bruyère ... elle fleurit en février" — during the 1830s. This now-extinct heath was cultivated by Mons Martine, "habile cultivateur fleuriste, rùe des Bourguignons, à Paris", and because the name was published by Utinet (1839) it could not be employed in any other sense. Thus Ross (1969) substituted the name *E. erigena*, and that remains valid for this heather today.

It is worth adding a comment here about the English names for this heather. As so often happens when an English name is needed, the Latin name will simply be translated. Thus for a long time this heather was called Mediterranean heath: it still often is, in both conversation and print. More recently, it has been dubbed Irish heath. The older moniker is often mistaken as meaning that the heath grows in, or came from, the region around the Mediterranean Sea. However, when Linnaeus employed the epithet "*mediterranea*", he surely intended the literal meaning of *mediterraneus*, in the middle of the land — inland. He annotated the specimen in Joachim Burser's herbarium (**UPS**), which was labelled "*In alpibus Austriae*", with the words "multiflora mediterranea", which can only be taken as meaning the inland "multiflora". Thus Mediterranean heath is a doubly inappropriate common name. Irish heath is the recommended name according to the Botanical Society of the British Isles (Dony *et al.* 1986; see also Scannell & Synnott 1987), yet it is not unambiguous, because that same name is often used (especially in North America) for Saint Dabeoc's heath, *Daboecia cantabrica* (Nelson 2004b).

IRISH HEATH IN THE WILD

On the Iberian peninsula, *Erica erigena* occurs as disconnected populations, some containing only a few hundred plants (as, for example, in the Sierra Nevada; Blanca *et al.* 2001: 132), in a scattering of mainly peripheral regions (Fig. 68). The southernmost populations are in Andalucía (Málaga, Cádiz and Granada) in southern Spain. The heath is also recorded in Galicia (see Fraga 1984b) in the far

Erica erigena **Painted by Deborah Lambkin**   PLATE 9

northwest; in a single, habitat on the north coast in Asturias, where I saw it in 2007; and in the east in Valencia (McClintock 1971; Mansanet *et al.* 1980; Fraga 1984b; Bayer & López González 1989; Bayer 1993; Bolòs & Vigo 1995). *E. erigena* occurs in south-western Portugal, including the vicinity of Lisbon: the Flemish botanist Charles de l'Écluse saw it flowering in December and January on the banks of the Tagus River (Rio Tejo) above Lisbon, and a passable illustration (labelled "Erica Corio folio III") was published first in his book *Rariorum aliquot stirpium per Hispanias observatarum historia* (l'Écluse 1576) and later in other works.

In western France, *Erica erigena* inhabits a very limited area of the Gironde, north of Bordeaux, in Médoc, perhaps extending over not more than 30 hectares in all of the seven stations reported (Besançon 1978b; Lahondère 2006). The French localities are approximately 400 kilometres from the nearest population at Río España in northern Spain and more than 1,100 kilometres from the southernmost one in Ireland.

In the west of Ireland, Irish heath occurs in counties Mayo and Galway only (Foss *et al.* 1987), between The Mullet in north-western Mayo and Errisbeg in south-western Connemara, County Galway, a range of approximately 100 kilometres. There is little conspicuous variation in flower colour in Ireland (Fig. 69); pure white flowers (*Erica erigena* f. *alba* (Bean) D. C. McClint.; McClintock 1984a: 191) are extremely rare (cf. Moore 1852, 1855, 1860), but shrubs with very pale purple flowers are occasionally

seen. The populations in Connemara generally seem to have paler flowers than plants from Mayo (see p. 122). Ecologically, *E. erigena* is remarkably adaptable. "The wettest parts of the bogs appear to be its favourite spots", observed Moore (1860) (cf. Fig. 70), and "when it grows on the sides of low hills, it is nearly or altogether confined to the margins of rills of water which descend from the high grounds ...".

Examination of pollen grains preserved in peat deposits within the region where *Erica erigena* occurs in western Ireland failed to reveal any pollen of this species in the oldest peat. Pollen of *E. erigena* was only recovered in samples close to the present–day land surface, indicating that, in the sites examined, this heather was a recent immigrant. Foss & Doyle (1988, 1990) suggested that seeds of the heather could have been imported from northern Spain during the mediaeval period. Recently conducted DNA studies support this thesis by indicating that plants in several Mayo populations are not genetically distinct from plants growing in La Coruña in north-western Spain. Kingston & Waldren (2006) argue that this "reflect[s] a recent separation of the populations".

Erica erigena is certainly an anachronistic plant when considered in the context of

Fig. 69. *Erica erigena* in bloom at Lough Furnace, County Mayo (E. C. Nelson).

Fig. 70. *Erica erigena* on wet blanket bog, Erris Peninsula, County Mayo (E. C. Nelson).

Ireland's native flora. As noted, these shrubs are in bloom when few other species are in flower and when most deciduous shrubs still have no leaves. The Irish heath is also an opportunist, colonising disturbed or vacant sites: the colony that became established on the newly exposed western shore of Lough Carrowmore (Fig. 71) after the water-level was reduced in the mid-1900s created a floral spectacle in the 1970s and 1980s. This was, incidentally, where several excellent cultivars were collected, including 'Irish Dusk' and 'Irish Salmon' (McClintock 1982a). More than a century earlier, this same population was described by David Moore (1860): "... on the shores of Carrowmore Lake [*E. erigena*] only grows where it is partially covered with water during the winter."

The first to publish the combination *Erica mediterranea* var. *hibernica* was apparently David Moore (1852). He described the plants from Ballycroy in County Mayo as "all of dwarf habit, with much darker coloured flowers than any I had previously seen, — characters which greatly enhanced their value from a floricultural point of view ...". Moore noted that the heather took "full possession of the ground, to the almost total exclusion of the other kinds of common heaths. The flowers are generally of a deep pink colour, and the plants grew from six inches to a foot tall" (Moore 1852). Given the description, Moore's varietal name is valid and predates by three years Hooker and Arnott's publication (in the seventh edition of *The British flora*; Hooker & Arnott 1855). However the first use of the epithet *hibernica* at varietal level was in 1838 by J. C. Loudon, under *Gypsocallis mediterranea*. Loudon (1838) made explicit references to descriptions published in *Supplement to English botany* (Hooker 1833) and J. T. Mackay's *Flora hibernica* (1836), so the name was validly published. Plants under this name were being sold at Rollisson's Tooting Nursery.

Erica erigena was discovered in Ireland in 1700 by the Welsh antiquary and naturalist Edward Lhuyd (c. 1659/60–1709) (see p. 66). However, he did not recognise the heather, because it was not in flower, and his record remained unpublished (Vines & Druce 1914; Mitchell 1975: 282). More

Fig. 71. An extensive population of *Erica erigena* on the western shore of Lough Carrowmore, County Mayo, with gorse (*Ulex europaeus* L.) in bloom, March 1983 (E. C. Nelson).

than a century later, in October 1829[2], James Townsend Mackay (see p. 240) of the College Botanic Garden, Dublin, collected a "strange" heath, also not then in bloom, on Errisbeg Mountain near Roundstone in Connemara; consequently, the credit for the discovery of this plant has generally been given to him (Mackay 1831, 1997; Nelson 1979a; Robinson 2006). In 1846, following the discovery of another remarkable heath, *E. ciliaris* (see p. 167), near Roundstone, Mackay (1997) wrote that the

> beautiful Spring flowering Heath ... was first found by me, on Erris-beg mountain ... in 1829, which though then out of flower I determined to be E. mediterranea ... which I had previously known in Gardens — I was led to look for it on Urris-beg [*sic*] by Mr John Nimmo ... . On the plant I brought to Dublin flowering the following Spring I found I was right in my conjecture[.]

Tim Robinson (1990, 2006) identifies the site where Mackay found *Erica erigena* as a "little hollow" in Gleann Mór, "on the north-east flank of the mountain ... for the place is called French Heath Tamhnóg by the very few people who have names for such out-of-the-way ... spots" (see Fig. 72). He explains (Robinson 2006) the origin of this strange, bilingual name as "a local misremembering of what some visiting botanist had to say about the remarkable heather that had overgrown it." That might have been Mackay, who inexplicably used the name "Corsican heath" (Mackay 1997) for *E. erigena*, although the species does not grow in Corsica.[3] In subsequent years, the heather was traced from Errisbeg northwards into Mayo (Mackay 1997) as far as the Mullet Peninsula. It is said that the Irish populations of *E. erigena* are the most extensive, yet the species is not uncommon in its Iberian sites.

Fig. 72. "French Heath Tamhnóg" in Gleann Mór on Errisbeg, Connemara, in summer. This is the locality to which John Nimmo directed James Townsend Mackay in 1829 (E. C. Nelson).

Erica erigena is not native in Britain, and its distribution pattern led Professor David Webb (1983) to comment that the distribution of *E. erigena* cannot be explained by the standard theories applied to disjunctions in the European flora. Perhaps it has, according to Webb, some "quite unexplained power of long-range distance dispersal, or else ... it underwent a severe fragmentation of area some time ago". Another possibility is that it is expanding its range with the assistance of human beings; in Ireland, at least, *E. erigena* is, as already pointed out, an opportunist, colonising old railway embankments, cutover bogs, and even lough shores laid bare by falling water levels (Fig. 71). One of its most remarkable colonies grows on the lazy beds of the derelict clachan on the southern slope of Mweelrea by the shore of Killary Harbour. Thus, it is not confined to pristine moorlands and unaltered mountainsides. Man's activities certainly assist its spread.

ETHNOBOTANY

If the novel theory that the Irish heath came to Ireland in mediaeval times, accidentally, as packing material or bedding for animals in a trading vessel (Foss & Doyle 1988, 1990), is correct, then in the distant past this heather was harvested for the purpose somewhere in the Iberian Peninsula. James Townsend Mackay recorded that John Nimmo, who directed him to the heather on Errisbeg, "... told me that the peasantry gathered there a strange kind of Heath with they used for fuel" (Mackay 1997: 34). Around the start of the twentieth century, half a century after the Great Hunger (the famine of 1845–1849), another visitor to Connemara reported that the leafy shoots were harvested to make into besoms, and also switches with which to "spray" potatoes with Bordeaux mixture to prevent infection by the blight fungus, *Phytophthora infestans* (Anonymous 1909; Scannell 1983; Robinson 1990, 2006): this latter use seems to be a unique invention of the people in Connemara.

Apart from its ornamental value, *Erica erigena* is a valuable source of pollen for bees, and of nectar for those insects with long enough mouthparts to reach the liquid (which probably precludes honey bees unless they puncture the bases of the corolla and "rob" the nectar).

CULTIVATION

Although the Irish heath is sometimes stated to have been brought into cultivation in Britain in 1765 by one Joshua Brookes (about whom nothing else seems to be recorded), William Aiton (1789) claimed it was in the University of Oxford Botanic Garden as early as 1648 (cf. Stephens & Browne 1658). However, given the confusing synonymy, there cannot be any certainty when it was first grown, and these early records may refer to *Erica carnea* or perhaps even *E. multiflora*, or none of these heathers (see p. 205).

There are, however, indisputable records of the Irish heath being grown in the early 1830s, soon after James Mackay found the population on Errisbeg in Connemara (Nelson 1979a, 1982a).

Erica erigena is hardy, although Bannister (1996) suggested that it is likely to be killed completely if the air temperature falls below −15°C. The Irish heath is also lime-tolerant, even though in Ireland it is often found in the wettest, and so presumably the most acidic, parts of blanket bogs, often partly submerged in winter. Around the lough shores, as at Lough Carrowmore, its roots are certainly well under water in winter. At the same time, *E. erigena* must also be salt-tolerant. It colonises hill-slopes that are exposed to ocean spray, as around Mallranny in County Mayo. At Playa de España in Asturias, the Irish heath grows on coastal cliffs within reach of salt-laden spray.

In gardens, the Irish heath will grow in any good garden soil and will also tolerate more harsh conditions including gravel. Untrimmed plants have reached over 3.6 m in cultivation (McClintock 1989e), so this can become a very substantial shrub and might be classed as a "tree" heath (see p. 197).

HYBRIDS

Erica erigena, like *E. carnea*, is not known to have formed any natural hybrids in the wild, although it was at first suggested that *E.* × *stuartii* (in its original state, 'Stuart's Original'; see p. 302) was a hybrid between *E. tetralix* and the Irish heath. As noted above (p. 115), in cultivation *E. erigena* and *E. carnea* have yielded the hybrid *E.* × *darleyensis* (see p. 316). Since it was found in Smith's Darley Dale nursery in Derbyshire during the late nineteenth century, *E.* × *darleyensis* has been produced artificially by Dr John Griffiths (Leeds, Yorkshire), Kurt Kramer (Edewecht-Süddorf, Germany), Mrs Ann Parris (Monmouthshire, Wales), Barry Sellers (Norbury, London), David Wilson (British Columbia, Canada), and perhaps others, including James Walker Porter (Carryduff, County Down, Northern Ireland).

Experiments with other couplings have not yielded viable seed. Kramer has tried using *Erica erigena* as a pollen- and a seed-parent without success: as pollen-parent with *E. arborea* (1986), *E. australis* (2002), *E. manipuliflora* (2001), *E. spiculifolia* (1992), *E. terminalis* (2000), *E. umbellata* (2004) and *E. vagans* (2001); and as seed-parent with pollen from *E. australis* (1997; this also failed for Wilson in 1999) and *E. lusitanica* (2002).

Barry Sellers (pers. comm., 23 October 2006) did succeed in raising seedlings from *Erica lusitanica* crossed with *E. erigena*; two with yellow foliage perished, but one with green foliage survived and bloomed (see pp. 213 & 346). The surviving clone has been given a cultivar name, 'Lucy Gena' (®E.2009:03) (Nelson 2010b).

Erica erigena R. Ross, *Watsonia* 7: 164 (1969). Type: "Connemara", *J. T. Mackay*: lecto. K (in Herb. Hooker); iso. BM! (Ross 1967: 71).

E. mediterranea sensu Thunb., *Diss. Bot. De Erica*: 9, 37, 56 (1785), nom. superfl., non L.

E. lugubris Salisb. *Prodromus* ... : 295 (1796), nom. superfl. quoad. descr.

E. carnea var. *occidentalis* Benth. in DC., *Prodromus* 7: 614 (1839).

E. carnea subsp. *occidentalis* (Benth.) M. Laínz, *Bol. Inst. Est. Astur.* **10**: 199 (1964).

E. mediterranea var. *hibernica* D. Moore, *Phytologist* 4: 597 (1852).

E. mediterranea var. *hibernica* Hook. & Arn. in Hooker, *British flora*, seventh edition: 266 (1855), nom. superfl.

E. hibernica (Hooker & Arnott) Syme, *Sowerby's English botany* **6**: 42–43, tab. 892 (1866), Ross (1967), nom. inval. non *E. hibernica* hortulanorum ex Utinet, *Annales de flores et de pomone*: 94 (1839).

Gypsocallis mediterranea var. *hibernica* J. C. Loudon: *Arb. & frut. brit.*: 1086–1088 (1838).

ILLUSTRATIONS. S. J. Roles, *Flora of the British Isles: illustrations* pt 2: tab. 939 (1960); S. Ross-Craig, *Drawings of British plants*, part 19: pl. 26 (1963); R. W. Butcher, *A new illustrated British Flora*: 109 (1961); J. Smythies, *Flowers of south-west Europe*: 287 (10) (Polunin & Smythies 1973); W. F. Walsh, *An Irish florilegium* **II**: plate 18 (Walsh & Nelson 1987); E. Sierra Ràfols, *Flora iberica* **4**: 488 (lam. 177) (1993).

DESCRIPTION. *Evergreen shrub* to ±3 m tall, erect, bushy, obligate reseeder; shoots puberulent especially between infrafoliar ridges and above petioles when young; infrafoliar ridges narrowing rapidly and not extending below next leaf whorl. *Leaves* in whorls of 4, narrow-lanceolate to linear, glabrous except for microscopic spicules around apex, flattened above, edge with slight cartilaginous rim; margins recurved, ± contiguous below, sulcus closed, linear. *Inflorescence* raceme-like, *flowers* in various shades of mauve to pale lavender, very rarely white, axillary, solitary or in clusters of 2–5 on very greatly reduced lateral shoots; *pedicel* ±2.5 mm long, ±0.3 mm in diameter, curved, usually tinged red, glabrous or sometimes sparsely tomentose on lower part (below bract and bracteoles) with simple, sometimes gland-tipped hairs, not markedly thicker below calyx; bract and bracteoles in a whorl ± at mid-point (sometime above), 0.5–1 mm long, broadly triangular, margins hirsute. *Calyx* lobes 4, fused at the based into a cup, truncate at base and broader than apex of peduncle; lobes c. 3 mm long, narrow-triangular, waxy, thicker in texture than corolla, and usually the same colour or paler, with slight central rib, thickened at tip and with a short sulcus. *Corolla* ±5 mm long, ±2.5 mm across, ovate-urceolate, glabrous, lobes ±1 mm long, ±1.5 mm across at base, overlapping alternately, usually spreading, rounded, margins sometimes irregularly toothed. *Stamens* 8, ±5.5 mm long, thus anthers partly emergent; *filament* straight, hooked at base and attached under the nectary, ±4.5 mm long, to 0.3 mm broad, narrowing only at base, white or pale pink, shading to brown at apex; *anther* ±1 mm long, basifixed, *thecae* c. 0.4 mm wide, surface ±smooth, pore occupying almost the entire lateral side. *Ovary* 4-locular, barrel- or bottle-shaped, green, glabrous, apex emarginate; c. 10–12 ovule per locule; *style* c. 6 mm, straight, terete, tapering slightly to apex which is not enlarged; *stigma* not enlarged. *Seed* 0.8–1.0 × 0.5mm (Nelson 2006a). n = 12 (Fraga 1983: 533–534; see Nelson & Oliver 2005b).

FLOWERING PERIOD. (January–)February–May(–June); Bayer (1993) indicates that *Erica erigena* can be in flower on the Iberian Peninsula as early as September.

DISTRIBUTION & ECOLOGY. Europe: Ireland (Counties Galway and Mayo); France (Médoc); Spain in western Galicia and northern Asturias (one site), in Levante and Andalucia; Portugal (central and southern). North Africa: the report from northern Morocco needs verification. **Maps:** Webb (1983:

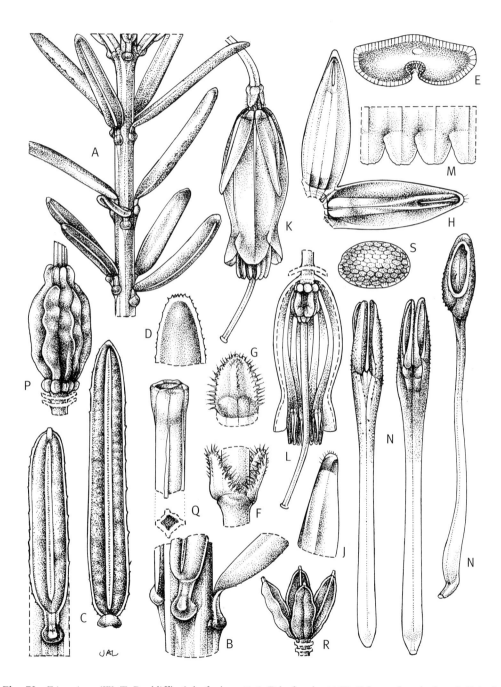

Fig. 73. *Erica erigena* 'W. T. Rackliff'. **A** leafy shoot × 6; **B** leaf nodes × 12; **C** leaves from back × 8; **D** leaf tip from front × 16; **E** section through leaf × 22; **F** bracteoles × 30; **G** bracteole × 32; **H** calyx lobes displayed × 11.5; **J** calyx lobe tip from side × 14; **K** flower from side × 8; **L** section through flower × 8; **M** corolla lobes × 6; **N** stamens from back, front and side × 14; **P** ovary and nectary × 22; **Q** style apex (stigma) from side with cross–section × 40; **R** capsule after dehiscence × 5.5; **S** seed × 28; from specimens grown by E. C. Nelson. Drawn by Joanna Langhorne.

158), Foss & Doyle (1988: 163), Bolòs & Vigo (1995); for detailed maps of Irish localities see Foss *et al*. (1987); for Galicia see Fraga Vila (1984b: 13). Fig. 68.

In Ireland, *Erica erigena* inhabits bogs and lough shores, stream beds and flushes. The heather occurs on the coast within a few metres of the ocean in several sites in counties Mayo and Galway. At Río España, Asturias, on the northern Spanish coast, shrubs of *E. erigena* thrive on coastal cliffs, growing in damp, shallow gulleys and on rock ledges, with *Schoenus nigricans*. *E. vagans* grows on the same cliffs. In Valencia, Spain, Irish heath grows on dolomitic soils with pH 7–8 and associates in some habitats with *E. terminalis* and *E. multiflora* (Mansanet *et al.* 1980).

Erica erigena is classified as a member of the Oceanic Southern-temperate element of the Irish and British floras, a category also assigned to *Daboecia cantabria*, *E. ciliaris* and *E. vagans* (Preston & Hill 1997).

ETYMOLOGY. Ross (1969) wrote: "Hooker & Arnott's epithet *hibernica*, referring to the Irish origin of their plant, is now replaced by *erigena*, which has the same significance" (cf. Morley & Scannell 1972; Morley 1974).

Erigena or Eriugena was a name used in the latter part of the first millennium AD, and until the twelfth century AD, to mean "of Irish birth" (O'Meara 1988). *Eriu* was an ancient name for Ireland. The Christian philosopher Johannes (or John) Scotus Eriugena, often simply called Erigena or Eriugena, is the most celebrated bearer of this name. *Eriugena* was formed, according to O'Meara (1988: 1 note 1), on the model of Virgil's *Graiugena* (*Aeneid* 3: 550) meaning "of Grecian birth".[4]

VERNACULAR NAMES. Irish heath, Irish heather, Mediterranean heath (English); tall spring heath (Krüssmann 1984); fraoch camogach (Irish); brezo de Irlanda (Spain); bruyère de l'ouest (French).

CULTIVARS. See p. 113 and Appendix I (pp. 376–377).

Notes

[1] Bentham (1839) wrote: "Varietas ab omnibus auctoribus uti species distincta mihi characteribus valdé incertis sejuncta videtur et intermedia numerosis cum E. carnea juncta (v. s. c. et sp.)"

[2] I have discussed the dating of Mackay's discovery elsewhere (Nelson 1979a; Mackay 1997) and concluded that he found it in 1830, based on the phrasing of his paper read to the Royal Irish Academy on 30 November 1830 (Mackay 1831a) and several other sources published at the time (for example, Mackay 1831b). However, in the long-unpublished 1846 lecture (Mackay 1997), the manuscript of which came to light in the early 1990s, Mackay reported that he had found the plant when it was not in bloom, adding, as quoted here, that he only confirmed the identification after the specimen flowered in Dublin "the following spring". In this light, allowing for several months to pass between discovery and confirmation, he must have been at Roundstone in October 1829. No other record of this visit has been traced; no other plants are attributed to it.

[3] Mackay was evidently misinformed, for he wrote (1831b) that "it is the principal heath of Corsica"

[4] There is no justification, given Ross's explicit reference, for substituting "harbinger of Spring" from ἦρι (*ēri*, early) and γένος or γενεά (*genos* or *geneā*, race, kind) as the meaning of the epithet, by analogy with *Erigenia* (cf. Scannell 1976a).

THE CORNISH HEATH AND ITS CONGENERS
Erica vagans, E. manipuliflora, E. multiflora, E. umbellata

This section treats three species, *Erica vagans*, *E. manipuliflora*[1] and *E. multiflora*, that have an entangled nomenclatural history. They are bewilderingly similar in superficial appearance and are, perhaps, related, although a study of the DNA of European and African species (McGuire & Kron 2005) inexplicably presents a complicated and confused picture that I have alluded to already (see p. 110). Other work, very recently published, on the microscopic anatomy of seed coats of *Erica* species (Fagúndez *et al.* 2010), suggests that while *E. vagans* and *E. manipuliflora* are closely allied, *E. multiflora* is distinct and separate; the cells of its seed coat have pitted inner walls, a unique character among European *Erica* species.

The species' distributions do not overlap, except on the Adriatic coasts of Italy, southern Croatia, Albania and north-western Greece where the ranges of *E. multiflora* and *E. manipuliflora* coincide. Otherwise, the species have separate areas of occurrence. *E. vagans* is the western species, ranging from Galicia to Cornwall with an isolated, enigmatic population in the northwest of Ireland. *E. manipuliflora* is the eastern species, principally occupying habitats on the Adriatic and Aegean coasts and islands and ranging through southern Turkey and as far as southern Lebanon; it also occurs on Cyprus. *E. multiflora* is largely confined to the western Mediterranean region, with an outlying station in Cyrenaica, northern Libya.

These species were grouped together in *Flora europaea* (Webb & Rix 1972) because of their "great morphological similarity". Decades previously, Richard Anthony Salisbury had suggested that *Erica vagans* and *E. manipuliflora* (as *E. verticillata* Forssk. non Bergius) should be placed in a separate genus, for which he coined the name *Gypsocallis* (most probably from Greek: γύψος, *gypsos*, limestone; καλός, *kalos*, beautiful[2]) because "they are both confined to Gypsaceous soil". Salisbury's opinion was brought into effect by Gray (1821) who placed only one species in *Gypsocallis*: *Erica vagans* (Cornish heath). David Don (1834) went further and included *E. multiflora*, *E. carnea*, *E. erigena* (as *E. mediterranea*), *E. manipuliflora* and *E. umbellata* in an enlarged *Gypsocallis*, a grouping that Salisbury would not have supported.[3] His brother, George, evidently agreed (G. Don 1834). Bentham (1839) relegated *Gypsocallis* to the rank of section. Webb & Rix (1972) suggested that *E. manipuliflora* "might be reduced to a subspecies" of *E. vagans* "on purely morphological criteria" but recognised that their separate geographic ranges and "usually rather distinct habit probably justify" their retention as separate species. In his publications, McClintock (1971a, 1989a, 1992a) maintained them as three distinct species — and they are so in my opinion, too.

They can be separated reliably by examining the anthers. In *Erica vagans*, these are tiny, only about half a millimetre long, and the two thecae that comprise each anther are clearly separate, even divergent. *E. multiflora*, which like *E. vagans* grows in the western part of Europe, has much larger anthers, around a millimetre and a half long, and the thecae are erect and parallel. The anthers of *E. manipuliflora* are intermediate in length, being around a millimetre long.

To this trio, I will append a fourth species, *Erica umbellata*, because it shares some characters with the other three, especially spurless (muticous), exserted anthers.

Cornish heath and related *Erica* Painted by Christabel King PLATE 10

LOWER LEFT: *E. vagans* **'Lyonesse'** ♀ (*E. vagans* f. *alba*). Floriferous, dome-shaped heather (±0.3–0.4 m tall × ±0.5–0.7 m spread); foliage rich green, glossy; flowers pure white, larger than in 'Kevernensis Alba', in long, tapering spikes; anthers pale golden-brown; August–October. (Painted 6 September 1987.)

Mr and Mrs D. F. Maxwell found several excellent heathers on the Lizard Peninsula, Cornwall, in 1923: 'Lyonesse' and 'Mrs D. F. Maxwell' (see below), along with *Calluna vulgaris* 'Mullion' (see p. 355). 'Lyonesse' was introduced by Maxwell & Beale Ltd, Broadstone, Dorset, in 1925. "Excelling in the size of its inflorescences and in the purity of its whiteness" was A. T. Johnson's (1932) assessment. AGM 1993 (Recommended 1999 by the Heather Society; withdrawn 2011). Lyonesse, according to legend, is a drowned kingdom situated somewhere off the coast of Cornwall and the Isles of Scilly. Some suggest that this legend, common to the Cornish and Breton people, may be a folk memory of the final stages of the inundation of the continental shelf after the last ice-age.

LOWER MIDDLE: *E. vagans* **'Valerie Proudley'** ♀ (*E. vagans* f. *alba*, *E. vagans* f. *aureifolia*). Slow-growing heather forming a compact dome (±0.15 m tall × ±0.25 m spread); young foliage golden-yellow, turning golden green, and yellow-green ("dull gold") in winter; flowers sparse, white (Fig. 78); September–November.

'Valerie Proudley' arose in 1965 as a sport on a "normal", white-flowered plant of *E. vagans* at Aldenham Heather Nursery, Watford, Hertfordshire. It was named by the late Brian W. Proudley (d. 2003) after his wife, and introduced by Aldenham Heather Nursery in 1968. Vickers (1976) noted that this is "a little gem in the right place".

This cultivar has a tendency to revert to plain green foliage; reversions should be removed carefully and destroyed. Despite this instability, the type specimen of *E. vagans* f. *aureifolia* came from a plant of 'Valerie Proudley' (see Miller & Nelson 2007). AGM 1993.

LOWER RIGHT MIDDLE: *E. vagans* (unnamed).

LOWER FAR RIGHT: *E. vagans* **'Mrs D. F. Maxwell'** ⬤ ♀ Floriferous, robust heather (±0.3 m tall × ±0.45 m spread); foliage dark green; flowers cerise to rose-pink (H6–H7), on red pedicels; August–October.

Douglas Fyfe Maxwell (1927: 60) is the best source for the history of this "striking" heather:

> It was found by my wife not many miles from Helston, that "quaint old Cornish town" where streams gush down the gutters of the precipitous main street. This find came as manna from Heaven to us; we considered it the reward of Diligence! After searching the downs for days, without finding any Heather of sufficient distinction to be worth while collecting, we had—at the time—turned our attention towards the absorbing if somewhat material occupation of gathering mushrooms, when my wife spotted a patch of Heather of an unusual colour on the field bank, not far from where she was standing.

They gathered cuttings, wrapped them in sphagnum and sent these back to the nursery in Dorset. "We made a pilgrimage to the shrine the following year and found the original plant making fine growth, in spite of the mutilation it had received the season before" (Maxwell 1927: 61). 'Mrs D. F. Maxwell' was introduced by Maxwell & Beale Ltd, Broadstone, Dorset, in 1925, making "its debút at Holland Park Autumn Show ... where it was accorded, by unanimous vote, an Award of Merit" (Johnson 1932).

A water-colour portrait of 'Mrs D. F. Maxwell' by Winifred Walker was first published in the nursery's 1926 catalogue, and was reprinted in *The low road* (Maxwell 1927: opp. p. 62) (Cleevely 1997). AGM 1993.

UPPER LEFT: *E. multiflora*. (Painted 7 November 1986.)

UPPER MID-LEFT: *E.* × *griffithsii* **'Heaven Scent'**. Hardy, vigorous, upright heather (±1 m tall × ±0.6 m spread); foliage dark grey-green; flowers noticeably fragrant, lilac-pink (H11), in long spikes; July–December.

The name (® no. 90, as *Erica manipuliflora*), a reference to its strong perfume, was suggested by Mrs Cherry Turpin in 1990 (Jones 1997). The clone was not a recent one, although its exact origins are obscure. On 16 March 1949, a hardy heather was sent to the nursery of Maxwell & Beale Ltd (Corfe Mullen, Dorset) from the Royal Botanic Gardens, Kew. Maxwell & Beale marketed the plants from 1951 under the name *E. verticillata* (which is illegitimate name for *E. manipuliflora*, see pp. 140, 146). Four decades later, A. W. ("Bert") Jones recognised that this handsome heather was identical with a new hybrid, involving *E. manipuliflora*, that had been raised by Dr John Griffiths, and so it was re-named. (Recommended 1999 by the Heather Society; withdrawn 2011)

Erica × *griffithsii* 'Heaven Scent' was included among the heathers painted by Christabel King as a representative of *E. manipuliflora* — its re-identification means there is no portrait of *E. manipuliflora* among the original watercolours.

'Heaven Scent' is attractive to butterflies, such as red admirals (Nelson 2006b).

7. ERICA VAGANS. Cornish heath.

This is the first heather than attracted my attention and was the subject of my earliest independent research while an undergraduate at the University College of Wales, Aberystwyth (1968–1971). The perplexing colony of the Cornish heath, known since the 1930s (Praeger 1938), on a remote hillside near my home in County Fermanagh, Northern Ireland, provided me with ample material for a project on ecology (Nelson 1971) and also led to an introduction to, and subsequent collaboration with, my late friend and mentor, Paddy Coker, to whom I dedicate this monograph. Paddy, among others (Webb 1954: 215–217; Coker 1968; McClintock & Rose 1970), had also been intrigued by this population, which most unusually comprises only white-flowered plants, and which is situated around 500 kilometres from the nearest populations on the Lizard Peninsula, Cornwall. In Fermanagh, at Carrickbrawn, *Erica vagans* inhabits a base-rich flush on a gently sloping hillside that otherwise is cloaked in blanket peat, covered with *Calluna vulgaris*. The colony occupies an area approximately 40 by 70 metres (for detailed description of the site and vegetation analyses, see Nelson & Coker 1974) (see Figs 2–4, pp. 5, 6).

In Cornwall, *Erica vagans* is almost confined to areas of the Lizard Peninsula that are underlain by serpentine (Turpin 1982a, 1984). Here the flower colour varies from pure white (*E. vagans* f. *alba* (G. Sinclair) D. C. McClint.; McClintock 1984a: 191) to rich pink, although the general colour is pale lilac (Clapham *et al.* 1962). The overwhelming majority of the wild-collected cultivars of Cornish heath have come from Cornwall, including such splendid clones as 'Mrs D. F. Maxwell' (Plate 10; Fig. 74) and 'Lyonesse' (see Plate 10).

Fig. 74 (left). *Erica vagans* 'Mrs D. F. Maxwell' (E. C. Nelson).
Fig. 75 (right). *Erica vagans* in Picos de Europa, with sticky flax (*Linum viscosum* L.) (E. C. Nelson).

The Cornish heath was known to English botanists including the Revd John Ray (1627–1705), who called it "Juniper, or Firre-leaved heath, with many flowers" and reported that the heather occurred "By the way-side going from Helston to the Lezard-point in Cornwal, plentifully" (Ray 1670: 101). Ray, who had travelled on the continent, also noted that he had seen the same plant in the south of France near Narbonne and Montpellier (Ray 1670): "Eandem observavi in Gallia Narbonensi eundo à Monspelio hìnc versus Pedenatium, indè versus montem Lupi." In fact, the heather he saw was more likely to have been *Erica multiflora*, because *E. vagans* does not frequent the extreme south-western corner of France.

Erica vagans, as almost universally understood through the nineteenth and twentieth centuries, at least in Britain, is a low-growing, late-summer flowering heather native to Cornwall (as the English name indicates). Cornish heath is most abundant in the western and central Pyrenees, but it peters out eastwards. From the Pyrenees, *E. vagans* ranges westwards through the northern part of Spain into Galicia (Bayer 1993) (Fig. 75). In France, beyond its headquarters in the south-west and in the western Pyrenees and their northern foothills where it is common (Dupont 1962: 117), there are only scattered populations. Inland colonies of *E. vagans* occur along the western and northern edge of the Massif Central between the Aveyron and Allier valleys. In Brittany, Cornish heath also inhabits only isolated habitats, around Golfe du Morbihan, and on some of the off-shore islands: it is common on Belle-Île, for example. The heath is also reported from Îles Chausey, Golfe de St Malo, Normandy (Dupont 1962).

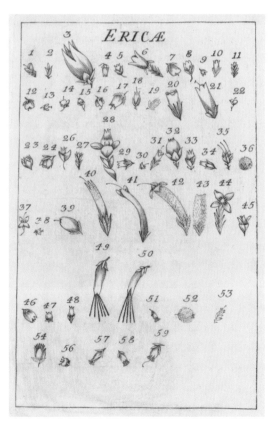

Carl Linnaeus named *Erica vagans* within a dissertation about the genus *Erica* that was defended on 19 December 1770 by Johan Adolph Dahlgren (1744–1797); the description is the sparsest possible: six words (Linnaeus 1770). There is also a tiny drawing of a flower crouching on a page among 58 others that could be a flower from *E. vagans* (Fig. 76). In *Mantissa plantarum altera*, issued in October 1771[4], Linnaeus gave a much more detailed description. He described *E. vagans* as having spreading leaves arranged in whorls of four ("*foliis quaternis patulis*"), solitary bell-shaped flowers with exserted anthers that lacked spurs, and added, "*Habitat in* Africa; *etiam* Tolosæ." Whereas "*Habitat in* Africa ..." certainly does not apply to the heather from the western parts of Europe, the one that we call Cornish heath, most of the rest of the description does indicate *E. vagans*. Although Williams (1910) pointed out that some of the traits mentioned in the *Mantissa* description "are characteristic only of the Greek plants", this can easily be explained. Linnaeus only

Fig. 76. Drawings of the flowers of species of *Erica* which accompanied the dissertation defended by Johan Adolph Dahlgren. The tiny figure numbered 58 was stated to represent *E. vagans*.

had access to a few — probably just two — specimens and with this exceedingly limited range of material, he did not distinguish between the western European and eastern Mediterranean plants.

In Linnaeus's own herbarium, now preserved in the Linnean Society of London, the specimen that is labelled *Erica vagans* is undeniably what we know as *E. manipuliflora*, a species that grows around the Adriatic and Aegean Seas and the eastern Mediterranean. However, there are clear indications in *Mantissa plantarum altera* that he also studied an important specimen in Joachim Burser's herbarium which today resides in the Botany Section of the Museum of Evolution in Linnaeus's own university at Uppsala, Sweden. In volume XXV of Burser's herbarium, on sheet number 45 (**UPS**) (Fig. 77), there are two specimens that are beyond doubt the western European *E. vagans*, and the accompanying label indicates that Burser collected the heather "Prope Tolosam GalloProvincialiae" — near Toulouse, Provence. Burser also made the comment (in Latin) that this heather agreed very much with the preceding one (which is *E. carnea*, also known as *E. herbacea*). I believe that it was Burser's suggestion (which we would not consider a correct judgement) that prompted Linnaeus to remark, in *Mantissa* (1771), that *E. vagans* was similar to *E. herbacea*.

Joachim Burser (1583–1639), a native of Saxony, was a medical doctor and keen botanist who became Professor of Botany and Medicine in Ritter-Academy, Sorø, Denmark, in 1625. He had been a pupil of Caspar Bauhin, and during the early 1600s, he travelled extensively in central and southern Europe, getting as far as the Pyrenees, where he collected many plants. Among the specimens surviving in his herbarium labelled as having come from the vicinity of Toulouse are ivy-leaved bellflower (*Wahlenbergia hederacea*), prostrate toadflax (*Linaria supina*), *Silene muscipula* and *Helichrysum serotinum*, as well as the Cornish heath.

When McClintock (1989b) proposed the conservation of the name *Erica vagans* with a new type specimen (a neotype collected in Cornwall; Jarvis 2007), he had evidently only studied the specimen in Linnaeus's own herbarium and he did not take into account the specimen in Burser's herbarium. Thus he was misled into stating categorically that "Linnaeus's *vagans* was not the plant of western Europe" (McClintock 1989b) — it was, undoubtedly, but mixed up with the other. Nor was it correct to remark that the "name *E. vagans* seems to have been (mis)used ... from very early times". There was certainly confusion between *E. vagans*, *E. manipuliflora* and *E. multiflora*, and Linnaeus's description of *E. vagans* did incorporate characters of *E. manipuliflora*.

Meanwhile, McClintock's proposal was accepted (Brummitt 1993), and *Erica vagans* remains irrevocably attached to the Cornish heath from western Europe, and *E. manipuliflora* to the whorled heath from the eastern Mediterranean.

Fig. 77. Joachim Burser's specimens of *Erica vagans*; preserved in the Museum of Evolution, Botany Section (UPS), Uppsala University, Sweden. The left-hand specimen has been mounted more or less vertically, but the sharply bent shoot-tips indicate that the branch was growing more or less horizontally. This may have given Linnaeus the idea that the heather was a creeping or wandering plant – hence *Erica vagans*.

ETHNOBOTANY

In Spain, various uses have been recorded for Cornish heath, including as kindling, for besoms, and as bedding and feed for domestic animals (Pardo de Santayana *et al.* 2002). *Erica vagans* is a valuable source of unifloral honey in France and Spain (Persano Oddo *et al.* 2004; Pardo de Santayana *et al.* 2002). If the natives of Cornwall ever used it for special purposes they have left no record of these, as far as I know.

VARIATION & CULTIVATION

The Cornish heath is one of the best of the late-flowering heaths, floriferous and colourful, and *Erica vagans* is tolerant of base-rich soils, especially those with high levels of magnesium (Small & Small 2001), so it does not need acidic conditions with a low pH. Thus, *E. vagans* is often recommended for gardens which have a neutral to slightly alkaline soil (Anderson 1965; Underhill 1971, 1990). However, plants grown in lime-rich conditions do not thrive (Jones 1977), remaining small and flowering sparsely.

The shrubs can be long-lived, "certainly thriving for well over half a century" (McClintock & Blamey 1998). They need to be carefully trimmed to keep them in good flowering condition, and if not pruned, they do tend to spread and sprawl; a plant in David McClintock's garden was 4.2 metres across after 25 years.

Propagation of cultivars by cuttings is easy. Seed may also be gathered and sown; seedlings will tend to have flowers in various shades of mauve.

White-flowered Cornish heath has been known since the early 1800s, at least — the nurseryman Conrad Loddiges of Hackney, London, had this variant for sale in 1820, and "white Cornish heath" was listed a few years later in the Duke of Bedford's heather collection at Woburn (Sinclair 1825). There is a reference to "white firleaved English heath" in Richard Weston's *Flora anglicana* (1775: 12); the name "fir-leaved heath" echoes that used by John Ray (1670). While Weston (c. 1733–1806), a Leicestershire thread-hosier and writer on matters agricultural and horticultural, employed the Latin name *Erica multiflora* [var.] *alba*, the plant referred to was surely what we know as *E. vagans*. Thus the "albino" variant, now called *E. vagans* f. *alba* (G. Sinclair) D. C. McClint., must have been found in Cornwall before 1775.

Some cultivars have yellow foliage (*E. vagans* f. *aureifolia* D. C. McClint.); 'Valerie Proudley' (Fig. 78; see also Plate 10) is the best example although not stable (McClintock 1988a: 62).

Fig. 78. Golden foliage is popular with many heather enthusiasts, providing additional colour tones especially in winter; this is *Erica vagans* 'Valerie Proudley' (E. C. Nelson).

Other teratologic variants include *Erica vagans* f. *anandra* P. G. Turpin that possesses tiny green florets composed of a duplicated calyx, an ovary and unusually long style, without a corolla and stamens. Major-General Turpin found this near Kynance Farm on the Lizard, Cornwall, in September 1977 (Turpin 1982a, 1982b, 1984; illustrated in McClintock & Blamey 1998: fig. 1). *E. vagans* f. *viridula* D. C. McClint. comprises plants in which the flowers are replaced by slender, elongated "leafy" branchlets; this was first collected in 1897 in France (Turpin 1982b). Occasionally such plants produce shoots bearing "normal" flowers: "twice I have had evidence of a whole raceme reverting to normal" (McClintock 1989c). The shoots that bear flowers do not also have the "leafy" branchlets. McClintock reported that he had seen plants on which the "normal" flowers were pink. There is one clone in cultivation under the name 'Viridiflora' (in McClintock & Blamey 1998: plate 1) (Fig. 79), and Dr John Griffiths (Leeds, Yorkshire) provided specimens for me to examine in October 2006. Griffiths's clone of 'Viridiflora' very occasionally produces very pale pink, almost white, flowers in which many of the stamens are grotesquely malformed and the corolla is usually duplicated and malformed. Turpin (1982b) reported that examples of *E. vagans* f. *viridula* were found by him on the Lizard in 1977 and 1979.

Fig. 79. *Erica vagans* f. *viridula* 'Viridiflora'; note the feathery side-shoots that replace the flowers (E. C. Nelson).

HYBRIDS

The well-known natural hybrid between *Erica vagans* and *E. tetralix*, called *E.* × *williamsii* (Williams's heath) is endemic to the Lizard Peninsula in Cornwall (see p. 311). It has also been created artificially with considerable success. David Wilson (British Columbia, Canada), using *E. tetralix* as the seed-parent, raised *E* × *williamsii* in 1985 — the reciprocal cross was not successful. Dr John Griffiths (Leeds, Yorkshire) also raised the hybrid using *E. vagans* as the pollen-parent.

One man-made hybrid involving *Erica vagans* is now a familiar garden plant — *E.* × *griffithsii* (see p. 319), raised by Griffiths, has *E. manipuliflora* as the other parent.

Griffiths obtained viable seed when he used *Erica vagans* as a pollen-parent on *E. ciliaris* and *E. terminalis*, but the seedlings soon perished; similarly, the reciprocal cross with *E. ciliaris*. Viable seed also was produced by *E. vagans* after Griffiths used *E. lusitanica* pollen. Kurt Kramer (Edewecht-Süddorf, Germany) records success when using *E. vagans* as the pollen-parent with *E. terminalis* as the seed-parent in 2002, and in 2003 with *E. spiculifolia*; in both instances the seedlings remain to be assessed.

Other species which have been tried unsuccessfully so far in artificial crossing with *Erica vagans* are *E. australis* (Griffiths), *E. carnea* (Griffiths, Kramer), *E. ciliaris* (Griffiths, Wilson), *E. cinerea* (Kramer), *E. erigena* (Griffiths, Kramer), *E. mackayana* (Griffiths), *E. multiflora* (Griffiths: non-viable seed was obtained), *E. sicula* subsp. *bocquetii* (Kramer) and *E. umbellata* (Kramer).

Erica vagans L., *Dissertationem botanicam de Erica*: 10, fig. 56 (1770); *Mantissa plantarum altera*: 230 (1771). Type: "Goonhilly Downs, Cornwall", *W. B. Turrill*, 12 July 1932: conserved neo. **K**! (McClintock 1989b; Jarvis 2007: 501).

E. multiflora Huds., *Fl. Angl., ed. 2*: 166 (1762) et auctt. plur. non L.

E. didyma J. Stokes, in Withering, *Botanical arrangement* **1**: 400 (1787, ed. 2).

E. vaga Salisb., *Prodromus stirpium in Chapel Allerton*: 294 (1796).

Gypsocallis vagans (L.) Gray, *Natural arrangement* **2**: 398 (1821).

ILLUSTRATIONS. J. Sowerby, *English botany* **1**: tab. 3 (1790); S. J. Roles, *Flora of the British Isles: illustrations* pt 2: tab. 940 (1960); S. Ross-Craig, *Drawings of British plants*, part 19: pl. 27 (1963); R. W. Butcher, *A new illustrated British Flora*: 110 (1961); J. Smythies, *Flowers of south-west Europe*: 287 (4) (Polunin & Smythies 1973); L. Denkewitz, *Heidegarten*: 185 (1987); E. Sierra Ràfols, *Flora iberica* **4**: lam. 182 (p. 501) (1993); M. Blamey, *Heathers of the Lizard*: plate 1 (McClintock & Blamey 1998).

DESCRIPTION. *Evergreen shrub* to 1 m tall, capable of regenerating form durable rootstock after fire or grazing; branches erect or decumbent; adventitious roots produced from shoots (in cuttings, rootlets emerge above the leaf nodes (Small 1974)). *Leaves* in whorls of 4 or 5, linear, ±10 mm long, <1 mm broad, glabrous except for microscopic spicules at tip, and some very short, simple hairs on petiole, flattened above, margins recurved, convex underneath, edges with vestiges of spicules (only visible under microscope), sulcus closed, linear. *Inflorescences* axillary, 1–8 flowers in each umbel on condensed lateral shoots, flowers in various shades of lavender, mauve or pink, very rarely white; numerous umbels forming a "spike"; *peduncle* ±8 mm long, glabrous, straight, slender, of equal diameter throughout; bract near base or ± ¹/₄ way up, and 2 bracteoles close to bract or distant, to ¹/₃ way up, bract ±1 mm long, bracteoles smaller, triangular-lanceolate, margins with simple hairs and some ± dendritic. *Calyx* unequally 4-lobed, ±1.3 mm long, lobes ±1 mm long, glabrous except for fringe of minute hairs on margins, sulcus open, 0.5 mm long. *Corolla* campanulate-cyathiform, ±3–5 mm long, lobes ± 1–1.5 mm long, usually erect, glabrous. *Stamens* 8, emergent; *filament* ±4 mm long, straight, slender, equal width, attached on side of ovary stipe below nectary; *anther* dark, muticous, thecae ±0.5 mm long, ±0.3 mm broad, erect, separate, diverging at apex, with large lateral pore towards apex. *Style* ± 4 mm long, erect, slender, tapering slightly; *stigma* not broader than style, ± 0.2 mm diameter, green or pink; *ovary* "pear-shaped", to 1.5 mm tall, ±1 mm diameter, distinctly stipitate; ± 8 ovules per locule, *nectary* dark surrounding ovary stipe, lobed. *Seed* 0.4–0.65 × 0.5 mm (see Nelson 2006a). n = 12 (Fraga 1983: 537; see Nelson & Oliver 2005b).

FLOWERING PERIOD. July–October.

DISTRIBUTION & ECOLOGY. Britain (Cornwall only, and there confined to the Lizard Peninsula; see Turpin 1984); Ireland (County Fermanagh only; see Nelson & Coker 1974); France (western half; Pyrenees); Spain (north only). **Maps:** Dupont (1962: 118, carte 9), Bolòs & Vigo (1995). Figs 4 and 81.

Erica vagans is a naturalised garden-escape even in Ireland and Britain (cf. Preston *et al.* 2002); for example, in Ireland it has long been known on the fixed dunes at Murlough Bay in County Down (the earliest report was in 1899, *fide* Hackney 1992), and since 1957 at Shane's Castle in County Antrim (Hackney 1992). The Cornish heath has also "jumped the garden fence" in the western USA (northern California (Ferguson 1991) and Oregon (Small 2004a)), in New Zealand (Sykes 1981), and in Madeira (McClintock 1994).

Erica vagans inhabits heathland, moors, pine and oak woodlands, and coastal dunes, usually on neutral or base-rich soils, including those overlying serpentine, ascending to 1,600 m altitude in Spain (Bayer 1993), and to 1,800 m in the Pyrenees. On the Lizard Peninsula, Cornish heath is

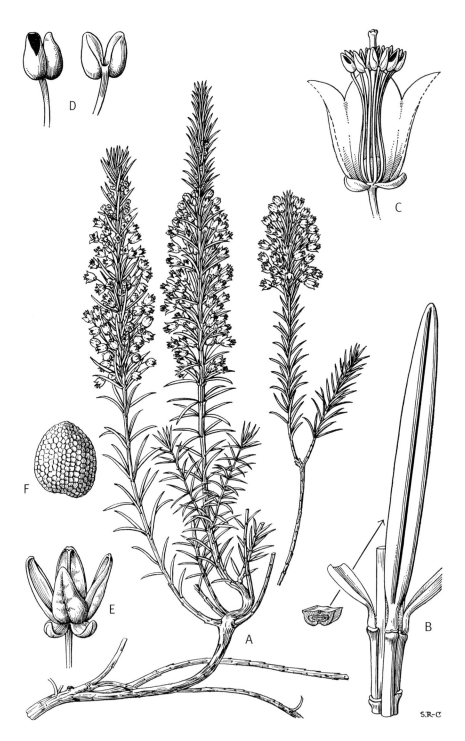

Fig. 80. *Erica vagans*. **A** part of a plant, and a flowering stem from a different one × 1; **B** stem and leaves, and section of a leaf × 8; **C** flower, part of corolla cut away × 8; **D** anther and part of filament in back and three–quarter front views × 16; **E** fruit, the persistant corolla and stamens removed × 8; F seed × 40. Drawn by Stella Ross–Craig.

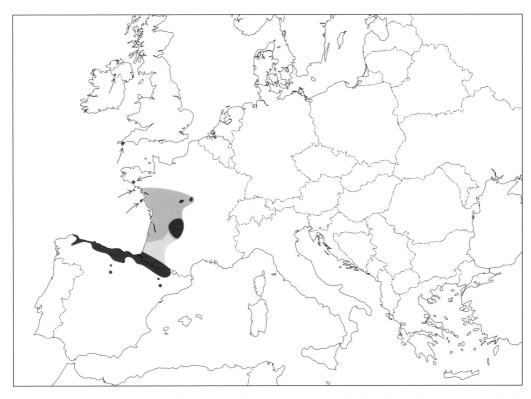

Fig. 81. Distribution of *Erica vagans*; for a more detailed distribution map of the Cornish heath's populations in Britain and Ireland see Fig. 4 (p. 6).

restricted to habitats with magnesium-rich soils of higher pH (Marrs & Proctor 1978). The heather is absent from Lizard habitats where the soil is deficient in calcium; all the Cornish sites where *E. vagans* grows were found to have soils containing substantial concentrations of exchangeable calcium (Marrs & Proctor 1978). In Ireland, the colony of *E. vagans* is restricted to a flush which is relatively rich in exchangeable calcium (Nelson & Coker 1974).

Erica vagans is classified as a member of the Oceanic Southern-temperate element of the Irish and British floras, a category also assigned to *Daboecia cantabrica*, *E. ciliaris* and *E. erigena* (Preston & Hill 1997).

ETYMOLOGY. Latin: *vagans* = creeping or wandering (see above).

The specimens that Joachim Burser collected near Toulouse around four centuries ago are very well preserved, and may explain why Linnaeus selected the epithet *vagans* for this species.

To anyone familiar with the habits of heathers, the left-hand specimen (Fig. 77, p. 134) shows clear signs that the main stem had been growing at a slant, possibly even almost horizontally, at least for a season, because the ends of the shoots — that year's new growth, in fact — are sharply angled at their bases. I suggest that this specimen, which *seems* to have been taken from a low, creeping shrub led Linnaeus to write "Caulis Empetrif fruticosus ...", meaning "with stem of crowberry or shrubby". Burser's own comment that the shrub was very like *Erica herbacea* — *E. carnea* is a low-growing, spreading heather — may have reinforced this view. Linnaeus, it should be remembered, had not seen *E. vagans* growing in its natural habitats and could have had no first-hand knowledge about its habit. Thus the idea that *E. vagans* was a creeping heather with wandering shoots may have arisen from a chance collection of a shoot that had become almost prostrate.

Vagans is the present participle of the verb *vago* (Gilbert-Carter 1950), which can mean to wander about, to ramble, to go to and fro, to roam, to range. Thus *vagans* can convey the sense that the heather is wandering, rambling, roaming or spreading. It may also mean widespread. Linnaeus might have intended the name to convey widespread, given the information he had about its distribution — "*Habitat in* Africa; *etiam* Tolosæ", it lives in Africa; also Toulouse. Sowerby (1790) commented:

> As we are assured by Dr Smith that it is undoubtedly the plant intended by Linnaeus, and which he called *vagans* (wandering), because found in so many different and remote countries, as Africa, France and other part of the south of Europe; to which we now add our own Kingdom, as a further apology for the name.

Even if Linnaeus intended the meaning as widespread, the vernacular names for *Erica vagans* in several European languages take up the theme of roaming: bruyère vagabonde, erica vagabonda, Wanderheide (*E. vagans* is not native in any Italian- or German-speaking region of Europe). "Wandering heath" has never been the vernacular name in England except on rather rare occasions such as in the epigraph of *Wandering heath* (quoted facing title page), a volume of short stories by the Cornish author "Q" (1895), *alias* Sir Arthur Quiller-Couch (1863–1944), who became Professor of English Literature at the University of Cambridge and whose legacy includes "the unsurpassed window on Cornwall and the Cornish in his fiction" (Smith 2004).

VERNACULAR NAMES. Cornish heath (English); bruyère vagabonde (French); erica vagabonda (Italian); Wanderheide, Cornwall Heide, Cornische Heide, Herumschweifende Heide (Dippel 1887; McClintock & Blamey 1998) (German); ainarra burusoila, txilar burusoila (Basque); rugón, caurioto, brezo rojo (Spanish; for other vernacular names used in Spain see Bayer (1993) and Pardo de Santayana *et al.* (2002)); fransk klockljung (Swedish).

CULTIVARS. See p. 131 and Appendix I (pp. 383–384).

8. ERICA MANIPULIFLORA. Whorled heath.

The whorled heath was first described in 1775 in *Flora ægyptiaco-arabica*, a posthumous work by the Swedish botanist Pehr Forsskål (1732–1763) who had collected the heath about Constantinople (otherwise Byzantium, modern Istanbul) (Hepper & Friis 1994; Meikle 1985: 1061). However, by the year of its publication, the binomial which Forsskål had chosen, *Erica verticillata*, had already been published by Peter Jonas Bergius (1730–1790) for, and it remains valid for, a handsome species from the Cape of Good Hope (which is now extinct in the wild but is the subject of a concerted re-introduction campaign (see Hitchcock 2003, 2006, 2007)). Although Forsskål's *E. verticillata* lingered in European literature into the twentieth century, the name is invalid under modern rules of botanical nomenclature, and the northern-hemisphere species is now named *E. manipuliflora*, a binomial published in 1802 by the English botanist Richard Anthony Salisbury (see Fig. 61, p. 94).

It is possible that the specimen in Carl Linnaeus' herbarium is a duplicate from Forsskål's collection although there is no evidence indicating that this is so.

The whorled heath was one of three *Erica* species which were observed and collected by John Sibthorp (1758–1796), and drawn by his artist-companion Ferdinand Lucas Bauer (1760–1826), during their travels through Asia Minor and Greece between May 1786 and March 1787 (Lack &

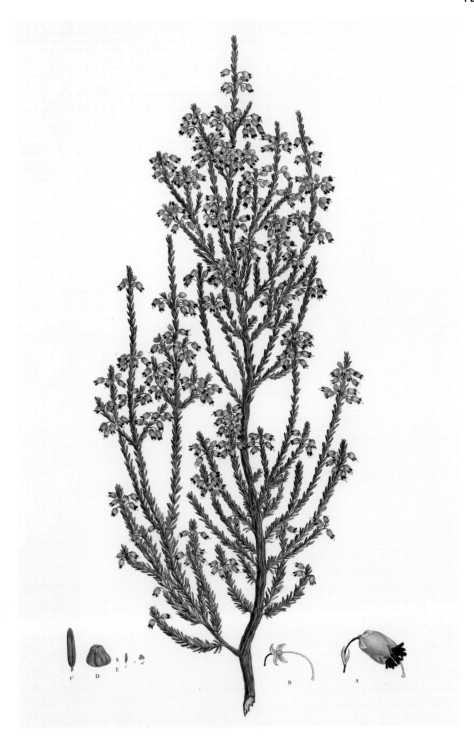

Erica manipuliflora   **Painted by Ferdinand Lucas Bauer**   **PLATE 11**

Published in J. Sibthorp & J. E. Smith, *Flora graeca sibthorpiana*, centuria 4, tab 352 (1823).

Mabberley 1999; Harris 2007; Nelson 2007g). They also visited Crete. Sibthorp's own herbarium specimens are in the Fielding-Druce Herbarium, Department of Plant Sciences, Oxford (**OXF!**), as are the original drawings which Bauer made in the field. According to *Florae graecae prodromus* (Sibthorp & Smith 1806), Sibthorp collected this species "In sylvis ad pagum *Belgrad*, prope Byzantium, et in insulâ Cretâ": Belgrad Forest, centred on the village of Belgrad, is a short distance to the west–north-west of Büyükdere (cf. Webb 1966b: 90–91, Nelson 2007g), and about 17 kilometres north of the centre of Istanbul. Thus Sibthorp, like Forsskål before him, collected specimens of *E. manipuliflora* on the European (western) side of the Bosporus. He gave specimens to friends, and from one such specimen *E. manipuliflora* was described by Salisbury (Nelson 2007g). Salisbury (1802) was explicit about the source of that specimen: "Juxta *Bujuchtar* a se lectam misit amicissimus Sibthorpe [*sic*]" – my very good friend Sibthorp sent it, collected by him near Bujuchtar. The toponym "Bujuchtar" is equated with the modern Turkish Büyükdere, as just mentioned (Webb 1966b).

Ferdinand Bauer's pencil sketches confirm that the travellers had seen *Erica manipuliflora* in Crete, a short while before they encountered it near Byzantium. One of Bauer's pencil sketches of this heather is on a folio (Ms 247 1 f. 111) that is clearly labelled "Candia", one of the names then used to denote the whole of Crete. The same folio also has on it sketches of several Cretan endemics, confirming the provenance of the heather. *E. manipuliflora* grows not far from the ancient centre of Hania (Χανιά, Canea) which was the base used by Sibthorp and Bauer — a painting by Bauer of the Venetian port of Hania survives (Lack 1996: 222; Lack & Mabberley 1999). The heather occurs today (Fig. 82) in an area that the botanists must have crossed to get from the town to Moni Tsangarólon, the monastery dedicated to the Holy Trinity, Ayía Triádha, which Bauer also sketched (see Harris 2007), and the peninsula beyond. There is an extensive, easily seen population to the south of Ayía Triádha through which the modern road passes. This particular sketch of *E. manipuliflora* was certainly a preliminary drawing for the plate that was published in *Flora græca sibthorpiana*. There is a second sketch among Bauer's drawings of a sprig and a flower of *E. manipuliflora* (Ms 247 4 f 98v) labelled "Asia" but it does not match anything on the published illustration. Thus the splendid plate of this heather (Plate 11) published in *Flora græca* (Sibthorp & Smith 1823) is principally based on Cretan material, rather than on specimens collected by Sibthorp, or sketches made by Bauer, in Turkey.

Fig. 82. *Erica manipuliflora* on Akrotiri not far from Hania, Crete; this population is close to Ayía Triádha which John Sibthorp and Ferdinand Bauer visited in the autumn of 1786 (E. C. Nelson).

Erica manipuliflora was also one of the plants that Ferdinand Bauer chose to decorate the splendid title-pages of *Flora græca*, arguably the finest botanical publication of all time. He paired a pink-blossomed sprig of whorled heath with a similar piece of white-flowered tree heather (*E. arborea*) above the oval frame for the title-page of *Centuria quarta* (Sibthorp & Smith 1823) (p. viii).

The geographical range of *Erica manipuliflora* is centred in the north-eastern quarter of the Mediterranean Sea. This species does not often stray far inland, remaining within the coastal periphery, and often within splashing distance of the sea. There are populations along the north-eastern, Black Sea coast of European Turkey. *E. manipuliflora* occurs on many Aegean islands, and on the mainland coasts of Greece and Turkey. On Cyprus there are small populations in the north, in woodland on limestone (Viney 1994a, 1994b), where it may have been known as koukoulóhorto (Holmboe 1914; Chapman 1949) although this seems to be a "book name" of quite general application and not a genuine Cypriot vernacular name for this particular heath.[5] I

Fig. 83. *Erica manipuliflora* in Crete (E. C. Nelson).

know the plant well in various localities on Crete, usually on calcareous substrates, and it is recorded on the island of Gavdos, and from the Karpathos group (Turland *et al*. 1993). The species ranges eastwards along the southern coast of Turkey, into western Syria and Lebanon. In the Adriatic region, again the whorled heath inhabits many of the Greek islands and adjacent mainland; I have seen *E. manipuliflora* on Paxos[6] around the port of Lakka. From the Ionian islands, the heather ranges north through Albania into Croatia, reaching its northern limit around Split (although there is an isolated record from Otok Molat, an island about 200 kilometres to the northwest: Domac 1963[7]). There are also populations in two localities in Apulia on the western side of the Adriatic, on the "heel" of Italy (Medagli & Ruggiero 2002).

Erica manipuliflora is a variable plant. In the course of a short walk on the Greek Aegean island of Skiathos, I collected a range of specimens including one that superficially resembled *E. scoparia* (and, being misled by its appearance, I originally labelled it with that name! **WSY**0077232). Some plants have white shoots, others have brown shoots. Some plants have a pubescent ovary (a character not noted in standard floras); in others the ovary is entirely glabrous. As Stevens (1978) stated, some plants have spreading foliage, while in others the leaves are erect and appressed to the stem. The variation seems to be random; there is no clear geographic trend. Thus I am not inclined to follow McClintock (1992a) who designated a subspecies, *E. manipuliflora* subsp. *anthura* (Link) D. C. McClint., "because it grows in areas away from typical *E. manipuliflora*. It is so distinct in its typical guise ..." (McClintock 1992a). The second-hand description in German

Fig. 84. *Erica manipuliflora* growing on coastal cliffs composed of conglomerate, near Loutro, Chora Sfakion, Crete. In this locality the plants are all tightly pruned, most probably by goats (E. C. Nelson).

(Anonymous 1845) contained within a brief report of a contribution made by Johann Heinrich Friedrich Link (1767–1851) at a meeting of the Gesellschaft Naturforschender Freunde zu Berlin on 17 December 1845 is quite vague, and there is apparently no extant specimen that can act as a type. The report mentioned a beautiful new *Erica* species, related to *E. vagans*, *E. multiflora* and *E. erigena*, that had been collected in the autumn near Split in Dalmatia (Anonymous 1845), which, as just noted, is at the northern extremity of the range of *E. manipuliflora*. To base a taxon on such insubstantial evidence is unwise, although it is certainly proper to treat *E. anthura* as a synonym of *E. manipuliflora*.

HYBRIDS

No natural hybrids are known, but *Erica* × *griffithsii* (see p. 319) is a chance hybrid between *E. manipuliflora* and *E. vagans*, which first occurred in Britain before 1939 perhaps in the Royal Botanic Gardens, Kew, and later (in the 1980s) was synthesised by Dr John Griffiths (Leeds, Yorkshire). Dr Griffiths's other success with *E. manipuliflora* as a parent was against *E. tetralix* — the resulting hybrid, also now a familiar garden plant, is named *E.* × *garforthensis* (see p. 334). Both David Wilson (British Columbia, Canada) and Kurt Kramer (Edewecht-Süddorf, Germany) have also synthesised this hybrid.

Other crossings have not yielded results: with *Erica australis* (Griffiths), *E. ciliaris* (Griffiths), *E. cinerea* (Kramer), *E. erigena* (Griffiths, Kramer), *E. mackayana* (Griffiths), *E. multiflora* (Griffiths, Kramer) and *E. terminalis* (Griffiths, Kramer).

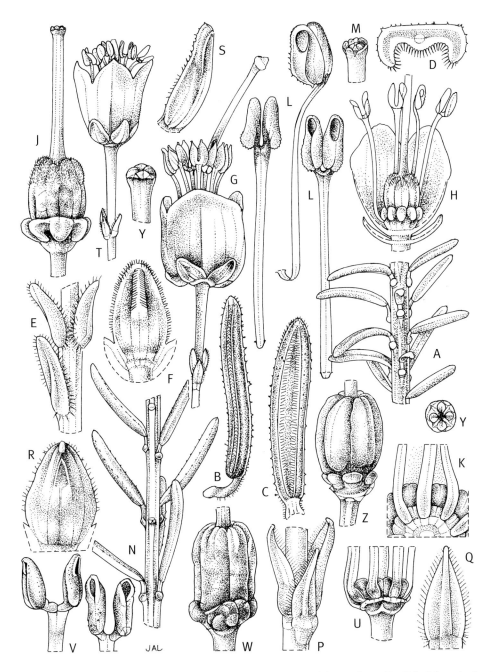

Fig. 85. *Erica manipuliflora* 'Aldeburgh'. **A** leafy shoot × 3.5; **B** leaf from side × 10; **C** leaf from back × 10; **D** transverse section of leaf × 25; **E** bract and bracteoles × 22; **F** calyx lobe × 22; **G** flower from side × 9; **H** longitudinal section of flower × 8; **J** gynoecium with nectary × 9; **K** nectary lobes and filament bases × 12; **L** stamens from back, side and front × 18; **M** style apex (stigma) from side × 16. *E. manipuliflora* (ex Dalmatia). **N** leafy shoot × 4.5; **P** bract and bracteoles × 24; **Q** bract × 22; **R** calyx lobe × 21; **S** calyx lobe from side × 21; **T** flower from side × 6; **U** nectary lobes and filament bases × 9; **V** anthers × 16; **W** ovary × 10; **Y** style apex (stigma) from side and above × 18; **Z** capsule × 14; from specimens grown by Allen Hall. Drawn by Joanna Langhorne.

Erica manipuliflora Salisb., *Trans. Linn. Soc. London* 6: 344 (1802) *nom. cons.* Type: "Juxta *Bujuchtar* a se lectam misit amicissimus Sibthorpe": neo. **OXF!** (Nelson 2007g).
E. verticillata Forsskål, *Flora ægyptiaco-arabica*, 210 (1775) non P. J. Bergius (1767).
E. forskalii Vitman, *Summa plantarum* **2**: 414 (1789) *nom. rej.*
E. anthura Link, *Flora* **23** (1): 190 (1845).
E. naematodes Link ex Nyman, *Bot.Not.* nos 8 & 9: 133 (1851).
E. vagans var. *anthura* Link ex Nyman, *Bot. Not.* nos 8 & 9: 133 (1851).
E. manipuliflora subsp. *anthura* (Link) D. C. McClint. *Yearbook of The Heather Society* **3** (10): 16 (1992a).

ILLUSTRATION. F. L. Bauer, *Flora graeca sibthorpiana*, centuria **4**: title-page & tab. 352 (Sibthorp & Smith 1823).

DESCRIPTION. *Evergreen shrub*, erect or spreading or prostrate (especially at high altitude), to 4 m tall; regenerating from lignotuber. *Branches* erect, arching or prostrate; shoots silvery white to brown, very minutely pubescent, sometimes becoming glabrous, with infrafoliar ridges. *Leaves* in whorls of 3 or 4, green, leathery, erect (sometimes appressed to stem) or spreading, long and linear-lanceolate to short and almost elliptical, ±3–9 mm long, to 1 mm broad, flattened above, appearing glabrous but often with spicules at apex and sometimes with numerous, minute, forward-pointing hairs especially near petiole; edges with irregular transparent cartilaginous border and forward pointing spicules; margins revolute, usually convex, contiguous underneath concealing abaxial surface, sulcus linear, fringed with minute hairs; petiole flattened above, convex on underside, margins and adaxial side with dense short hairs. *Inflorescence* composed of several to many axillary umbels: flowers 1–5 in axillary umbels on very short shoots, in varying shades of mauve or pink, or rarely white (see *E. manipuliflora* f. *albiflora* below, p. 147); peduncle to 10 mm long, glabrous, slender, erect or arching; bract and 2 bracteoles, lanceolate, ±0.7 mm long, margins with fringed of simple hairs, position varies but usually in lower part of peduncle. *Calyx* comprising 4 entirely free sepals overlapping at base, or 4-lobed with unequal lobes fused at base and calyx truncate at base; ±1 mm long, glabrous, lobe/sepal margins with fringe of minute hairs, tip swollen with short sulcus. *Corolla* campanulate to cyathiform, ± 3.5–4.5 mm long, ± 3.5 mm diameter, glabrous, lobes ±1–1.5 mm long, to 2 mm across at base, erect. *Stamens* 8, emergent; *filament* ± 3.5–4.5 mm long, mainly white, sometimes tinged red at apex below anther, glabrous, erect, straight, hooked at base and attached underneath the nectary, with a slight sigmoid bend at apex; *anther* ± 0.8–1 mm long, ±0.3–0.5 mm wide, muticous, *thecae* separate, diverging towards apices, not fused but joined by filament apex, papillose; pore elliptical, lateral, towards apex of theca. *Ovary* ±1–1.5 mm tall, 4-locular, ±21 ovules per locule; glabrous or with sparse, minute hairs towards apex; lobed nectary encircling base; *style* ±5.5 mm long, erect, glabrous; *stigma* capitellate, slightly broader than apex of style, ±0.3 mm across.

FLOWERING PERIOD. August–November.

DISTRIBUTION & ECOLOGY. Europe: Italy (Apulia), southern Croatia, Montenegro, Albania, Greece (including Crete, and the Ionian and Aegean islands), European Turkey. Asia: Turkey, northern Cyprus (Meikle 1985; Viney 1994a), Syria, Lebanon (Post & Dinsmore 1923). **Maps:** Hansen (1950: 20 as *E. verticillata*), Browicz (1983), Viney (1994a: Cyprus only), Medagli & Ruggiero (2002: Italy only). Fig. 86.

Erica manipuliflora grows on limestone, conglomerate, serpentine and schistose rocks, and on sandstone and chalky rocks (Post & Dinsmore 1923), in pine forest as an understorey, and in phrygana, maquis and garrigue, ascending to 2,000 m in Crete (Turland *et al.* 1993) and to 2,200 m in Sandras Dağ, south-western Turkey (McClintock1991a). At high altitudes in Crete and Turkey

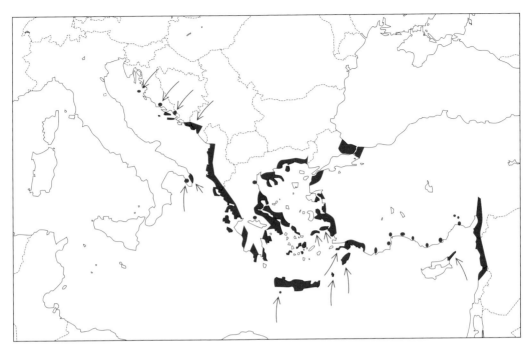

Fig. 86. Distribution of *Erica manipuliflora*.

the shrubs are prostrate, but they can also be spreading to prostrate at low altitude.

A report (Pignatti 1982) of *Erica manipuliflora* from several localities in Sicily arose from mis-identification of *E. multiflora*; although the error was corrected many years ago, several Italian Internet sites (accessed February 2007) perpetuate it.

ETYMOLOGY. Latin: *manipulus* = bundle, handful; *flos* = flower.

VERNACULAR NAMES. whorled heath (English) — derived from the rejected binomial *Erica verticillata* Forsk.; erica pugliese (Italian); χαμορείκι[6] (hamoríki), κουκουλόχορτα[5] (koukoulóhorto) (Greek); Büschelheide (German).

VARIANTS AND CULTIVARS

The whorled heath is not common in cultivation, although it not as uncommon as it used to be. Until the 1980s there was only one cultivar available commercially, namely the hardy clone 'Aldeburgh' (see Appendix I, p. 380). With the single exception of 'Olivia Hall' (see p. 148), which came from Antalya in southern Turkey, all the other named clones were obtained from the eastern coasts of the Adriatic: 'Cascades' came from Croatia (then Yugoslavia) in 1988; 'Corfu' (® no. 67) and 'Don Richards' (® no. 66) were collected on Corfu by Don Richards about 1972; A. G. Small obtained 'Ian Cooper' (® no. 45) and 'Korčula' on Otok Korčula, Croatia (then Yugoslavia), during 1978. No Aegean clones are recorded in cultivation.

While Stevens (1978) described that the corolla of *Erica manipuliflora* was "white or pink", which could be taken as implying parity between these variants in Turkish and Aegean populations, white-flowered plants seem to be very rare (cf. McClintock 1984a). I do not recall seeing any shrubs of the whorled heath with white flowers in populations that I have seen in Greece and Turkey.

'Olivia Hall' has extremely pale pink, sometimes almost white flowers (Joyner 2007), and its dark anthers indicate that it is capable of producing anthocyanin. Despite this, Mrs Hall's pressed specimen (**WSY**0096182) was designated by McClintock (1988c) as the holotype of *E. manipuliflora* f. *albiflora* (Miller & Nelson 2008). The recently-named, wild-collected clone 'Bert Jones' is similar, with almost white flowers and dark anthers (Nelson 2011a)

Hall (2002) reported that stems and shoot tips of both 'Korčula' and 'Aldeburgh' were damaged by periods of low air temperatures in central England during the winter of 2000–2001.

9. ERICA MULTIFLORA. Many-flowered heath.

This lovely heather is little grown in gardens in northern Europe and North America because it has a reputation for being frost-sensitive and difficult to propagate. In it natural habitats, *Erica multiflora* develops a woody, underground rootstock (lignotuber) from which it will regenerate after fire (Fig. 87) has destroyed the above-ground stems. The young flowering shoots that result can be spectacularly laden with blossom (Fig. 88), suggesting that occasional hard pruning of cultivated plants would be beneficial in stimulating flowers. Old stems are never so profuse with blossom.

In the wild, plants can be found that are both floriferous and colourful ranging from rich pink to pure white[8] (McClintock 1984a: 191). Other variants have been described and named, mainly at varietal level, but none of these appears to warrant retention; they were differentiated using such variable characters as corolla size, shape and colour, the length of the pedicels, and the length and width of the leaves. References in the early nineteenth century to double-flowered plants — presumably bearing flowers with duplicated corollas — cannot be verified.

Fig. 87. *Erica multiflora* regenerates very rapidly after its shoots, but not its underground lignotuber, were destroyed by a fire about six months earlier; northern Mallorca, Balearic Islands (E. C. Nelson).

Fig. 88 (left). Regenerating shoots of *Erica multiflora* tend to be very floriferous for a few years. Here, the heather is growing with carob (*Ceratonia siliqua* L.) which also regenerates vigorously after a fire: Mallorca, Balearic Islands (E. C. Nelson).

Fig. 89 (right). *Erica multiflora* in the wild; note the long, slender peduncles: Mallorca, Balearic Islands (E. C. Nelson).

Erica multiflora has an extraordinary distribution pattern. Populations occur on the northern periphery of the Mediterranean Sea, in eastern Spain, southern France and Italy, as well as on the Dalmatian coast (Mullaj and Tan 2010; Tan *et al.* 2010). It occurs also on the Balearic islands, on Sardinia, Sicily and Malta. On Corsica, *E. multiflora* is restricted to the northern extremity, Cap Corse. *E. multiflora* ranges eastwards along the north African coast from Morocco into north-western Tunisia. Then, skipping out the next thousand or so kilometres, the heather reappears in northern Libya, in Cyrenaica, an area which Brullo & Guglielmo (2001) characterise as "an island surrounded in the north by the Mediterranean Sea and in the south by the Libyan Desert". Many species occur there which are otherwise absent from north-eastern Africa, including *E. sicula*, one of the Pentaperan heaths (see pp. 290–300), and the endemic Libyan strawberry tree, *Arbutus pavarii* Pamp.

Nowhere does its range overlap that of *Erica vagans*, although the two species do occur in moderately close proximity in south-western France. There is overlap with *E. manipuliflora* in southern Italy (Medagli & Ruggiero 2002) and in the coastal regions of southern Croatia, Albania (see Mullaj and Tan 2010) and north-western Greece (see Tan *et al.* 2010).[9]

As noted above (p. 129), in the treatment of European *Erica* species published in *Flora europaea* (Webb & Rix 1972), *E. multiflora* is placed in the cluster of three species possessing "great morphological similarity", with *E. vagans* and *E. manipuliflora*. The same species were included

Fig. 90. *Erica multiflora* in the wild; Mallorca, Balearic Islands (E. C. Nelson).

Fig. 91. Mature plants of *Erica multiflora* in scrubby vegetation on limestone; the plants are in flower but they also have many withered, rust-coloured flowers: Mallorca, Balearic Islands (E. C. Nelson).

in *Gypsocallis* by the Don brothers (D. Don 1834; G. Don 1834), and in section *Gypsocallis* by George Bentham (1839). To Salisbury, the Dons and others, *E. multiflora* had more in common with *E. carnea* and *E. erigena*. Indeed, as already recorded (p. 119), Linnaeus gave *E. carnea* in one of its guises the name "*multiflora mediterranea*". DNA studies which are not underpinned by voucher specimens and so cannot be checked, are equivocal about *E. multiflora*, placing it close to *E. terminalis* and *E. manipuliflora* (McGuire & Kron 2005). Certainly *E. multiflora* differs from *E. carnea* and *E. erigena* in the morphology of its stamens, more than it differs from *E. vagans* and *E. manipuliflora*.

Formally described and named by Linnaeus (1753), *Erica multiflora* was known to earlier botanists and was one of the heathers included by Bauhin & Cherler (1650: 356) in *Historia plantarum universalis*; the accompanying woodcut, a poor representation of the species, was chosen as the lectotype of the species (Jarvis & McClintock 1990).

ETHNOBOTANY

Erica multiflora is a valuable source of honey in Spain (like *E. vagans*) and Croatia (Persano Oddo *et al.* 2004). In Spain it also was used to make besoms, and as kindling "although the flowers, when burnt, produce a characteristic sparking, said to send people to sleep" (Pardo de Santayana *et al.* 2002)!

HYBRIDS

No natural or man-made hybrids involving *Erica multiflora* have been confirmed, although possible hybrids have been raised. Seedlings from *E. terminalis* pollinated by Kurt Kramer (Edewecht-Süddorf, Germany) with *E. multiflora* remain alive. Dr John Griffiths (Leeds, Yorkshire) raised seedlings using *E. tetralix* as the seed-parent, but they were lost before it could be established if they were of hybrid origin. Other attempts, each one unsuccessful, have been made artificially to cross this species with *E. manipuliflora* (Griffiths, Kramer) and *E. spiculifolia* (Kramer) — in each instance *E. multiflora* was the pollen source. Griffiths did obtain non-viable seeds from *E. multiflora* when he used pollen from *E. manipuliflora* and from *E. vagans*.

Erica multiflora L., *Species plantarum* 354 (1753). Type: icon. "Erica foliis coros multiflora": J. Bauhin & J. H. Cherler (1650: 356)! (Jarvis & McClintock 1990; Jarvis 2007: 500).

ILLUSTRATIONS. H. C. Andrews, *Coloured engravings of heaths* **2**: tab. 111 (1801); H. C. Andrews, *The heathery* **3**: tab. 175 (1807); J. Giroux, *Erica multiflora ...* (1934); J. Smythies, *Flowers of south-west Europe*: 287 (1) (Polunin & Smythies 1973); E. Sierra Ràfols, *Flora iberica* **4**: lam. 182 (p. 501) (1993).

DESCRIPTION. *Evergreen shrub*, erect, to ±0.3–1.5(–2.5) m, regenerating from lignotuber. *Branches* upright, rigid, sparingly branched. *Leaves* thick and leathery, spreading, in whorls of (3–)4 or 5, linear, ±10 (–15) mm long, 1–1.5 mm broad, often slightly curved and so sickle-shaped, glabrous but with spicules along edge, acute; margins revolute, closely contiguous and parallel, very densely tomentose, sulcus narrow; petiole ±1 mm long, ±1 mm broad, flattened, tomentose on upper surface, glabrous underneath. *Inflorescence* variable, white to pink; flowers in axillary clusters of (1–)4, often clustered towards the tips of the shoots, but also lower down; peduncle ±10 mm long, ±3 mm in diameter, terete, glabrous; bract and 2 bracteoles close together, about ¼ way up from base of petiole, approximately the same size and shape, ±1 mm long, translucent, broadly triangular-lanceolate, fringed with very short hairs. *Calyx* 4-lobed, fleshy and waxy, white or pink; lobes unequal, 1–2 mm long, obovate, acute, glabrous, tips

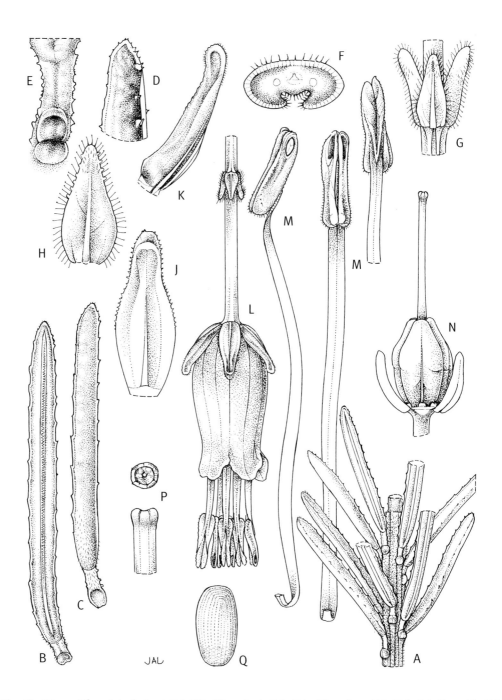

Fig. 92. *Erica multiflora.* **A** leafy shoot × 5; **B** leaf from back × 6; **C** leaf from front × 6; **D** leaf tip from side × 24; **E** leaf base × 22; **F** transverse section of leaf × 27; **G** bract and bracteoles × 22; **H** bracteole × 27; **J** sepal × 22; **K** sepal from side × 22; **L** flower from side × 9; **M** stamens from side, front and back × 18; **N** gynoecium with two sepals × 6; **P** style apex (stigma) from side and above × 10; **Q** seed × 52; from specimens grown by Allen Hall. Drawn by Joanna Langhorne.

coloured cream or red, with very short, inconspicuous sulcus, margins not hirsute; base truncate, ±1 mm across, much broader than top of peduncle. *Corolla* 4-lobed, broadly ovoid, lobes erect (or patent). *Stamens* emergent, 8; *filament* ±3.5 mm long, <0.2 mm broad, flat, straight, linear and not altering in width; *anther* dark brown, fixed laterally a little above base, muticous, *thecae* ±1.5 mm long, ±0.5 mm across, papillose, lobes erect, remaining parallel, fused towards base, pore lateral, oval, ±0.5 mm long, near apex of theca. *Ovary* barrel-shaped, green, glabrous, apex emarginate; *style* straight, terete, tapering slightly, glabrous; *stigma* not expanded; *nectary* at base, nectar profuse. *Seed* ±0.5–0.7 mm, ovate.

FLOWERING PERIOD. August–January (–March); Bayer (1993) indicated that plants can occasionally be in flower through the year.

DISTRIBUTION & ECOLOGY. Europe: eastern Spain (Murcia to Giron), Balearic Islands (Bayer 1993); southern France (including northern tip of Corsica) (Giroux 1934; Aubert 1967); Italy (including Lampedusa, Sardinia and Sicily); Malta and Gozo; southern coastal Croatia; Albania; north-west Greece. North Africa: Morocco (Ball 1877; Fennane *et al.* 1999; Fennane & Ibn Tattou 2005); Algeria (Quézel & Santa 1963); Tunisia; Libya (Siddiqi 1978). **Maps**: Hansen (1950: 20), Aubert (1967: 10), Bolòs & Vigo (1995: 36). Fig. 93.

Erica multiflora "est une des rares Ericacées véritablement calcicoles" (Giroux 1934: 39): it occurs in rocky places and on cliffs, to about 550 m altitude in France (Aubert 1967) and to 900 m altitude in Spain (Bayer & López González 1989), in woods, scrub and thickets, usually on calcareous soils (Bayer & López González 1989; Webb & Rix 1972).

ETYMOLOGY. Latin, *multiflorus* = many-flowered.

VERNACULAR NAMES. Many-flowered heath (English); bruyère à nombreuses fleurs (French); leħjet ix-xiħ (Maltese); bruc d'hivern, brezo, bruguera, sepell, petorra (for other names used in Spain see Bayer 1993; Pardo de Santayana *et al.* 2002); cipell, xipell (Balearics); queiró, queiroga, torga (Portuguese); Vielblütige Heide (German).

CULTIVARS. See Appendix I (p. 380).

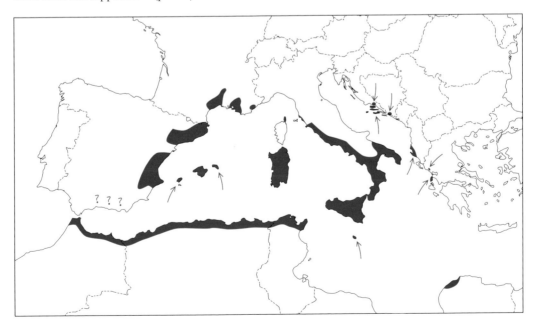

Fig. 93. Distribution of *Erica multiflora*.

10. ERICA UMBELLATA. Dwarf Spanish heath.

Erica umbellata grew for a decade and more in a old ceramic sink, out of doors in The Fens on the Cambridgeshire/Norfolk border, and it flourished without anything but rudimentary care and with frequent neglect (Nelson 1999b) (Fig. 95). The plant was the original of 'David Small', distinguished by its richly coloured (vivid amethyst or heliotrope) flowers. Every May it blossomed profusely: "... *Erica umbellata* is a species which links up in flower with the April and mid-June flowering varieties" (Slinger 2007). Allen Hall grew plants outdoors in south-eastern England and found the species "perfectly hardy" (Hall 1999: 20): pink- and white-flowered plants "gave splendid displays of colour in April and May, and plenty of seed later on". *E. umbellata* survived "without a blemish" the very severe winter of 1962–1963 in north-western England (Chapple 1965). Dr Keith Lamb grew the same heather in his garden in central Ireland, and he also reported it was easy and undemanding, and hardy; late-spring frosts in 1998 which damaged other plants (such as *Bergenia* (Saxifragaceae)) did not affect *E. umbellata* (Lamb 1999). An additional benefit is that this species of heather is apparently lime-tolerant (Lamb 1999). I am sure that the secret of success is excellent drainage, a suggestion seemingly confirmed by Hall (1999); when grown in water-retentive soils *E. umbellata* is not so tolerant of cold. Our ceramic sink was never watered even in summer. Dr Lamb grew his plant in a rockery. Yet *E. umbellata* is almost unknown to gardeners despite having been introduced to cultivation in Britain in 1787.

Richard Salisbury (1796) grew this heather and decided he did not like Linnaeus' name for it, so he published *Erica lentiformis*: this was Salisbury's common practice when he thought a name was somehow inappropriate.[10] Remarkably, that is the only synonym for *E. umbellata*! Most other European heaths have been encumbered with many more, by Salisbury and others.

In the wild there is noticeable variation in the size and colour of the corolla; plants with pure white flowers (*E. umbellata* f. *albiflora* D. C. McClint.) have been recorded since the early nineteenth century (McClintock 1984a: 191) and there is one clone in cultivation, 'Anne Small' (see p. 382). Sometimes plants untypically possess short, included stamens; this variant was collected in 1848 by Friedrich Martin Josef Welwitsch (1806–1872) at Villa Nova de Milfontes, Alentejo, Portugal. Not infrequently, plants with anther-less stamens are also encountered — whether these are functionally female and fertile, capable of producing viable seeds, is not known, but they cannot produce pollen. Johan Martin Christian Lange (1818–1898) described and named the anther-less plant as *E. umbellata* var. *anandra*[11] (Lange 1863).

It is very doubtful whether there is merit in retaining *Erica umbellata* subsp. *major* (Cosson & Bourg) Pinto de Silva & Teles (1971: 229–230), described simply as having a larger corolla and less-exserted stamens, given the substantial variation that is evident throughout the species's range. However, Portuguese botanists treat it as an endemic variety of significance in the context of the flora of the Algarve (Costa *et al.* 1998).

With its spurless (muticus), emergent anthers, *Erica umbellata* may be related to *E. multiflora*, and perhaps to *E. vagans* and *E. manipuliflora*, but it differs from those species in having a strictly terminal, umbellate inflorescences on the main branches and leafy lateral branches, and more recaulescent bract and bracteoles adjacent to the calyx. Bentham (1839), following Salisbury's unpublished scheme (see Appendix II, p. 387), placed *E. umbellata* in section *Pyronium* by itself, whereas David Don (1834), and his brother (G. Don 1834) included it in *Gypsocallis* which, as noted previously, also contained *E. multiflora*, *E. vagans*, *E. manipuliflora*, as well as *E. carnea* and *E. erigena*. Until all these species have

Erica umbellata   **Painted by Franz Andreas Bauer**   PLATE 12

Published in F. A. Bauer, *Delineations of exotick plants cultivated in the Royal garden at Kew*, tab. 5 (1796).

Fig. 94. *Erica umbellata* near Ronda, southern Spain (© M. Rix; reproduced by courtesy).

been subjected to rigorous DNA analysis supported by voucher herbarium specimens, speculating on the position of any of them is probably futile.

The first person to record *Erica umbellata*, common on untilled ground beyond Lisbon and flowering in December, was the Flemish botanist Charles de l'Écluse (Carolus Clusius) during his stay in the Iberian Peninsula from the autumn of 1564 into the spring of 1565. December seems a little early for *E. umbellata* to be in bloom — *Flora iberica* (Bayer 1993) records the flowering period as February to July (April to July according to Bayer & López González (1989)). During 1982 we saw the species in full bloom towards the end of July, perhaps a little past its prime, at Cabo Villano, northwest of Santiago de Compostela in Galicia. Be that as it may, a water-colour painting made for l'Écluse, the basis for the woodcut published in *Rariorum aliquot stirpium per Hispanias observatarum historia* (l'Écluse 1576)[12], is among the portraits of heathers in *Libri picturati* preserved in Krakow (see p. 166), and Ramón-Laca & Morales (2000) are confident of the identification.

Erica umbellata is recorded as having been introduced into English gardens in 1787 by Messrs Lee & Kennedy, Vineyard Nursery, Hammersmith. A short time afterwards, Franz Andreas Bauer, "Botanick Painter to His Majesty", a consummate botanical artist like his younger brother Ferdinand, Sibthorp's companion, prepared a portrait of the heather (Plate 12), using a plant grown at Kew, for *Delineations of exotick plants* (Bauer 1796: tab. 5); the date 1 January 1793 is engraved on the published plate.

Around this time, Henry Andrews also painted *Erica umbellata* for his book *Coloured engravings of heaths* (tab. 143, 1804).

ETHNOBOTANY

Erica umbellata is a valuable source of unifloral honey in Spain (Persano Oddo *et al.* 2004), and the stumps have been used to make charcoal (Pardo de Santayana *et al.* 2002).

CULTIVATION

In 1926, when shown at the Royal Horticultural Society in London, *Erica umbellata* received an Award of Merit, but the species has always been scarce in British and Irish gardens, and very few nurserymen propagate it. Somehow this species has gained an undeserved reputation for not being "sufficiently hardy to recommend out of hand" (Slinger 2007)": even so, Leslie Slinger admitted that "where it does grow well it is a helpful little shrub."

Propagation is said to be easier from seed (Hall 1999) than from cuttings, but the named clones can only be propagated by taking cuttings. Seed gathered from garden plants in late summer (Sellers 1999: 25) germinates readily. Hall (1999: 20) noted that all the seedlings he raised from both white- and pink-flowered parents had pink flowers.

Experiments conducted in north-western Spain suggest that the origin of *Erica umbellata* clones may have some effect on their capacity to root, plants from coastal populations rooting better than inland plants (Iglesias-Díaz & González-Abuín 2004). 'David Small' derives from a coastal habitat; it was among plants collected at Cabo Villano, A Coruña, on 21 July 1982.

In the wild *Erica umbellata* is confined to siliceous, acidic habitats (Bayer 1993), but, as noted, in cultivation it is regarded as lime-tolerant (Lamb 1999).

Fig. 95. *Erica umbellata* 'David Small'; cultivated at Outwell, Norfolk (E. C. Nelson).

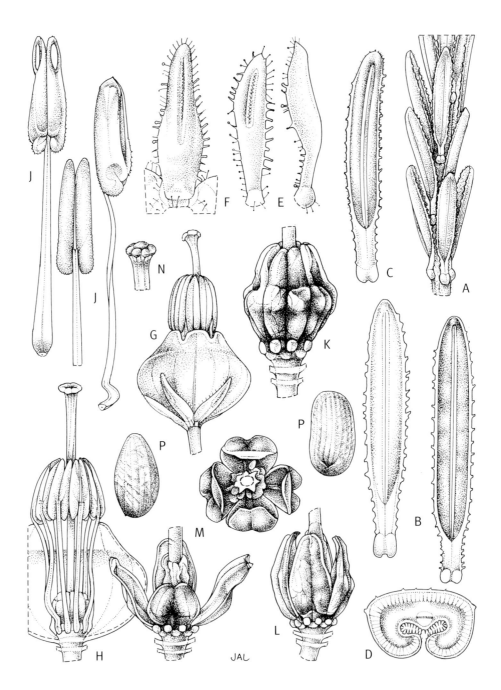

Fig. 96. *Erica umbellata* 'Anne Small'. **A** leafy shoot × 7; **B** leaf from back and front × 17; **C** young leaf from back × 17; **D** transverse section of leaf × 48; **E** bracteoles from back and side × 14; **F** sepal × 25; **G** flower side view × 8; **H** stamens and gynoecium × 12; **J** stamens from front, back and side × 18; **K** ovary × 18; **L** dehisced capsule × 9; **M** dehisced capsule from side and above × 8; **N** style apex × 28, **P** seed from front and side × 40; from specimens grown in RHS Garden Wisley, no. W 2001 3921. Drawn by Joanna Langhorne.

HYBRIDS

There are no records of natural hybrids involving *Erica umbellata*, nor has it yielded any through artificial cross-pollination. Both Dr John Griffiths (Leeds, Yorkshire) and Kurt Kramer (Edewecht-Süddorf, Germany) have used this species as a pollen- and as a seed-parent. Griffiths did get viable seeds from several of the crosses, but the seedlings were always identical with the seed-parents which were *E. ciliaris*, *E. cinerea* and *E. tetralix*. The same occurred when he placed *E. carnea* pollen on *E. umbellata*; the reciprocal cross was also a failure. Other failed couplings were with *E. erigena* and *E. lusitanica*, both employed as pollen-parents. Kramer's crosses all failed: with *E. arborea*, *E. carnea*, *E. spiculifolia* and *E. vagans* using *E. umbellata* pollen, and with *E. arborea*, *E. carnea*, *E. erigena* and *E. lusitanica* as the pollen-parents.

Erica umbellata L., *Species plantarum*: 352 (1753). Type: "*Habitat in* Lusitania", *P. Löfling*: lecto.: **LINN** 498.97! selected by Ross (1967) (see Jarvis & McClintock 1990; Jarvis 2007: 501). *E. lentiformis* Salisb. *Prodr. stirpium Chapel Allerton*: 294 (1796).

ILLUSTRATIONS. F. A. Bauer, *Delineations of exotick plants*, tab. 5 (1796); H. C. Andrews, *Coloured engravings of heaths* **2**: tab. 143 (1804); H. C. Andrews, *The heathery* **2**: tab. 99 (1805); J. Smythies, *Flowers of south-west Europe*: 287 (11) (Polunin & Smythies 1973); J. Tyler, *Yearbook of The Heather Society* **2001**: iv.

DESCRIPTION. *Evergreen shrub*, to 0.5–0.6 m in height, rarely to 1 m, bushy, without lignotuber; roots capable of extending as deep as 2 m (Silva & Rego 2004); shoots upright, slender, hirsute especially when young with very short, often gland-tipped hairs. *Leaves* in whorls of 3, small, ±2–4(–5) mm long, ±0.5 mm broad, linear, green, ± flat above, convex and glabrous below, slightly curved, ascending or (like those immediately below the umbels) erect and appressed to stems, edges hirsute with simple, sometimes gland-tipped hairs becoming ±glabrous; margins recurved, contiguous underneath, sulcus linear, closed. *Inflorescence* a terminal umbel of (1–)3–6 florets; *pedicel* slender, erect, pubescent with short erect hairs, occasionally gland-tipped; *bract* and 2 *bracteoles* recaulescent, close to the calyx; bract ± 1.5 mm long, ±0.3 mm, bracteoles smaller, leaf-like, green, linear, margins with few gland-tipped hairs, sulcus linear, extending almost to base. *Calyx* of 4, ± separate, unequal, green sepals, resembling bracts, with simple, non-glandular, and gland-tipped hairs on margins, linear , ±2 mm long, ± 0.5 mm across, sulcus closed, linear, extending from tip almost the entire length of each lobe. *Corolla* ±3–5(–7) mm long, ± 3.5 mm in diameter, glabrous, turbinate to urceolate, lobes erect, concave (so not reflexing). *Stamens* 8, emergent; *filaments* ±3.5 mm long, flat, white, erect, glabrous; *anther* ±1.3 mm long, muticous, very dark (more or less black), *thecae* ±elliptical in lateral view, ±0.4 mm across, erect, parallel, surface papillose. *Style* emergent, ±5 mm long, slender, terete, glabrous; *stigma* capitellate, broader than apex of style; *ovary* ±1 mm tall, ±0.7 mm diameter, cylindrical-ovate, glabrous, walls thin, sclerified; ±12 ovules per locule. n = 12 (Fraga 1983: 536–537; see Nelson & Oliver 2005b).

FLOWERING PERIOD. February–July.

DISTRIBUTION & ECOLOGY. Europe: Spain, Portugal (Bayer 1993). North Africa: northern Morocco only (Ball 1877; Fennane *et al.* 1999; Fennane & Ibn Tattou 2005). Fig. 97.

Erica umbellata inhabits open heaths and scrub, open woodland and coastal sands, always on siliceous, acidic soils, and ascends to about 1,500 m altitude. It is not a "re-sprouter", and so after fire must re-establish its populations from seed. In adverse conditions, such as severe drought, many seedlings may perish, putting the species at risk of local extinction (Quintana *et al.* 2004).

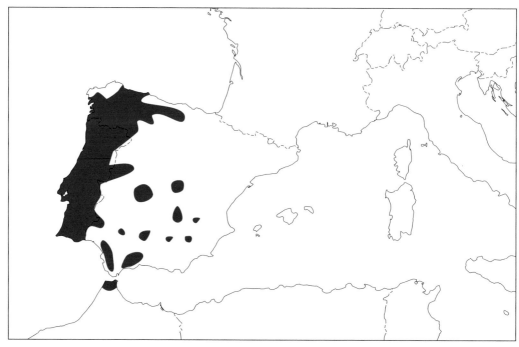

Fig. 97. Distribution of *Erica umbellata*.

ETYMOLOGY. Latin: *umbellatus* = umbellate, forming an umbel.

VERNACULAR NAMES. Dwarf Spanish heath, Portuguese heath (English); mogariza, quirola, quiruela, brecina (Spanish; for other names used in Spain see Bayer 1993; Pardo de Santayana *et al.* 2002); queiró (Portuguese); Doldenheide (German).

CULTIVARS. See Appendix I (pp. 382–383).

Several unnamed clones or seedlings are undoubtedly in cultivation in Britain (see, for example, Hall (2002)) but only two cultivars have been selected and named. A clone with yellow-green foliage was propagated in 1980 but subsequently lost.

NOTES

[1] This name was threatened by *E. forskalii* Vitman (1789), but a proposal (Nelson & Wulff 2007b) to conserve *E. manipuliflora* Salisb. (1802) against this earlier, little-used name was approved (Brummit 2009: 287–288).

[2] George Don (1834: 800) derived *Gypsocallis* from γύψος and κάλλιστος (*kallistos*, most beautiful). Philip Oswald (pers. comm., February 2007) commented that there seems no reason to suggest that *Gypsocallis* stemmed from the superlative κάλλιστος (*kallistos*). Though the adjective καλός has only one lambda (λ), most derivatives have *kalli-*; Liddell & Scott (1890) stated that "kalo- is later and less common". Some words ending in -καλλις are classical, and Dioscorides has ἡμεροκαλλίς, *Hemerocallis*. We do not have Salisbury's own explanation of *Gypsocallis*. Gray (1821: 398) gave it the name "Gypsum-heath".

[3] Salisbury suggested, in the hitherto unpublished manuscript letter referred to earlier (see p. 387; Appendix II), that *E. multiflora* should form a separate genus with *E. carnea* and *E. erigena*, for which he coined the name "Tesquaria" (from Latin *tesqua, -orum*, "rough places, wild regions, wastes, steppes" (Lewis 1915): "Such places were sacred to the gods …" (Lewis & Short 1890)). This view was not followed by later botanists, and Salisbury's generic name remained unpublished.

[4] The publication date means that *Erica vagans* was not first described and named in *Mantissa altera plantarum*, although this is often stated (e.g. McClintock 1989b).

[5] The derivation is obscure: *koukoúla* is a hood, cowl or tea-cosy, *koukoúli* is a silk-worm cocoon and *hórto* is grass (but often other plants when in a compound word) (P. H. Oswald, pers. comm., 28 July 2007).

[6] On Paxos, according to Sands (1991), the heather's name is *hamoríki*. The modern Greek name χαμορείκι or χαμορίκι, meaning low-growing heather, is ultimately derived from χαμαί (*chamai*), on the ground, and ἐρείκη (*ereikē*), heather.

[7] Flora Croatica database (http://hirc.botanic.hr/fcd/) accessed 2 February 2007.

[8] The name *E. multiflora* f. *alba* (Regel) D. C. McClint. is invalid because the basionym *E. multiflora* var. *alba* Regel is a later honomyn of *E. multiflora* var. *alba* Weston (1775) which is most probably a synonym of *E. vagans* var. *alba* Sinclair (1825).

[9] Flora Croatica database http://hirc.botanic.hr/fcd/, accessed 2 February 2007.

[10] Sir James Edward Smith (1809) commented on Salisbury's proclivity to change names: "Mr. Salisbury, who has paid much attention to [*Erica*], has given an arrangement of 246 species … with often expressive, if not classical, specific characters, but with an unlicenced change, and frequent perversion, of names." (See also Stearn 2007.)

[11] In *Flora iberica* (1971: **4**: 499) this variety is given the rank of *forma* because within the protologue (Lange 1863: 223) used the term *forma*: "a forma typica recedens corolla majore, cylindrico-urceolata (nec globoso-campanulata), antheris castratis."

[12] L'Écluse's *Erica Coris folio V*; the same woodcut was printed in Thomas Johnson's revised edition (1633) of John Gerard's herbal, but not in the original 1597–1598 edition (Nelson 1998). Johnson's name for his twelfth *Erica* was "*Erica ternis per intervalla ramis* Heath with three branches at a joint", and he gave a translation of l'Écluse's text: "growing in untilled places in Portingale above Lisbone, where it floured in December". I suggested that the illustration was made from a dried specimen of *E. cinerea* (Nelson 1998: 51), but the evidence of the original water-colour (Ramón-Laca & Morales 2000) is fully accepted.

THE CILIATED, OR DORSET, HEATH

Erica ciliaris

The individual flowers of *Erica ciliaris* are unique in the context of hardy European heaths. The mature corolla is pendent, at least one centimetre long, and markedly zygomorphic, with a narrow, obliquely angled mouth (Figs 98–100). The included anthers, which are relatively small and are positioned just inside the corolla mouth, lack spurs. While the flowers of other northern-hemisphere *Erica* species are not strictly actinomorphic, none shows the degree of zygomorphy that *E. ciliaris* displays. In addition, the flowers are arranged in an erect raceme. Richard Salisbury considered placing the species in a separate genus to be named "Loxomeria" (from Greek λοξος, *loxos*, oblique, slanting; μερις, *meris*, a part: evidently an allusion to the corolla), which opinion Bentham (1839) followed, employing Salisbury's otherwise unpublished name at section level. Thus the general view has been that *E. ciliaris* occupies an isolated position within the northern species of *Erica*, not closely allied to the others.

The Dorset heath's long, curved corolla with a narrow, "pinched" mouth, precludes most bees from acting as pollinators, as they do not have mouth parts of sufficient length to reach the nectar, which is the principal "reward" offered by Dorset heath (Herrera 1987), at the base of the blooms. These insects would not get any "reward" for their attention to the heather's flowers. (In Britain the buff-tailed humble bee has been observed puncturing the corolla at the base and thus is a nectar thief, not a pollinator.) Long-tongued insects such as moths and butterflies are more likely to be able to access the nectar. However, Rose *et al.* (1996) state that visits by insect pollinators "such as honey-bees and butterflies" are rare. Thus the conventional description of pollination in *Erica ciliaris* lays emphasis on self-pollination (autogamy), at least in British populations (Rose *et al.* 1996). Rose *et al.* (1996) also suggested that thrips "may be important" (as in *E. tetralix* in the Faeroe Islands, see p. 235), adding that when plants are growing in closely packed colonies, wind-pollination (anemogamy) "may occur … at the later stages of floral development when the style protrudes out from the corolla." However, a somewhat different situation is indicated by studies in Spain (Herrera 1987). When flowers were "bagged" to prevent insects from visiting them, fruit (and seed) production virtually ceased suggesting that *E. ciliaris* is not self-pollinated, and that thrips are also not of any significance. Furthermore, included anthers and a long, tubular corolla are unlikely to have evolved in response to wind-pollination.

The insect visitors of *Erica ciliaris* in southern Spain have been recorded in another paper by Javier Herrera (1988). Every week for two years Herrera visited a four-hectare heathland plot that had Dorset heaths growing on it, as well as ling and *E. scoparia* (besom heath), within the Doñana National Park in the south-west of Spain. He observed and recorded the insects visiting the flowers of the indigenous shrubs, spending 160 minutes watching *Calluna* (noting 134 insects), and 235 minutes on *E. ciliaris*. Sixty-four insects were seen visiting the Dorset heaths, principally bees (Anthophoridae, Halictidae) and a few butterflies (Lycaenidae) (see Table 6 for the names of the insects).

Table 6

Insects recorded as visiting the flowers of Dorset heath (*Erica ciliaris*) and ling (*Calluna vulgaris*) in Doñana National Park, Spain (extracted from Herrera 1988: Appendix).

	proboscis length	*Erica ciliaris*	*Calluna vulgaris*
DIPTERA			
Phthiria sp.	2 mm	■	■
Calliphoridae (not indentified)			■
Muscidae			■
Eristalis tenax L.	5 mm	■	■
Eristalodes taeniops Wied.			■
Erisyrphus balteatus De Geer			■
Lathyrophthalmus quinquelineatus Fabr.	3 mm		■
Sphaerophoria scripta L.			■
HYMENOPTERA: APOIDEA			
Amegilla fasciata F.	11 mm	■	
Amegilla quadrifasciata Vill.		■	
Epeolus fallax Mor.			■
Ceratina cucurbitina Rossi			■
Ceratina cyanea K.		■	
Xylocopa cantabrita Lep.		■	■
Apis mellifera L.	6 mm		■
Colletes caspicus subsp. *dusmeti* Nosk.			■
Colletes succintus L.			■
Lasioglossum immunitum Vach.	3 mm	■	
Lasioglossum prasinum Sm.			■
Megachile maritima K.		■	
Megachile pilidens Alfken		■	
HYMENOPTERA: NON-APOIDEA			
Odynerus sp.			■
Cerceris arenaria L.			■
Mellinus arvensis L.			■
LEPIDOPTERA			
Aricia cramera Eschscholtz			■
Syntarucus pirithous L.	6 mm	■	■
Gegenes sp.		■	

Dorset heath and bell heather   Painted by Christabel King   PLATE 13

UPPER LEFT: *E. ciliaris* **'Stapehill'**. Dense, low, spreading bush (±0.25 m tall × 0.45 m spread); foliage dark green; flowers pale lilac (H3) at base, flushed lavender (H3) at tip; August–October. Some descriptions of this clone suggested that flowers have a creamy white base (Maxwell & Patrick 1966), but Vickers (1976) recorded the colour as lilac. 'Stapehill' was introduced by C. J. Marchant, Stapehill, Wimborne, Dorset, by 1954; Chapple (1964) stated that Marchant "raised" it.

UPPER LEFT CENTRE: *E. ciliaris* **'Mrs C. H. Gill'** ♈ Low, spreading shrub (±0.2 m tall × 0.45 m spread); foliage dark green; flowers "rich deep red" (crimson H13), in erect spikes; July–October. "One of the best ... and a first class garden plant", according to D. F. Maxwell (Maxwell & Patrick 1966). 'Mrs C. H. Gill' was found near Wareham, Dorset, and introduced by Maxwell & Beale Ltd, Broadstone, Dorset, in 1927. The cultivar was named after Mrs Gill of Thirsk, Yorkshire, described by Maxwell as "a great little lady who loved her garden, and was so proud of her heather" (Maxwell & Patrick 1966: 124). Johnson (1932) wrote: "... the bold, erect spikes bear flowers of a full-toned rosy-red." AGM 1993. (Recommended by the Heather Society; withdrawn 2011).

CENTRE: *E. cinerea* **'Heatherbank'**. Upright heath (±0.3 m tall × ±0.5 m spread); foliage dark green; flowers amethyst (H1) at first, then becoming bicoloured with white at base; June–September. (Painted 11 October 1985, cultivated by Major-General Turpin.) This spontaneous seedling was found in Mrs Margaret Bowerman's garden, Champs Hill, Coldwaltham, Surrey, in 1977, the name being registered in 1982 (® no. 23).

UPPER RIGHT: *E. cinerea* **'C. D. Eason'** 🏵️ ♈ Floriferous, neat, spreading heather (±0.25(–0.35) m tall × ±0.5 m spread); foliage dark green; flowers bright magenta (H14), in long, dense spikes; June–September. (Painted 31 August 1985, cultivated at the Royal Botanic Gardens, Kew.)

Charles Douglas Eason (c. 1890–1972), an Australian, was chief propagator at Maxwell & Beale's nursery, Broadstone, Dorset. John Eason (1993), his eldest son, recalled walking with his father on the moors that then surrounded Broadstone:

> It was on one of these walks, among the wild heathers, that we noticed one of the cinereas had a bright magenta colour. He took it back to the nursery, made a lot of cuttings, and then waited to see if the flowers were the same colour as on the parent plant. He showed it to Mr Beale who said he had better name it after himself: hence *Erica cinerea* 'C. D. Eason'.

Maxwell & Beale Ltd introduced 'C. D. Eason' in its 1929 catalogue. A water-colour portrait of 'C. D. Eason' by Winifred Walker (1882–1966) was published in the nursery's catalogues between 1933 and 1972 (Cleevely 1997). One of three outstanding cultivars in trials at the Royal Boskoop Horticultural Society in the early 1990s (van de Laar 1998). AGM 1993.

LOWER LEFT: *E. cinerea* (wild).

LOWER CENTRE: *E. cinerea* **'Katinka'**. Compact, upright bush (±0.3 m tall × ±0.6 m spread); foliage dark blue-green, glaucous; dark flowers beetroot (H9); June–October. (Painted 22 July 1985, cultivated by Major-General Turpin.)

The very dark flowers do not show up well against the dark foliage of this Dutch cultivar, which is similar to 'Velvet Night'. H. J. Weber found 'Katinka' (a "fantasy" name) in 1960 at Plantsoenendienst Gemeente, Driebergen-Rijsenburg, Netherlands, and introduced the cultivar in 1968.

LOWER RIGHT: *E. cinerea* **'Domino'** (*E. cinerea* f. *alba*). Neat (when trimmed), spreading plant (±0.2–0.35 m tall × ±0.75 m spread); foliage dark green in spring, becoming green in summer; flowers arrayed in long spikes; corolla white; calyx and flower-stalk "dark, almost ebony in colour. The contrast was very telling" (Maxwell 1927); July–October. (Painted 21 July 1985, cultivated by Major-General Turpin.)

Another of Maxwell & Beale's 1929 introductions, found by Douglas Fyfe Maxwell in 1927 on the moorland near Broadstone, Dorset, England:

> During this summer I found a deep-coloured form of *E. cinerea*, with one shoot from it bearing white flowers. ... There was not a cutting over a quarter-of-an-inch long to be found on the white-flowered branch; nevertheless, a few have now been rooted, and if after a thorough test this variety does not revert to the type—which I think is as unlikely as it would be deplorable—it will be launched into the world of Heather fanciers under the name of ... *Domino*." (Maxwell 1927: 52.)

Johnson (1932) applauded 'Domino' as "a decided acquisition, a Heath of singular attractiveness and not lacking in vigour."

11. ERICA CILIARIS. Dorset heath.

The Dorset heath, long known as ciliated heath or fringed heath (before it was dubbed Dorset heath in the late 1890s[2]), was among the heathers observed by the Flemish botanist Charles de l'Écluse (Carolus Clusius) during his prolonged stay on the Iberian Peninsula in 1564–1565. In *Rariorum aliquot stirpium per Hispanias observatarum historia*, published in 1576, l'Écluse described and illustrated many of the plants he had found, using the notes he had taken and drawings made in the field, as well as his dried specimens. His eighth *Erica*, which bloomed in October, was *E. ciliaris*; he saw it in Portugal and elsewhere. The water-colour that was the basis for a wood-cut printed in l'Écluse's book (Fig. 11, p. 15) survives in the famous *Libri picturati* collection of sixteenth-century natural history illustrations, now in the Biblioteka Jagiellóniska, Kraków, Poland (vol. **20** f. 31v) (Ramón-Laca & Morales 2000). The wood-cut was reused several times for illustrating other books, including John Gerard's *The herball or generall historie of plantes* (1597), and Thomas Johnson's "very much enlarged and amended" edition of 1633 (Nelson 1998). However neither Gerard not Johnson seem to have been aware that the "Challice Heath" (an odd, inexplicable moniker) occurred as a native plant in England.

"C. Bauhin, mistakenly, calls it *anglica*, which has given rise to the idea of its being an English plant, but it is not." Thus William Curtis (1800) drew attention to an apparent enigma, without, it seems, seeking an answer. Frederic Williams (1910) and David McClintock (1966a, 1980a) followed suit. Was the Dorset heath the plant which was named *Erica hirsuta anglica*, hairy English heather, in Caspar Bauhin's ΠΙΝΑΞ *theatri botanici* (1623: 486) and which was described in more detail in Jean Bauhin's *Historia plantarum* (Bauhin & Cherler 1651: 358) as Linnaeus (1753) believed? In fact, the hairy English heather was the much more familiar and common cross-leaved heath, *E. tetralix* (Nelson 2008a).

The standard history of *Erica ciliaris* gives the date of introduction into English gardens as about 1773 (Aiton 1789; Curtis 1800), and the heather was not admitted as a species native to Britain until after 1828, when the Revd John Savery Tozer (c. 1790–1836), Curate of St Petrock's in Exeter, sent a specimen collected on "a bog near Truro" to Dr Robert Greville (Lindley 1829). Sir Charles Lemon MP (c. 1784–1868), a notable gardener of Carclew, Cornwall, claimed to have known it earlier, because it grew on parts of his estate.

However, Richard Weston (c. 1732–1806) of Leicester (see p. 135), a prolific writer on matters agricultural and horticultural, seems to have been the first to report *Erica ciliaris* as a native English plant. In the first volume of his four-volume work, *Botanicus universalis et hortulanus*, published in 1770, there is a listing of heathers. Weston (1770: 99) glossed "ciliaris" with "rough-leaved English heath". There can be little doubt he meant to indicate it was a native species, because towards the end of the same volume, under the title "The trees, shrubs, and fruits, native of Great-Britain and Ireland", he listed the species again, without the then-superfluous adjective "English" (Weston 1770: 329). I am confident credit should go to Weston, even though he does not give any of the expected details of where, when and by whom it was gathered. Five years later, he also employed "Rough-leaved English heath" as the plant's English designation in his *Flora anglicana* (Weston 1775: 12).

Erica ciliaris was not found in Dorset until 1833, when the Revd Robert Blunt (1808–1884; later to take the name Dalby) gathered specimens on Corfe and Wareham heaths (Bowen 2000). There is even a suggestion that the heather may not be a wholly native plant in the county: "some could have been introduced with *Pinus pinaster* from west France" about 1800 (Bowen 2000). In August 1998, *E. ciliaris* was discovered in an isolated half-hectare of wet heath and mire at an altitude of 210 m on the Blackdown Hills. The site, on the Somerset/Dorset border, is the most northerly one in England where the Dorset heath is considered to be truly native (Edgington 1999).

In western Europe, *Erica ciliaris* has a distinctly Atlantic distribution, ranging southwards from Brittany through the west of France, mainly relatively close to the coast, into northern Spain, and then through Portugal and southern Spain, crossing the Straits of Gibraltar into northern Morocco (Ball 1877; Fennane & Ibn Tattou 2005). The species was described by Linnaeus in *Species plantarum* using a specimen collected near Oporto in Portugal between 27 July and 7 August 1751 (McClintock 1980a) by Pehr Löfling (1729–1756), a Swede who had been one of his pupils at Uppsala University.

ERICA CILIARIS IN IRELAND

On 14 September 1846, an amateur naturalist Thomas Fleming Bergin (d. 1863), by profession an engineer and the Secretary to the Dublin and Kingston Railway Company, gathered specimens of Dorset heath in the west of Ireland. Six days later, he donated three living plants to the Royal Dublin Society's Botanic Gardens, Glasnevin (Eager *et al.* 1978; Nelson 2008f). Bergin also pressed numerous voucher specimens: he presented several to James Townsend Mackay on 10 October 1846 (**TCD!**). The discovery of this heather in Ireland was announced by Mackay soon afterwards, at a meeting of the Royal Irish Academy in November 1846 (Mackay 1997; see also Nelson 1979a; Eager *et al.* 1978).

Fig. 98 (left). *Erica ciliaris* in Connemara, where the flowering season is in autumn; in background (left) are the discoloured shoot tips of the hybrid heather, Praeger's heath, *E.* × *stuartii* (see p. 301) (E. C. Nelson).

Fig. 99 (right). *Erica ciliaris* in bloom in July at Castriz, Galicia, north-western Spain, with *E. erigena* (E. C. Nelson).

Erica ciliaris   **Painted by Wendy F. Walsh**   **PLATE 14**

Author's collection (unpublished); this is the Connemara plant.

On Thursday 26 August 1852, Professor John Hutton Balfour of the Royal Botanic Garden, Edinburgh, reported that "on the road to Clifden ... on the left hand side of the road in hollow ground we gathered Erica ciliaris"[1], and he had specimens to prove this. Since then, the Connemara records have been subject to some considerable doubt. Numerous botanists searched the bogs between Clifden and Roundstone for the Dorset heath, without success. In 1898, Colgan and Scully suggested that "Bergin was the victim of an imposition and that Balfour's specimen was somehow mislabelled", and so the species was taken off the list of the Irish flora. Then to everyone's surprise, in 1965 Michael Lambert came across a small colony at the edge of the "Bog Road" between Roundstone and Ballyconeely, a few kilometres west of Craiggamore (Webb 1966a), and the heather still grows at that site (Walsh *et al*. 1983; Curtis 2000) (Fig. 98).

Bergin's herbarium specimens (as noted by Webb 1966a) all possess glandless hairs, and they all have a mixture of long and short hairs on the stem below the inflorescence, a very noticeable character not mentioned by previous commentators (Webb 1966a; McClintock 1968, 1972). Their apparent uniformity suggests that the population found by Bergin was clonal. One sheet (**TCD**!), attributed to Mackay, has a mixture of specimens on it; three have gland-tipped hairs. The specimen collected on 9 April 1965 by Professor David Webb and Michael Lambert (**TCD**!) is eglandular but unlike Bergin's material, lacks the distinctive mixture of short and long hairs on the stems just below the lowest flower, suggesting that this does *not*, as widely supposed, represent the same clone from which Bergin's specimens were gathered. However another explanation can be advanced: that the colony is not clonal and, thus, not uniform in such characteristics as vesture and glands.

FOSSIL HISTORY

The Dorset heath has a long history in Ireland, even if the present colony is regarded as dubious, the result of a deliberate introduction. Flowers and seeds of *Erica ciliaris* have been identified in an "organic deposit beneath clay and gravel of a subsequent glaciation" found near Gort, County Galway (Jessen *et al*. 1959; Godwin 1975). Thus, providing that the identification is correct, the Dorset heath was growing in western Ireland during what is termed the Gortian interglacial, which occurred around 428,000 to 302,000 years BP (see Coxon & Waldren 1995). Growing thereabouts at the same time were *E. cinerea,* and *E. tetralix*, bell heather and cross-leaved heath respectively, and, surprisingly, *E. scoparia* (besom heath) (see p. 177; Jessen *et al*. 1959). These heathers must have been wiped out when the area was covered by the glacier that buried the organic deposit containing their fossilised remains.

Did the Dorset heath return to Ireland after the glaciers melted and the ice retreated in the present post-glacial era? Or, like *Erica scoparia*, did it not manage to make its way north again, until assisted by a "forger of Nature's signature"? I tend to the view that Nature alone has been at work.

VARIATION AND CULTIVATION

As noted, wild plants can lack the glandular tips that are very often present on the long cilia that occur on the edges of the leaves and the calyx lobes and on the stems. The type specimen in Linnaeus' herbarium does not have gland-tipped cilia (McClintock 1980a). Plants entirely lacking cilia have been reported from France, and so have glabrescent ones.

Other wild-reported variants include plants with pure white flowers (*Erica ciliaris* f. *albiflora* Sauvage & Veilex; McClintock 1984a: 191), as well as ones with bicoloured flowers, this variant being

Fig. 100. *Erica ciliaris* in cultivation, northern California (E. C. Nelson).

represented in cultivation by 'David McClintock', which was found near Carnac in Brittany. Plants with yellow or golden foliage (*E. ciliaris* f. *aureifolia* D. C. McClint.; McClintock 1988a: 61) are also reported, but are very uncommon — 'Aurea' came from near Wimborne in Dorset in the 1920s.

Differences in corolla size and shape are also on record. The plant long known as *Erica* "*maweana*" (Backhouse 1882) was originally thought to be a separate species but is now considered merely a cultivar. James Backhouse had obtained this "New, distinct hardy heath" from George Maw (1832–1912), who only vaguely recalled collecting it, probably in Portugal; it was described as forming "much branched, but compact bushes, [and] bearing showy clusters of large bright purplishcrimson blossom." 'Mawiana' was illustrated in *Curtis's botanical magazine* (tab. 8443) in 1912. Plants grown under the name 'Mawiana' (which is the correct orthography, not 'Maweana': see Nelson 2003b) today do not tally with the early descriptions.

Erica ciliaris will grow in lime-free soils and can be propagated easily from cuttings or seed; cultivars must be propagated vegetatively.

HYBRIDS

Erica × *watsonii* (*E. ciliaris* × *tetralix*; see pp. 306–311), found in England, where it is abundant, and in France and northern Spain, where it is rather uncommon, is the only naturally occurring hybrid involving *E. ciliaris*. This has also been artificially created by Dr John Griffiths (Yorkshire, England), Kurt Kramer (Edewecht-Süddorf, Germany) and David Wilson (British Columbia, Canada) (see p. 306).

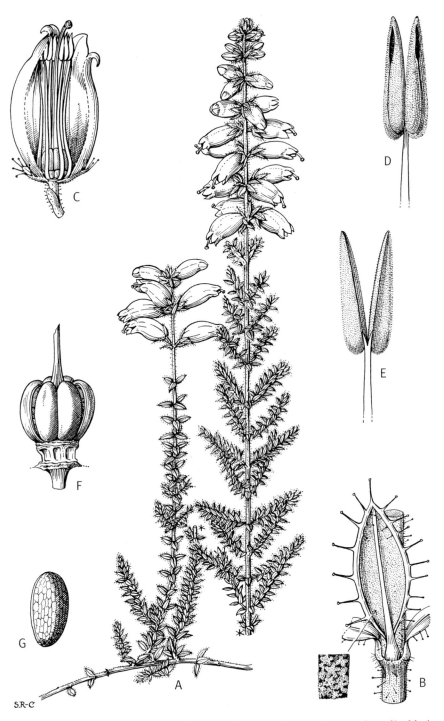

Fig. 101. *Erica ciliaris.* **A** part of a plant × 1; **B** stem and leaves, and part of lower surface of leaf further enlarged × 12; **C** flower: a calyx lobe, part of a corolla, and one stamen removed × 4; **D** anther and part of filament in front view × 16; **E** anther and part of filament in back view × 16; **F** fruit: calyx, persistant corolla and stamens removed × 8; **G** seed × 40. Drawn by Stella Ross-Craig.

Other attempts to cross-fertilise *Erica ciliaris* artificially have not been especially productive; pollen-parents tried have included *E. carnea*, *E. cinerea*, *E. erigena*, *E. lusitanica*, *E. mackayana*, *E. manipuliflora*, *E. sicula* subsp. *bocquetii*, *E. terminalis*, *E. umbellata* (all Griffiths) and *E. vagans* (Griffiths, Wilson). Griffiths has also used *E. ciliaris* to pollinate *E. australis*, as well as *E. mackayana*, *E. manipuliflora*, *E. terminalis* and *E. vagans*. Only some of these yielded seed; often this was not viable or, when viable, plants identical to the seed-parents were produced. Other seedlings perished before any judgement could be made about their state.

Erica ciliaris L., *Species plantarum*: 354 (1753). Type: "*Habitat in* Lusitania", *P. Löfling*: lecto. **LINN** 498.64! (McClintock 1980a; Jarvis 2007: 498).

ILLUSTRATIONS. S. Edwards, *Curtis's Bot. Mag.* **14**: tab. 484 (1800); M. Smith, *Curtis's Bot. Mag.* **138**: tab. 8443 (1912); S. J. Roles, *Flora of the British Isles: illustrations* pt 2: tab. 935 (1960); S. Ross-Craig, *Drawings of British plants*, part 19: pl. 25 (1963); R. W. Butcher, *A new illustrated British Flora*: 107 (1961); L. Denkewitz, *Heidegarten*: 154 (1987); W. F. Walsh, *An Irish florilegium* pl. 16 (Walsh *et al.* 1983); M. Blamey, *Heathers of the Lizard*: plate 1 (McClintock & Blamey 1998).

BIOLOGICAL FLORA OF THE BRITISH ISLES. *J. Ecol.* **84**: 617–628 (1996).

DESCRIPTION. *Diffuse, bushy, evergreen shrub* 0.2–0.5(–1–2) m tall, capable of regenerating rapidly from underground roots; shoots erect or ascending, branching; variously with gland-tipped or eglandular hairs. *Stems* hirsute with mixture of long hairs and very short, simple, often glandular hairs; without infrafoliar ridges. *Leaves* in whorls of 3, ±3–5 mm long, ±1 mm broad, lanceolate to ovate, upper surface flattened, densely and minutely pubescent, sometimes glabrescent, with long cilia spaced around edge and at tip, margins recurved underneath around tip leaving abaxial surface exposed; abaxial midrib glabrous or minutely pubescent underneath; *petiole* hirsute or glabrous, edges fringed with hairs. *Inflorescence* a terminal, ±unilateral, elongated, erect raceme, ±40–120 mm long; *flowers* ±15–45, zygomorphic, pendent; *pedicel* curved, ±2 mm long, hirsute with short hairs and cilia; *bract* near base of pedicel, leaf-like, green, lanceolate, ±2.5 mm long, minutely pubescent, with long marginal cilia; 2(–3) *bracteoles* almost recaulescent, similar to but smaller than bract. *Calyx* composed of 4(–5) free, lanceolate, green sepals, with long cilia along margins, also with fringe of simple hairs; sulcus open, ±0.5 mm long. *Corolla* glabrous, elongate ovoid-urceolate to ellipsoid-urceolate, ±11 mm long, mouth asymmetrical, oblique, narrow; lobes ±1 mm long, erect. *Stamens* 8, ±9 mm long, included, slightly declinate; *filament* ±7.5 mm long, ±0.3 mm across at base, straight, hooked at base and attached under the nectary, tapering upwards; *anther* ±1.8 mm long, muticous, *thecae* elongated, erect, parallel, free, papillose on outer surfaces, inner surfaces smooth; pore lateral, toward apex, oval to elliptical, to 0.6 mm long. *Style* glabrous, curving slightly, ±8 mm long, ±0.4 mm diameter at base, tapering upwards; *stigma* capitate, broader than style apex, ±0.7 mm across, emergent; *ovary* ±1.5 mm long, cylindrical, 4-locular, ribbed, emarginate, 45–50 ovules per locule; *nectary* lobed, ±0.5 mm wide. *Seed* ±0.3–0.5 × 0.2–0.3 mm, ovate (Nelson 2006a); capsules can contain 60–80 seeds (Rose *et al.* 1996). n = 12 (Fraga 1983: 534; see Nelson & Oliver 2005b).

FLOWERING PERIOD. May–October(–December). The small colony of Dorset heath in Connemara is later flowering than might be expected (Nelson 1982b); it can bear fresh blooms and buds as late as November when other heathers in the area, including *Erica mackayana* and *E.* × *stuartii*, have long since ceased blooming.

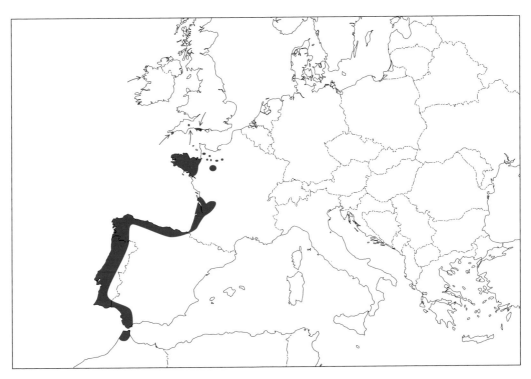

Fig. 102. Distribution of *Erica ciliaris*.

DISTRIBUTION & ECOLOGY. Europe: western half of Portugal; common through northern Spain, north of Cordillera Cantabrica and in scattered localities in south-western Spain; western France especially in the south-west and in Brittany (Dupont 1962: 197, 199); south-western England (Dorset, Somerset, Devon, Cornwall); Ireland (Connemara but perhaps not indigenous: cf. Curtis 2000). North Africa: only in northern Morocco (Fennane & Ibn Tattou 2005). Some of the populations in Britain may not be indigenous — Bowen (2000: 122) suggested that while *Erica ciliaris* "has been here [Dorset] since early times, some could have been introduced with *Pinus pinaster* from west France *c.* 1800."
Maps: Hansen (1950: 44), Rose *et al.* (1999: 618), Preston *et al.* (2002); more detailed maps are provided by Chapman (1975) and Chapman & Rose (1994) for some of the English populations, and by Fraga Vila (1984b: 13) for Galicia. Fig. 102.

Fresh seeds germinate rapidly, while storage in dry conditions for several months does not affect the capacity of the seeds to germinate. Light is required for germination (Rose *et al.* 1996). The ability of *Erica ciliaris* to resprout from roots gives it the capacity to recover from trampling and other damage to its aerial parts.

The Dorset heath inhabits moist heathland and moorland, rarely open woodland, on acid soils, usually at low altitude. *Erica ciliaris* will prefer the drier hummocks when it grows in bogs. In Britain it can be found at around 400 m above sea-level on Dartmoor (Preston *et al.* 2002), while in Spain, *E. ciliaris* occurs up to 800 m, and very rarely to 1,800 m (Bayer 1993). This species is a member of the Oceanic Southern-temperate element of the British and Irish floras (Preston & Hill 1997); *E. erigena*, *E. vagans* and *Daboecia cantabrica* also are assigned to this category.

ETYMOLOGY. Latin: *cilium* = eye-lash; fringed with hairs.

VERNACULAR NAMES. Dorset heath, ciliated heath (English); bruyère ciliée (French); Wimperblättrige Heide (Dippel 1887), Wimpernheide (German); carapaça, urze-carapaça, cordões-de-freira, lameirinha (Portuguese); argaña, agaña, carroncha, moguerica (Spanish; for other vernacular names used in Spain see Pardo de Santayana *et al.* (2002)).

CULTIVARS. See p. 165 and Appendix I (pp. 369–370).

Notes

[1] J. H. Balfour ms diary 26 August 1852; original ms in Royal Botanic Garden, Edinburgh.

[2] The first instance which I can trace of this name in print is in *The garden* on 30 April 1898 as the caption of a painting by Henry George Moon (1857–1905) and in the short accompanying article (see Nelson 2008a). Although that article is not signed, I suspect it was written by William Robinson, owner of *The garden*. Significantly, within the sixth edition of his *The English flower garden* (p. 514), published on 1 July 1898, Robinson also gave the name Dorset heath for *Erica ciliaris* (McClintock 1982c; Nelson 2008a) — in the fifth edition (Robinson 1897) none of the heathers has an English name.

Whereas horticultural authors were quick to follow Robinson's lead (for example Weathers 1901: 581 ("*E. ciliaris* (Dorset Heath). A native of Dorset, Cornwall, and parts of Ireland ..."); Fitzherbert 1903: 36 ("Other pretty species are the ... Dorset Heath, *E. ciliaris* ..."): Cook 1908: 235 ("*E. ciliaris* (Dorset Heath). Although in smoky and foggy places, such as London, this Heath is not always satisfactory, in the purer air of the surrounding ...")), botanists did not adopt the new name so readily.

THE WIND-POLLINATED BESOM HEATHS
Erica scoparia, E. azorica, E. platycodon

I have deliberately headed this section in the plural because there are three rather similar, but distinguishable, heathers that fall within the compass of *Erica scoparia*, the archetypal besom heath (also called broom heath or green heather). Whether they deserve the rank of separate species, or of subspecies within *E. scoparia*, is a debatable topic and one likely to remain unresolved until detailed DNA studies are undertaken on all the insular and continental populations concerned. We must bear in mind, however, the perplexing indications from recent DNA studies (McGuire & Kron 2005; Pirie *et al.* 2011) that morphologically similar heathers may not have a recent common ancestor and so may not be as closely related as their superficial similarities suggest.

The besoms heaths are relatively widespread; they range over three archipelagos in Macaronesia, the European mainland and parts of north-western Africa, as well as islands in the western Mediterranean, the Balearics, Corsica, and Sardinia. The "true" besom heath, *Erica scoparia*, grows on the European mainland as far north as central France, and is also recorded in northern Italy, the Iberian Peninsula (Bayer 1993), northern Morocco (Fennane & Ibn Tattou 2005), Algeria (Quézel & Santa 1963) and Tunisia. On the Azores, the local counterpart has long been called *E. azorica*, and a remarkable plant it is, too, a tree as tall as any other tree heath that I have seen. Another besom heath, named *E. platycodon*, also potentially a small tree, inhabits just three of the Canary Islands — El Hierro, the westernmost, Tenerife and La Gomera (Bramwell & Bramwell 1974; Hohenester & Welss 1993). What may be the same heath — it certainly is a besom heath — is found on Madeira and its satellite island, Porto Santo (Sjögren 1972, 2001; McClintock 1989d; Hood 2006); this can be called *E. platycodon* subsp. *maderincola*.

Given that Portuguese and Spanish botanists regard the Macaronesian plants as distinct at species level from the continental ones, I have decided to follow their lead, although in some ways this is unsatisfactory, especially because there have been no detailed comparative investigations involving all the populations. In cultivation, they do appear different and can be told apart. Professor David Webb (1972b; see also Webb & Rix 1972) took a broad view of this species: "Except in its much shorter calyx and slightly shorter corolla the Azorean plant does not seem to differ constantly from the continental one in other characters, so that subspecific status seems appropriate." He did not discuss or distinguish the Madeira populations, nor those on the Canary Islands (which, anyway, were outside the scope of *Flora europaea*). More than a decade later, whilst arguing in support of specific rank for *Daboecia azorica* (see p. 82), Webb (*in litt.* to D. McClintock, 9 June 1983) admitted that *Erica azorica*, which he had relegated to a subspecies of *E. scoparia*, was a "borderline case, and I <u>did</u> dither over it ...", and he told McClintock that "if you [wish] to reinstate <u>Erica azorica</u> as an independent species, I would have less objection" than to the demoting of *Daboecia azorica*.

The besom heaths have several, probably significant, characteristics in common. Firstly, they are wind-pollinated. The flowers have exserted and greatly expanded stigmas (see note p. 105) that resemble miniature parasols or toadstools (see Hall 1997); the diameter of the stigma is several times greater than that of the distal portion of the style (Fig. 103). No other northern *Erica* has such an enlarged stigma, although the wind-pollinated African species do (see Rebelo *et al.* 1985; Oliver 2000). Ironically, a poorly labelled specimen of a besom heath was once mistaken for a Cape heath and led to a phantom Cape species.[1] Secondly, the flowers do not have nectaries.

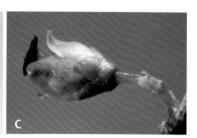

Fig. 103. Close-up photographs of individuals flowers of besom heaths (© Allen Hall, reproduced by kind permission): A. *Erica azorica*; B. *E. platycodon*, C. *E. scoparia*.

Thirdly, they all have relatively small flowers that are produced in vast quantities. Herrera (1987) estimated that within a square metre of a southern Spanish heathland, the shrubs of *Erica scoparia* produced, when blooming was at its peak, about 57,000 flowers, *Calluna* around 16,500, whereas *E. ciliaris* only produced 52 flowers in that same area.[2] Fourthly, they all have included stamens and their anthers lack spurs. They produce colossal quantities of pollen — 92,000 grains per flower, which is four times as much as *Calluna* (Herrera 1987: Table 1)! Fifthly, the leaves have a similar gross morphology, only found in one other northern species, namely *E. terminalis*. The midrib on the underside of the leaf is broad, hairless and elevated. While the leaf margins are recurved, they do not overlap the midrib but, rather, their edges remain parallel with the elevated midrib on either side of it. Thus the underside of the leaf is furrowed by two parallel, longitudinal grooves — other northern *Erica* species have only a single abaxial sulcus because the leaf margins overlap the midrib and meet above it.

How may they be distinguished? The distinctions are mainly quantitative. The flowers of *Erica platycodon* are larger than those of the other species, with larger stamens that are twice the size, with anthers that are about 1 mm long. The filaments are geniculate, with a sigmoid bend towards the apex so they nestle on the ovary. The flowers of *E. azorica* are the smallest, and the stamens are proportionately shorter, but the anthers are proportionately broader, about as broad as long (Table 7).

Table 7

Comparison of stamen characterisitcs in the besom heaths.

	E. platycodon	*E. scoparia*	*E. azorica*
anther (theca) length	±0.85–1.2 mm	±0.6 mm	±0.4–0.65 mm
anther (theca) width	±0.4–0.6 mm		±0.3–0.4 mm
filament length	>0.9 mm	±0.85 mm	±0.6–1.1 mm

Given that the stamens are very small, 2 mm long at most, the proportions are significant when the species are compared. The foliage of *Erica azorica* tends to be finer than that of the other species; again, when seen side by side in gardens, the differences are discernible, even if hard to describe.

Studies of the seeds of the besom heaths suggest that the continental populations can be distinguished from the Macaronesian ones by the external morphology of the seeds. Fagúndez &

Izco (2003) examined small samples of seeds from ten populations in the Azores, Madeira, Tenerife (Canary Islands), Spain and Portugal. They found that the mean length of the seeds ranged from 0.52 mm to 0.57 mm, the seeds from continental populations being the smallest. More significant differences occurred in the microscopic structure, especially in the boundaries of the cells of the seed coat (testa). In the besom heaths from the islands these boundaries are channelled, whereas in the continental plants they were not channelled but "totally anastomosing" (Fagúndez & Izco 2003: figs 4 & 5). Thus, Fagúndez & Izco (2003) concluded that there are "four subspecies within *Erica scoparia*".

POLLINATION

The markedly broadened, peltate apex of the style (Fig. 103), coupled with the lack of a nectary, undoubtedly indicates that the besom heaths are wind-pollinated: as just noted, there are southern African species with similarly enlarged style apexes that are wind-pollinated (cf. Rebelo *et al.* 1985; Oliver 2000). Pollen is copiously produced and freely shed by all the besom heathers. Writing about the Azorean species in his famous book *Plants, seeds, and currents in West Indies and Azores* (1917), Henry Brougham Guppy (1854–1926) noted that "when men or cattle brush against the branches in June dense clouds of pollen are given off." The flowers apparently offer no other reward to attract animal pollinators, and, indeed, they are also small and tend not to be brightly coloured, so the individual blossoms are also relatively inconspicuous. While some other northern species, notably *Erica arborea* and *E. lusitanica*, may sometimes be wind-pollinated, it is suggested, their functioning nectaries, as well as fragrance, indicate that they also have retained the capacity to attract insects. Only the besom heaths and the Balkan heath (*E. spiculifolia*), among northern heathers, have neither nectar nor (to humans at least) fragrance.

Pollen alone, however, may be reward enough. Bees need and, of course, collect pollen, although there are no published observations of pollen-gathering from this heath, as far as I can tell. *Erica scoparia* is reported to be a valuable source of honey: McClintock (1990b) noted this, quoting the German botanist Dr Hermann Schacht's account of Madeira and Tenerife (1859: 95). Persano Oddo *et al.* (2004) reported that *E. scoparia* yields "abundant" unifloral honey in Spain. Honey is a complex product for which nectar is the essential ingredient. Any honey containing significant amounts of *E. scoparia* pollen must have been made with nectar obtained from other plants that were blooming at the same time — these could have been other heather species. Neither Bayer (1993) nor Pardo de Santayana *et al.* (2002) make any mention of honey being produced from *E. scoparia* in Spain, while Allen Hall (1997), then living at Cheam in Surrey, England, observed that "I have yet to see a bee on the flowers of my plant".

FOSSIL EVIDENCE

The uniquely distinctive flowers of *Erica scoparia* (in a wide sense), with the relatively huge parasol-like stigma, provide some of the best evidence that heathers have migrated and retreated as the climate of the northern hemisphere has fluctuated between relatively warm and relatively cold periods during the Pleistocene (the stage covering about 1.8 million years and concluding at the end of the last "ice age").

Fossilised leaves, flowers and seeds (taken together, much better evidence than solitary pollen grains) found near Gort, County Galway, in western Ireland, extracted from a peat deposit, which had been covered by glacial till and so dates from before the last glaciation, have been identified as belonging to various *Erica* species. The peat was formed during a warm period known as the Gortian Interglacial (equivalent, most agree, with the Hoxnian) that lasted for about 120,000 years, commencing around 420,000 years BP. Among the preserved plant debris were (Jessen *et al.* 1959: 25–29): the tips of several shoots of *Calluna vulgaris* (ling), as well as some "decayed" flowers; flowers, seeds and probably also leaves of *E. ciliaris* (Dorset heath, see p. 169); a decayed capsule with a portion of the corolla of *E. cinerea* (bell heather); several leaves and several decayed flowers identified as *E. tetralix* (cross-leaved heath); and six seeds which were referred to *Daboecia cantabrica*. The most remarkable heather finds were described by Jessen and his colleagues (1959) as follows:

> From the upper layers came eleven small flowers of *Erica*-type conspicuous by an exserted style and a big peltate or bowl-shaped stigma, and also a considerable number of quite similar flowers that had lost the style. On the pedicel, up to 1.5 mm. long, only traces of the bract have been seen. Calyx 1.2–1.3 mm. long, connate below with four nerved, broad, obtuse lobes; corolla broadly four-lobed, coriaceous, and narrowly campanulate, 1.7–1.8 mm. long; stamens awnless and included — eight were counted in one of the flowers lacking the style, in the other dissected flowers five was the maximum number, episepalous as well as epipetalous. ... Filament when entire up to 1.5 mm. long, anthers 0.8–1.1 mm. long, separated in upper half and subacute at the top, open by a slit of nearly their full length; tetrads, 35–43μ in diameter. Of the ovary ... only single valves were seen, up to 1.5 mm. long, having a low median ridge on the inner side corresponding to similar ridges on the ovoid central column, 0.9–1.5 mm. long; style up to 1.6 mm. long, quadrangular at the base and jointed to the central column; stigma up to 0.9 mm. broad when expanded.

They concluded that, apart from differences between the sizes of some seeds extracted from the peat-preserved flowers and modern flowers, there was "no doubt" the Gort fossils matched "[*Erica*] *scoparia* only" (Jessen *et al.* 1959: 28). The excellent drawings that accompanied their paper

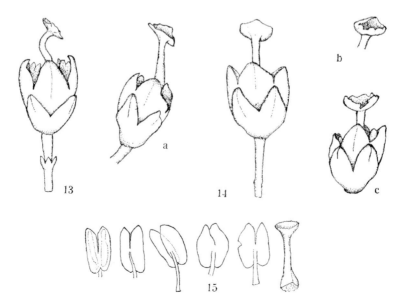

Fig. 104. Drawings of the flowers recovered from interglacial peat deposits at Gort, County Galway; reproduced from Jessen *et al.* (1959) by permission of the Royal Irish Academy.

(also reproduced in Godwin 1975) (Fig. 104) convince me that it was a besom heath. Data on the size of the seeds of *E. scoparia*, available in various publications, seem to coincide with theirs. The Gort seeds were ovoid to ellipsoid, 0.6–0.93 × 0.3–0.57 mm. The ranges of modern measurements for *E. scoparia* (summarised by Nelson 2006a) are 0.4–0.65 × 0.2–0.35 mm. Thus, apart from the fact this heath seems to have produced larger seeds, beyond reasonable doubt we may conclude that a variety[3] of the besom heath occurred in south-western County Galway during the Gortian interglacial and grew alongside heathers that are familiar in Connemara nowadays. A corollary is that the climate pertaining when the peat formed was not greatly different from the present climate of the region — warm, mild and equable.

There is just one question — which of the three besom heaths flourishing today most closely resembles the Gort fossils? Could one of the Macaronesian taxa provide a better match? Jessen *et al.* (1959) compared the Gort material with *Erica scoparia* from the far south of Spain; they did not also compare the Irish fossils with either the Canarian or Azorean populations (*E. platycodon* or *E. azorica* respectively). This is an intriguing puzzle that no one seems to have bothered to pursue. Could its resolution provide some clues to why *Daboecia* and a besom heath coexist today on the Azores?

ETHNOBOTANY

As the Latin name indicates, this is — or was — much used in the making of besoms and brooms in places where it is native. An extension of this is the manufacture of ornaments in the form of rabbits (Fig. 105), the fur represented by a shaped "besom" of heather twigs.

Like tree heather (*Erica arborea*), *E. scoparia* is also useful for firewood and kindling, and for various other purposes. *E. scoparia* is made into woven fencing mats or screens (Fig. 106).

Fig. 105 (left). Rabbits made from besom heath (*Erica scoparia*) on display in a shop in Madeira; these were not locally manufactured but had been imported (E. C. Nelson).

Fig. 106 (right). Besom heath screening fence in Mallorca, Balearic Islands; there was also a very small quantity of *Calluna vulgaris* in the woven fencing so this was evidently imported too (E. C. Nelson).

CULTIVATION

All three besom heaths, *Erica scoparia*, *E. azorica* and *E. platycodon* (including *E. platycodon* subsp. *maderincola*) are represented in cultivation. They are not spectacular plants, given that they have small, often green flowers. However, they bloom so profusely, so the effect of the tiny flowers *en masse* can be dramatic. None has been reported to reach tree-like dimensions in gardens — even the Azorean plant remains a bushy shrub. All require lime-free to neutral soil. They seem to be hardy in Britain and Ireland, although *E. platycodon* subsp. *platycodon* (from Tenerife) did not thrive in the southeast of England during the long, cold winter of 1995–1996 (Hall 1997) and in central England its shoot tips were damaged by persistent low air temperatures in the winter of 2000–2001 (Hall 2002).

12. ERICA SCOPARIA (*E. scoparia* subsp. *scoparia*). Besom heath.

Erica scoparia L., *Species plantarum*: 335 (1753). Type: "*Habitat* Monspelii, *in* Hispania, & Europa *australi*": lecto. **LINN** 498.17! (Jarvis & McClintock 1980; Jarvis 2007).
E. leptostachya Guthrie & Bolus, *Flora capensis* **4** (1): 217 (1905).

DESCRIPTION. *Evergreen shrub* to 1–2.5(–4) m tall, older plants tree-like in some habitats; regenerating from a lignotuber, with roots that can extend to 1.4 m deep (Silva *et al.* 2002, 2003). *Leaves* linear, ±4–7(–10) × 0.6–1.0 mm, in whorls of 3 or 4, spreading or ascending, when young pubescent, becoming glabrous, margins revolute, furrowed underneath. *Inflorescences* numerous and crowded on shoots, appearing to form a raceme; individual inflorescence an umbel of 1–3 flowers on very reduced lateral branchlets; *pedicel* green, glabrous, ±3 mm long, often curved, with bract and 2 bracteoles at $^1/_3$ of way along; bract and bracteoles similar, ± 0.3 mm long, green, margins with minute hairs. *Calyx* green, closely appressed to corolla, ±1.5 mm long, with 4 erect, equal lobes, ±1 mm long, base abruptly truncate (at least twice the diameter of the pedicel apex). *Corolla* ± 2–2.5 mm long, campanulate, greenish or rarely tinged with red; lobes 4, erect, ±1–1.5 mm long, acute. *Stamens* 8, included or rarely partly exserted; *filament* ±0.8 mm long; *anther* ±0.6–0.8 mm long, dorsifixed, muticous, thecae erect, fused, parallel, muticous, pore proportionately large. *Ovary* glabrous, ovate, ±1 mm tall; *style* ±1–2 mm long, cylindrical; *stigma* peltate, discoid, slightly concave, to 1 mm in diameter, usually dark red at anthesis. *Capsule* 1.3–1.8 mm tall, ovoid, glabrous. *Seeds* 0.4–0.65 × 0.2–0.35 mm (Nelson 2006a), boundaries of anticlinal walls not channelled (see Fagúndez & Izco 2003). n = 12 (Spanish specimens: Fraga 1983: 538; see Nelson & Oliver 2005b).
FLOWERING PERIOD. December–June (–July).
DISTRIBUTION. European mainland: southern and south-western France including Corsica; north-western Italy and Sardinia; Spain including Balearic Islands; Portugal. North Africa: Morocco (Ball 1877; Fennane & Ibn Tattou 2005), Algeria, Tunisia. **Maps:** Bolòs & Vigo (1995: 35). Fig. 108.

Erica scoparia is naturalised in one locality on the West Frisian Islands in the Netherlands (van Oostrom & Reichgelt 1956), and also near Brest in Brittany (Dizerbo 1974): in both instances the heather is said to have been introduced during the Second World War. Besom heath is also recorded as naturalised in Tasmania (Rozefelds *et al.* 1999; Baker 2005).

The species was also listed among introduced aliens found in 1996 close to the buildings of La Possession research station on the sub-Antarctic Crozet Island (Frenot *et al.* 2001, 2005), but no specimens have been seen to verify if it was the continental plant. The shrubs flowered in 1998, 1999

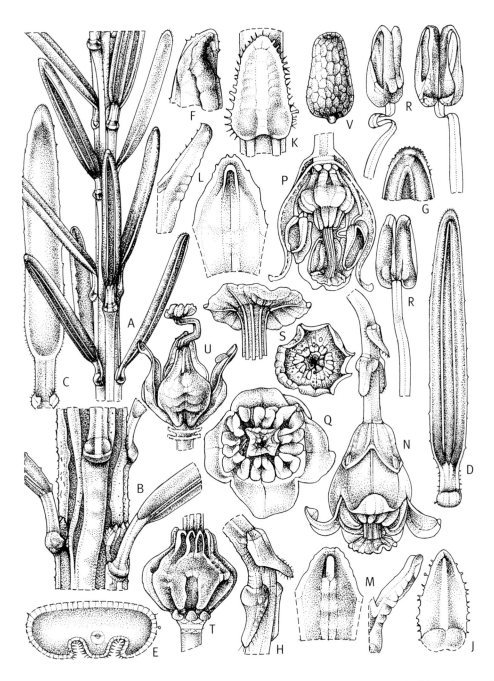

Fig. 107. *Erica scoparia.* **A** leafy shoot × 6; **B** leaf nodes × 20; **C** leaf from below × 14; **D** leaf from above × 14; **E** section through leaf × 30; **F** leaf tip from side × 34 ; **G** leaf tip from back × 34; **H** bract and bracteole × 26; **J** bract × 54; **K** bracteole × 54; **L** calyx lobes from front and side × 26; **M** alternate calyx lobes from front and side × 26; **N** flower from side × 12; **P** section through flower × 14; **Q** flower from below × 14; **R** stamens from side, front and back × 20; **S** stigma from side × 32 and above × 27; **T** ovary × 30; **U** capsule from side × 14; **V** seed × 46; from specimens grown by Allen Hall. Drawn by Joanna Langhorne.

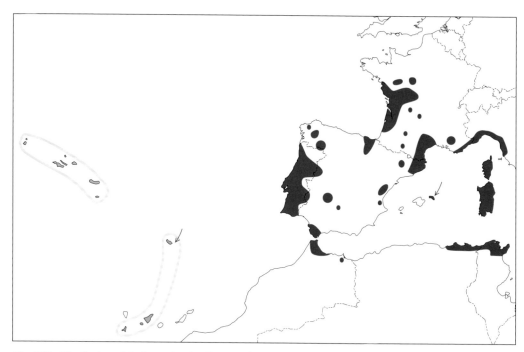

Fig. 108. Distribution of the besom heaths: *E. azorica* (green); *E. platycodon* subsp. *platycodon* (blue), subsp. *maderincola* (blue, arrowed) and *E. scoparia* (pink).

and 2000, but subsequently this "very small site, 1 or 2 m² in area, was destroyed by rubble deposits" and the heather was obliterated (Yves Frenot, pers. comm., 26 August 2005).

ETYMOLOGY. Latin: *scoparius* = a broom (for sweeping) made from twigs, a besom.

VERNACULAR NAMES. besom heath, broom heath, green heath(er) (English); bruyère à balai (French); scopa (Italian); brezo de escobas (Spanish; for other vernacular names used in Spain see Pardo de Santayana *et al.* (2002)); moita-alvarinha, urze-das-vassouras, urze-durázia, vassoura (Portuguese); Feinzweigige Heide (Dippel 1887), Strauchheide (German).

CULTIVARS. No cultivar, assignable with certainly to this species, has been selected and named.

13. ERICA AZORICA (*E. scoparia* subsp. *azorica*). Azores tree heath, Azores heath.

The Azorean besom heath, *Erica azorica*, is a protected species under Appendix 1 of the Bern Convention and Appendix 2 of the European Union's Habitat and Species Directive (see Dias *et al.* 2004). A veritable tree heather, at least 10 metres tall (although accurate measurements are not easy to come by) (Fig. 109), *E. azorica* is common on all of the islands in the archipelago, often abundant and dominant. The Azorean taxon is an important pioneer, colonising species. For example, at the western tip of Faial, *E. azorica*, almost exclusively, has colonised substantial areas close to the site of the volcanic eruption of 1957–1958 which destroyed the village of Capelinhos and the nearby lighthouse, adding almost two and a half square kilometres to the island's area (Scarth & Tanguy 2001) (Fig. 110). On Pico, the heather has invaded abandoned "semi-natural pasture" at 815 m above sea level (Ribeiro *et al.* 2003). Numerous seedlings of the heath can also be seen on road-cuttings, for example, on São Miguel. However, *E. azorica* also occurs in much more mature vegetation, including species-rich woodland. In some places, such as on Terceira, the heather forms forest with *Juniperus brevifolia*. In this community, protected under European habitat protection legislation, the trees can be 5 metres tall (Dias *et al.* 2004), and associated species are the Azorean holly (*Ilex perado* subsp. *azorica*) and such ferns as bracken (*Pteridium aquilinum*), hard fern (*Blechnum spicant*) and Tunbridge filmy fern (*Hymenophyllum tunbrigense*).

As a tree, it is a remarkable plant, with bare trunks and bare, erect stems, and foliage in dense, terminal tufts (Figs 109 & 112). These agglomerations of foliage give the trees a tiered appearance somewhat reminiscent, to my mind, of some of the species of *Eucalyptus* in Australia. Guppy (1917: 395) reported trees of *Erica azorica* "twenty-five feet [tall] and a diameter of eleven or twelve inches" (about 7.5 m tall and 0.3 m diameter) but also commented that at that time "except when specially preserved it would be difficult to find on Pico trees more than twenty

Fig. 109. A mature tree of *Erica azorica* exceeding 10 m in height growing by the old track above Caldeira de Cima on the northern side of São Jorge, Azores (E. C. Nelson 1993).

Fig. 110. *Erica azorica* and the giant reed (*Arundo donax* L.) recolonising the 1957–1958 volcanic deposits which destroyed the village of Capelinhos, Faial, Azores (E. C. Nelson).

Fig. 111 (left). Low scrubby growth of *Erica azorica*, with *Calluna vulgaris* and *Daboecia azorica*, on the middle slopes of Pico, Azores (E. C. Nelson).

Fig. 112 (right). Dense growth of older trees of *Erica azorica* in successional forest, northern side of Pico, Azores (E. C. Nelson).

feet high and more than thirty years old ...". Tall trees do exist on several islands at present (Fig. 112); for example, in "an old successional forest" in the reserve at Mistério da Prainha on Pico, "individuals of *E. azorica* are mostly emergent, 8 to 11 metres tall trees" (Ribeiro *et al.* 2005).

The first botanically minded person to report the Azorean besom heath was Johann Georg Adam Forster (1754–1794), who observed *Erica azorica* on Faial in the middle of July 1775 during the last leg of Captain James Cook's second circumnavigation. While HMS *Resolution* was anchored at Horta, Forster botanised, recording this heather as *E. scoparia* (Forster 1787), which led Guppy (1917: 491) to suggest erroneously that the species "has not since been recorded on Fayal but from other islands ...". Francis Masson must have seen it, too, two years later in 1777, but he left no written account. George Bentham was aware that the Azorean plant was somewhat different from the besom heath of mainland Europe, naming it *E. scoparia* var. *parviflora*. Hewett Cottrell Watson (1804–1881) recorded the besom heath during his visit to the archipelago aboard HMS *Styx* in the summer of 1842, when he also saw *Daboecia azorica* and *Calluna vulgaris* (Watson 1843, 1844). Christian Ferdinand Friedrich Hochstetter (1787–1860) was the first to regard it as a species distinct from the continental European plant, noting that it was taller, more tree-like, "*caule elatiori saepius vere arboreo*", with very numerous flowers and deeply bifid anthers. He reported specimens to 15 feet tall (Seubert 1844). Hochstetter proposed the name *E. azorica* and this was published by Seubert.

Although Francis Masson collected around 60 different lots of seed while he was in the Azores in 1777, he did not, it seems, gather any of the besom heath, and the Azorean plant was probably not introduced into cultivation in Britain until 1939, when Collingwood Ingram (1880–1981) — best known as "Cherry" Ingram because of his deep interest in flowering cherries — obtained material (Ingram 1970: 100) "on a small island within sight of where banana and orange trees were being cultivated". This rather ambiguous reference seems to relate entirely to São Miguel, as Ingram does not mention visiting any other island in the archipelago.

> Among the plants I collected in the Azores and which were successfully brought to England
> ... were a number of species which have proved, against all expectation to be perfectly
> hardy in my part of Kent. Who would have anticipated that a plant growing at sea-level
> in a island where even in mid-winter the temperature has never been known to fall below
> 40°F. would be unaffected by a temperature as low as 16°F. — twenty four degrees below
> that minimum? Yet that is exactly what had occurred with plants of the Azorean tree heath
> – *Erica azorica* ...

Established in Ingram's garden, the Azorean besom (tree) heath was unharmed by a "moderately
severe" frost that killed (by bark-split) several plants of *Calluna* that he had unwittingly also collected
in the Azores.

Ingram's introduction of *Erica azorica* is perhaps the plant most frequently seen today in cultivation,
although other plants raised from seed obtained later certainly exist, or have existed — I recall a
specimen growing on the Rock Garden in the National Botanic Gardens, Glasnevin, during the late
1970s and 1980s. Don Richards and David McClintock brought more material to England in 1974
(McClintock 1990b).

A notable characteristic of these Azorean heaths in cultivation in Ireland and Britain is that the
fine foliage appears to spiral along the stems and the shoots themselves to undulate — this is most
obvious in a shrub established in the heather garden at the Royal Horticultural Society's Gardens,
Wisley, which can only be described as lolloping along the ground like an unruly, shaggy, green
monster (Fig. 114, p. 186). I have not seen anything like it on my several visits to the Azores; all the
plants are erect, disciplined shrubs. Collingwood Ingram (1970: 199) described the plant he collected
as a "hardy tree heath with insignificant flowers but with attractive, dark green, fuzzy foliage."
McClintock (1990b: 56) characterised this heather well: "This I find delightful with its distinctive
curling waves of foliage, embellished in early summer by myriads of chestnut-coloured flowers,
admittedly small, but so numerous as to be decorative."

Fig. 113. Trackway, lined with trees of *Erica azorica*, leading from slopes of Pico to Magdalena, Pico, Azores (E. C.
Nelson).

Fig. 114. *Erica azorica* at the Royal Horticultural Society's Garden, Wisley, Surrey (E. C. Nelson).

Hochstetter (Seubert 1844) described two varieties, one with greenish, and the other with rosy flowers. In the field both variants are seen; plants that are under stress for whatever reason tend to produce intensely red flowers (Fig. 115). The corolla is more likely to be red on the sunlit side in most plants.

The most important question concerning *Erica azorica*, given its very close resemblance to *E. scoparia*, is whether it was present in the Azores before human settlement. If it was not there before the Portuguese discovered Santa Maria in 1432, then, like *E. erigena* in western Ireland, it may have been brought to the Azores accidentally by man. The evidence (unpublished; J. van Leeuwen, pers. comm., April 2006) from peat deposits indicates that three different members of the Ericaceae, with distinct pollen grains, were ancient inhabitants — *Erica scoparia*-type pollen, as well as pollen identified as coming from *Calluna* and *Vaccinium*, are present in deposits approximately 2,500 years old. The pollen curves for both *Calluna* and *Erica* show clear increases about 500 years ago: in other words, one impact of human settlement on the islands was an expansion of *Calluna* and *Erica*.

Fig. 115. *Erica azorica* in bloom; Pico, Azores (E. C. Nelson).

How *Erica* "*scoparia*" reached the Azores cannot be determined from pollen grains preserved in peaty sediments — but it is not the only plant of European origin on these volcanic islands, which would have been devoid of land-dwelling plants when they emerged from the mid-Atlantic Ocean. With *Calluna*, *Daboecia* and *Vaccinium* (all Ericaceae), *Erica* somehow bridged the ocean gap of around 1,300 kilometres.

Remarkably, the Azorean population of *Calluna vulgaris* has not noticeably deviated, in terms of morphology, from its continental congeners to form a new species, whereas the species of the other genera of Ericaceae differ enough to rank by themselves as endemics, unique to the archipelago: *Vaccinium cylindraceum* and *Daboecia azorica*, for example.

There is another curious fact about *Erica azorica*. Given that this species grows naturally on the islands of Flores and Corvo, the two westernmost islands in the archipelago, *E. azorica* is the only species in the genus that is native in the western hemisphere when that is defined in terms of continental plates. Flores and Corvo lie to the west of the Mid-Atlantic Ridge, on the North American tectonic plate. The other islands are to the east of the ridge, rising from the Azores Platform, in a region where the boundaries between the African and the Eurasian plates are not at all clear (Scarth & Tanguy 2001).

Fig. 116. "St Miguel ass"; sketch of a donkey laden with besom heath from J. & H. Bullar's *A winter in the Azores* ... (1841).

ETHNOBOTANY

Tutin (1953) noted that the heather provides excellent wood for fuel and that it was burnt green or dry. Bullar & Bullar (1841) enlivened their account of winter in the Azores with a sketch of a donkey laden with heather (Fig. 116). When Hewett Cottrell Watson ascended Pico at the beginning of July 1842, he and his companions halted on the upper slopes: "I should guess the spot ... [was] about 5000 or 5,500 feet in elevation. ... We formed our beds with green bundles of the *Erica*; and having made a good fire with the dead and dry branches of the shrubs, we passed the night more comfortably than the preceding night on the deal boards in Mr. Daubeny's house" (Watson 1843: 405).

Erica timber was also used for building and other constructions (Sjögren 1973), so that old trees are very scarce. On São Jorge, where a few very tall trees survived in 1993, I saw *E. azorica* trunks and stems used to make fences and gates.

Guppy (1917: 398) reported the use of *Erica azorica* by the inhabitants of the Azores in the early twentieth century:

> ... it is only used for cooking their food, the foliage serving as fodder, the leafy branches as litter in their stables, and the branches of the Ling ... and Tree-Heath ... as brushwood. The procuring of these materials seems to be one of the principal occupations of their lives. On the lower slopes of the great mountain of Pico one meets all through the year a constant string of men, women and bullock-drawn carts carrying loads of *Erica azorica*, *Calluna vulgaris*, *Myrica faya*, *Laurus canariensis*, *Ilex perado*, etc. ...

Guppy (1917: 376) also reported that the slopes of Pico "are often marked by linear copses of the Tree-Heath, which present a variety of strange patterns, that look from a distance like huge hieroglyphs on a light green ground" (Fig. 113). A landowner, by trimming the heather, achieved two results: "he obtains shade for his cattle in summer and shelter from the cold winds in winter" and by allowing *Erica azorica* trees to remain "near his boundary lines, and ... allow[ing] them to propagate themselves only on the borders of his property, ... his land is partially enclosed by a living tree-fence".

A unique use of *Erica azorica* was in fiction. It was the model for a fictitious heather named *E. hybrasiliensis* with which Margaret Elphinstone cloaked her "near-mythical" island of Hy Brasil (Elphinstone 2002, 2003).

Erica azorica Hochst. ex Seub., *Flora azorica*: 40 (1844). Type: "C. Hochstetter no. 30": lecto. not traced.

E. scoparia subsp. *azorica* (Hochst. ex Seub.) D. A. Webb, *Bot. J. Linn. Soc.* **65**: 259 (1972); *Flora europaea* **3**: 8 (1972).

E. azorica α *chlorantha* Hochst. ex Seub., *Flora azorica*: 40 (1844).

E. azorica β *erythranta* Hochst. ex Seub., *Flora azorica*: 40 (1844).

E. scoparia var. *parviflora* Benth., *Prodromus* **7**: 692 (1839).

DESCRIPTION. *Evergreen shrub* or, rarely, *tree* to ±11 m in height. *Leaves* thin, lamina and petiole covered with forward-pointing spicules, ± 3.5 mm long, ±1.5 mm broad, with pair of furrows underneath. *Inflorescences* numerous and crowded on shoots, appearing to form a raceme; individual inflorescence an umbel of 1–3 flowers on very reduced lateral branchlets; *pedicel* green, glabrous, ±2 mm long, curved, with bract and 2 bracteoles at mid-point; bract and bracteoles similar, ± 0.3 mm long, green, margins with minute hairs. *Calyx* green, ±1.2 mm long, with 4(–5) erect, equal lobes (when 5-lobed, one lobe much narrower), ±1 mm long, closely appressed to corolla, and tapering, not truncate, at the base. *Corolla* pale to dark chestnut-brown with white lobes or greenish with pink tinge, campanulate, glabrous, <3 mm long, with 4 equal, erect, triangular lobes ±1 mm long. *Stamens* 8, included; *filament* straight, ±0.6–1 mm long, very slender, 0.1 mm wide; *anther* brown, ±0.4–0.6 mm long, 0.4 mm across at base, 0.5 mm across at top (in frontal view), muticous, thecae fused in lower half, pore ±0.2 mm long. *Ovary* green, glabrous, ovoid, ±1 mm tall; *style* very short, ±1.5 mm long, tinged red towards apex; *stigma* deep red-purple, ±2.3 mm diameter, ±1.5 mm tall, conical (so that style and stigma together are mushroom-like), with 4 raised carpel lobes at summit. *Seeds* ± 0.57 mm long, boundaries of anticlinal walls channelled (see Fagúndez & Izco 2003).

FLOWERING PERIOD. April–June.

DISTRIBUTION & ECOLOGY. Azores (endemic) on all islands, in both natural and anthropogenic habitats, along roadsides, in ravines, on slopes, in moist places, reaching around 1,500 m altitude. Fig. 108.

Erica azorica is the dominant species in the tree layer of the woodlands above 600–750 m on the islands, with *Juniperus brevifolia* (Tutin 1953) as a companion. The *Ericetum azoricae* (Tutin 1953)

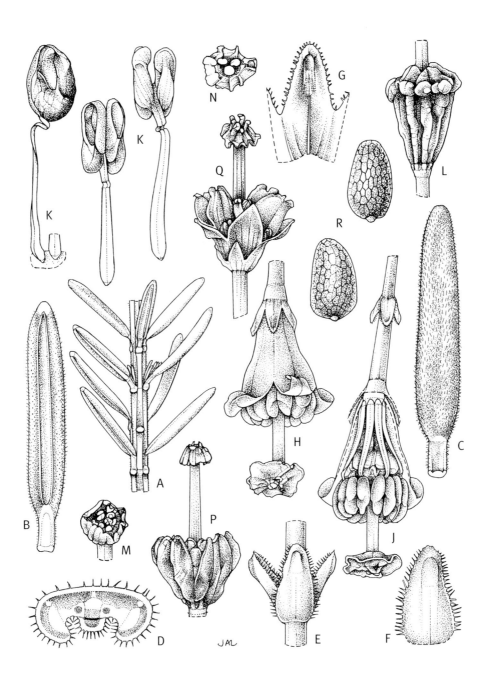

Fig. 117. *Erica azorica*. **A** leafy shoot × 6; **B** leaf from below × 12.5; **C** leaf from above × 13.5; **D** section through leaf × 34; **E** bract and bracteoles × 54; **F** bract × 65; **G** calyx lobe from back × 34; **H** flower from side × 13; **J** section through flower × 18; **K** stamens from side, front and back × 29; **L** ovary × 27; **M** style apex (stigma) from side × 26; **N** old style tip (stigma) from above × 26; **P** capsule from side × 18; **Q** open capsule with persistent calyx and corolla × 12; **R** seed from front and side × 24; from specimens grown by Allen Hall. Drawn by Joanna Langhorne.

— the vegetation zone dominated by *E. azorica* — extends to 1,500 m altitude. Above that, *Calluna vulgaris* becomes the dominant shrub, with *Daboecia azorica* (Guppy 1917; Tutin 1953).

One notable feature of the woodlands dominated by *Erica azorica* is a rich epiphyte flora, including many bryophytes, filmy ferns (*Hymenophyllum* spp.) and other ferns, as well as seedlings of the heather (Fig. 118). It is a remarkable experience to see young shrubs of *E. azorica* growing as epiphytes upon a mature tree, probably even their own parent.

Entomologists working on the herbivorous insects in the canopy of the *Erica azorica*-dominated forest have found that the heather supports more than 50 insect species and that "the number of herbivore species per individual tree crown was higher for *E. azorica* than for any other host" (Ribiero *et al.* 2005).

VERNACULAR NAMES. Azores heath, Azores tree heath (McClintock 1980b); urze (Azores).

CULTIVAR. I believe that the cultivar called *Erica scoparia* 'Minima' is a selected form of the Azores heath. It may indeed be the plant introduced by Collingwood Ingram. It has minute anthers that are remarkable because the thecae are entirely fused and there is often no septum between them.

Fig. 118. A seedling of *Erica azorica* growing epiphytically on a mature tree of the same heather; Pico, Azores (E. C. Nelson).

14. ERICA PLATYCODON (*E. scoparia* subsp. *platycodon* & *E. scoparia* subsp. *maderincola*)

THE CANARIAN BESOM HEATH

The Canarian besom heath was distinguished from the continental one, but only as a variety, in 1844 by Philip Barker Webb (1793–1854) and Sabin Berthelot (1794–1880) in their work about the plants of the Canary Islands. Noting that the Canarian plants were in all parts larger, they opined that the differences were not sufficient to make it a species. That was in the same year that Hochstetter described the Azorean species; Webb & Berthelot (1844) gave Hochstetter's name as a synonym of their *E. scoparia* var. *platycodon*. In 1993, Salvador Rivas-Martínez and his colleagues decided that the Canarian besom heath deserved to be treated as a separate species, although they did this without any explanation (Rivas-Martínez *et al.* 1993). Commenting on this, Professor Rivas-Martínez (pers. comm., 17 February 2005) noted that *E. scoparia* in southern Europe never grows into a tree, reaching 7 m in height, as does the heath on the Canary Islands. Moreover, the "leaves, bark, flowers, anthers and pollen are quite different and species status seems to us more adequate" (S. Rivas-Martínez, pers. comm., 17 February 2005). Fagúndez & Izco (2003) noted that the seeds of the two Tenerife populations that they studied were smaller and more rounded that those from continental populations of *E. scoparia*, with "irregular, oval or round" surface cells, while the seeds collected from plants in the Azores and Madeira were more elongated with tetragonal surface cells.

A difference in habit, which can be modified by environmental factors or even human usage, is not justification for specific status, and it is possible that the continental plants could attain tree-like proportions if allowed to grow unfettered. Regarding the bark of the Canarian besom heath, Antonio Machado (1976: 371) provided this comment: "The loose and hanging bark of *Erica scoparia* constitutes a unique habitat, housing the richest bark-fauna. In addition to its proper fauna, numerous terricolous species climb through it ... Sifting at *E. scoparia* ('tejo' in Spanish) is a gratifying expirience [*sic*] for any entomologist."

The differences in foliage have been noted by others, too: McClintock (1989d) wrote, "what I find noticeable ... is the way the sparser longer leaves stand out from the stems, nearly as much as in" the plants from Madeira and Porto Santo. As for the flowers, these are certainly larger then those of the European besom heath.

Erica platycodon occurs on only three of the Canary Islands: Tenerife, La Gomera (Fig. 119), and El Hierro. The heather inhabits a zone between 300 m and 1,200 m where the vegetation is dominated by *Myrica fayal* (fayal) and *Erica arborea* (tree heather). It is not known whether the Canarian besom heath once grew on the other islands, like tree heather (*E. arborea*) which Webb & Berthelot (1844) recorded clothing the cliffs at Peñitas de Chache on Lanzarote and in copses in the gorge at Ermita de Nuestra Señora de la Peña on Fuerteventura; *E. arborea* has not been recorded from either island in recent decades so is either very rare or, more probably, extinct on them (Bramwell & Bramwell 1974; Hohenester & Welss 1993; Chilton 1999).

Fig. 119. *Erica platycodon* subsp. *platycodon* on La Gomera, Canary Islands (E. C. Nelson).

THE MADEIRAN BESOM HEATH

A similar heath occurs on Madeira and Porto Santo (Hansen 1969; Sjögren 1972; McClintock 1994; all as *E. scoparia*). McClintock (1989d) distinguished it as a fourth subspecies of *Erica scoparia*, and Rivas-Martínez *et al.* (2002) have maintained it as a subspecies, but of the Canarian besom heath — *E. platycodon* subsp. *maderincola*, although again without providing any justification for this in print: "it is smaller and thinner, up to 4 m [tall] and 20 cm diameter, and the leaves and corolla are small" (S. Rivas-Martínez, pers. comm., 17 February 2005).

This besom heath is widespread on Madeira (Sjögren 1972; McClintock 1994; Jardim & Francisco 2000), from the coast, especially on the northern side of the island, to at least 1,300 m altitude (Sjögren 1972). It is much less common on Porto Santo, being confined to Pico Branco at the extreme east of this much smaller island (Jardim *et al.* 1998; Hood 2006). As well as being widespread, *Erica platycodon* subsp. *maderincola* is dominant over much of the higher parts of Madeira almost to the exclusion of other plants, such as, for example, on the plateau between Fonte da Pedra (1,022 m) and Pico das Urzes (1,418 m) — urze is the Portuguese word for heather — at the western side of Paúl da Serra (Fig. 120). In a few places, such as at Pico das Urzes, *E. arborea* also occurs but the besom heath is present in far greater abundance (Fig. 121). Sjögren (1972) noted, however, that on the highest ground (above 1,500 m altitude) *E. arborea* is the more abundant heath.

Erica platycodon subsp. *maderincola* acts as a pioneer species: seedlings and small plants are frequent on roadside cuttings. Natural regeneration of forest on Madeira commences with colonisation by the heathers *E. arborea* and *E. platycodon*, along with *Vaccinum maderense* (Sjögren 1972: 72). The besom heath seems to be primarily a re-seeder, reluctant to re-sprout from a rootstock after fire.

Fig. 120 (left). *Erica platycodon* subsp. *maderincola* near Pico das Urzes, Madeira (E. C. Nelson).

Fig. 121 (right). *Erica platycodon* subsp. *maderincola* (foreground) with *E. arborea* near Pico das Urzes, Madeira (E. C. Nelson).

ETHNOBOTANY

On Madeira, besom heath is still occasionally used to make besoms: I saw a street-cleaner with such a besom in late December 2006 in Ribeira Brava. The stout stems, cut into suitable lengths, of *Erica platycodon* are also used, as the stems of *E. azorica* are in the Azores, to make fences, especially along the sides of *levadas* (the contour-following irrigation channels). In the far north-west of the island in the vicinity of Seixal, Ribeira da Janela and Porto Moniz, fences are made in an entirely novel manner. Pieces of *E. platycodon* at least a metre long, complete with the foliage and fine twigs, are cut and then stuck into the ground upright to form a dense windbreak of dead stems. It does not matter that these die, and I saw piles of recently cut heather waiting to be used at Ribeira da Janela. The dead twiggy stems, closely packed, form very effective hedges, protecting vines and other crops from the strong north winds that batter this part of Madeira.

Fig. 122. Garden plots hedged with windbreaks composed of dead *Erica platycodon* subsp. *maderincola* in the vicinity of Ribeira da Janela on the northern coast of Madeira (E. C. Nelson).

CULTIVATION

It is not known when *Erica platycodon* came into cultivation in Britain, and even today it is only to be found in specialised collections. It is possible that plants came to England from the Canary Islands in the 1690s, among consignments of "trees" and seeds sent to Samuel Doody, Curator of the Chelsea Physic Garden, London, by Thomas Simmons (see p. 206) (Francisco-Ortega & Santos-Guerra 1999).

In recent decades, several attempts have been made to propagate and introduce yellow-foliaged variants, but they either turned green in cultivation, or have been lost. 'Lionel Woolner' (Woolner 1974; Proudley & Proudley 1974) from Las Mercedes (Sierra de Anaga), Tenerife, is believed to be extinct, while a second propagation from the same place, named 'Mercedes Gold', turned an indistinguishable, ordinary green. *E. platycodon* subsp. *maderincola* 'Levada Gold' (® E. 2010-08) is a recent selection from Madeira, propagated by David Edge, with golden yellow foliage in winter.

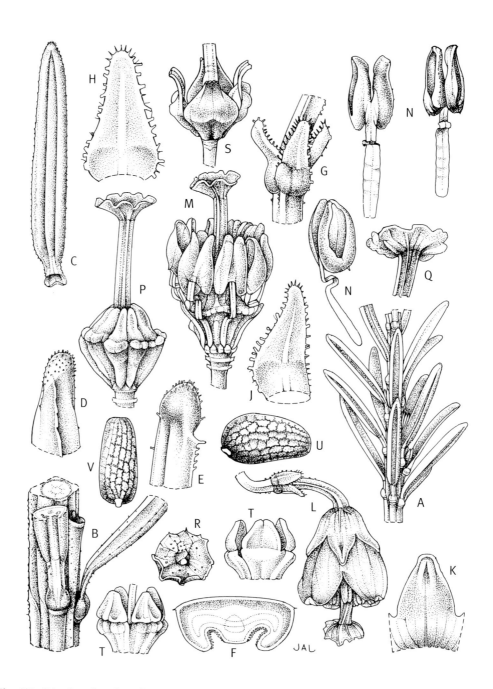

Fig. 123. *Erica platycodon* subsp. *platycodon*. **A** leafy shoot × 4; **B** leaf nodes × 7; **C** leaf from back × 6; **D** leaf tip from side × 14; **E** leaf tip ¾ view × 24, **F** transverse section through leaf × 22; **G** bract and bracteoles × 26; **H** bract × 48; **J** bracteole × 50; **K** calyx lobe × 20; **L** flower from side × 12; **M** ovary and stamens × 12; **N** stamens from front, side and back × 20; **P** gynoecium × 12; **Q** stigma from side × 18; **R** stigma from above × 18; **S** ripe capsule × 20; **T** old capsule from side × 14; **U** seed from side × 34; **V** seed from front × 34; from specimens grown by Allen Hall. Drawn by Joanna Langhorne.

Erica platycodon (Webb & Berthel.) Rivas Mart.,Wildpret, del Arco, O. Rodr., P. Pérez, García Gallo, Acebes, T. E. Díaz & Fern. Gonz., *Itinera geobotanica* 7: 355 (1993).

E. scoparia var. *platycodon* Webb & Berthel., *Phytogr. canar.* **3** (3): 14 (1844). Type: "Tejo Canariensium" (not traced).

E. scoparia subsp. *platycodon* (Webb & Berthel.) A. Hansen & G. Kunkel, *Cuad. bot. canar.*: 26–27, 79 (1976).

subsp. **platycodon**

ILLUSTRATIONS. Bramwell & Bramwell (1974): 77, fig. 56 (line drawing); Hohenester & Welss (1993): fig. 55 (colour photograph).

DESCRIPTION. *Evergreen shrub*, or very rarely tree to 6 m. *Leaves* apparently glabrous, but when young with minute hairs at apex and a few gland-tipped hairs along edge, ±6 × 1 mm, patent, with pair of furrows underneath; petiole sparsely pubescent when young. *Calyx* green, cup-shaped, base truncate, ± 2 mm long, with 4 equal lobes, ±1 mm long, triangular, not overlapping, glabrous. *Corolla* greenish white to deep red with paler lobes, campanulate to ±3 mm long, ±0.2 mm diameter, 4-lobed; lobes, ±1.2 mm long, overlapping minutely at base. *Stamens* 8, included; filament with pronounced sigmoid bend at apex below anther so that the stamens seem to rest on the rim of the ovary; *anther* ±1–1.2 mm long, 0.6–0.7 mm broad (frontal view) and 0.6 mm across (in lateral view), muticous, pore proportionately very large, 0.7 mm long; *filament* ±1 mm long. *Ovary* turbinate, ±2.5 mm high, glabrous, green, prominently ribbed; *style* ±3 mm long, cylindrical; *stigma* disc-shaped, concave, ±1.5 mm diameter, deep red (the same colour as the immature anthers). *Seeds* ± 0.53 mm long, boundaries of anticlinal walls channelled (see Fagúndez & Izco 2003).

FLOWERING PERIOD. February–May

DISTRIBUTION. Canary Islands only; presently reported on El Hierro, La Gomera (to 1,200 m) and Tenerife (to 800 m) (Bramwell & Bramwell 1974; Hohenester & Welss 1993). Fig. 108.

ETYMOLOGY. Greek: *platy* = wide, broad; *codon* = bell.

VERNACULAR NAME. Techo (Bramwell & Bramwell 1974).

subsp. **maderincola** (D. C. McClint.) Rivas Mart., Capelo, J. C. Costa, Lousã, Fontinha, R. Jardim & M. Seq., *Itinera geobotanica* **15** (2): 701 (2002).

E. scoparia subsp. *maderincola* D. C. McClint, *Yearbook of The Heather Society* **3** (7): 35 (1989). Holo. "ex Madeira in horto Bracken Hill, Platt. Kent", *D. McClintock*, 31 May 1980. **WSY**0100339! (see Miller & Nelson 2008).

E. scoparia subsp. *scoparia* sensu Webb & Rix, *Flora europaea* **3**: 8 (1972).

ILLUSTRATIONS. Jardim & Francisco (2000); R. Hood, *Heathers* **3**: 51 (2006).

DESCRIPTION. *Evergreen shrub* or small tree to 4 m (Jardim & Francisco 2000); stems glabrescent, bark turning grey, with prominent infrafoliar ridges. *Leaves* 10–12 mm long, spreading, in whorls of 3 or 4, appearing glabrous but covered with minute hairs (visible under microscope), straight or curving upwards, furrowed underneath. *Calyx* green, conical, tapering into pedicel at base, ±0.8 mm long, 4-lobed, lobes erect, triangular, not overlapping at base. *Corolla* greenish pink, ovoid, 2–3 mm long, to ±1.3 mm diameter, 4-lobed, lobes erect, not overlapping at base. *Stamens* 8, included, ±2.5 mm long; *filament* curving with an inconspicuous sigmoid bend at apex or straight, ±1.5 mm long, broadest at base, ±0.25 mm across, tapering upwards narrowing to ±0.1 mm; *anthers* brown, muticous, thecae

± 0.8–0.9 mm long, ±0.4 mm across at base, with obovate pore ±0.7 mm long on lateral surface, fused, broadening towards base. *Style* cylindrical, ±2 mm long; *stigma* discoid to saucer-shaped; *ovary* ovoid-turbinate, ±0.8 × 0.75 mm, tinged red, glabrous, more or less tapering into style. *Seeds* ± 0.55 mm long, boundaries of anticlinal walls channelled (see Fagúndez & Izco 2003).

FLOWERING PERIOD. April–June.

DISTRIBUTION & ECOLOGY. Madeira and Porto Santo only (Fig. 108, arrowed). *Erica platycodon* subsp. *maderincola* occurs at sea-level on the northern coast of Madeira, and is recorded also from the highest peaks (on Pico Ruivo, 1,850 m; *fide* Sjögren 1972). Generally, on Madeira, this species is more abundant at lower altitudes than *E. arborea* and it is also more frequent than tree heather in the "cloud zone vegetation". On high-altitude grazing land, *E. platycodon* also is the dominant species — in contrast, Sjögren (1972) noted, *E. arborea* has "frequently been eliminated by grazing, cutting and burning" from the over-grazed grassland communities above 1,100 m altitude.

The Madeiran besom heath is drought-tolerant, and can be found colonising steep slopes and cliffs, as well as volcanic deposits, and sandy and gravelly soils (Sjögren 1972).

ETYMOLOGY. Toponym; from Madera (Madeira), and *-incola* = an inhabitant.

VERNACULAR NAME. Urze durazia.

CULTIVARS. 'Madeira Gold' (see Appendix I: p. 381), originally described as having yellowish green foliage. Consequently it provided the holotype for *Erica scoparia* f. *aureifolia* D. C. McClint. (McClintock 1988a). However, the foliage colour is not golden or even yellowish green. 'Levada Gold' (® E. 2010-08), introduced in 2010 by Forest Edge Nursery, Wimborne, Dorset, is also from Madeira. It has yellow-green foliage and may be more reliable and distinctive.

NOTES

[1] *Erica leptostachya* Guthrie & Bolus, *Flora capensis* **4** (1): 217 (1905) (see also Oliver 1994b: 27–28): the type specimen was among *Erica* specimens associated with William McNab (1780–1848) (see p. 47), now in the herbarium of the Royal Botanic Gardens, Kew (holo: *ex herb. McNab 405a, 109*; **K**; iso **BOL**) but its original source is not known.

[2] The data provided by Herrera (1987) for flower number were as follows:

Calluna vulgaris	16,495 ± 4,827 (2)	
Erica ciliaris	52 ± 23 (7)	
Erica scoparia	56,902 ± 13,699 (5)	

where flower number indicates the mean number in individual plants at peak bloom per m², followed by the standard error; number of plants is indicated in parentheses (Herrera 1987: Table 1, p. 71).

[3] Jessen *et al.* (1959) gave the name *Erica scoparia* var. *macrosperma* to the fossils they extracted from the Gort peat. They provided the detailed description in English, quoted on p. 178, as required by the *International code of botanical nomenclature* (Vienna Code), Art. 36.1 (McNeill *et al.* 2006), so the varietal name was validly published. It cannot be synonymous with *Bruckenthalia spiculifolia* as reported by McClintock (1991c). For further discussion see pp. 177–179.

TREE HEATHERS
Erica arborea, E. lusitanica

All heathers are woody plants and, not surprisingly for a vast genus, they display a diverse range of habits, from low, almost herbaceous, subshrubs to substantial trees. None of the northern species fits into the "almost herbaceous subshrub" category; all are bushy shrubs or trees possessing woody stems and branches.

The distinction between a shrub and a tree is not easily defined, and a particular species of heather may not fit neatly into one or other category. It is important to bear in mind that "tree" and "shrub" refer to modes of growth, not to botanical classes of plants (Mitchell 1982). Definitions vary as demonstrated by the three sources (chosen at random) that I quote here: for example Mitchell (1982) regards a tree as a woody plant that has "the ability to grow over 6 m tall on a single stem", whereas Allaby (1998) defines a tree as a "woody plant that can grow more than 10 m tall". While they may disagree about the height a plant must reach before being classed as a tree, Mitchell (1982) and Allaby (1998), as well as Coe (1985), who gave no attainable dimension, agree that a tree possesses a "readily recognised" single trunk. Woody plants of lesser height with numerous branches at ground-level are termed shrubs (Coe 1985; Allaby 1998). Trees can be transformed into shrubs: cut down a tree of *Erica arborea* leaving the rootstock undamaged, and numerous shoots, which may then form a bushy thicket of stems, will spring from the rootstock.

Erica arborea, the archetypal tree heather, can form a substantial small tree. I have seen examples on the Atlantic islands — on La Gomera in the Canary Islands, and on Madeira — and there are

Fig. 124 (left). Tree heath, *Erica arborea*, on La Gomera, Canary Islands (E. C. Nelson).
Fig. 125 (right). Multi-stemmed tree heath, *Erica arborea*, on Madeira (E. C. Nelson).

reliable reports of trees of tree heather in Spain, the Arabian peninsula and elsewhere. Yet *E. arborea* is not the only northern *Erica* species which, allowed to reach full maturity, unmolested by grazing animals or human beings seeking firewood, can form a tree. Accepting 6 metres (approximately 20 feet) as the height to be attained, the other indisputable tree-heaths are the besom heaths, especially *E. azorica* and *E. platycodon* (see pp. 182–196).

The two species paired here are easily confused at first glance, but they may not be closely related (cf. Fagúndez & Izco 2010). They both bloom in the late winter and spring and have small white flowers in profusion. They can quickly be distinguished when not in bloom by the composition of the indumentum on the stems — in *Erica arborea* plumose hairs (these are multicellular, multiserrate hairs;

Tree heaths   Painted by Christabel King   PLATE 15

LEFT TO RIGHT:

E. arborea 'Albert's Gold' 🌿 ♀ (*E. arborea* f. *aureifolia*). Tall, bushy shrub (±2 m tall × ±0.8 m spread); foliage bright yellow in winter and spring, changing to golden green in summer; flowers sparse, white, slightly fragrant; March–May.

Albert S. Turner, The Hamlet, Hall Green, Birmingham, noticed this as a sport on a plant of *Erica arborea* var. *alpina* ('Alpina') (see below) in 1971, and it was propagated, named (® no. 9) and introduced by Denbeigh Heather Nurseries, Creeting St Mary, Ipswich, Suffolk, about 1975. AGM 1993.

Turner was a keen plantsman, raising seedlings of *Calluna* and conducting his own selection programme, aiming for "coloured foliage varieties, hoping to find one with double flowers" (Turner 1973). *Calluna* 'Hamlet Green', a distinctive cultivar with unusual foliage that is orange, yellow and green in winter, and 'Mauve Mist' with grey foliage and mauve flowers, were seedlings he found. He pondered this conundrum: "By careful selection could the red pigment in a purple-flowered plant be eliminated, even after many generations, to produce a blue-flowered heather?"

E. arborea 'Alpina' ♀ (*E. arborea* var. *alpina*). Tidy, compact shrub when young, yet reported to have reached at least 2.6 m in height and 7.6 m in width at the Royal Botanic Gardens, Kew, in about 50 years (Bean 1950), and in 1979 there was a specimen twice as tall, 17 ft (5.2 m), at the Royal Horticultural Society's Gardens, Wisley (McClintock 1989e); foliage vivid green; flowers small, white, closely packed in cylindrical spikes; pedicels white or green (never tinged pink); mid-May (cf. Turpin 1981). (Painted 21 May 1985; cultivated at the Royal Botanic Gardens, Kew.)

This plant came from relatively high altitude (c. 1,400 m asl) in Serranía de Cuenca in central Spain, and was collected by Georg Dieck, Zöschen, Germany, in 1892. By his own account, he only got to find this heather, which was growing above the tree-line with *Erica australis* in a very wild and unapproachable place, after he had won the favour of the most powerful of the local brigands, one Don Antonio de Torriz. (Dieck (1899: 85) used the phrase "allmächtigen Räubers", omnipotent robber, to characterise Don Antonio.) The Cuenca tree heather was propagated and introduced around 1899. Dieck (1899) stated that the foliage was light green and the habit was slim like a poplar ("pappelschlanke"). The Kew plant, noted above, was hardly poplar-slim.

There is a lot of uncertainty about the number of clones that were originally propagated by Dieck, but it is now considered improbable that more than one clone is represented in present-day gardens. Dieck's plant is hardy, hardier than clones of *Erica arborea* from other sources. AGM 1993.

Erica arborea var. *alpina* has yielded one sport, 'Albert's Gold' (see above), and is the seed-parent of *E.* × *afroeuropaea* and *E.* × *krameri*.

E. lusitanica (unnamed) plant cultivated by Major-General P. Turpin and shown at the Royal Horticultural Society on 19 February 1985. (Painted 20 February 1985.)

E. × veitchii 'Gold Tips' 🌿 ♀ Bushy, low shrub (±0.65 m tall × 0.6 m spread); foliage bright green when mature, the young shoots tipped golden yellow in spring; flowers white; March–June. (Cultivated in the Royal Botanic Gardens, Kew.)

'Gold Tips' arose as a chance seedling in Maxwell & Beale's nursery, Broadstone, Dorset, England, before 1948 (Turpin 1979). When introduced by Maxwell & Beale Ltd, the clone was named *E. arborea* 'Gold Tips'. AGM 1993.

see Oliver 2000: 26, fig. 2) and simple (unbranched) hairs are present, whereas in *E. lusitanica* there are no plumose hairs. In bloom, they may be separated by the presence or absence of hairs on the filaments of the anthers: *E. lusitanica* has hirsute filaments, a characteristic very rarely mentioned in descriptions (cf. Hansen 1950; Fagúndez 2006; Fagúndez & Izco 2007), while in *E. arborea* they are hairless.

15. ERICA ARBOREA. Tree heath, tree heather.

This remarkable heather, which can be spectacular in flower, probably has the most extensive natural range in the entire genus, including the species native in tropical Africa. *Erica arborea* has a latitudinal range close to 60° east–west and a longitudinal one of about 50° north–south, both equating to around 5,000 kilometres. Only *Calluna vulgaris* can match this, as indicated elsewhere (see p. 39).

Given its range in Europe, including the coastal regions and islands of the Mediterrean Sea, *Erica arborea* was most probably known to early Greek naturalists including Theophrastus (372–287 BC) who came from Eresos, a village on the Aegean island of Lesbos — also, perhaps, the home of the famous lyric poet Sappho (fl. 610–570 BC). Theophrastus would have known *E. manipuliflora* as it is also a native plant on the island. When he lived in Athens, presiding over the Peripatetic School, as Aristotle's successor, Theophrastus would probably also have been familiar with the same heathers because they would then have occurred within the city's hinterland. The Roman naturalist Gaius Plinius Secundus (Pliny the Elder) (AD 23–79) surely knew *E. arborea* as well.

Erica arborea occurs as far west as the Canary Islands (Fig. 124) and Madeira (Sjögren 1972; Fig. 125) in Macaronesia. It inhabits the Iberian Peninsula and extends eastwards through southern France into Italy, the Balkan states, Bulgaria and Greece. On the southern shores of the Mediterranean, tree heather occurs in northern Africa from Morocco to Tunisia. While present on the Balearic Islands, Corsica, Sardinia, Sicily (in a very restricted area) and Crete (Figs 126 & 127), *E. arborea* is not found on Cyprus or Malta, although extermination, as on Fuertaventura and Lanzarote in the Canary Islands, could explain these absences. From the Bosporus, north-western Turkey, tree heather extends eastwards, close to the southern shore of the Black Sea, as far as the region of Karabük; its

Fig. 126 (left). *Erica arborea* in flower, Mylia, Crete (E. C. Nelson).
Fig. 127 (right). Flowers of *Erica arborea*; Mylia, Crete (E. C. Nelson).

occurrence here is discontinuous, with seemingly isolated populations near Sinop, and farther to the east between Ordu and Trabzon. Populations of *E. arborea* are known on the eastern side of the Black Sea in Abkhazia, in the northern Caucusus, where the species is listed as rare. North of that, between Sochi and Tuapse, in the south of Russia, the tree heather reaches it eastern limit in Europe.

There are populations of *Erica arborea* on the Arabian Peninsula. The first person to report the tree heath there was the Finnish botanist Pehr Forsskål (1775) (see p. 140) who listed it as "ERICA *scoparia*" adding "floribus albis" — *E. scoparia* does not have white flowers. He was very probably confused by Linnaeus's treatment of *Erica* in *Species plantarum* in which the descriptions are sometimes incorrect (see p. 94)! Forsskål's report was evidently the basis for the indication of "E. scoparia" in the south-western quarter of Arabia on Professor Joakim Frederik Schouw's map (1823) for the genus *Erica* (Fig. 128), one

Fig. 128. Distribution of the genus *Erica*, including *Calluna* (as *E. vulgaris*), as known to Joakim Frederik Schouw in the early 1820s.

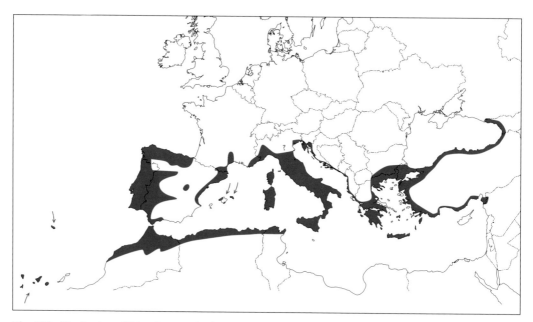

Fig. 129. Distribution of *Erica arborea* north of the Tropic of Cancer; populations in Arabian Peninsula and north-east Africa are not shown.

of the first maps published showing plant distribution patterns (Nelson 1993b). In Saudi Arabia, tree heather is a relict species and is rare and vulnerable, the small populations occurring in scattered localities at high altitudes, in the cloud zone, where it can form trees more than 10 m tall — there is a herbarium specimen in the Royal Botanic Gardens, Kew, from Asiv Province at 2,500 m altitude labelled "found almost exclusively in canyons and ravines, tree 15–18m tall". In Yemen, there are substantial communities of *E. arborea* in the Jebel Melhan mountains, at altitudes above 2,000 m; there it varies greatly in stature, according to annotations on herbarium labels, from a "rock plant" to a small tree.

As well as bring present in the north-west of the continent, *Erica arborea* inhabits several widely separated locations in Africa north of the Equator. Perhaps the most remarkable population of the tree heather is on Emi Koussi, a huge dormant stratovolcano known to have erupted in the Holocene in the Tibesti Mountains; at 3,415 m altitude, it is the highest mountain in the Sahara. *E. arborea* trees are reported at altitudes between 2,500 m and 3,000 m on Emi Koussi (Bruneau de Miré & Quézel 1959: 66): "Il s'agit en général d'énormes individus atteignant 5 à 6 mètres de hauteur et un diamètre de 30 à 40 centimètres à hauteur d'homme. Les jeunes pieds, quoique tour à fait exceptionnels, ne sont pas absolument absents, et au moins trois ont été observés."

The species is also reported in Sudan (Hedberg 1957), Ethiopia (Hedberg & Hedberg 2003), and Somalia (Hedberg & Hedberg 2003; Nepi 2006), and in Congo (Hedberg 1957). In Ethiopia, *Erica arborea* occurs in the mountains above about 2,200 m altitude (Hedberg & Hedberg 2003). Tree heather inhabits the "alpine" zone of other mountains in tropical east Africa, in Uganda, Kenya and Tanzania, between 1,600 m and 4,500 m altitude (Bentje 2006), including Mount Kenya and Mount Kilimanjaro which is about its southern limit, a little below 3°S latitude (Fig. 129). (Reports of *E. arborea* in southern Tanzania are probably due to confusion with another species.)

With so extensive a geographical range, it is not unexpected that *Erica arborea* is variable in vegetative and floral morphology, yet Pichi-Sermolli & Heiniger (1953) considered that there was no significant distinction between African (particularly Ethiopian) populations and European ones with continuous variation in many characters. Recent studies of DNA, which probed the ancestry of living populations from the extremities of the species' range, show they are certainly related: *E. arborea* appears to be monophyletic (Pirie *et al.* 2011; pers. comm.). The most recent common ancestor of its extant populations was found in tropical East Africa (rather than Europe) suggesting that the species has invaded Europe and Macaronesia on two different occasions (Désamoré *et al.* 2011). Indeed, we should now think of *E. arborea* as an African species (see Pirie *et al.* 2011: figure 2), its fragmented pattern of distribution best explained by migrations out of eastern Africa followed by extinctions in northern Africa when the Sahara expanded.

LONGEVITY & REGENERATION

The very extensive geographical range displayed by *Erica arborea* may have a lot to do with the capacity of individual plants to survive for long periods of time. Even if the leafy shoots, larger branches and the main trunk are destroyed by fire or grazing animals, or are removed for firewood, a well-established tree heath has a chance to regenerate from its underground lignotuber (Mesléard & Lepart 1989) (Fig. 130). This means that a single plant could live for decades, if not centuries, even though its leaves are reported to have an individual life span of just two and a half years, and the stems only about 30 years (Mouillot *et al.* 2002). There does not seem to be any information available on the longevity of individual plants. This ability to regenerate is shared with other members of the Ericaceae, including *Arbutus unedo* (strawberry tree), about which Professor David Webb (1948) made this telling comment: "... observations ... in Kerry make it apparent that the individual trees are here so long-lived as to be almost immortal ... a fair proportion of the Irish *Arbutus* trees are sprung from seed which ripened over a thousand years ago." The same must be true, surely, about

Fig. 130. Exposed rootstock of *Erica arborea*; Madeira (E. C. Nelson).

old plants of tree heath, although research in Corsica suggests that the lignotuber declines in its ability to produce new shoots as it ages (Mesléard & Lepart 1989).

Other northern heathers have lignotubers and can regenerate from these after fire, including *Erica australis* (Cruz & Moreno 2001; Silva *et al.* 2003), *E. lusitanica* (Silva *et al.* 2002, 2003), *E. manipuliflora* (Kazanis & Arianoutsou 2004; Pausas & Paula 2005), *E. multiflora* (Lloret & Lopez-Soria 1993; Vila & Terradas 1995), and *E. scoparia* (Silva *et al.* 2002, 2003; Paula & Ojeda 2006). Some species can regenerate, too, from root-stocks, but they could not be said to have lignotubers; these include *E. ciliaris* (Rose *et al.* 1996; Gallet & Rozé 2002), *E. cinerea* (Bannister 1965), *E. mackayana* (Webb 1955), *E. tetralix* (Bannister 1966), *E. vagans* (Turpin 1984), *Daboecia cantabrica* (Woodell 1958), and *Calluna vulgaris* (Gimingham 1972).

However, *Erica arborea* also colonises habitats by means of seedlings. In Mediterranean forests, such as that dominated by *Quercus suber* (cork oak) in southern Spain, *E. arborea* and *E. scoparia* are the dominant species in the soil seed bank, as well as being present in the forest understorey (Díaz-Villa *et al.* 2003). Tree heath will invade abandoned fields in Corsica, and so has spread widely during the past century (Mesléard & Lepart 1991). In undisturbed laurel forests (characterised by species of *Laurus*, *Ocotea*, *Persea* and *Apollonias*) in the Canary Islands, almost three-quarters of the seeds in the soil seed-bank comprised *E. arborea* which will behave as a pioneer species if the forest is destroyed; Bramwell & Bramwell (1974: 171) stated that the tree heath was "ecologically aggressive replacing felled laurel species in many places." While the heather seeds will not germinate in shade, if a gap is created in the canopy, they will sprout quickly (Fernández-Palacios & Arévalo 1998). That fresh seeds of tree heath need light for germination was also demonstrated by Valbuena & Vera (2002).

ETHNOBOTANY

Undoubtedly this wide-spread heather has numerous uses. It is a valuable source of honey in the Mediterranean region (Persano Oddo *et al.* 2004). Isolated trees on the summits of El Hierro (Canary Islands) capture fog, dripping abundant quantities of fresh water which collects in cisterns (Marzol-Jaén 2011). In Ethiopia and many other places, the wood is used as fuel (Siebert & Ramdhani 2005). In Spain branches are used to "thatch" the roofs of huts and grain stores, and utensils such as spoons (Pardo de Santayana *et al.* 2002), and even necklaces, jewel cases and musical instruments are carved from the wood. Long-lasting rustic fences can be made from branches too (Fig. 132).

The woody rootstocks of the ancient tree heaths are excavated for turning into the bowls of tobacco-pipes (Everett 2000a): it is said that the best wood comes from the lignotubers ("burls") of trees of *Erica arborea* that are at least a century old, and the preferred source remains Corsica. There are various tales about the "invention" of wooden tobacco-pipes, familiarly called brier pipes. The *Oxford English dictionary*[1] defines this use of "brier, briar, *n.*" thus:

> The White Heath (*Erica arborea*), a native of the south of France, Corsica, etc., the root of which is extensively used for making tobacco-pipes (introduced into England about 1859); also a pipe of this wood. So brier-root, brier-wood; brier-wooder (*nonce-wd.*), a smoker of a brier pipe.

According to the *OED* the earliest instances of the use of brier in connection with tobacco pipes both come from the *Tobacco trade review* which was launched by William Reed early in 1868. Advertisers used the phrases "Heath Pipe: in Bruyer Wood" and "Bruyer Pipes", and thus brier pipe — sometimes spelled briar — apparently came from the French *bruyère*[2] (heather), and it is not linked to the similar word meaning a prickly, entangling shrub such as a rose or bramble.

Fig. 131. Rustic fence of tree heath branches on a walking route in Madeira; most of the scrub up-slope is composed of *Erica arborea* (E. C. Nelson).

There are, however, earlier instances to be found of the use of brier/briar in connection with species of tobacco pipes. In *The Times* (of London) on 5 June 1861 (p. 15), Messrs Davis & Johnstone announced an auction of the "fancy stock" of a bankrupt tobacconist "comprising meerschaum, briar root, and other pipes". On 7 November 1863, again according to *The Times* (5 November: 16), Messrs Furber & Price were to hold an auction of the "Stock in Trade" of another tobacconist, including "briar-root pipes". It is note-worthy that both these advertisements referred to briar-root pipes. That the term briar-root/briar-root was common currency in London by the mid-1860s is also shown by reports of the activities of local burglars and pickpockets, as reported by *The Times* of 17 October 1864 and 22 March 1866 (Nelson 2008b).[3] Similar advertisements appeared in, for example, *The Irish Times*, established in Dublin in March 1859, Edward Keevil's Wholesale Pipe Warehouse, a Dublin emporium, even made a poor anagram out of the names, advertising "RAIRB TOOR SEPIP. — BRIAR ROOT PIPES at Continental Prices..." (Nelson 2009c). Albert Einstein's brier pipe is a prized exhibit[4] of the Smithsonian Institution's National Museum of American History!

CULTIVATION

Various authoritative sources, starting with William Townsend Aiton (1811: 402–403), allude to *Erica arborea* being cultivated in England in the 1650s. This may not be so, for they rely for this early date on a demonstrably unreliable entry in a seventeenth-century catalogue of the University of Oxford Botanic Garden. Stephens & Browne (1658: 59) listed "Erica major flore albo ... great white flowred Heath" among the heathers then cultivated in Oxford. Those names were used in John Gerard's *Herball* (1597), where the accompanying description suggests that Gerard was most probably referring to white-blossomed bell heather, *E. cinerea* (see Nelson 1998: 43): "Of this kind there is another sort with white flowers, but seldome found or seen, unless here and there a plant amongst the other sort." However, Stephens & Browne (1658) were undoubtedly referring not to the original 1597 edition of Gerard's book, but to Thomas Johnson's revised edition of 1633, because the page number that they cited, 1381, is for it (the page has the illustration of "Erica maior flore albo Clusij"). Johnson (1633) had revised Gerard's text, providing a new description based on Charles de l'Écluse's *Rariorum aliquot stirpium per Hispanias observatarum Historia* (1576): "This [heath], saith *Clusius*, which is the largest that I haue seene, sometimes exceeds the height of a man ... it hath plentiful store of floures growing all alongst the branches, ... well smelling, white and beautifull." The accompanying woodcut actually depicts the Irish heath, *E. erigena* (Nelson 1998; Ramón-Laca & Morales 2000), while the text refers to *E. lusitanica* (Ramón-Laca & Morales 2000), not to *E. arborea* (as I had suggested, Nelson 1998). In other words, nothing can safely be concluded about the history of *E. arborea* in English gardens from the list of heathers in Stephens & Browne's 1685 catalogue.

On the other hand, *Erica arborea* may have been introduced from one of its remoter outposts in the late seventeenth century, if documents among the papers of Sir Hans Sloane record successful transport of live plants. "Trees from the Canaryes received from Tho: Simmons 1694" listed "Bresto. Ericæ species", and "Seeds from yᶜ Canaryes Collected by Tho: Simmons" has the same (Francisco-Ortega & Santos-Guerra 1999): the Spanish name for any heather is "brezo". Simmons (who has not been identified) sent these to Samuel Doody (1656–1706), Curator of the Society of Apothecaries' Physic Garden at Chelsea, so there is a chance that *E. arborea* from the Canary Islands was cultivated there in the late 1690s. Given his use of the plural — "Ericæ" – there is also the real chance that Simmons sent *E. platycodon*, too.

There seems no doubt that tree heath was cultivated by the middle of the following century; the elder Aiton attributed its introduction to "about 1748", and certainly by the late 1780s there must have been a plant at the Royal Garden in Kew (Aiton 1789).

The principal variant in cultivation is the relatively dwarf, bushy plant variously named *Erica arborea* var. *alpina* or 'Alpina' (see p. 199). There are also cultivars with gold or yellow foliage, *E. arborea* f. *aureifolia* D. C. McClint. (McClintock 1988a: 61).

HYBRIDS

No naturally occurring hybrids involving *Erica arborea* have been confirmed beyond doubt, although there have been occasional, repeated claims that plants intermediate between *E. arborea* and *E. lusitanica* have been detected in Spanish habitats (Bayer 1993; Fagúndez 2006) — these would represent *E.* × *veitchii* (for discussion see p. 322). Other alleged crosses are generally dismissed as incorrect: *E. arborea* × *scoparia* (in fact, queried by the botanists who reported it (Rivas Goday & Bellot 1946: 151)) and *E. arborea* × *umbellata* for which Rivas Goday & Bellot (1946: 152) published the name *E.* × *lazaroana* (cf. Bayer 1993: 506; Pizarro Domínguez 2007[5]).

In recent decades, attempts have been made, some successfully, to create artificial hybrids involving *Erica arborea*. Kurt Kramer (Edewecht-Süddorf, Germany) produced *E.* × *oldenburgensis* (see p. 346) using pollen from *E. carnea*. The reciprocal cross failed for him and also for David Wilson (British Columbia, Canada). Kramer has also tried using *E. arborea* as the seed-parent with *E. australis*, *E. cinerea*, *E. erigena* and *E. umbellata*, and as a pollen-parent on *E. australis*, *E. cinerea*, *E. spiculifolia* and *E. umbellata*, all without viable seed being produced. Barry Sellers (Norbury, London) also failed to get viable seed from *E. australis* after cross-pollination with *E. arborea*.

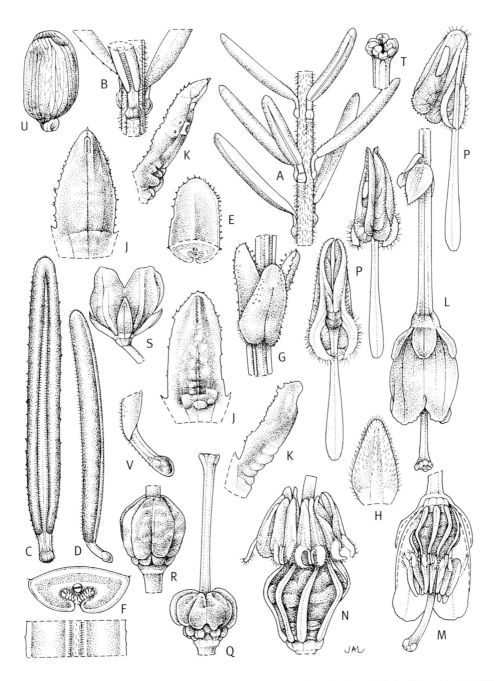

Fig. 132. *Erica arborea*. **A** leafy shoot × 10; **B** leaf node × 12; **C** leaf from back × 14; **D** leaf from side × 12; **E** leaf tip × 27; **F** transverse section of leaf with detail of leaf back × 47; **G** bracts and bracteole × 21; **H** bracteole × 14.5; **J** calyx lobes from back × 28; **K** calyx lobes from side × 28; **L** flower from side × 8; **M** longitudinal section of flower × 11; **N** stamens and ovary × 15; **P** stamens from back, front and side × 24; **Q** gynoecium × 12.5; **R** capsule × 6; **S** dehisced capsule with persistent calyx × 6; **T** style apex (stigma) × 27; **U** seed × 54; **V** petiole and leaf base × 14; from specimens grown in RHS Garden Wisley, no. W 1988 5552. Drawn by Joanna Langhorne.

On the other hand, Kramer succeeded in obtaining the first known hybrid involving a northern heath and a Cape heath, using the northern *Erica arborea* as a seed-parent; the pollen came from *E. baccans*, and the handsome progeny is named *E.* × *afroeuropaea* (see p. 337). He also got seedlings from *E. arborea* cross-pollinated with *E. curvirostris*, but these were lost. Other Cape heath pollen-parents were unsuccessful, including plants identified as *E. baueri*, *E. blandfordii* and *E. formosa*.

Erica arborea L., *Species plantarum*: 353 (1753). Type: "Monspelii"; lecto. **UPS**: Herb. Burser, vol. XXV: 42 (Jarvis & McClintock 1990; Jarvis 2007: 497).

E. caffra L., *Dissertatio de Erica*, 9 (1770).

E. scoparia Thunb., *Dissertatio botanica de Erica*, 48 (1785), non L.

E. procera Salisb., *Trans. Linn. Soc.* **6**: 328 (1802).

E. elata Bubani, in Hoffmansegg & Link, *Flora Portugaise* **1**: 411 (1809)

E. acrophya Fresen., *Flora* **31**: 604 (1838).

E. riojana Sennen & Elías, *Bol. Sociedad Iber. Ci. Nat.* **28**: 177 (1930).

E. lazaroana Rivas Goday & Bellot, *Anales Jardín Bot. Madrid* **2** (6): 152 (1946).

Arsace arborea (L.) Fourr., *Ann. Soc. Linn. Lyons* **17** (new series): 113 (1869).

Ericodes arboreum (L.) Kuntze, *Revisio genera plantarum* **II**: 389 (1891)

ILLUSTRATIONS. F. L. Bauer, *Flora graeca sibthorpiana* **4**: tab. 351 (Sibthorp & Smith 1823); J. Smythies, *Flowers of south-west Europe*: 287 (3) (Polunin & Smythies 1973); E. Sierra Ràfols, *Flora iberica* **4**: lam. 183 (p. 503) (1993).

DESCRIPTION. *Evergreen tree* to 7–15(–18) m tall, or *bushy shrub*, with a lignotuber (Mesléard & Lepart 1989); much branched; young shoots hirsute with eglandular or glandular plumose hairs, and with short, simple, eglandular or gland-tipped hairs. *Leaves* in whorls of 3, ±(2.5)–5–7 mm long, linear, upper surface flattened, sparsely hirsute with forward pointing spicules; edges with short hairs, sometimes with glandular tips; margins revolute, sulcus linear, underneath rounded (keeled), and sparse small hairs at tip, *petiole* hirsute. *Flowers* white or very pale pink, terminal on short leafy shoots, in umbels of 2–4; *pedicel* ±4 mm long, terete but broadening at apex below calyx, curving, glabrous, translucent, with bract and 2 bracteoles towards the base; bract narrow-lanceolate, ±1 mm long, hirsute with very short hairs along margins and sulcus; bracteoles smaller and broader in proportion, without obvious sulcus. *Calyx* cup-shaped, base truncate ±1.5 mm long, 4-lobed, lobes ±1 mm long, apex thickened, margins with short hairs, some gland-tipped. *Corolla* cyathiform-campanulate, ±3 mm long, ±2.5 mm diameter, 4-lobed, lobes erect, ±1 mm long, margins very minutely toothed, in bud lobes overlapping alternately. *Stamens* 8, erect, ±3 mm long, inserted at base of ovary, visible at mouth of corolla but not exserted; *filaments* with sigmoid bend towards top, broadest at base tapering upwards, glabrous; *anthers* ±1 mm long, elliptical in lateral view, with obovate pore, thecae erect, fused at attachment to filament, red when young; spurs present, ±0.4 mm long, pointing downwards, white. *Style* ±3.5 mm long, erect, broadest at base, tapering to apex, glabrous; *stigma* capitate, pink; *ovary*, obovoid, narrowest at base, ±1 mm tall, ±1 mm diameter, glabrous, ± 25–30 ovules per locule; *nectary* prominent, nectar profusely produced. *Seed* 0.4–0.6 × 0.2–0.2 mm (Nelson 2006a). n = 12 (Fraga 1983: 538; see Nelson & Oliver 2005b).

FLOWERING PERIOD. February–May (–July).

DISTRIBUTION & ECOLOGY. Europe: Madeira (Sjögren 1972; McClintock 1994); Portugal eastwards to Georgia and southern Russia on the eastern shore of the Black Sea. Africa: Canary Islands (Bramwell & Bramwell 1974), Morocco (Ball 1877; Fennane & Ibn Tattou 2005) to Tunisia, Chad, Congo,

Sudan, Eritrea, Ethiopia, Somalia, Kenya, Uganda, Tanzania (Beentje 2006). Asia: Turkey (Black Sea coast), Saudi Arabia, Yemen. **Maps:** Hansen (1950: 39), Browicz (1983: 11), Niedermaier (1985: Abb. 4), Bolòs & Vigo (1995: 35). Fig. 129 (p. 202).

Tree heath is found in garrigue, maquis, phrygana and open woods, usually on acidic, siliceous soils but also on limestone (in Somalia; see Nepi 2006), and attains 4,500 m altitude in east Africa (Beentje 2006).

Erica arborea has been reported as naturalised in a few places in England, at Budleigh Salterton, Devon, and at Brook in Surrey. It has also escaped from cultivation on Jersey in the Channel Islands (Clement & Foster 1994). The tree heather is a naturalised alien in Australasia: in Australia (New South Wales (Hosking *et al.* 2003), South Australia, Tasmania (Rozefelds *et al.* 1999), Victoria) and New Zealand (Sykes 1981).

ETYMOLOGY. Latin: *arbor* = a tree, hence *arboreus* = tree-like.

VERNACULAR NAMES. Tree heather, tree heath, white heath (English); bruyère blanche, bruyère arborescente, bruyère en arbre, bruyère des Maures (French); Baumheide (German); αρείκι, ρείκι, ρείκι δενδρώδες (Greek); brezo blanco, bruc de la flor blanca (Spain, for other names in use in Spain, see Bayer 1993; Pardo de Santayana *et al.* 2002); betouro, queiroga, quiróga, torga, urze, urze-arbórea, urze-branca, urze-molar, urze mollar (Portuguese); trädljung (Swedish).

CULTIVARS. See p. 199 and Appendix I (p. 363).

16. ERICA LUSITANICA. Portuguese heath, Spanish heath.

Often confused with *Erica arborea* (tree heath), this species can be distinguished clearly by two characters, both easily detectable using a good hand-lens: there are no branched hairs on the stems, but there are hairs, tiny, simple ones, on the lower portions of filaments of the stamens.

Erica lusitanica (Portuguese heath) is native on the Iberian Peninsula, but its distribution there is disjunct (Fagúndez & Izco 2007; see discussion below). This heather is found throughout Portugal (except the extreme north-west), extending eastwards into western and some central provinces of Spain. There are also populations on the northern side of the Cordillera Cantabrica in north-central and north-eastern Spain, ranging into south-western France as far north as Arcachon (to 44°30'N, *fide* Webb & Rix 1972; Besançon 1978a), but this species is absent from north-western, central and south-eastern Spain (Fagúndez & Izco 2007).

However, by being a prolific seed producer, as well as being capable of regenerating from rootstocks and roots, the species has demonstrated a propensity to become a pest far from its native habitats: in New Zealand, it is a notifiable weed; Australia especially Tasmania; California; Brittany; and also south-western England. I recently received a report that *Erica lusitanica* has escaped from cultivation in Hawai'i. *E. lusitanica* has become naturalised in Dorset, on Lytchett Heath at Poole (Hemsley 1905; Anonymous 1905a; Clement & Foster 1994). At one period, the early 1920s, it was even stolen by the cartload, presumably for sale as "lucky white heather". The heather was deliberately planted "in rough ground just outside the flower-garden" about 1876 and started producing seed "about 1880"; by 1905 "between one and two acres of ground are covered with thousands of bushes", some over two metres tall including one measuring 11ft (3.35 m) (Anonymous 1905a). In Cornwall, *E. lusitanica* has also become naturalised (Clement & Foster 1994); Thurston (1930) reported that the heather was planted "many years ago at Doublebois railway station, where hundred of plants are now growing, and flowering about February."

Fig. 133. *Erica lusitanica* (bright green shrubs) as a naturalised weed in northern California, USA (E. C. Nelson).

I have seen *Erica lusitanica* regenerating from established rootstocks along road sides in northern California (north of Fortuna) (Fig. 133); seedlings were frequent in the area (see below), too. And, in a former "white heather" nursery in Cornwall (see below), the heather also has regenerated from buried roots.

Research in New Zealand provides some of the reasons for *Erica lusitanica*'s success as a weed. A single mature shrub, between four and six years old, with around twenty flowering branches, can produce about nine million seeds every year (Mather & Williams 1990). While a proportion of these will germinate soon after release, around half to three-quarters of those seeds remain dormant in the topmost layers of soil. The buried seed-bank thus established can contain almost half a million seeds per square metre, and these may remain viable for four years — but of course they are replenished each year as long as the heather blooms. Germination is stimulated by fire, among other factors, and given that *E. lusitanica* can also resprout from underground rootstocks after fire, it is superbly well adapted to survive, and to expand its territory, once it gains a foothold. With few natural "enemies" likely to retard its progress, *E. lusitanica* could be unstoppable in many places.

HISTORY

Erica lusitanica was not distinguished by a separate name until 1799, when Karl Asmund Rudolphi (1771–1832), a Swedish-born German physician and naturalist, called sometimes the "father of helminthology", published a discussion of the various heaths known by the name *E. scoparia*. He applied the name *E. lusitanica* to the species that Carl Thunberg had called *E. scoparia* in his *Dissertatio botanica de Erica* (1785). However, Rudolphi's work was published in a rather obscure periodical and was not well-known, so for many decades after 1834, the Portuguese heath was listed as *E. codonodes*, a name published by John Lindley (1834). Although Rudolphi (1799) wrote that he had received specimens from Lisbon, no specimens associated with him can be found: there is none in the Museum of Evolution in Uppsala (**UPS**), where Thunberg's herbarium is preserved.

Charles de l'Écluse (Carolus Clusius) had seen this heather near Lisbon during the winter of 1564–1565, and it was painted for him; the water-colour survives in the famous *Libri picturati*, now in the Biblioteka Jagelloniska, Kraków, Poland (Ramón-Laca & Morales 2000), and the woodcut made from this was printed in several works, including his own *Rariorum aliquot stirpium per Hispanias observatarum historia* (l'Écluse 1576). The engraving was also used incorrectly in John Gerard's *Herball* (1597) to illustrate an entirely different heather: Gerard's description points to *Erica cinerea*.[6]

IBERIAN POPULATIONS

During his research on the morphology of seeds of Spanish *Erica* species, Fagúndez noticed that there was a difference between seeds gathered from populations of *E. lusitanica* in north-eastern Spain and those from populations in the south-west (Fagúndez & Izco 2007). The mean length of the seeds from the six north-eastern populations exceeded 0.37 mm, whereas the south-western seeds were smaller, the mean length being 0.36 mm or less. The difference in mean seed length suggested to Fagúndez that two subspecies might be involved, and he sought further differences between the populations of *E. lusitanica*. Other characteristics that tend to separate the plants from the two regions are flower shape, which is always difficult to define, and size, with the north-eastern plants generally having a shorter corolla (3–4.5 mm long) with more of the style exserted (extending 1.5–2 mm beyond the corolla). Corolla length and style exsertion are linked — a shorter corolla means more exposure of the style when the length of the gynoecium remains more or less constant. A further character is also ill-defined and somewhat confused; in the north-eastern plants the filaments have sparse, minute hairs only on the lower half, and mostly at the base. Apart from the hairiness of the filaments, all these characteristics could be subject to environmental influences — even the size of seeds may not be constant and may be affected by differences, both regional and seasonal, in climate, especially rainfall, or soil fertility. The differences need to be tested by cultivation experiments to eliminate environmental effects.

Another notable, and surprising, distinction is that shrubs of Portuguese heath in the north-eastern populations are attacked by insects which cause galls, but galls have not been reported from the south-west of the Iberian Peninsula. Fagúndez & Izco (2007) suggested that the galls they have found on the plants in north-eastern Spain could be linked with the larvae of the gall-midge *Myricomyia mediterranea* (F. Löw, 1885) (Diptera; Cecidomyiidae) because the same insect has been identified as inducing galls on *Erica arborea* (Frutos 1986).[7] However, this is a most unreliable difference; galls on *Erica* species, and the insects that cause them, need much more thorough botanical and entomological investigation.

In trying out the "key" provided by Fagúndez & Izco (2007) on cultivated plants, I found that these have hairs only on the lower half of the filaments, suggesting that the plants growing in England at present all belong to *Erica lusitanica* subsp. *cantabrica*, the north-eastern subspecies. However the other distinctions — corolla length and style exsertion — were equivocal, and it is noteworthy that the seeds of cultivated plants are substantially larger (around 0.4 mm long; see Hall 2006) than the wild-collected seeds according to the dimensions given by Fagúndez & Izco (2007).

Data in Huckerby *et al.* (1972) also suggests the seeds of *E. lusitanica* often can be substantially longer (0.4–0.65 mm; mean 0.48 mm). These data tend to suggest that environmental factors are implicated in the apparent differentiation.

While clear geographic separation — disjunction — does lead to populations becoming distinct, the differences between the plants of *Erica lusitanica* growing in the disjunct populations on the Iberian Peninsula are not especially marked; there is substantial overlap. Much more research, including detailed DNA studies of the populations, is needed before the two subspecies proposed by Fagundéz & Izco (2007) can be accepted with absolute certainty. And yet it may be that gall-inducing insects are very particular, discriminating between two plants that look the same to us — insects can often be excellent "taxonomists", indicators of characteristics that cannot be seen by the human eye.[8] It may be informative that in *E. carnea*, some cultivars (clones) are susceptible to gall-midges while others apparently are not, even when "growing cheek by jowl" (McClintock 1990c).

CULTIVATION

Erica lusitanica is an attractive garden shrub, reaching over 3.5 m in height (supported). Allen Hall (2002) recorded that plants can attain 1.8 metres in four years. Vigorous plants are usually smothered in white flowers from late in autumn until late in spring. The buds are often tinged pink (Fig. 134). The Portuguese heath is tolerant and moderately hardy; old, well-established plants will stand more frost than young seedlings.

Fig. 134. *Erica lusitanica* 'Sheffield Park' flowering in late September 2006 at the Royal Horticultural Society's Garden, Wisley, Surrey (E. C. Nelson).

The only marked variation in cultivation is yellow foliage (*Erica lusitanica* f. *aureifolia* D. C. McClint.; McClintock 1988a: 62); 'George Hunt' is the only named yellow clone, and was the first heather to be protected by plant breeders' rights, in 1972

Portuguese heath was planted on a commercial scale in south-western England during the 1920s at Rowan's White Heather Nursery, at Kernock in the Tamar Valley, Cornwall. The stock (cuttings) came from the keen gardener and Rector of Ludgvan, the Revd Arthur Townshend Boscawen (c. 1862–1939), who was "one of the pioneers of the Cornish fruit, flower and vegetable-growing industry" (Desmond 1977). The plantation survived into the twenty-first century. At Kernock, 10 hectares of heather was harvested for its white flowers, for sale between October and May as "lucky white heather", especially for Saint Andrew's Day (30 November) and Burns' Night (25 January). The plantation was grubbed out several years ago, so that white heather is no longer produced at Kernock. Afterwards *Erica lusitanica* regenerated from broken roots that remained buried on the site.[9]

NOMENCLATURE AND HISTORY IN ENGLISH GARDENS

The name "Erica lusitanica" was printed in several English horticultural publications during the early 1770s. The first appearance seems to have been in William Malcolm's *A catalogue of hot-house and green-house plants* (1771), where it was specified also as "Portugal upright red heath" (see Harvey 1988). Malcolm had a nursery at Kennington Turnpike in South London, and his 1771 catalogue was one of earliest to employ Latin binomials as the primary names for listing his stock (see Harvey 1983: 9). These names were repeated in Weston's *The English flora: or, a catalogue of trees*, one of the components of his *Flora anglicana* (1775). In neither Malcolm's nor Weston's catalogue was the Latin binomial accompanied by a description, only the phrase "Portugal upright red Heath" which evidently was intended as an English name — the phrase suggests that Malcolm and Weston meant *Erica australis*, because *E. lusitanica* never has red flowers. It is surprising that neither David McClintock nor Dr John Harvey commented on the use of this name. Consequently, Harvey (1988) was misled into dating the introduction of *E. lusitanica* as 1771, although there is no reliable evidence for it being cultivated in English gardens at this period.

HYBRIDS

The chance, nursery-raised hybrid *Erica* × *veitchii* arose when *E. lusitanica* and *E. arborea* crossed in Veitch's Exeter nursery during the late 1800s. As noted above (p. 206), there are occasional reports that this hybrid occurs in the wild in Spain — the most recent one is discussed on p. 322.

Erica lusitanica has been used in modern cross-breeding programmes by Dr John Griffiths (Leeds, Yorkshire, England), Kurt Kramer (Edewecht-Süddorf, Germany) and Barry Sellers (Norbury, London, England). Griffiths obtained viable seed when using pollen from *E. lusitanica* on the following species: *E. carnea*, *E. ciliaris*, *E. erigena* and *E. vagans*, but the seedlings were lost and no decision could be made about their hybrid status. The reciprocal crosses involving *E. ciliaris* and *E. erigena* as pollen-parents were both failures for him. Griffiths also reported obtaining non-viable seed from *E. cinerea*. Kramer's unsuccessful seed-parents were *E. carnea*, *E. cinerea*, *E. erigena* and *E. spiculifolia*. Remarkably, he has succeeded in raising seeds from *E. lusitanica* when pollinated with *E. carnea*, and the progeny do appear to be hybrids. In 1996, Sellers (see pp. 125 & 349) successfully raised three seedlings after cross-pollination of *E. lusitanica* with *E. erigena* but only one survived to flowering stage. Examination of the flowers and foliage demonstrated clearly that the plant was intermediate between the parents. This clone has been given the cultivar name 'Lucy Gena' (® E.2009:03) (Nelson 2010b) but it remains to be seen whether it will be garden-worthy.

Erica lusitanica Rudolphi, *J. Botanik* [Schrader] Bd. 2 (pt 4): 286 (1799), non Malcolm, *Cat. hot-house & green-house plants* (1771) nom. nud. sine descr.[10] Type: not traced, "aus Lissabon".

E. scoparia Thunb., *Dissertatio botanica de Erica*: 48 (1785), non L.

E. polytrichifolia Salisb., *Trans. Linn. Soc.* **6**: 329 (1802).

E. codonodes Lindl., *Edwards's Bot. Reg.* **20**: tab. 1698 (1834).

E. lusitanica subsp. *cantabrica* Fagúndez & Izco, *Nordic J. Bot.* **24**: 390 (2007).

ILLUSTRATIONS. *Edwards's Bot. Reg.* **20**: tab. 1698 (1834); M. Smith, *Curtis's Bot. Mag.* **131**: tab. 8018 (1905); E. Sierra Ràfols, *Flora iberica* **4**: lam. 183 (p. 503) (1993); J. Smythies, *Flowers of south-west Europe*: 287 (9) (Polunin & Smythies 1973); B. Keel, *Illustrations of alien plants of the British Isles* (Clement *et al*. 2005: 128).

DESCRIPTION. *Evergreen shrub* attaining 1(–3.5–4.5) m in height, regenerating from a lignotuber, with roots that can extend at least 1.6 m deep (Silva *et al*. 2002, 2003); young shoots hirsute with simple, smooth hairs of varying lengths, longest ± 0.25 mm long. *Leaves* in whorls of 4 (sometimes in 3s), linear with edges parallel or lanceolate and narrowing slightly towards tip, to 7 × 0.5 mm, light to dark green, glabrous but with microscopic spicules at tip and along edges; margins revolute, contiguous; sulcus extending from tip to petiole, closed, lower leaf surface completely concealed. *Inflorescences* usually numerous and crowded towards ends of shoots, appearing to form elongated panicles: 1–4 flowers in each terminal umbel at tip of short, leafy lateral shoots; pedicel curving or straight, glabrous, with small, translucent, narrowly triangular bract about ¼ way from base, and 2 similar bracteoles about middle; bract and bracteoles c. 1 mm long, margins ciliate. *Calyx* ± conical, tapering into pedicel, c. 1 mm long, with 4 triangular lobes c. 0.5 mm long, white, waxy, thicker in texture than corolla; lobes with sulcus c. 0.25 mm long, margins ciliate towards apices. *Corolla* white, often tinged pink in bud, narrow-campanulate to tubular-obconical, c. 5 mm long, glabrous, lobes ± 1 mm long, erect or slightly recurved. *Stamens* 8, included; *filaments* white, c. 2.5 mm long, flattened, strap-shaped, c. 0.2 mm broad but narrowing towards top, completely encircling the base of the ovary, curving outwards in lower part and with pronounced sigmoid curve at top, with microscopic hairs at base extending about ⅓ way up on sides and surfaces; anthers dark brown, c. 0.8 mm long, dorsifixed about ⅓ way up; with 2 paler (white to tan), hirsute, narrowly-triangular, downward-pointing spurs attached about ⅓–½ way along; thecae fused in lower part, ± ovate, surface minutely papillose; pore oval, ± half size of theca. *Ovary* vase-shaped to ovoid, stipitate, 8-ribbed, 4-locular, c. 1.5–2 mm long, c. 1 mm in diameter, widest above middle, green or tinged red, glabrous, with nectary ring at base (not as prominent as in some species), nectar copious; apex slightly emarginate; 15–25 ovules per locule; *style* c. 2.5 mm long, terete, straight, white, stained red and slightly expanded towards apex; *stigma* emergent, tinged red, shallowly cup-shaped. *Seed* (0.25–)0.4–0.65 mm long, ellipsoidal.

FLOWERING PERIOD. December–March(–May). In cultivation in England *Erica lusitanica* generally blooms between December and March or April, but can have flowers as early as late September.

DISTRIBUTION & ECOLOGY. Europe: much less widespread than descriptions and reports suggest and occurring in widely scattered and relatively small pockets; in western and southern Portugal; south-western and north-eastern Spain, but absent from central, north-western and eastern Spain; south-

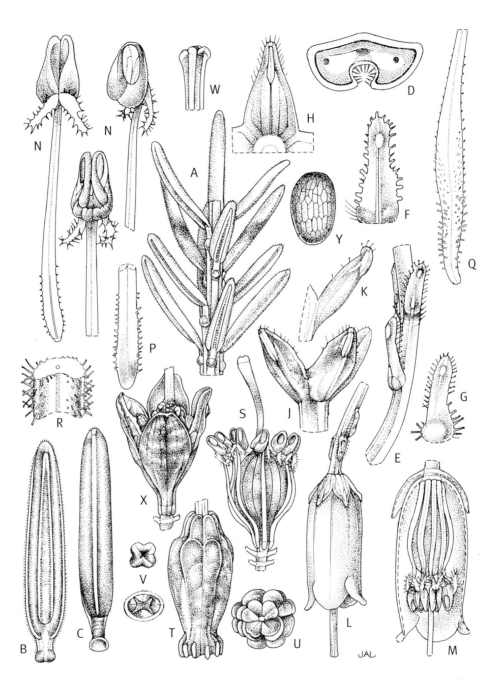

Fig. 135. *Erica lusitanica* 'Sheffield Park'. **A** leafy shoot × 7.5; **B** young leaf from back × 15; **C** mature leaf from back × 15; **D** transverse section of leaf × 43; **E** peduncle with bract and bracteole × 39; **F** bract × 34; **G** bracteole × 34; **H** calyx lobe × 9; **J** side view of calyx × 9; **K** calyx lobe from side × 9; **L** flower × 7; **M** longitudinal section of flower × 9; **N** stamens from back, front and side × 34; **P** base of filament × 26; **Q** glandular filament × 26; **R** section of filament with gland-tipped hairs × 54; **S** gynoecium and stamens × 9; **T** ovary × 15; **U** ovary from above × 15; **V** style apex (stigma) from above × 37; **W** style apex (stigma) from side; **X** dehisced capsule ×12; **Y** seed × 14; from specimens grown by Allen Hall and RHS Garden Wisley, no. W 1996 0006A. Drawn by Joanna Langhorne.

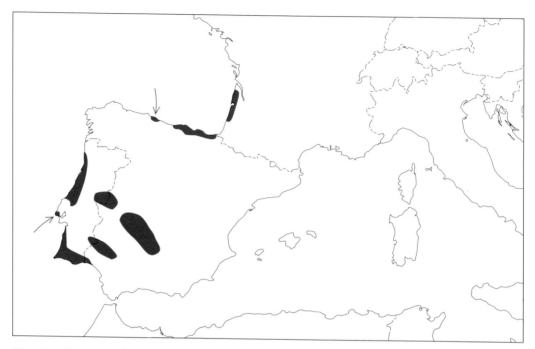

Fig. 136. Distribution of *Erica lusitanica*.

western France (Gironde). **Maps:** Hansen (1950: 37), Dupont (1962: 229, carte 52), Fagúndez & Izco (2007). Fig. 136.

The populations in Portugal and south-western Spain were placed in *Erica lusitanica* subsp. *lusitanica* by Fagúndez & Izco (2007), while those in north-eastern Spain and France were segregated into *E. lusitanica* subsp. *cantabrica* Fagúndez & Izco (see discussion above, p. 211).

As noted, *Erica lusitanica* is naturalised in several places including southern England, and potentially is a serious threat in most areas, including Tasmania (Rozefelds *et al.* 1999), New Zealand (Sykes 1981), and California.

ETYMOLOGY. Latin toponym: *Lusitania*, Portugal.

VERNACULAR NAMES. Portuguese heath is the name recommended by English authorities, including McClintock (1980b), Stace (1991), and Clement & Foster (1994) (and is also used in German, Portugisiche Heide). However, in other English-speaking countries, the name Spanish heath is used, particularly New Zealand (e.g. Sykes 1981; Mather & Williams 1990) and Australia (Tasmania, New South Wales, Victoria and South Australia). In Iberia, the names used are more or less those applied to *Erica arborea*: brezo blanco, brezo (Spain; see Bayer 1993); urze-branca, urze-de-Portugal, queiroga, torga (Portuguese).

ETHNOBOTANY. This is frequently used as a substitute for white-flowered *Calluna vulgaris* when "lucky white heather" is required (see, for example, photograph in McClintock 1970b), although it had nothing whatever to do originally with that particular tradition (see pp. 50–51).

CULTIVARS. See Appendix I (p. 379). There are several named cultivars in commerce including 'George Hunt' and 'Sheffield Park' (Fig. 134).

NOTES

[1] brier, briar, *n. The Oxford English dictionary*. Second edition. 1989. *OED Online*. Oxford University Press. (http://dictionary.oed.com/cgi/entry/50027501 accessed 25 February 2007).

[2] The French word *bruyère* can be traced back to the twelfth century and is said to derive from the Latin *brucaria*, meaning a field of heather, which, in turn, is derived from mediaeval (tenth century) Latin *brucus* (see Cazalbou 2000).

[3] There are also earlier instances of "brier" recorded from the United States: the first seems to be *The brier-wood pipe*, the title of verses by the Irish-born writer Charles Dawson Shanly (1811–1875), published on 6 July 1861 in *Vanity fair*.

[4] See http://americanhistory.si.edu/collections/object.cfm?key=35&objkey=35 (accessed 25 February 2007).

[5] Pizarro Domínguez (2007) recently illustrated the "author's original voucher" specimen (**MAF** 20627), which looks like plain *Erica arborea*, and provided an "Abbreviata diagnosis vel character essentialis" for *E. × lazaroana* that appears to represent a direct translation into Latin of Rivas Goday & Bellot's original Spanish description (1946: 152).

[6] In my paper describing Gerard's *Herball* (Nelson 1998), I mistook the engraving as a depiction of *Erica arborea*. Ramón-Laca & Morales (2000), benefiting from study of the original water-colour in Krakow, corrected that misidentification (see also Ramón-Laca *et al.* 2005).

[7] Other gall-midges (Diptera: Cecidomyiidae) are reported to cause galls on *Erica arborea*, and on other species of *Erica*, but most records (and especially that of Fagúndez & Izco 2007) are not supported by rigorous taxonomic studies of the gall-midge species involved, nor of their host ranges, and cannot be accepted as authoritative (K. M. Harris, pers. comm., 3 March 2007).

All of the known gall-midge galls on *Erica* are small, apical, "artichoke" galls, composed of leaf-like scales. Each contains one or more larvae during the brief period of development, and gall-midge species from two different lineages (*Myricomyia* and *Dasineura / Wachtliella*) are involved.

Dauphin & Aniotsbehere (1993), in an illustrated account of galls on plants in France, recorded *Myricomyia mediterranea* (Löw) from *Erica arborea* and *E. scoparia* (but not *E. lusitanica*), and noted a second species of gall-midge, *Dasineura zimmermanni* (Tavares), on *E. arborea* from the Iberian Peninsula. They also recorded *Dasineura ericaescopariae* (Dufour) from *E. arborea, E. carnea* and *E. scoparia*; *Dasineura broteri* (Tavares) on *E. ciliaris*; and *D. elegans* Tavares on *E. umbellata* in Portugal. Another species, *Wachtliella ericina* (Löw), is recorded as abundant in Europe on *E. arborea* and *E. carnea* (Harris 1966; Shaw & Sutherland 1966; Skuhravá & Skuhravy 2003). Further study is needed to clarify this complex and should ideally involve DNA sequencing of the gall-midges to establish host ranges and possible synonymies.

Thus, while galls are reliably reported on more than a dozen species of *Erica*, the insects involved are not authoritatively identified. The heathers known to bear galls are *Erica arborea* §, *E. australis* ‡, *E. carnea* ‡, *E. ciliaris* §, *E. cinerea* ‡, *E. erigena* ‡, *E. lusitanica*, *E. mackayana* (Cabo de Peñas, Asturias, Spain, 10 July 2007; *leg* E. C. Nelson. **WSY**0077265), *E. manipuliflora* ‡, *E. multiflora* ‡, *E. scoparia* §, *E. tetralix* (Bannister 1966), *E. umbellata* ‡ and *E. vagans* § (‡ see McClintock 1990c; § = galls collected, mainly in northern Spain, by McClintock, Nelson & Small in 1982: see McClintock 1983). (My thanks to Dr K. M. Harris for his assistance and comments.)

The other insect species noted by Dauphin & Aniotsbehere (1993) are not gall-midges: *Eriococcus devoniensis* (Green) is a scale insect (Coccidae) originally described from specimens collected in 1896 at Budleigh Salterton, Devon, on *Erica cinerea* (http://www.sel.barc.usda.gov/catalogs/eriococc/Eriococcusdevoniensis.htm accessed 2 March 2007), while *Nanophyes niger* (Waltl) is a weevil (Coleoptera, Curculionidae).

[8] Portuguese populations of *Erica lusitanica* and *E. arborea* both harbour the same species of mealybug, *Cucculococcus arrabidensis* (Neves), formerly thought to belong to a separate genus, *Lusitanococcus* (Marotta & Franco 2001).

[9] Richard Harnett (Kernock Park Plants), pers. comm., 2006.

[10] See commentary under "Nomenclature and history in English gardens", p. 213. The use of "*Erica lusitanica*" by nurserymen as early as 1771 was not noted by Nelson & Small (2000: **3**: 50) where 1832 is given as the date of introduction into Britain of *E. lusitanica* Rudolphi. Underhill (1990) has "about 1800" for this date, presumably following Synge (1956).

SOUTHERN HEATH

Erica australis

17. ERICA AUSTRALIS. Southern heath.

The southern heath is a handsome plant with relatively large flowers (Fig. 137), among the largest produced by the hardy European *Erica* species, blooming from early spring into summer. The size of the individual flower is variable, as is the colour of the corolla, from pale pink through various shades of lilac-pink; plants with pure white corollas are recorded in the wild (*Erica australis* f. *albiflora* Vicioso) and have been long known in cultivation ('Mr Robert'). Benito Cebrian (1948: 53) ventured that white-blossomed plants were frequent in "el tipo [the type variety]" and in *E. australis* var. *aragonensis* (see below). The recently named 'Holehird White', a sport from 'Holehird', is not a pure white; the stigma is red, and the corolla turns very pale shell-pink (H16) as it ages. *E. australis* is not as frequently planted in gardens in Ireland and Britain as it might be, having a reputation for not being reliably hardy; it will not tolerate shade, and is prone to damage by wind and snow, perhaps because it grows too lushly. However, seed is readily produced (see below), and spontaneous seedlings are reported from gardens in England and the western USA (Metheny 1977).

In the wild, on the Iberian Peninsula and in northern Morocco, *Erica australis* is confined to acidic soils, in open woodland or scrub (Fig. 138), and is recorded as reaching 1,600–2,000 m altitude. This heather is a source of honey in Spain (Persano Oddo *et al.* 2004). There is substantial variation reported within Spain (Bayer 1993), especially with regard to habit and the spurs on the anthers, with a number of taxa designated as distinct: *E. australis* var. *occidentalis* (Merino) Benito Cebrián (1948: 53), a reputedly smaller shrub from the extreme north-west of Spain; *E. australis* var. *aragonensis* (Willk. & Lange) Coutinho (1913), a hardier plant from montane habitats; and *E. australis* subsp. *bethurica* Ladero (1970) which was stated to be intermediate between the "type" and *E. australis* var. *aragonensis*. Various studies on Spanish populations (for example, Castroviejo 1982; Fraga Vila 1982, 1984a; Fagúndez & Izco 2004a) indicate that this variation is not geographically restricted, and that these taxa merge into one another, so that their retention as distinct subspecies or varieties is not warranted.

This species was known to Charles de l'Écluse (Carolus Clusius), the Flemish botanist, who found it during his travels in Portugal, between Lisbon and Coimbra, ten miles from Lisbon near the Tagus River (Rio Tejo), in the early winter of 1564–1565. He named it *Erica Coris folio II*: it had purplish flowers, grew in the same place as the first of his heathers, *Erica Coris folio*, and blossomed at the same time, in November and December. The first of the heathers was most probably *E. lusitanica* (cf. Ramón-Laca & Morales 2000); it has more frequently been identified as, or equated with, *E. arborea* (for example by Linnaeus). The original painting of *Erica Coris folio II*, which was the prototype for the wood-cut (Fig. 8, p. 14) that was reproduced first in l'Écluse's *Rariorum aliquot stirpium per Hispanias observatarum historia* (1576), and reprinted many times, is in the collection of botanical and zoological paintings called *Libri picturati*, now in Biblioteka Jagelloniska, Kraków, Poland (Ramón-Laca & Morales 2000).

Linnaeus described and named *Erica australis* in *Dissertationem botanicam de Erica* (1770); the type specimen was sent to him by Clas Alströmer (1736–1794) from Spain (Jarvis & McClintock 1990: 518, Jarvis 2007: 498).

Fig. 137. *Erica australis* in flower; cultivated in National Botanic Gardens, Glasnevin, Dublin (E. C. Nelson).

Fig. 138. Dense scrub composed almost entirely of *Erica australis* at near Puerto de Piedrasluengas, Picos de Europa, Spain (E. C. Nelson).

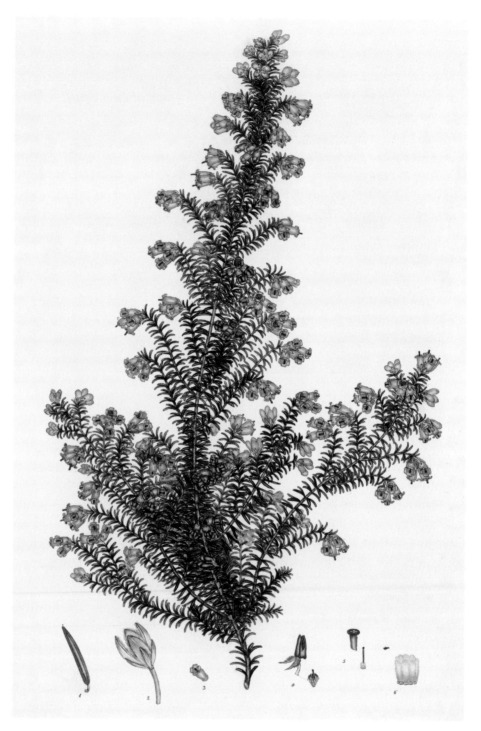

Erica australis   **Painted by Henry C. Andrews**

PLATE 16

From Henry C. Andrews's *Coloured engravings of heaths*, 3: tab. 150 (1805).

HISTORY IN CULTIVATION

Erica australis is stated to have been introduced into cultivation in Britain by George William Coventry (1722–1809), sixth Earl of Coventry, in 1769 (Aiton 1789; Bean 1914), the year before it was named and described by Linnaeus (1770). Lord Coventry's estate, Croombe Park, in Worcestershire, had been created from "as hopeless a spot as any in the island" (quoted in Hadfield 1985: 213) by Lancelot "Capability" Brown, who laid out the park and designed the house. Coventry was evidently a very keen plantsman, for he is credited with a number of important introductions, including *Koelreuteria* and *Chimonanthus fragrans* from China. As well as *E. australis*, which was probably grown in a conservatory to start with, Lord Coventry is acknowledged as being the first to cultivate *E. carnea* in 1763. His interest in heaths, and doubtless his patronage of London nurseries, was marked more than four decades later, when one of the Cape heaths was named after him: *E. coventryi*. Messrs Lee & Kennedy of Hammersmith, famous for their Cape heaths, gave this species its name as early as 1808, and it was illustrated by Henry C. Andrews in the last volume of his sumptuous *Coloured engravings of heaths* (1824: **4**, tab. 226). Andrews noted that the *Erica* had "received its specific title in honour of the Earl of Coventry". Conrad Loddiges of Hackney, another major supplier of Cape heaths, added that the "late Earl of Coventry [was] ... a kind patron of our establishment in its earliest years" (Loddiges 1820: **5**: tab. 423).

Richard Salisbury (see below) grew the Spanish heath under several different names: labelled *Erica protrusa* (Salisbury 1796), it was in his Yorkshire garden at Chapel Allerton, but perhaps under glass, by 1796.

SEEDS AND ELAIOSOMES

In the context of European hardy heathers, *Erica australis* is unique — its seeds have a natural protuberance called an elaiosome (Fig. 139), a fact noticed more than two centuries ago by Richard Salisbury (Nelson 2007b), whose name recurs frequently in this monograph. An elaiosome (from ἔλαιον (*elaion*), oil, properly olive oil) is defined as "a structure on the surface of a seed that secretes and stores oil, usually as an attractant to ants, which assist in seed dispersal" (Allaby 1998): dispersal of seeds by ants is termed myrmecochory. In some plants, elaiosomes are very conspicuous, sometimes brightly coloured; in *E. australis*, they are rather inconspicuous, except under a microscope.

Apart from Salisbury's brief comment (discussed below) in 1802, there was no published account of the presence of elaiosomes on the seeds of any *Erica* species before 1998, when *E. cedromontana* was described and named (Oliver & Oliver 1998: 269–272). That Cape heath produces a tiny, 0.5 × 0.25 mm, "ellipsoid, slightly flattened, smooth, shiny, orange [seed], with a small white elaiosome". (The seeds of *E. australis* are 0.4–1.1 mm long; see Nelson (2006a).) Oliver & Oliver (1998) suggested that "the possession of an elaiosome ... would appear to indicate that the seeds are distributed by ants", and they speculated that the "selfsame ants" were involved in pollination of *E. cedromontana*. Discussing the seeds of the entire subfamily Ericeae, Oliver (2000: 52) noted that "like the fruits the seeds have

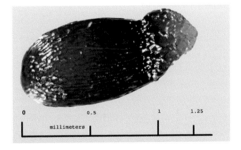

Fig. 139. The seed of *Erica australis* showing the protuberance called an elaiosome (© Allen Hall; reproduced by courtesy).

never been investigated in … any detail". He continued: "A few have been noted with elaiosome-like appendages which feature has never before been recorded in the family." However, no list of the African *Erica* species possessing elaiosomes has been compiled and published.

In the past decade, several students of *Erica* have noticed the elaiosome on the seed of *Erica australis*. In mid-March 1996, the late Inge Oliver made a series of carefully annotated sketches of *E. australis*, using material collected in southern Spain by Dr Fernando Ojeda, among which she drew a seed with a "white elaeosome [*sic*]" (see Volk *et al.* 2005). More recently, Fagúndez & Izco (2004a) described the elaiosome (caruncle; see below) as "a conspicuous mass of enlarged cells continued from the testa cells. It has a fleshy aspect and a yellowish color …". Scanning electron microscope photographs show seeds with shrunken, seemingly damaged, hilar regions: in one photograph, however, the details of the enlarged cells of the caruncle are seen clearly. Commenting on the possibility that seeds of *E. australis* are dispersed by ants, Fagúndez & Izco (2004a) noted that whereas myrmecochory has "never been reported for this species", they had "observed an intense removal of seeds by ants in field populations …".

Richard Salisbury was certainly a careful and accurate observer who in the 1790s and early 1800s had minutely studied around 250 different species of *Erica*, a remarkable piece of research that would not be repeated until George Bentham undertook work for Augustin-Pyramus de Candolle's *Prodromus* in the late 1830s. Salisbury's publications include his monograph on *Erica* that was read before the Linnean Society of London on 6 October 1801. This monograph was published at the end of May 1802, and in it Salisbury succinctly described 246 species, naming 53 new ones and renaming many others. One of the European species which he renamed (for the second time!) was *E. australis*: in the Linnean Society monograph, he used the name *E. pistillaris* (Salisbury 1802: 368). Salisbury's (1802) Latin diagnosis for *E. pistillaris* reads:

> *E. pedunculis foliolis gemmaceis obsitis: corolla 3lineari, laevi; tubo curvulo, infundibuliformi; limbo recurvo.* [Erica with peduncles covered with budded leaflets; with the corolla ¼in long, smooth, with the tube somewhat curved and funnel-shaped and the limb recurved.] …
> *Semina ad hilum, appendiculam fungosam exserunt.*

The final Latin sentence may be translated as "At the hilum, the seeds put forth a small, spongy appendage" — an elaiosome. This meaning is made clear in the letter, already discussed (pp. 98–99), which was written before late December 1819 probably to John Edward Gray. In this letter, Salisbury set out his views about the classification of *Erica* and even provided brief diagnoses for several new genera (see Appendix II, p. 387). One of his proposed segregates was a genus to be called "Tylosporum" (see below), comprising a single species — "E. Australis L. this Portugal shrub grows 10 feet height & differs from all others in its seeds which have a Caruncula as large as a Mimosa seed."

There are two significant points about Salisbury's remarks. A caruncle is "an excrescence at or about the hilum of certain seeds": an elaiosome is a caruncle. Like many other botanical epithets that Salisbury intended to publish, "Tylosporum" was derived from Greek: τύλος (*tylos*), a callus, lump or swelling, and σπορά (*spora*), a seed. While Salisbury's new generic name was never published, under *Erica australis* Bentham (1839: 666) noted: "*Hæc species Salisburio sect. propriam Tylosporam constituit* [This species was put by Salisbury in its own section Tylospora]". Bentham did not pursue this, nor did he describe the seed of *E. australis*.

Bentham's comment was taken up more than a century later by Dr Irmgard Hansen (*olim* Degener), who decided that *Erica australis* did merit segregation into a separate section within *Erica*. However, Hansen (1950) does not seem to have realised the significance of the name that Salisbury

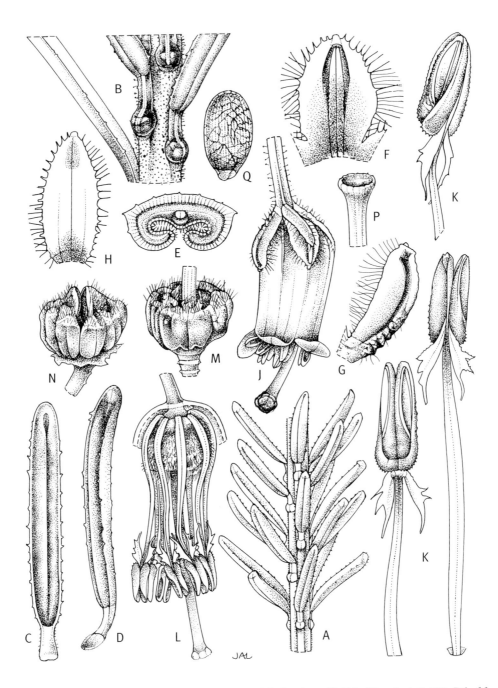

Fig. 140. *Erica australis* 'Mr Robert'. **A** leafy shoot × 5.5; **B** leaf nodes × 20; **C** leaf from back × 19.5; **D** leaf from side × 12; **E** transverse section of leaf × 54; **F** calyx lobe from front × 8; **G** calyx lobe from side × 8; **H** bracteole × 18; **J** flower from side × 6; **K** stamens from front, back and side; **L** gynoecium and stamens × 17; **M** ovary × 8; **N** capsule × 8; **P** style apex (stigma) × 14; **Q** seed from side × 24; from specimens grown by Allen Hall. Drawn by Joanna Langhorne.

had coined, and her protologue of Section *Tylospora* Salisb. ex I. Hansen makes no reference to the presence of the elaiosome. Like Bentham and the majority authors, Hansen (1950) seems to have regarded the seeds of *Erica* species as having little or no taxonomic value.

In autumn 2004, I obtained some seeds collected from plants cultivated by Allen Hall in his garden at Loughborough, Leicestershire. Although only six seeds from *Erica australis* 'Riverslea' were then available, one clearly showed an intact pale yellow elaiosome. Subsequently, having gathered many more seeds, both Allen Hall and I have examined hundreds of seeds, all of which had elaiosomes in place (Hall 2006). The elaiosome in *E. australis* varies in colour from almost white to golden brown. It seems to be fragile and does not appear to have a long natural "shelf-life", at least in the relatively cool, damp conditions prevailing in Britain. Examination of seeds extracted from old flowers suggests that the elaiosome quickly decays, and observations on seeds collected in Spain and England seem to show fungal hyphae associated with the caruncle.

ECOLOGY

Erica australis is a facultative "re-sprouter" (Cruz *et al.* 2003). Plants develop an underground lignotuber from which they can regenerate vigorously after fire (see, for example, Cruz & Moreno 2001). Coupled with this, it has also been shown that the species contains chemicals that will inhibit germination (Carballeira 1980) and also retard growth of other plant species (Ballester *et al.* 1979). A lot of research time has been expended on studying this species's capacity to regenerate from seed, its responses to fire and the physiology of resprouting. Yet, strangely, nowhere can I find any reports that indicate the botanists who carried out the research programmes have been aware of the elaisome attached to the seed, and that ants are probably very important in seed dispersal.[1]

After a fire, *Erica australis* will readily produce new shoots from the lignotuber, but a small number of seedlings may also appear. In an interesting study of heathland in León, northern Spain, dominated by *E. australis*, Valbuena *et al.* (2000) found that there was a very substantial seed-bank in the top layer of soil; their soil samples were from plots 20 × 25 cm and 5 cm deep. They calculated that before an experimental fire there were more than 2,200 seeds of this heather capable of germinating in every square metre. Not many of these germinated — the heather resprouted — and the fire evidently killed about half of the seeds, leaving fewer than 1,200 viable seeds in each square metre. Experiments undertaken

---

Erica australis **and others**   Painted by Christabel King   PLATE 17

LOWER LEFT: **E. australis 'Mr Robert'** ♔ (*E. australis* f. *albiflora*). Upright bush (±1.8–2.5 m tall × 0.85 m spread); foliage paler green than other cultivated clones; flowers white, tubular-bell-shaped, anthers pale tan ("golden") apparent at the corolla mouth; April–June. (Painted 15 May 1985.)

   "Transcendently beautiful" (Maxwell & Patrick 1966), 'Mr Robert' was collected at Algeciras in southern Spain by Robert Williams (after whom it was named), son of J. C. Williams, Caerhays, Cornwall, England, in 1912. Second-Lieutenant Williams (28 years old), 2nd Battalion of the Grenadier Guards, was killed in action on 8 October 1915 at Loos. AGM 1993.

LOWER CENTRE & RIGHT: **E. umbellata** (unnamed) (right: anthers included; painted 3 July 1986; cultivated by David Small.)

UPPER LEFT: **E. australis** (unnamed).

UPPER RIGHT: **E. sicula** (source not recorded). (Painted 2 May 1986.)

by Cruz *et al.* (2003) corroborate those findings. Seeds of *E. australis* heated for five minutes at 150°C subsequently failed to germinate, whereas those heated to 100°C for the same time did sprout. Seeds kept for one hour at 70°C also germinated. Cruz *et al.* (2003) also found that exposure to moderately high temperatures almost doubled the percentage germination, and they extrapolated from these results to suggest that it was "likely that only those seeds buried at a certain depth could survive a fire".

Closer examination of the data published by Valbuena *et al.* (2000) shows that the seeds of *Erica australis* which perished were those in the top 2 cm of the soil profile. Before the fire the upper 2 cm of soil contained around 340 seeds per m²; fire killed two-thirds of these seeds. The lower layer, more than 2 cm deep, contained only about 110 seeds before the fire, and almost the same number after the fire. Valbuena *et al.* (2000) commented that "no difference ... was observed" in the number of seeds capable of germination within the 2–5 cm layer before and after their experiment. My interpretation of these figures is that the insulating effect of the soil is a significant factor in the survival of the seeds of *E. australis* — the seeds that are "buried" survive. The corollary is that seeds taken underground by ants will have a better chance of surviving. Whether the presence of an oil-rich elaiosome affects the chances of survival of a seed, whether it is more combustible because of the elaiosome, has not been investigated.

Additional, fascinating results of the work by Luz Valbuenca and his colleagues suggest that germination of *Erica australis* seed "is favoured" by storage (Valbuena & Vera 2002). They also found that seed subjected to cold pre-treatment germinated earlier than seeds not so treated although there was no significant difference in the percentage germination.

In contrast, the same experimental fire broke the dormancy of many *Calluna* seeds. Before the fire, a square metre contained only about 90 seeds capable of germinating, yet after the fire that increased to around 700 *Calluna* seeds per square metre. The authors observed that "neither seedlings nor sprouts of *Calluna vulgaris* had appeared ... 10 months after the fire" (Valbuena *et al.* 2000).

CULTIVATION AND VARIATION

Southern heath demands a lime-free, acidic (or neutral) soil for successful cultivation. It is also prone to damage, especially by snow and wind. Propagation of the named cultivars must be by vegetative means, layering or cuttings. *Erica australis* can be raised from seeds, which when freshly gathered, germinate readily as fire is not needed to break dormancy; indeed, as just noticed, Valbuena *et al.* (2000) demonstrated that fire can kill the seeds of this heath.

The only variant recognised in horticulture is the form with white flowers, *E. australis* f. *albiflora* (McClintock 1984a: 191).

Recently a "semi-double"-flowered clone has been named: 'Trisha' (® E.2007:01; see Nelson 2008e), which has malformed flowers with up to nine corolla and calyx lobes, was a seedling raised by Kurt Kramer (Edwecht-Süddorf, Germany) (for description see pp. 363–364).

HYBRIDS

There are no reports of any natural hybrids of *Erica australis*, but the species has featured in recent breeding experiments in Britain, by Dr John Griffiths (Leeds, Yorkshire) and Barry Sellers (Norbury, London), and in Germany by Kurt Kramer (Edwecht-Süddorf). None of the attempted crosses produced viable seed, although Hall (1992: 48) reported seeing "a young plant which was a cross between *E. australis* and *E. arborea*" in Kramer's nursery in March 1991. Among the species used as pollen- and/or seed-parents, as well as *E. arborea* (Kramer, Sellers), were *E. carnea* (Griffiths, Kramer), *E. ciliaris* (Griffiths), *E. cinerea* (Kramer), *E. erigena* (Griffiths, Kramer), *E. lusitanica* (Griffiths), *E. manipuliflora* (Griffiths), *E. spiculifolia* (Kramer), *E. terminalis* (Griffiths), *E. tetralix* (Griffiths) and *E. vagans* (Griffiths).

Kramer has also experimented, without success, pollinating *Erica australis* with pollen from African species; the two tried were *E. blandfordii* and *E. patersonii*.

Erica australis L., *De Erica*: 9 (1770). Type: *C. Alströmer* 80; lecto. **LINN** 498.50! (Jarvis & McClintock 1990: 518; Jarvis 2007: 498).

E. umbellata Asso, *Syn. Stirp. Arag.*, 49 (1779), non L.

E. protrusa Salisb., *Prodromus stirpium … Chapel Allerton*, 293 (1796).

E. pistillaris Salisb., *Trans. Linn. Soc.* **6**: 368 (1802).

E. aragonensis Willk. & Lange, *Linnaea* **25**: 46 (1852); *E. australis* var. *aragonensis* (Willk. & Lange) Coutinho, *Fl. Portugal*, 463 (1913)

E. occidentalis Merino, *Bol. Soc. Esp. Nat. Hist.* **2**: 67 (1902); *E. australis* var. *occidentalis* (Merino) Benito Cebrián, *Brezales y brezos*, 53 (1948).

E. australis subsp. *bethurica* Ladero, *Anales Inst. Bot. Cavanilles* **27**: 89–91 (1970).

Ericodes australe (L.) Kuntze, *Revisio genera plantarum* **II**: 966 (1891)

ILLUSTRATIONS. H. C. Andrews, *Coloured engravings of heaths* **3**: tab. 150 (1805); H. C. Andrews, *The heathery* **2**: tab. 52 (1806); J. Smythies, *Flowers of south-west Europe*: 287 (7) (Polunin & Smythies 1973); P. Millan, *Anales Inst. Bot. Cavanilles* **27**: 90 (1970); E. Sierra Ràfols, *Flora iberica* **4**: lam. 181 (p. 498) (1993).

DESCRIPTION. *Evergreen shrub* to 1.5(–2.5) m tall; capable of resprouting from underground lignotuber, with roots extending to 3.5 m deep (Silva *et al.* 2003); shoots densely tomentose, covered with simple and plumose hairs, and some gland-tipped hairs. *Leaves* in whorls of 4, linear, to 7 mm long, ±1 mm wide, green, ascending, glabrous, flattened above, with parallel edges which are conspicuously transparent and cartilaginous, margins recurved and contiguous underneath, sulcus linear, closed, adaxial surface "keeled"; petiole flattened, with long, shaggy hairs on edges. *Inflorescence* terminal on leafy lateral shoot, flowers in 4s, sometimes with subsidiary whorls; the buds resin-coated and very sticky; *peduncle* hirsute (sometimes with long, shaggy hairs) toward top; bract and bracteoles recaulescent towards the apex of pedicel, like calyx segments, to 2 mm long, pale green to dark red. *Calyx* with 4 unequal, free sepals, ±2.5 mm long, cartilaginous or fleshy and markedly thickened at apex, in bud the outer pair overlapping the inner, with central green or reddish portion and broad, colourless, transparent sides, keeled, sulcus extending from tip ± 1 mm long, with long shaggy hairs on the "keel" and margins, margins with hairs and often beaded with resin, with gland at apex. *Corolla* ± 8 mm long, ±3 mm diameter, tubular-campanulate, but somewhat lop-sided and slightly curved, 4-lobed at mouth, lobes erect or recurving, rounded, to 2 mm long. *Stamens* 8, included; *filaments* ± 6.5 mm long, hooked at base and attached under nectary, curving, with or without a pronounced sigmoid bend below anther, white, glabrous, c. 0.2 mm broad, tapering only very slightly towards apex; *thecae* brown to tan, ± 1.5 mm long, erect, not fused except at point of junction with filament; *spurs* pale, ±1 mm long, pointing downwards, divergent, outer margins irregularly lobed or fimbriate. *Style* stout, glabrous, straight, 9 mm long, 0.8–1.0 mm in diameter at base which is deeply inserted into top of ovary, tapering to 0.6 mm at apex below stigma, white or tinged pale pink towards the apex; *stigma* expanded, about same diameter as base of style (0.8–1.0 mm), red or green; *ovary* barrel-shaped, 1.5–2.0 mm tall, hirsute, green, with c. 20 ovules/locule,

Fig. 141. Distribution of *Erica australis*.

with nectary ring at base; *nectar* copious. *Seeds* (0.4–0.6)–1.1 × 0.3–0.5 mm (see Nelson 2006a), brown with conspicuous pale elaiosome (Fig. 140); 15–45 seeds per capsule (Cruz *et al.* 2003). n = 12 (Fraga 1983: 536; see Nelson & Oliver 2005b).

FLOWERING PERIOD. (October–) January–May.

DISTRIBUTION. Europe: Spain, to 2,000 m altitude and mainly in the western half but also occuring in the north-east; Portugal. North Africa: northern Morocco (Ball 1877; Fennane & Ibn Tattou 2005) ascending to about 1,450 m altitude. **Maps:** Benito Cebrián (1948: 58), Hansen (1950: 33); for Galicia see Fraga Vila (1984b: 18). Fig. 142.

ETYMOLOGY. Latin: *australis* = south.

VERNACULAR NAMES. Southern heath, Spanish heath (English); brezo rubio, brezo colorado, brezo rojo (Spanish; for other vernacular names used in Spain see Bayer 1993; Pardo de Santayana *et al.* 2002); chamiça, urgeira, urze-vermelha (Portuguese); Spanische Heide (German).

CULTIVARS. see pp. 225–226, and Appendix I (pp. 363–364).

NOTES

[1] Valbuena *et al.* (2001: 25, Table 1) indicated that seed dispersal in *Erica australis* was by wind (anemochory). This seems to be based on an observation reported by Valbuena *et al.* (2000: 164) that "seeds germinating from withering flowers of *Erica australis* ... [confirm] the importance of flowers dispersed by the wind and which retain seeds inside. This may be important for the dispersion distance of this species ...".

THE CROSS-LEAVED HEATHS

Erica tetralix, *E. mackayana*, *E. andevalensis*

The three species included in this section have their leaves arranged in whorls of four, like a cross, and have prominent spurs on the anthers. They are *Erica tetralix*, cross-leaved heath, *E. mackayana* (more familiar as *E. mackaiana*), Mackay's heath, and *E. andevalensis*, which being a native of south-west Spain and the adjacent part of Portugal does not have an English name.

Erica andevalensis is morphologically similar to *E. mackayana*, but as with the winter-blooming heaths *E. carnea* and *E. erigena*, these two species have different habits and habitats, and do not coexist anywhere. McClintock (1989a) relegated *E. andevalensis* to the status of a subspecies of *E. mackayana*, but the plant's ecological propensities are so extraordinary and so different from those of any other species of *Erica*, including the southern African ones, that its status as a distinct species should be firmly reasserted. It is possible that further investigations, including DNA studies, which have not been conducted as far as I can determine, will alter this view. Spanish and Portuguese botanists maintain *E. andevalensis* as a separate species, and their opinions are here respected.

For many decades into the mid–1950s, *Erica mackayana* was thought perhaps to have arisen as a hybrid between *E. tetralix* and *E. ciliaris*. Peter Gay (1955a), for example, referred to this as the "old problem", noting that while *E. mackayana* "appears to be a hybrid" between those heathers, "it differs greatly from those [plants] which we know to be hybrids"; in other words, it differed from *E.* × *watsonii*. In the late 1920s Margaret Smith studied the morphology and anatomy of the leaves of *E. tetralix*, *E. mackayana*, and also of the peculiar plant then called *E. crawfordii* (cf. Nelson 1995b), and showed that the leaf anatomy of cross-leaved and Mackay's heaths was quite different. In particular, the large upper epidermal calls in *E. mackayana* showed, with rare exceptions, a clear transverse septum that divided the cells in half; this septum was not present in *E. tetralix* but was seen in about half the equivalent cells in such plants as *E.* × *stuartii* (Smith 1930, as *E. praegeri*) and in *E. crawfordii* which are deemed to be true hybrids. Smith (1930: 205) concluded that *E. mackayana* was "even more distinct from *E. tetralix* than has been thought."

The natural hybrid between *Erica tetralix* and *E. mackayana* — *E.* × *stuartii*, Praeger's heath — has long been known to grow in Ireland where these two species are sympatric. Unbeknownst to English-speaking botanists interested in *Erica*, Praeger's heath had also been collected in Asturias, northern Spain, during the early 1990s (Díaz González & García Rodriguez 1992), and more recently the plant also has been found in Galicia in north-western Spain (Fagúndez 2006). There is no chance of either of these species producing a natural hybrid with *E. andevalensis*, but these have been created by artificial pollination in Germany (by Kurt Kramer) and Britain (by Dr John Griffiths).

18. ERICA TETRALIX. Cross-leaved heath.

Cross-leaved heath has the distinction of being the most northerly-occurring species of *Erica* (Fig. 143). The accepted limit of its natural range is in Nord-Trøndelag, north-western Norway, at 68°10'N latitude (Fægri 1960: 63), almost two degrees north of the Arctic Circle (66°33'N). However, the species has recently been recorded by Solveig Gamst even further north at Skarpenes (69°44'N), northeast of Tromsø, although there is some doubt about whether it is native or was

Cross-leaved heath and related *Erica* spp. Painted by Christabel King PLATE 18

LOWER LEFT: *E.* × *stuartii* '**Stuart's Original**' (® E.2007:03; formerly 'Stuartii'; see McClintock 1979a, 1982a; 'Charles Stuart' **rejected** (*Yearbook of The Heather Society* **1998**: 72)). Low, spreading shrub (±0.25 m tall × ±0.25 m spread); young shoots red, maturing to greyish green; flowers slender, tubular, small, white to shell-pink (H16) at base shading to beetroot (H9) at tip; July–October. Fig. 198 (p. 305).

A. T. Johnson (1928: 81) praised this as "an engaging little shrub, distinct enough and pretty enough for any collection". This most peculiar clone was once thought to represent a hybrid between *Erica mackayana* and *E. erigena*, but in the 1970s when a sport was noticed, its correct identity was determined (McClintock 1979a). There were several consequences, including the changing of the binomial for the hybrid *E. mackayana* × *tetralix* from *E.* × *praegeri* to *E.* × *stuartii*. A further nomenclatural change is necessitated under the *International code of nomenclature of cultivated plants*, so that the familiar cultivar name 'Stuartii' has been replaced by 'Stuart's Original' (® E.2007:03). This clone was gathered on 11 August 1890 in Connemara, County Galway, by Dr Charles Stuart (1825–1902), Churnside, Duns, Berwickshire; it was propagated and soon established in cultivation at the Royal Botanic Garden, Edinburgh. In the summer of 2000 at Craiggamore, Connemara, I found and photographed (Fig. 197, p. 305) a small plant with flowers indistinguishable from those of Stuart's clone (Nelson 2001a).

LOWER RIGHT: *E. mackayana* '**Dr Ronald Gray**' (*E. mackayana* f. *eburnea*). Low, spreading, but not tidy heather (±0.1 m tall × ±0.35 m spread); foliage green; flowers white, sparse, small; July–October.

Dr Ronald Gray (1880–1966), a keen heather gardener, of Hindhead, Surrey, found this as a sport on 'Lawsoniana' in 1962 (McClintock 1984a; McClintock (1966b) gave the date as 1964; the cultivar was introduced by Maxwell & Beale Ltd, Broadstone, Dorset, in 1966. For many years this was the only white-blossomed clone of Mackay's heath and, like its parent 'Lawsoniana' (see Nelson 2000d), it is a rather poor thing. 'Dr Ronald Gray' has been superseded by the splendid 'Shining Light' (see Fig. 224, pp. 379–380).

CENTRE LEFT: *E.* × *stuartii* '**Irish Lemon**'. Neat, spreading heather (±0.2 m tall × ±0.5 m spread); young shoots bright yellow, eventually maturing to green; flowers large mauve (H1/H2), usually in a single terminal umbel, but occasionally several axillary inflorescences appear to form a short raceme; May–September.

David McClintock collected this on moorland by Lower Lough Nacung, County Donegal, on 26 August 1966; this clone was originally called simply "number 1" (McClintock 1982a). The bright yellow of the young shoots persists until flowering commences, and the colours of foliage and flowers clash. 'Irish Lemon' was introduced by John F. Letts and B. W. Proudley, Foxhollow Heathers, St Briavels, Gloucestershire, by 1971. They also introduced 'Irish Orange' (with orange-coloured young shoot-tips) which was collected on the same occasion at Lough Nacung. AGM 1993.

RIGHT CENTRE: *E. mackayana* '**Plena**' (*E. mackayana* f. *multiplicata*). Low, neat, spreading heather (±0.15 m tall × ±0.4 m spread); foliage dark green; flowers magenta (H14), each one "double" due to the corolla containing numerous petaloid staminodes and additional corollas; July–October.

The first person to collect a double-flowered variant of Mackay's heath in Connemara was the English-born naturalist and co-author of *Cybele hibernica* (1866), Alexander Goodman More (1830–1895); he collected herbarium specimens during 1869 (McClintock 1982a) but, as far as is known, he did not gather material for propagation. The plant as presently known in cultivation derives from Dr Frank C. Crawford's collection, also in Connemara, on 5 August 1901: this was once named *Erica* "crawfordii", Crawford's heath (for discussion of this apparently invalid name see Nelson 1995b). Remarkably, similar, if not identical, plants have been seen more recently in Connemara – in 1965 by Father Brennan and in September 1969 by Dermot Burke (Scannell & McClintock 1974; McClintock 1982a; Nelson 1995b, 2000e).

UPPER LEFT: *E. tetralix* '**L. E. Underwood**'. Compact shrub (±0.2 m tall × ±0.35 m spread); foliage grey-green; flowers cerise (H6) when open, buds salmon (H15); June–September.

Described as having "silver foliage and terra-cotta buds which open to iridescent apricot flowers" (Maxwell & Patrick 1966). This was found about 1937 "... on the common at West End, Woking" (Maxwell & Patrick 1966).

UPPER CENTRE: *E. tetralix* 'Alba Mollis' (*E. tetralix* f. *alba*). Compact shrub with upright stems (±0.2 m tall × ±0.3 m spread); foliage grey-green with distinctive silver sheen especially when young, the long cilia on the edges of the leaves without glandular tips; flowers white; June–October.

An old cultivar, probably: this name has been in use since at least the 1860s when it was listed (as "Alba Moles") by James Smith of Darley Dale, Derbyshire. The cultivar so-named received an Award of Merit in 1927. The anthers do not produce pollen, but the flowers are fertile (Griffiths 1985), so this clone has been much used in artificial breeding programmes as a seed-parent. AGM 1993.

UPPER RIGHT: *E. mackayana* (unnamed).

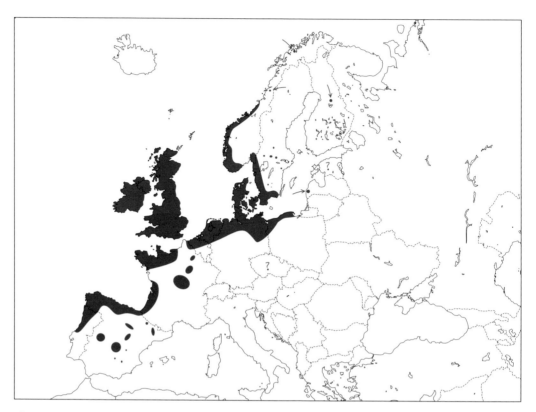

Fig. 142. Distribution of *Erica tetralix*.

introduced (Gamst & Alm 2003). From Nord-Trøndelag, *E. tetralix* ranges south through western Norway, spreading eastwards into south-western Sweden. There is a small, isolated population at Kuhmo in north-eastern Finland (64°12'40"N. 29°29'30"E), where *E. tetralix* is a protected species (Mäkinen & Tiikkainen 1966; Sallinen 2007). Other European states where it is rare and protected are Lithuania, Latvia and the Czech Republic, all in the north-eastern part of its range.

Cross-leaved heath is a familiar and common plant on wet heaths and bogs in Ireland (Fig. 143) and Britain, reaching an altitude of 880 m in Coire Etchachan, Aberdeenshire, Scotland (Preston *et al.* 2002). In south-central England, the species has declined during the last half century, due largely to habitat loss (Preston *et al.* 2002), but it is still abundant in suitable habitats in the far south-west. In southern Britain, *Erica tetralix* is also to be found in drier heathlands on nutrient-deficient soils. *E. tetralix* is classified as a member of the Suboceanic Temperate element of the flora of Britain and Ireland; the constituent species have a relatively even distribution pattern within the islands without a concentration along the western seaboards (Preston & Hill 1997).

Erica tetralix ranges through western Europe southwards into the Iberian Peninsula. It ascends to 2,200 m in the French Pyrenees. In Spain, it has a distinctly northern (Fig. 144) and western pattern of distribution, being absent from southern and eastern parts. Portuguese populations of cross-leaved heath are largely confined to the northern half of the country; reports that *E. tetralix* occurs in Algarve are incorrect and probably due to confusion with the recently recognised *E. andevalensis* (see p. 250).

Fig. 143. *Erica tetralix* on raised bog on the Cork-Kerry border in south-western Ireland (E. C. Nelson).

Fig. 144. *Erica tetralix* in Picos de Europa, northern Spain; here the heather was growing in a place where the underlying rock was limestone (see Nelson 1996a) (E. C. Nelson).

The cross-leaved heath displays substantial variation in the hairiness of its foliage. Sometimes plants with almost glabrous foliage are found, especially on the coast — they appear dark green. Others have very densely hirsute leaves so that the whole plant is silvery white in appearance; I saw plants like this in the southern foothills of Picos de Europa at Cervera de Pisuerga, Palencia, northern Sain, during 2007 (Canovan *et al.* 2008: 49). Some have cilia that lack the prominent, usually red, glandular tips. There is also substantial variation in the hairiness of the inflorescence (Fig. 145). A named variant with leaves in whorls of five (*Erica tetralix* var. *quinaria* Guffroy) is said to be more common in wetter areas (Brough & McClintock 1978), and in some plants the leaves can be almost spirally arranged. Some plants have flowers arranged evenly in a star- or starfish-like umbel rather than a lopsided one; McClintock (1986) employed this as the prime characteristic for *E. tetralix* f. *stellata* D. C. McClint., encompassing 'Pink Star' and 'Helma', and also wild-collected examples. Occasionally the inflorescence seems to be elongated and raceme-like (*E. tetralix* f. *racemosa* D. C. McClint.) (McClintock 1982b).

Various other, impermanent teratological conditions have been recorded, including flowers with deeply cleft, not bell-like, corollas (Sigerson 1871). The cultivar 'Mary Grace' produces ugly, ragged flowers, at least when it begins to bloom. This aberration may be due to mites, which cease activity in mid-summer, but Dr Arthur Massee (1899–1967), who examined the clone, could never find any trace or evidence of them (McClintock 1966b). The Dutch cultivar 'Stikker' demonstrates another abnormality; its flowers develop normally to full size but the lobes at the apex of the corolla do not separate and recurve, so that the mouth remains closed and bud-like. Another anomaly is the replacement of the flowers by elongated, catkin-like shoots composed of numerous, tiny, leaf-like bracts; one clone in cultivation, which originated in the wild in the Netherlands, has very silvery "catkins" and is named 'Meerstal' (® E.2005.06; Baron 2006).

Fig. 145. *Erica tetralix*: a now-extinct clone collected at Brandonas de Arriba near Santiago de Compostela in northwestern Spain by David McClintock, Dr Charles Nelson and David Small in 1982 and named 'Arriba' (® no. 151) (E. C. Nelson).

POLLINATION

"When flowering *Calluna* and *Erica tetralix* grow together it is remarkable that the majority of bees turn to *Calluna*, while *Erica* offers its flowers almost in vain, although they are fragrant much longer and also contain abundant nectar." So wrote Else Hagerup and Olaf Hagerup (1953) in a classic paper about the pollination of heathers. They suggested that in some habitats *E. tetralix* is pollinated by thrips, and they questioned the efficacy of pollination by other insects, including bumblebees: "Visits by bumble-bees, butterflies and flies are so rare that they are no more than accidental chances without any quantitative value." The Hagerups' work has been partly corroborated and partly contradicted in a paper by Hanne-Dorthe Haslerud (1974) in which the same two heather species were observed in Norway. Haslerud noted that where *E. tetralix* and *Calluna* "grew together some bumblebees preferred *Erica*." The majority of the bumblebees caught on cross-leaved heath were *Bombus pascuorum*, which has a proboscis 10–12 mm long, long enough to reach into the corolla of *E. tetralix* and sip the nectar. The pollen loads that the bees carried contained very high percentages of *E. tetralix* pollen, showing the insects were also collecting the heath's pollen, and so Haslerud (1974) concluded that "pollination by bees must be a possibility."

As for thrips, which Hagerup & Hagerup (1953) had found in the flowers of *Erica tetralix*, all winged females, Haslerud (1974) noted that there were no thrips at all in the buds of plants at Stavanger, although each of the open flowers contained as many as four insects. At the start of the flowering season, the thrips "carried only two pollen tetrads, later up to 12." Pollination by thrips is possible, she concluded. Other simple experiments, including bagging flowers to prevent insects getting to them, led to other conclusions: self-pollination can take place, but wind-pollination "seems rather improbable", even "unimportant" (Haslerud 1974: 215). While there were punctures in the *Erica* flowers, Haslerud (1974) opined that "nectar thieving was apparently unimportant in the syndrome of bumblebee-pollination."

Haslerud (1974) concluded with this comment: "the effectivity of insect-pollinators may vary a great deal from one year to another."

As noted, flowers are frequently robbed of nectar: an insect such as a honey bee (*Apis mellifera*) will puncture the corolla near the base where the nectar is held, and will sip the nectar through the hole without effecting pollination (see Bannister 1966; Edwards 1996; Proctor *et al.* 1996).

There is a lot about pollination in heathers that is mysterious and unexplained. We blithely assume that all plants produce pollen, and that all pollen is viable. Yet, the fine white-flowered cultivar 'Alba Mollis' does not produce any pollen (Griffiths 1985), and when Professor David Webb examined the pollen of *Erica tetralix* specimens from Portacloy in the extreme northwest of County Mayo he found it was "entirely abortive" (Lamb 1964) leading him to mistake these unusual plants, that have darker, almost glabrous foliage (Lamb 1964), for *E.* × *stuartii* (cf. Nelson 2005a) which they certainly are not. Like almost every other aspect of heathers, pollen, evidently, can be unexpectedly variable.

GLAND-TIPPED HAIRS

I know of no explanation of the purposes of the gland-tipped hairs that are a prominent feature of many *Erica* species, including, among the northern hemisphere heathers, *E. tetralix*, *E. mackayana* and *E. ciliaris*. They seem to be analogous to the gland-tipped hairs that cover the leaves of the insectivorous sundews (*Drosera* spp.) which often frequent the same bogs as cross-leaved heath, but *E. tetralix* is not an insect-trapping plant and as far as is known the fluid secreted on to the glandular tips of these hairs does not have a digestive action. One possible purpose for these hairs, which are often very sticky, is dispersal, and there is a tantalising indication of this in a recently published paper titled "Estimating adhesive seed-dispersal distances: field experiments and correlated random walks" by Mouissie *et al.* (2005). The research team set out to estimate the distance that "adhesively dispersed seeds" could travel using a sheep dummy and a cattle dummy. In the cases of ling (*Calluna vulgaris*) and cross-leaved heath (*E. tetralix*), they noted that it was not the individual seeds that adhered to fur but the "flower heads or even whole branches".

CULTIVATION

Although typically an inhabitant of the wetter part of moors and peat bogs, *Erica tetralix* will grow well in any good, non-alkaline garden soil; it does not need to be cultivated in peat. In cultivation, cross-leaved heath forms a low, compact shrub, generally with greyish green foliage that contrasts well with the white or pink flowers. Several cultivars with yellow foliage (*E. tetralix* f. *aureifolia* D. C. McClint.; McClintock 1988a: 62) are available, including 'Ruth's Gold' and 'Swedish Yellow'. Also available are cultivars with white flowers (*E. tetralix* f. *alba* (Aiton) Braun-Blanq.; McClintock 1984a: 191). Cross-leaved heath is hardy, having a maximum frost-resistant of about −20°C (Bannister 1966, 1996), although the tolerance of individual cultivars may depend on their origins.

HYBRIDS

Erica tetralix is one of the parents of three naturally occurring hybrids (see pp. 301–315): *E.* × *stuartii* (with *E. mackayana*: Ireland, Spain); *E.* × *watsonii* (with *E. ciliaris*: England, France, Spain); and *E.* × *williamsii* (with *E. vagans*: endemic to Cornwall). All three also have been produced by manual cross-pollination of the parent species.

Cross-leaved heath is also involved in another recently created and named artificial hybrid: *Erica* × *garforthensis* (with *E. manipuliflora*) was first raised by Dr John Griffiths (Leeds, Yorkshire) in 1983 (see p. 334), using *E. manipuliflora* as the pollen-parent. David Wilson (British Columbia, Canada) has also succeeded with this cross. For Wilson and Griffiths, as well as Kurt Kramer (Edewecht-Süddorf, Germany), the reciprocal crossing failed.

Using *Erica tetralix* as a seed-parent, Griffiths has been able to effect the following crosses: with *E. andevalensis* (seedlings were hybrids), *E. carnea* (seedlings were hybrids but were weak and soon died) and *E. multiflora* (seedlings lost before they could be assessed). Using *E. tetralix* pollen, he was less successful: attempts to cross-pollinate *E. australis*, *E. carnea*, *E. maderensis* and *E. manipuliflora* all failed. With *E. cinerea* and *E. terminalis*, the seedlings resembled the seed-parent. Kramer's attempts to use *E. tetralix* as a seed-parent all failed, but he did succeed using *E. tetralix* pollen in obtaining seedlings with *E. andevalensis* (one clone named 'Pat Turpin' is in cultivation) (Hall 1992: 48; Small 1992), and *E. spiculifolia*.

Recalling William Herbert's failed attempt to get Cape heath pollen to pollinate *Erica tetralix* (Herbert 1822; see p. 327), it is noteworthy that Kramer failed to get any viable seeds when he tried cross-pollinating *E. tetralix* and *E. mammosa* in 1986.

Erica tetralix L., *Species plantarum*: 353 (1753). Type: "*Habitat in* Europæ *paludibus cespitosis*": lecto. **LINN** 498.24! (McClintock 1980a; Jarvis 2007: 501).
E. glomerata Salisb., *Prodromus stirpium … Chapel Allerton*: 293 (1796).
E. botuliformis Salisb., *Trans. Linn. Soc.* **6**: 369 (1802).

Eremocallis glomerata (Salisb.) S. F. Gray, *Natural arrangement of British plants* **2**: 398 (1821).
Ericodes tetralix (L.) Kuntze, *Revisio genera plantarum* **II**; 967 (1891).
Tetralix septentrionalis E. Meyer, *Preuss. Pflanzengatt.*: 100 (1839).

ILLUSTRATIONS. J. Curtis, *Brit. Entomol.* **1**: tab. 13 (1824); S. J. Roles, *Flora of the British Isles: illustrations* pt 2: tab. 933 (1960); S. Ross-Craig, *Drawings of British plants*, part 19: pl. 22 (1963); R. W. Butcher, *A new illustrated British Flora*: 105 (1961); L. Denkewitz, *Heidegarten*: 181 (1987).

BIOLOGICAL FLORA OF THE BRITISH ISLES. *J. Ecol.* **54**: 795–813.

DESCRIPTION. *Evergreen shrub*, stems erect or semi-prostrate, or very rarely completely prostrate: in open, relatively dry habitats forming a much-branched bushy shrub to 0.5–1 m tall, with upright branches; in wet, boggy habitats, a straggling, subshrub forming clumps of shoots, lower portions of the branches semi-prostrate, often buried and producing adventitious roots. *Shoots* densely hairy with crinkled short hairs intermingled with some long, simple hairs; *internodes* ± equal except below inflorescence where they get progressively longer. *Leaves* in whorls of 4 (very rarely to 6), 3–8 mm long, 0.6–1.5 mm across, spreading in lower part of shoot but becoming erect and parallel to shoot below inflorescence, lanceolate to linear, floccose on upper surface, with long, gland-tipped cilia ± evenly spaced along edge and a single gland-tipped hair at apex, margins revolute, contiguous near apex, but sulcus widening and opening towards petiole, lower margin of even width, abaxial leaf surface with long hairs, sulcus margins with dense short, even hairs, midrib hirsute below; petiole flat, floccose. *Inflorescence* terminal, an umbel, usually lopsided, with ± 9 flowers; *pedicel* densely woolly, c. 3.5 mm long, bract about ²/₃ way along, leaf-like, c. 3 mm long, elliptical, green, very woolly on edge and outer (lower) surface, sulcus open, broadening toward base, c. 1.5 mm long; inner (upper) surface glabrous, with very long, gland-tipped, marginal cilia, especially near tip; bracteoles like bract but slightly smaller, elliptical-spathulate, thickening at apex. *Calyx* with 4 free, triangular-lanceolate sepals, to 2.5 mm long, to 1 mm across, green, like bract and bracteoles, woolly on outer surface, glabrous on inner, with long, gland-tipped cilia (very rarely without cilia). *Corolla* usually pink, darker on sunlit side shading to white, or rarely pure white, ±7 mm long, ±5 mm in diameter, obovoid-urceolate to tubular-urceolate, glabrous except on apical lobes outside (in buds, the hairs on lobes are clearly visible but the recurved lobes conceal these); lobes 4, recurving. *Stamens* 8, ±5 mm long, erect, included; *filament* glabrous, curving, hooked at the base and attached under the ovary, with pronounced sigmoid bend towards apex; *anther* ± 1 mm long, lateral, with two very narrow-triangular white spurs (spur margins very minutely irregular) at junction with filament, ±0.7 mm long; *thecae* ±1.0–1.2 mm long, to 0.5 mm across at base, triangular-elongate, prognathous; pore obovate, ±5 mm long. *Ovary* barrel-shaped, c. 1.5 mm long, densely covered with long silky hairs, emarginate; with prominent nectary ring; 40–50 ovules (Szkudlarz 2001) per locule; *style* ± 6 mm long, ±0.5 mm in diameter at base, erect, straight, glabrous, tapering upwards but apex expanded, *stigma* broader than style, ±0.7 mm in diameter. *Seeds* 0.3–0.43 × 0.2–0.33 mm, brown, ovate (Szkudlarz 2001; Nelson 2006a). n = 12 (see Fraga 1983; Nelson & Oliver 2005b).

FLOWERING PERIOD. May–September(–November).

DISTRIBUTION & ECOLOGY. Europe only; Ireland, Britain, Norway, Sweden, Finland, Latvia, Lithuania, Poland, Germany, Czech Republic, Denmark, Netherlands, Belgium, France, Spain, Portugal. Reports that *Erica tetralix* occurs naturally on the Faeroe Islands and Iceland are erroneous (Bannister 1966). **Maps:** Fægri (1960), Bannister (1966: 797), Bolòs & Vigo (1995: 33). Fig. 143 (p. 232).

In bogs, wet and dry heaths, open woodland, on acid soil; very rarely on limestone but then

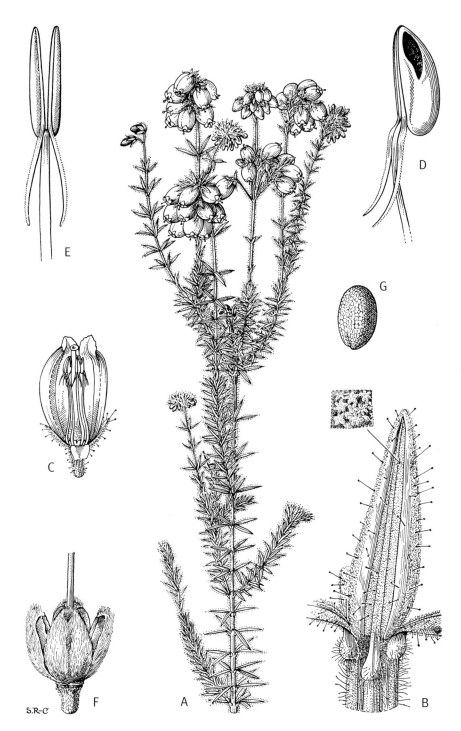

Fig. 146. *Erica tetralix.* **A** part of flowering plant × 1; **B** stem and leaves, and part of lower surface of leaf further enlarged × 12; **C** flower – part of calyx lobe and part of corolla cut away × 4; **D** anther and part of filament, in side view × 20; **E** anther and part of filament, in back view × 20; **F** fruit – the calyx and the persistent corolla and stamens removed × 8; **G** seed × 40. Drawn by Stella Ross-Craig.

always on acidic soil overlying the base rock (for example in Picos de Europa, Asturias, Spain (Nelson 1996a), and in The Burren, County Clare, Ireland (Brodie & Watts 2001)); Davies (1984) suggested that the ability of *Erica tetralix* to grow on limestone heaths was due, at least in part, to the evolution of physiologically adapted races but their existence has not been demonstrated.

Erica tetralix is naturalised in eastern North America: USA (recorded from localities in Maine, Massachusetts, New Hampshire, North Carolina, Ohio, West Virginia[1]).

ETYMOLOGY. Τετράλιξ was the name of a thistle-like plant, known to Theophrastus, that grew in the summer (Hort 1916).[2] In any case, it cannot have been cross-leaved heath as that does not occur anywhere within Theophrastus' likely range of knowledge, despite its connection to this particular *Erica* species by eighteenth-century botanists.

Carl Linnaeus (1753) took up, as he often did, a name which had been applied to this plant by an earlier botanist, in this case the German Heinrich Bernhard Rupp (1688–1719), who is better known by the Latinised version of his surname. "ERICA spuria sive TETRALIX" was listed in the 1726 edition of Ruppius's *Flora ienensis*; although Linnaeus made no reference to this work under his entry for *Erica tetralix*, Ruppius's (1726: 31) synonymy indicates that he intended the cross-leaved heath (see Jussieu 1828). It is noteworthy that Linnaeus did not use this epithet for any other plant described and named in *Species plantarum*.

VERNACULAR NAMES. Cross-leaved heath, bog heather (English); cross-leaf heath (USA); fraoch naoscaí (Irish); kellokanerva (Finnish); Glockenheide, Moorheide, Gemeine Glockenheide, Sumpfheide (Dippel 1887) (German); bruyère quaternée, bruyère à quatre angles, bruyère tétragone (French); ainarra lauhostoa, txilar lauhostoa (Basque: *fide* http://www.jpdugene.com/; accessed 3 February 2007); margariça, urze-peluda (Portuguese); klockljung (Swedish).

On the Scottish name "cat-heather" and its use for *Erica tetralix*, see under *E. cinerea* (p. 364).

An old name for cross-leaved heath was rinze-heather (see, for example, Ramsay 1808: 496); its use was probably confined to Scotland. Under "rinse", the *Oxford English dictionary*[3] explains the noun as meaning: "A small bundle of twigs (esp[ecially] of heather) used for cleaning out pots or other vessels. Hence **rinse-heather**, the variety of heather used for making this." An alternative spelling is "ringe-heather": "*Ringe*, a whisk or small besom, made of heath. *Ringe-heather*, Cross-leaved Heath" (Jamieson 1808[3]).

CULTIVARS. See p. 230 and Appendix I (p. 382).

19. ERICA MACKAYANA. Mackay's heath.

Although there is controversy about the status of three, Ireland is generally considered to have six native species of *Erica*. Only two, *E. tetralix* (cross-leaved heath, see above) and *E. cinerea* (bell heather, see p. 257), are widespread throughout the island. *E. vagans* (Cornish heath, see p. 132), which some botanists consider to be an introduction, is confined to a single population in County Fermanagh (Nelson & Coker 1974), and the spring-flowering *E. erigena* (Irish heath, see p. 119), almost certainly a long-naturalised introduction (Foss & Doyle 1987), is only found in counties Galway and Mayo, where it is particularly abundant in the region to the north of Clew Bay (Foss *et al.* 1987). *E. ciliaris* (Dorset heath, see pp. 167–169) is restricted to Connemara, and the tiny population of perhaps a dozen plants was recently (Curtis 2000) categorised as introduced. The remaining species is *E. mackayana* (Mackay's heath) (Figs 147–150) — on the unfamiliar spelling of the botanical name see p. 249 below.

Mackay's heath was discovered in western Ireland (Figs 147–150) and northern Spain by different individuals almost simultaneously in 1835 (Nelson 1979a). William McCalla, a young Scottish schoolmaster whose home was in Roundstone, Connemara, County Galway, recognised that it was distinct from *Erica tetralix*. The Spanish populations were first encountered on 27 June 1835 in the Sierra del Peral, Asturias, between Grado and Candamo by Michel-Charles Durieu de Maisonneuve (1796–1878) (Gay 1836; Lacaita 1929; McClintock 1983a). In a way, it is a shame that the plant was eventually named after James Townsend Mackay, Curator of the College Botanic Gardens, Ballsbridge, Dublin, and not McCalla, but McCalla, who was mainly interested in seaweeds, had upset the Irish botanical fraternity through some unstated misdemeanour (Nelson 1981c). In fairness, Durieu de Maisonneuve had an equal right to the heather's name — yet, obsequiously, the "meticulous" Jacques Étienne Gay (1786–1864) thought "*hookeri*" (after Professor William Jackson Hooker of the University of Glasgow) was more appropriate (Lacaita 1929: 256).

Fig. 147. *Erica mackayana* at Craiggamore, near Roundstone, Connemara (E. C. Nelson).

McCalla considered this heather to be an undescribed species (Mackay 1997; Nelson 1979a). In September 1835, he showed living examples to a visiting English botanist, Charles Cardale Babington (1808–1895), who collected specimens. Babington wrote a description of the species which he read to the Linnean Society of London on 1 December 1835, proposing "biformis" as the epithet. However, on the same day, Professor Hooker (1835), published a note about the heather, having received specimens from Mackay, and proposed naming it *Erica* "mackaii" in his honour. Babington was not pleased. Eventually the two botanists agreed, and in the summer of 1836 Babington formally named the new heather *Erica mackaiana* "in accordance with a suggestion of Dr. Hooker's" (Babington 1836), within Mackay's *Flora hibernica* (1836), which was published in Dublin on 15 June 1836 (Nelson & Parnell 1992).

The *locus classicus* for *Erica mackayana* in Ireland is Na Creagaí Móra, meaning the big crags and anglicised as Craiggamore, a low knoll (201 ft altitude, so sometimes referred to as "Hill 201") to the south of the so-called "Bog Road" which links Toombeola and Clifden (Robinson 2006: 38–39; Scannell 1975; Nelson 1978). I know the area well and have spent many days tramping across the spectacular bogland looking for this heather and mapping its distribution (Nelson 1981a). Mackay's heath has a substantial range in this part of Connemara, extending at least 5 kilometres east and west and around 8 kilometres north and south. My work was supplemented in the early 1980s by Ian Small[4], who extended the area southwards and westwards.

Mackay's heath also grows in four other separate localities, namely: a small population, containing a few hundred plants, near Carna (More 1874; Webb & Scannell 1983; Scannell 1985), also in Connemara; on blanket peat around Lower Lough Nacung (Fig. 150), at the base of Errigal Mountain, County Donegal, within about 200 metres of the lough margin (Webb 1954: 188–189); a small colony on blanket peat south of Bellacorick, County Mayo, about 80 kilometres north of Roundstone (van Dooslaer 1990); and another small population on the Iveragh Peninsula, north-east of Cahirciveen, County Kerry (Nelson 2005a). The populations in counties Mayo and Kerry are the most recently discovered ones; Lieveke van Doorslaer found the Bellacorick one in 1989, and David Edge chanced upon the Iveragh Pensinsula colony during 2003.

Almost without exception, in Ireland *Erica mackayana* is accompanied by its hybrid with *E. tetralix*, *E.* × *stuartii* (see pp. 301–306). The exceptions, reported by Ian Small[4], were two isolated patches to the west of Errisbeg; *E.* × *stuartii* was not encountered within half a mile (±0.8 km) of either.

In northern Spain, *Erica mackayana* has a range of around 300 kilometres, from near Santander westwards towards Santiago de Compostela and La Coruña (Fraga & Nelson 1984; Fraga 1984b). However it does not extend south of the Cordillera Cantabrica, being restricted to localities within about 35 kilometres of the northern coast. Although *E. tetralix* is also widespread in the region west of the Pyrenees, the two species rarely co-exist. Dr Maria Fraga Vila, who studied the ecology and distribution of *Erica* species in Galicia, reported four localities where they could be found growing in reasonably close proximity in that province (see map in Fraga & Nelson 1984). We saw them growing within the same patch of gorse (*Ulex europaeus*) near Covadonga in Asturias during July 2007. Thus, unlike in the Irish populations, *E.* × *stuartii* is not a constant cohabitant in Spain — it is, however, not absent as we have so long thought, being recorded from at least five localities in Asturias (Díaz González & García Rodriguez 1992) and, most recently, in another in Galicia (Fagúndez 2006) (see p. 303).

Fig. 148. *Erica mackayana* on raised bog near Roundstone, Connemara; the Twelve Bens in the distance (E. C. Nelson).

Seed is produced by Spanish populations of *Erica mackayana*, but the species apparently does not often set seed in Ireland, although fertile pollen is produced. It is possible that self-incompatibility is the cause, because Irish plants yield seed when grown alongside Spanish plants (McAllister 1996). Thus in Ireland, *E. mackayana* reproduces nearly always by vegetative means. However, reproduction by seed must occur, at least very occasionally, in Connemara given the amount of variation visible to the naked eye in the size and shape of the corollas — the plants are not all the same. Differences in habit were obvious when a random selection of material from throughout the species's Connemara range was propagated and then grown together at the National Botanic Gardens, Glasnevin (Nelson 1989a).

In 1955, Professor David Webb suggested that the Irish populations of *Erica mackayana* comprised a single "biotype". Because it reproduced in Ireland almost exclusively by layering, "populations over a considerable area consist of a single clone" (Webb & Scannell 1983: 130; see also McClintock 1972), McAllister (1996) postulated that the small colonies in Ireland were "reduc[ed] ... to self-sterile clones." Thus, in any particular area it is likely that the individual clumps of Mackay's heath all belong to a single clone, and furthermore clones are likely to monopolise quite substantial territories.

The Irish plants of *Erica mackayana* have a remarkable ability to propagate vegetatively; shoots pressed into peat quickly produce roots, and even small portions of roots are capable of regenerating new plants (McAllister 1996). This is most obvious on the face of abandoned turbary banks where new shoots or clusters of shoots can be seen. *E. tetralix* (cross-leaved heath) will not be found performing in the same manner but *E.* × *stuartii* (Praeger's heath) shares this faculty with its other parent, *E. mackayana*.

Another consequence of this reproductive strategy is that distinctive, individual clones of *Erica mackayana* may persist for decades, if not centuries and even longer. The repeated finding of the double-flowered plant, first collected near Craiggamore in 1869 (McClintock 1972, 1982a; Scannell & McClintock 1974), most likely demonstrates this. Further evidence for the persistence of individual clones of Mackay's heath and its hybrid was the chance finding in 2000 of a small plant with flowers indistinguishable from the cultivar long known as 'Stuartii' (see Plate 18, p. 230) (Nelson 2001a).

HISTORY OF ERICA MACKAYANA IN IRELAND

The extraordinary distribution pattern of *Erica mackayana* has long demanded some explanation. Why is it found only in western Ireland and northern Spain? It is one of a suite of plants — the so-called "Lusitanian" species (see, for example, Praeger 1934) — that link the floras of Iberia and Ireland. Modern terminology is rather different: *E. mackayana* is now regarded as a member of the Oceanic Temperate element (Preston & Hill 1997) of the flora of Ireland and Britain, which also includes bell heather (*E. cinerea*).

Erica mackayana, unlike *E. erigena*, has a long record as a native species in Ireland. Leaves identified as coming from *E. mackayana* are recorded from various interglacial deposits in Ireland. Jessen (1948) found such leaves in deposits from Gort, County Galway, in association with fruits (capsules) of *Rhododendron ponticum*: "evidence that the history of the so much debated Irish "Lusitanians" goes far back in time" (Jessen 1948: 175). Seeds (four specimens from one sample) identified as belonging to *E. mackayana* were found in some "vegetable debris" extracted from a karstic cavity in County Monaghan; these cannot be accurately dated "but the flora is unquestionably Quaternary, not Tertiary" (Mitchell *et al.* 1980). Along with the *E. mackayana* seeds, were seeds of *E. tetralix* and *Calluna vulgaris*; cones from Scots pine; holly, yew and juniper leaves; and fruits of alder and birch.

Erica mackayana   **Painted by Wendy F. Walsh**   **PLATE 19**

Adapted from W. F. Walsh, R. I. Ross and E. C. Nelson, *An Irish florilegium*, pl. 15 (1983).

Figs 149 and **150.** *Erica mackayana* on abandoned turbary (peat-cutting) banks close to Craiggamore, Connemara, County Galway and in a similar man-made habitat close to Lough Nacung, County Donegal (E. C. Nelson).

These finds, especially the fossilised leaves, are not unique. Peat of interglacial age from Fulga Ness in the Shetland Islands also yielded *Erica mackayana* and with it *Daboecia cantabrica* (Birks & Ransom 1969), indicating that some of the heathers of southern affinity had migrated much further north during interglacial stages than their present distribution patterns would lead us to expect.

Studies[5] of the "genetic fingerprints" of Irish and Spanish populations of *Erica mackayana* suggest that the populations "have been separated for a considerable time" and that the separation "is due to postglacial migration" (Kingston & Waldren 2006), but not too much reliance can be placed on these conclusions.

The debate continues about whether plants like Mackay's heath, St Dabeoc's heath and the other "Lusitanian" plants that inhabit mainly western parts of Ireland (see eg. Nelson 1979b; Coxon & Waldren 1995) survived in some ice-free, mild refuge on the edge of the continental shelf (Mitchell & Watts 1979) through the last glacial period, or whether the refuge was much further to the south in what is now Iberia. After the ice retreated, the plants were able to migrate from their refuges and re-colonise areas from which they had been evicted by glaciers and tundra. That they could, and did, migrate northwards when the climate was mild is shown by the Shetland finds. That they probably did so in the present interglacial seems to be indicated, but not proved, by Kingston's studies (Kingston & Waldren 2006).

CULTIVATION & VARIATION

Erica mackayana was introduced into cultivation from Connemara late in 1835, within a few weeks of Babington's visit to Connemara (see Nelson 1979: 292–293).[6] Mackay's heath can be cultivated in any limefree soil, and the cultivars are easily propagated by cuttings. Several Irish clones of *E. mackayana* are in cultivation, reflecting variation within and between the isolated native populations. One, from Lough Nacung, called 'Donegal' (McClintock 1982a), has a more open habit, larger leaves and flowers than the clone called 'Connemara' (alleged, without any evidence, to have come from County Galway). 'Errigal Dusk' (® no. 33), also from Lough Nacung, has barrel-shaped amethyst (H1) flowers in August and September, and dark green foliage. Gay (1955a) had reported that the Donegal plants tended to have more flowers per umbel and longer pedicels than the Connemara ones, and that leaf size and colour differed, and this seems borne out by cultivated plants. As noted, cultivars with "double" flowers are grown, particularly 'Plena' (see below) and 'Maura' (heliotrope (H12) flowers, July–September).

Figs 151–154 (left to right). In coastal heathland on Cabo de Peñas, Asturias, northern Spain, *Erica mackayana* is much more variable that in Ireland. **151.** "Normal" Mackay's heath with pink flowers; **152.** Mackay's heath with *E. cinerea* (bell heather; left) and gorse (*Ulex europaeus*); **153.** "Normal" and white forms of Mackay's heath, with *E. ciliaris* (Dorset heath; top left); **154.** Mackay's heath with pure white flowers (*E. mackayana* f. *eburnea*) (E. C. Nelson).

No Spanish clones were reported in cultivation in Ireland or Britain until our visit to northern Spain in July 1982. Although material was gathered from many populations in the region, only two have merited naming: 'Galicia', with deep magenta flowers from July to October, and good, dark green foliage; and 'Shining Light' (see p. 379) (Nelson 1989b).

Within *Erica mackayana*, three variants have been named. McClintock (1983b) named and described *E. mackayana* f. *eglandulosa* based on material from Carna, County Galway, which, contrary to the protologue, is not entirely eglandular. A second, comprising those plants that occur in the wild and in cultivation in which the stamens have been replaced by numerous petaloid structures is designated *E. mackayana* f. *multiplicata* E. C. Nelson (1995b). The extreme example of this is the so-called double-flowered clone named 'Plena' which, as just noted, was first collected in Connemara during 1869 by A. G. More, and subsequently in August 1901 by Dr F. C. Crawford. Also included is 'Maura', gathered at Carna in 1970 by Miss Maura Scannell (McClintock 1982a; Scannell & McClintock 1974).

White-blossomed plants are *Erica mackayana* f. *eburnea* E. C. Nelson & D. C. McClint. (Nelson & McClintock 1984; McClintock 1984a: 191). They are uncommon, yet not infrequent, in the wild throughout northern Spain. In fact, this *forma* is more abundant in Spain than published accounts imply. According to Bayer (1993), the flowers in *E. mackayana* are "rosada o purpúrea, ocasionalmente albina" (Figs 151–154). At Cabo de Peñas, Asturias, several dozen white-flowered plants occur in a relatively small patch of coastal heathland, and when in bloom they are especially prominent, standing out from among the "normal" rose-pink-flowered shrubs.

HYBRIDS

The facility with which Mackay's heath forms a hybrid with *Erica tetralix* (cross-leaved heath) has already been noted, and *E.* × *stuartii* is familiar not just to Irish botanists but also more generally to gardeners (see p. 305).

The hybrid has been created artificially — Professor David Webb succeeded (see p. 329) in obtaining seedlings from *Erica tetralix* pollinated by *E. mackayana*. Both Dr John Griffiths (Leeds, Yorkshire) and David Wilson (British Columbia, Canada) have more recently obtained hybrid seedlings from this coupling; for Griffiths, the reciprocal cross yielded viable seeds but the seedlings were lost before they could be studied, while Wilson's reciprocal cross failed.

Dr Griffiths has ventured further with *Erica mackayana* as a pollen- and as a seed-parent. He failed to raise hybrids when *E. ciliaris* and *E. manipuliflora* were pollinated by Mackay's heath, and the reciprocal cross with *E. ciliaris*. Interestingly, he obtained viable seed from *E. mackayana* pollinated by *E. andevalensis*, but the seedling were lost before any conclusion could be reached about them. *E. vagans* was another failed pollen-parent. Kurt Kramer (Edewecht-Süddorf, Germany) has tried cross-pollinating *Erica spiculifolia* with *E. mackayana* without success.

Erica mackayana Bab. in J. T. Mackay, *Fl. hibernica*: 181 (1836). Type: 'Craigha Moira Connemara', *C. C. Babington*, 2 September 1835: lecto: **CGE!** designated by Sell (1990).
E. mackaii Hook., *Companion to the botanical magazine* **1**: 158 (1835), nom. provis., non rite publ.
E. tetralix var. *mackaiana* (Bab.) Loudon, *Fruticetum & arboretum britnannicum* **2**: 1079 (1838).
E. ciliaris subsp. *mackaiana* (Bab.) Moore & More, *Contrib. Cybele hibernica*: 183 (1866).

ILLUSTRATIONS. S. J. Roles, *Flora of the British Isles: illustrations* pt 2: tab. 934 (1960); S. Ross-Craig, *Drawings of British plants*, part 19: pl. 23 (1963); R. W. Butcher, *A new illustrated British Flora*: 106

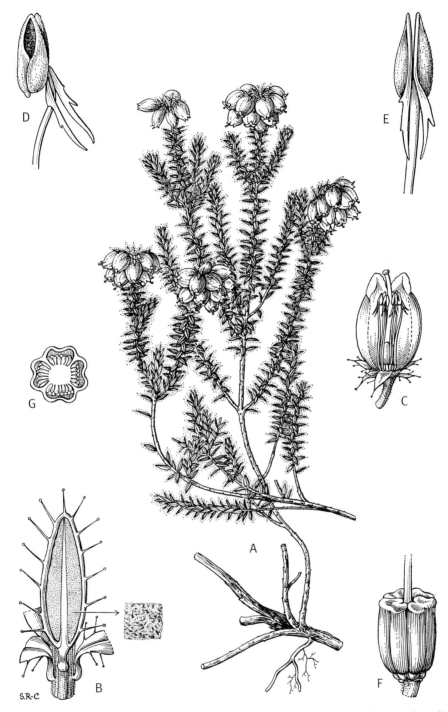

Fig. 155. *Erica mackayana.* **A** flowering stems × 1; **B** stem and leaves × 12, and part of lower surface of leaf further enlarged; **C** flower, part of corolla and one stamen cut away × 4; **D** stamen in side view × 20; **E** stamen in back view × 20; **F** immature fruit and base of style, the persistent parts of the flower removed × 16; **G** transverse section of immature fruit × 16. Drawn by Stella Ross-Craig.

(1961); J. Smythies, *Flowers of south-west Europe*: 287 (8) (Polunin & Smythies 1973); L. Denkewitz, *Heidegarten*: 175 (1987); W. F. Walsh, *An Irish florilegium* pl. 15 (Walsh *et al*. 1983); E. Sierra Ràfols, *Flora iberica* 4: lam. 178 (p. 490) (1993).

BIOLOGICAL FLORA OF THE BRITISH ISLES. *J. Ecol*. **43**: 319–330 (1955).

DESCRIPTION. *Evergreen, bushy shrub* to 0.3(–0.9–1.5) m tall, branches ascending, or sprawling; *shoots* pale brown, with dense covering of very short, simple straight hairs and occasional gland-tipped cilia; infrafoliar ridges not apparent. *Leaves* in evenly spaced whorls of 4, always spreading more or less horizontally, obovate, flat, upper surface dark green, glabrous, margins and apex with long (usually) gland-tipped cilia ± evenly spaced, margin slightly recurved given very narrow glabrous rim below; sulcus open exposing abaxial surface, midrib green, glabrous, lower surface is covered with white tomentum; petiole curved, glabrous, expanded at base and usually stained red-brown. *Inflorescence* a terminal umbel, with ±10 flowers; pedicel red, curving, sparsely pubescent with very short, simple hairs, and fewer, longer gland-tipped hairs; bract $3/4-2/3$ way along, leaf-like, green; bracteoles closely pressed to calyx; bract and bracteoles spathulate, ±2 mm long, >0.5 mm broad, leaf-like, glabrous except for several (±8), long (>1 mm) marginal, gland-tipped cilia. *Calyx* of 4 unequal, free sepals, ±3.5 mm long (including apical cilium c. 1 mm), ±0.5 mm broad, lanceolate, glabrous, with long, gland-tipped hairs on margins (lowermost hairs have much shorter stalks), sulcus white, resembling an inverted "V". *Corolla* urceolate, to 7 mm long, pale to dark pink, usually darker on sunlit side shading to white, or pure white; limb glabrous, lobes with sparse, simple hairs on outer surface (visible in buds); lobes ±1 mm long, rounded, recurved after anthesis. *Stamens* 8, included, to 5.5 mm long; filaments ± straight but at maturity becoming curved with sigmoid bend towards apex; *anthers* lateral, with 2 narrow, ± linear spurs at junction with filament, as long as, or longer than, thecae; thecae ovate, ±1 mm long, sometimes markedly prognathous, pore oval, ±0.75 mm long. *Ovary* ribbed, 4-locular ±1.5 mm long, glabrous, emarginate, very numerous ovules/locule; with prominent nectary ring at base; *style* ±5 mm long, straight, terete, tapering slightly, glabrous; *stigma* capitate, dark. Seed 0.3–0.45 × 0.2–0.33 mm (Nelson 2006a). n = 12 (Fraga 1983: 535; see Nelson & Oliver 2005b).

FLOWERING PERIOD. July–September (–November).

DISTRIBUTION & ECOLOGY. Europe only; western Ireland (counties Kerry, Galway, Mayo, Donegal); northern Spain. **Maps**: Ireland — see Nelson (1981a, 2005a); Spain — see Nelson & Fraga (1984). Fig. 156.

In Ireland, Mackay's heath inhabits damp heaths and bogs, abandoned turbary banks (the place where turf had been cut) (Figs 150 & 151), and the margins of loughs and ditches, where it is often associated with *Erica tetralix*. It ascends to about 1,500 m altitude in Spain (to 1,680 m: *fide* Bayer & López González 1989), but never exceeds an altitude of 100 m in Ireland. In northern Spain it is a very prominent and dominant component of the species-rich coastal heathland (Figs 152–155), growing with *E. ciliaris, E. cinerea, E. vagans, Calluna* and *Daboecia cantabrica*. Only very occasionally in Spain does *E. mackayana* occur within the same habitat as *E. tetralix*.

ETYMOLOGY. Eponym. Babington (1846) wrote: "in honour of Mr. J. T. Mackay, the eminent botanist to whom we owe the discovery of *E. mediterranea* [= *E. erigena*, see p. 123] in Ireland." James Townsend Mackay (1775–1862), a noted Scottish horticulturist and field botanist ,was curator of Trinity College Botanic Garden, Ballsbridge, Dublin, Ireland.

In *Flora hibernica*, published on 15 June 1836, Babington employed what has always been regarded as an intentional latinisation of Mackay's surname, notionally removing the last letter ("y") and appending the termination -*iana*: thus *mackaiana*. This is the form of the epithet that has been employed in Irish botanical literature for 170 years (see also Babington 1836).

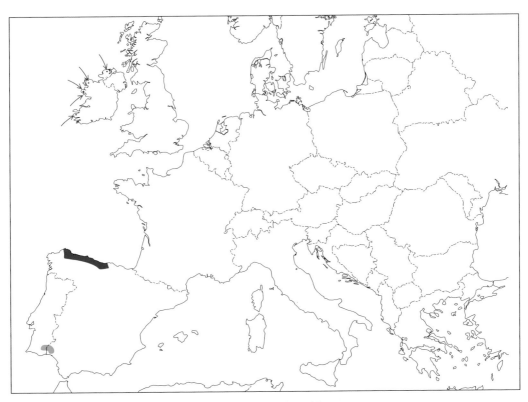

Fig. 156. Distribution of *Erica mackayana* (pink) and *E. andevalensis* (blue).

In 2006, a new edition of the *International code of botanical nomenclature (Vienna code)* was issued, containing a revised Article 60.7 (McNeill *et al.* 2006: 107–108) that lays out new guidelines for the formation of names from personal names (eponyms) where implicit Latinisation of those names occurred. The revised article states (and it is the exception that is significant):

> When changes in spelling by authors who adopt personal, geographic, or vernacular names in nomenclature are intentional Latinisations, they are to be preserved, except when they concern (a) only the termination of the epithets ... or (b) changes to personal names involving (1) omissions of a final vowel or final consonant or (2) conversion of a final vowel to a different vowel, for which the final letter of the name is to be restored.

One example given is relevant to the case of *Erica mackaiana/mackayana*, as follows: "... *Hypericum* "buckleii" ... commemorate[s] ... S. B. Buckley. The implicit latinisation ... Buckleius ... [is] not acceptable under Art 60.7. The name [is] correctly ... *H. buckleyi* ... " (McNeill *et al.* 2006: 107, *ex. 15*).

Thus *Erica mackaiana*, after J. T. Mackay, must be changed to *E. mackayana*.

Of course, this change is annoying — books and nurserymen's catalogues, labels in gardens and garden centres, international registers and databases, all have to be amended, although probably at small expense beyond great irritation (Nelson 2007c).

VERNACULAR NAMES. Mackay's heath (English); fraoch Mhic Aoidh (Irish); carroncha (Spanish); Spanische Moorheide, Mackays Heide (Dippel 1887) (German).

CULTIVARS. See pp. 230, 245 and Appendix I (pp. 379–380).

20. ERICA ANDEVALENSIS

Originally described from Huelva in the southwest of Spain (Cabezudo & Rivera 1980; Márquez *et al.* 2004) where it occurs around pyrites mines and along associated rivers and streams, *Erica andevalensis* has recently been reported to the west of Huelva, across the international frontier, in the south-eastern corner of Portugal (Capelo *et al.* 1998), growing in remarkably similar habitats at Mina de São Domingos, Mértola, Baixo Alentejo.

The plant was recorded in the nineteenth century from south-western Spain as *Erica tetralix*, which it resembles, particularly in the arrangement of four leaves in a whorl, but from which it differs in possessing a glabrous ovary, although in *Flora iberica* (Bayer 1993: 493) it is stated that the ovary can have hairs on the apex. In 1989, McClintock decided unilaterally that the Huelva plant was better considered as a subspecies of *E. mackayana*, and he published the required new combination, *E. mackayana* subsp. *andevalensis* adding my name to his as the authority without any consultation. I have never considered the case for merging *E. mackayana* and *E. andevalensis* proven. They occur at opposite extremities of the Iberian Peninsula, around 600 kilometres apart. In this case, despite acknowledged morphological similarities, I prefer to retain them as separate species. The extraordinary ecology of *E. andevalensis*, and the probable physiological consequences, argue for separate status. As said before, until detailed, reliable DNA analyses are available for European species at population level, the status of the three cross-leaved heaths remains a matter of redundant speculation.

THE COPPER-FLOWER

The mines of the south-western Iberian Peninsula have been worked for many centuries. Since Roman times, at least, copper, silver and gold have been exported from the region, and it is sometimes suggested that the reference in *The Bible*, in the First Book of Maccabees 8, to the gold and silver mines of Spain refers to this region. One consequence of the long history of mining of the toxic, oxidised mineral ores is that many localities in Huelva and south-eastern Portugal are highly contaminated. The adjective "lunar" can be applied with impunity to the massive open-cast mines at Río Tinto (Fig. 157), reputed to be the oldest in the world. Belén Márquez García has recently likened the "inhospitable" landscape of these mines, in which the soil pH can be as low as 2, to that of Mars (Márquez García *et al.* 2006). Almost the only plant found growing on the mines, their enormous spoil heaps, and the poisoned stream-beds leading to the main rivers, such as Río Odiel, is *Erica andevalensis*. Along the river's margins, dense thickets of *E. andevalensis* occur (Figs 158 & 159), the shrubs rooted in gravel and sand into which lethal, iron-stained water percolates (McClintock 1983a; Nelson *et al.* 1985). At Mina de São Domingos, Mértola, Portugal, the heather grows around pools of water that is characterised as "enriched with sulphuric acid" (Capelo *et al.* 1998).

Since the mid-1980s, *Erica andevalensis* has become a much-studied plant — indeed one of the most intensely investigated of all the heaths — and its exceptional ability to thrive in soils that have very high concentrations of copper, zinc and iron has been documented (Nelson *et al.* 1985; Asensi *et al.* 1999; Stewart & Anderson 2005). This heather has one of the highest tolerances of copper of any known plant. Thus *E. andevalensis* is one of a small number of so-called "copper flowers", an excellent indicator of copper mineralisation. It has the capacity to "tolerate high metal levels by excluding" the toxic minerals from its tissues (Stewart & Anderson 2005). In a recent paper, Rodríguez *et al.* (2007) have drawn attention to the possibility of using *E.*

Fig. 157. Open-cast mine at Río Tinto in south-western Spain; the only plant growing on the slopes is a heather, *Erica andevalensis* (E. C. Nelson, July 1982).

Fig. 158. Río Odiel with *Erica andevalensis* shrubs up to 1 m tall; the shrubs will be submerged when the river is in spate (E. C. Nelson, July 1982).

andevalensis for revegetating places where the soil is contaminated with heavy metals, such as old mines. However much this is desirable, the uncontrolled or uncritical use of *E. andevalensis* outside its native habitats would be very unwise given the propensity of some heathers (for example, *Calluna vulgaris* or *E. lusitanica*) to become pest plants.

Other aspects of this extraordinary heather have also been studied. Its seeds germinate best after chilling (Aparicio 1995): only a small proportion (5–20%) of unchilled, dormant seeds germinated in Aparicio's experiments. This suggests that in the wild the freshly shed seeds remain dormant and will only germinate after a prolonged period of wet, cool weather. Pollination studies showed that *Erica andevalensis* is self-compatible yet also out-crossing, and depends on insects or wind to transfer pollen from plant to plant (Aparicio & García-Martín 1996).

On an entirely different topic, *Erica andevalensis* was used in Huelva by the local people as a urinary antiseptic. Extracts of the flavonoids contained in its leaves and shoots were made, and these have been shown to be effective in preventing gastric ulcers in rats (Ruiz *et al.* 1996). Subsequent work has shown that triterpenoids extracted from *E. andevalensis* acted to retard cell division (mitosis) in the root tips of onions (*Allium cepa*), and, extrapolating from this result, Spanish pharmacologists have found that ursolic acid from the heather had "pronounced activity against ... three [human] cancer cell lines", suggesting it is a potential antitumor agent (Martín-Cordero *et al.* 2001). Previous studies had included investigations of the antibacterial (Toro Sainz *et al.* 1987) and diuretic (Toro Sainz *et al.* 1988) properties of phenolic acids found in *E. andevalensis*.

CULTIVATION & VARIATION

In July 1982, with David Small and David McClintock, I visited the extraordinary habitat of the species in southern Spain, and we later described these in *The Kew magazine* (Nelson *et al.* 1985). We collected cuttings from a random selection of plants, including five white-flowered shrubs (*Erica andevalensis* f. *albiflora* E. C. Nelson & D. C. McClint.; Nelson &

Fig. 159. The shrubbery on the river gravels is composed exclusively of *Erica andevalensis*, Río Odiel, Huelva, south-western Spain (E. C. Nelson, July 1982).

Fig. 160. *Erica andevalensis*. Watercolour by Wendy F. Walsh (unpublished, author's collection).

Fig. 161 (left). *Erica andevalensis*; Río Odiel, Huelva, south-western Spain (E. C. Nelson, July 1982).

Fig. 162 (right). *Erica andevalensis* f. *albiflora* 'Blanco de Odiel', collected in July 1982 at Río Odiel, Huelva, south-western Spain (E. C. Nelson).

McClintock 1984; McClintock 1984a), and so *E. andevalensis* was introduced into cultivation in Ireland and Britain from Huelva. The cuttings from various sites were inserted into a compact mist-propagation unit installed in the back of David Small's car, and by the time we returned to England many had already rooted (a rooting-time of less then three weeks). Initially the young plants were grown on at Denbeigh Heather Nursery, Creeting St Mary, near Ipswich. Later, some of the plants were transferred to the National Botanic Gardens, Glasnevin, Dublin, and planted out in a trial bed in the nursery. The plants grew well, forming erect, open shrubs about 0.5 m in height. Their rather gawky, twiggy habit was not attractive, however, and while the clones were not damaged in Dublin by frosts, this species does not have much merit as a garden subject. The mauve-pink flowers (Fig. 161) are never abundant and showy; the white-flowered plant, 'Blanco de Odiel' (*E. andevalensis* f. *albiflora*) (Fig. 162), is more notable but has the same erect, twiggy habit.

HYBRIDS

Although so recent an introduction, *Erica andevalensis* has figured in the cross-breeding experiments of both Dr John Griffiths (Leeds, Yorkshire) and Kurt Kramer (Edewecht-Süddorf, Germany). Both have raised hybrids. Griffiths has used *E. andevalensis* only as a pollen-parent: he raised hybrid seedlings from *E. tetralix*; the seedlings obtained from *E. mackayana* were lost before they could be assessed, while the seed from *E. terminalis* was not viable. Kramer has used *E. andevalensis* only once, and unlike Griffiths, it was as a seed-parent, with pollen from *E. tetralix* 'Foxhome'; a single cultivar, 'Pat Turpin', resulted (McClintock 1992a; Small 1992; Hall 1992). Kramer also succeeded in back-crossing this with both *E. andevalensis* and *E. tetralix* (Hall 1992).

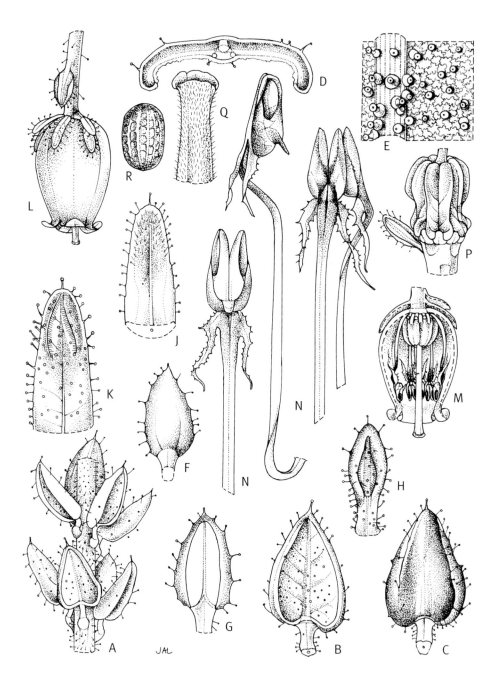

Fig. 163. *Erica andevalensis*. **A** leafy shoot × 6; **B** leaf from below × 7.5; **C** leaf from above × 7.5; **D** section through leaf × 22; **E** indumentum from back of leaf × 24; **F** bract from above × 7; **G** bract from below × 7; **H** bracteole from below × 11.5; **J** sepal from above × 17; **K** sepal from below × 18; **L** flower from side × 7; **M** section through flower × 7; **N** stamens from front, side and back × 20; **P** ovary × 15; **Q** style tip (stigma) from side × 21; **R** seed from side × 36; from specimens grown by David Edge. Drawn by Joanna Langhorne.

Erica andevalensis Cabezudo & J. Rivera, *Lagascalia* **9** (2): 224 (1980). Type: Zalamea la Real, Huelva, Spain, 21 July 1979, *Cabezudo & Rivera*: holo. **SEV** (41555); iso. **MA**, **MAF**.

Erica tetralix sensu Willkomm, *Supplementum prodromi florae hispanicae*, 136 (1893), non L.

Erica mackayana subsp. *andevalensis* (Cabezudo & J. Rivera) D. C. McClint. & E. C. Nelson, *Bot. J. Linn. Soc.* **101**: 282 (1989)

ILLUSTRATIONS. *Lagascalia* **9** (2): 225 (1980); E. Sierra Ràfols, *Flora iberica* **4**: lam. 179 (p. 492) (1993); Bayer & López González 1989; R. Tavera Mendoza, p. 120 in Blanco *et al.* (1999).

DESCRIPTION. *Evergreen, erect, twiggy shrub* to ±1(–2) m tall, to 0.65 m across. *Shoots* ascending, rather brittle, brown, with very short indumentum of simple non-glandular hairs with a few gland-tipped hairs interspersed. *Leaves* in whorls of 4(–5), internodes on young shoots ±1.5 mm long, on older shoots to ±5–7.5 mm long; ovate, to ±5.5 mm long, to ±2.5 mm across, glabrous and dark green above, margins revolute, with gland-tipped cilia around edges (very rarely without marginal cilia), underneath white-tomentose, with numerous minute hairs except on midrib. *Inflorescence* terminal, umbellate, composed of (4–)10–12(–17) flowers, ±1 cm across; July(to September–October); *pedicels* red, ±4–6 mm long, with short hairs, often gland-tipped; *bract* leaf-like, 2–3 mm long, at about mid-point; *bracteoles* immediately below calyx, narrow, leaf-like; flowers dark pink (rarely white); *calyx* with 4 free sepals, leaf-like, narrow-lanceolate, ±1.5 mm long, ±0.5 mm wide, with very short non-glandular hairs especially on tip on both inside and outer side, and short gland-tipped hairs especially on edges (rarely ± glabrous); sulcus ±0.5 mm long, open; *corolla* tubular-urceolate, to ±6(–7) mm long, to ±3–4 mm across, lobes revolute. *Stamens* 8, included, ±4.5 mm long; *anther* dark (or pale) brown, 1 mm long, sometimes markedly prognacious, with 2 slender, white spurs, (0.3–)0.5–1 mm long; *filament* ± 3.5 mm long, with pronounced sigmoid bend at apex below anther. *Stigma* emergent, capitate; *style* ±5 mm long; *ovary* ±1.4–2 mm tall, barrel-shaped, glabrous or with very sparse hairs on summit (*fide* Bayer 1993), brown or green; 10–30 ovules per locule (Aparicio 1999). *Seed* ±0.4 mm long, ovoid to ellipsoidal. n = 12 (Aparicio 1994; see Nelson & Oliver 2005b).

FLOWERING PERIOD. June–November.

DISTRIBUTION & ECOLOGY. South-western Iberian Peninsula only (Fig. 156, p. 249); Spain: Andalucia: Huelva; principally in the catchment of Río Odiel from Alonso and Tharsis in the west, to San Telmo, and east to Río Tinto and Nerva (Márquez *et al.* 2004); Sevilla; catchment of Río Guadiamar (Pérez Latorre *et al.* 2002). Portugal: Baixo Alentejo; Mina de São Domingos, Mértola. On river margins and stream beds, in gravel and sand which often are heavily contaminated with toxic minerals; on abandoned mines, spoil heaps, and roadsides. The species may be the only plant in a habitat. **Maps:** Aparicio (1999), Márquez *et al.* (2004).

Erica andevalensis has been gazetted by the provincial government of Andalucia as an endemic species in danger of extinction (Aparicio 1999; Márquez *et al.* 2004).

ETYMOLOGY. Toponym, from the region of Andévalo, Huelva, Spain.

VERNACULAR NAME. Brezo de Andévalo (Bayer & López González 1989).

CULTIVAR. Only one clone, 'Blanco de Odiel' (Figure 162), has been given a cultivar name to date; see Appendix I (p. 363).

Notes

[1] See URL http://plants.usda.gov/java/profile?symbol=ERTE4; accessed 29 April 2007.

[2] The name, in its Latin form, also occurs within a discussion of thistles in Pliny's *Historia naturalis* (Book XXI. LVI. 94):

> *Carduus et folia et caules spinosae lanuginis habet, item acorna, leucacanthos, chalceos, cnecos, polyacanthos, onopyxos, helxine, scolymos. chamaeleon in foliis non habet aculeos. est et illa differentia quod quaedam in his multicaulia ramosaque sunt, ut carduus, uno autem caule nec ramosum cnecos. quaedam cacumine tantum spinosa sunt, ut erynge. quaedam aestate florent, ut tetralix et helxine. ...*

> The thistle has both leaf and stem covered by a prickly down, and so have acorna, leucacanthos, chalceos, cnecos, polyacanthos, onopyxos, helxine, scolymos. The chamaeleon has no prickles on its leaves. There is however this difference also, that some of these plants have many stems and branches, the thistle for instance, while the cnecos has one stem and no branches. Some are prickly only at the head, the erynge for instance; some, like tetralix and helxine, blossom in summer. ... (see Jones 1961: 230, 231).

Thus, *tetralix* was, it seems, probably a sort of thistle that flowered in summer, and it has been equated by Hort (1916: **2**: 480) and Jones (1961: cf. index to volume **7**) with the yellow star-thistle (or St Barnaby's thistle), *Centaurea solstitialis*, a biennial knapweed, with bracts possessing a very long terminal spine, common throughout the Mediterranean, and now very occasionally naturalised in Britain (Clement & Foster 1994; Stace 1991; Blamey & Grey-Wilson 1993).

[3] *Oxford English dictionary* (Second edition. 1989) (*OED Online*. URL http://dictionary.oed.com/cgi/entry/50207111 accessed 31 May 2008).

[4] I. Small [c. 1983]. "Connemara project." Unpublished report (quoted from a copy in the present author's possession).

[5] Unfortunately, the study by Kingston & Waldren (2006) is flawed, because insufficient attention was given to ensuring that the samples of "*Erica mackayana*" from Connemara and Donegal did not include *E.* × *stuartii*. By courtesy of Dr Kingston, in June 2006 I examined some of the voucher specimens (**TCD**) for this study and found that the Donegal specimens, without exception, represented the hybrid (populations "Ireland C" and "Ireland D"; see Kingston & Waldren 2006: Appendix 1). It is doubly unfortunate that the voucher specimens for the Spanish populations could not be traced, so that some explanation could be found for Kingston & Waldren's curiously ambiguous statement (2006: 154) that "the herbarium collections made at the same time as the tissue samples were collected ... [showed] that the outlying Spanish population fell within the range of variation found in *Erica tetralix*."

Given that the samples designated as *Erica mackayana* probably also included material that should have been identified as *E. tetralix* or *E.* × *stuartii* (*E. mackayana* × *tetralix*) the study's conclusions are dubious.

[6] Ninian Niven, Curator of the Royal Dublin Society's Botanic Gardens, Glasnevin, obtained plants of *Erica mackayana* "for the use" of the Glasnevin Botanic Gardens during his visit to western Ireland between 19 August and 2 September 1836: "this I am disposed to consider a distinct species after seeing it growing, as it does, confined remarkably to one tract of bog, and mixed up with [*E.*] "Tetralix."" (Niven 1836; see Nelson & McCracken 1987: 91). Niven also brought back *E. erigena* and white-flowered *Daboecia cantabrica*.

THE BELL HEATHERS

Erica cinerea, E. maderensis, E. terminalis

When George Bentham was working on *Erica* and other Ericeae for Augustin-Pyramus de Candolle's encyclopaedic *Prodromus systematis naturalis regni vegetabilis*, he was aware of a heather collected on Madeira that was similar to, but not identical with, *E. cinerea*, the bell heather common through western Europe. Bentham decided the Madeiran plant was a variety of the much more widespread species. In 1904, the German botanist Joseph Friedrich Nicolaus Bornmüller (1862–1948) concluded that the isolated bell-heather population on Madeira deserved to be regarded as a separate species. No detailed, comparative study of *E. maderensis* and *E. cinerea* has been conducted.

Erica terminalis (Corsican heath) is also included here for several reasons. It is one of the few species that is known to be capable of producing a viable hybrid with *E. cinerea*, albeit not in the wild, only following artificial cross-pollination. Three decades ago, McClintock (1976, 1981) suggested that *E. maderensis* might be related to *E. terminalis*, and while this hypothesis is unproven and probably wrong, there are a number of similarities among the three heathers.

Bell heather (*Erica cinerea*) and Corsican heath (*E. terminalis*) are familiar to gardeners in western Europe and North America, the former more so than the other. Madeira heath (*E. maderensis*) is a collector's plant; it is not at all well known to gardeners and is not available commercially at present.

21. ERICA CINEREA. Bell heather, grey heath.

Bell heather (Fig. 164) is a favourite garden plant displaying a wide range of foliage and flower colours, the most diverse in the genus, it is claimed with justification, only matched in diversity by *Calluna vulgaris*. Golden, yellow, pale green, dark bronze-green, as well as "ordinary" green, are the foliage shades available, while flowers can be in numerous tones of lilac, pink, mauve, crimson and purple, as well as pure white (*E. cinerea* f. *alba* (Aiton) D. C. McClint.; McClintock 1984a: 191). Foliage colours are not constant and static — they may be seasonal and alter with aspect, and colours may also vary as the individual leaves age and mature. Plants with yellow or golden foliage have been designated as *E. cinerea* f. *aureifolia* D. C. McClint. (McClintock 1988a: 62). There are also some cultivated clones that might be termed monstrosities, plants with the corolla split into ragged segments (McClintock 1980c), or with the flowers replaced by clusters of coloured (or green) bract-like structures (the so-called bracteomaniac or "wheatear" bell heathers; McClintock 1980c: 189).

This multifarious variation in cultivated selections is not surprising given that *Erica cinerea* is one of the more wide-ranging heathers, occurring principally on the north-western Atlantic fringe of Europe. Bell heather is recorded from Viðoy, the northernmost of the Faeroe Islands (its northern extremity, Enniberg, is situated at 62°23'N lat., 6°34'W long.[1]); this heather, with *Calluna vulgaris*, was present on the archipelago before human settlement (Hannon *et al.* 2001). From the Faeroes, bell heather ranges from south-western, coastal Norway southwards through the Shetland and Orkney archipelagoes, Ireland and Britain, and from north-western Germany into the Low Countries and much of France (although it is absent from eastern and south-eastern parts, east of the Rhône valley). *E. cinerea* occurs in a very restricted part of southern Liguria in north-western

Fig. 164. *Erica cinerea*; a pale flowered plant (E. C. Nelson).

Italy (Serra 1966). It also ranges through northern Spain and into southern Portugal. There is a remarkable isolated population growing in evergreen oak woodland at Cap Rosa in Algeria (approx. 36°N lat., 8°E long.) (Quézel & Santa 1963); specimens I have seen do not appear to differ from European ones, although Réné Charles Joseph Ernest Maire (1878–1949) designated it as *E. cinerea* var. *numidica* (Maire 1931; Quézel & Santa 1963: 721). The species' north–south range is around 3,200 kilometres, about 26° longitude.

To confuse the geographical pattern, *Erica cinerea* is apparently naturalised on Madeira: in the mid-1980s both Dr J. R. Press and M. J. Short collected specimens at Parque de Queimadas (around 880 m altitude) (**BM!**), and there is also a record dating from the 1860s. These are most probably escapes from cultivation. Reports of *E. cinerea* from Corsica (Hegi 1927; Bayer & López González 1989) and even Transylvania (Bayer & López González 1989; Bayer 1993) are erroneous. Bell heather has become naturalised elsewhere; in North America, on Nantucket Island; in Australia (South Australia: Robyn Barker, pers. comm., 2005); and New Zealand (Healy 1944).

Erica cinerea has distinct ecological propensities (Bannister 1965). A typical habitat provides freely draining acidic soil (Fig. 165), and little or no shade; although in some places, and especially in south-western Europe, bell heather is found as an understorey shrub in open pine and oak woodlands. The soil can be peaty; bell heather often colonises the tops of abandoned turbary banks in cut-over bogs in Ireland. It occurs on mineral soils, too. Being intolerant of lime in the soil, bell heather does not inhabit regions where the bedrock is limestone or chalk, although there are exceptions. It will thrive when the surface layers of soil are leached and acidic as in limestone heath overlying chalk, or when peat develops over limestone and there is no seepage of limerich water into that peat, as in The Burren of County Clare in western Ireland (Nelson & Walsh 1991) (Fig. 166).

Erica cinerea   **Painted by Wendy F. Walsh**   **PLATE 20**

Published in E. C. Nelson & W. F. Walsh, *The flowers of Mayo* (1995).

Fig. 165. *Erica cinerea* growing on a hag of peat which is relatively free-draining; with *Calluna vulgaris* in bud behind; near Pettigo, County Donegal (E. C. Nelson).

Fig. 166. *Erica cinerea*; The Burren, County Clare (E. C. Nelson).

Bell heather is essentially also a lowland species intolerant of hard frost and snow. In Ireland and Britain, *Erica cinerea* rarely if ever occurs above 800 m altitude; its maximum altitude is probably around 790 m (2,600ft) on Purple Mountain, County Kerry, where the shrubs are very stunted (Scully 1916: 185; Preston & Hill 1997; Preston *et al.* 2002). On the Iberian Peninsula, it reaches 1,450 m (Bayer 1993), while in Liguria, the maximum altitude attained by *E. cinerea* is only about 450 m (Serra 1966).

Erica cinerea is abundant and ubiquitous in western Ireland, forming a principal component of the vegetation on rocky, coastal hillocks and on mountain slopes. In Connemara (western County Galway) (see Nelson 2001b: 7) and on Exmoor (west Somerset and north Devon), bell heather is principally associated with the dwarf, summer-flowering western gorse, *Ulex gallii*, the pair combining in blossom to create a splendid tapestry of brilliant purple and yellow.

ETHNOBOTANY

Erica cinerea has a number of uses, as well as being an excellent garden shrub. The most significant nowadays is as a valuable source of nectar that yields a dusky, aromatic honey (Pardo de Santayana *et al.* 2002; Persano Oddo *et al.* 2004). In Spain, the flowers and foliage have been used medicinally and as a source of dye (Pardo de Santayana *et al.* 2002).

CULTIVATION

Bell heather is easily grown in well drained, lime-free soil. Cultivars must be propagated vegetatively by layering or cuttings.

HYBRIDS

Although several putative, wild-occurring hybrids of *Erica cinerea* have been reported, none has been proven; they have all been phantoms, with the actual parentage not including this species. Focke (1881) suggested that *Erica mackayana* was the progeny of *E. tetralix* × *cinerea*, while *E.* × *williamsii* (see p. 310) was at first claimed as a hybrid between *E. cinerea* and *E. vagans*. Neither suggestion is accepted nowadays.

Attempts by horticulturists to produce hybrids of *Erica cinerea* by artificial cross-pollination have mostly been unsuccessful — those who have tried include the German nurserymen Georg Arends, Lothar Denkewitz and Kurt Kramer, and the English plantsman, Dr John Griffiths. The only species to yield fertile seed when crossed with bell heather are *E. terminalis* and *E. spiculifolia*. Arends (1951) was the first to achieve the cross with *E. terminalis*. His achievement was repeated by Kramer (see p. 327), but only when *E. terminalis* was the seed-parent — the hybrid has been named *E.* × *arendsiana* (Nelson 2007e). Kramer also reports that he obtained viable seeds in 1987 from *E. spiculifolia* pollinated with *E. cinerea* — the reciprocal cross, made in 1994, failed.

The failed attempts have included the following species as seed- and/or pollen-parents: *Erica arborea* (Kramer), *E. australis* (Kramer), *E. carnea* (Griffiths, Kramer), *E. ciliaris* (Griffiths), *E. erigena* (Griffiths), *E. manipuliflora* (Kramer), *E. tetralix* (Griffiths, Kramer, Denkewitz), *E. umbellata* (Griffiths, Kramer) and *E. vagans* (Kramer). Kramer failed also to get viable seeds from *Erica maderensis* when pollinated by *E. cinerea*, yet Dr John Griffiths records success and that the seedlings showed hybrid characteristics although they were not kept because they had no merits.

Erica cinerea L.[2], *Species plantarum*: 352–353 (1753). Type: "*Habitat in* Europe *media*": lecto. **LINN** 498.46! (Jarvis & McClintock 1990; Jarvis 2007: 498; see also Jarvis 1992: 563).
E. mutabilis Salisb., *Trans Linn. Soc.* **6**: 369 (1802).

ILLUSTRATIONS. J. Curtis, *Brit. Entomol.* **1**: tab. 35 (1824); S. J. Roles, *Flora of the British Isles: illustrations* pt 2: tab. 937 (1960); S. Ross-Craig, *Drawings of British plants*, part 19: pl. 24 (1963); R. W. Butcher, *A new illustrated British Flora*: 108 (1961); L. Denkewitz, *Heidegarten*: 157 (1987); W. F. Walsh, *Flowers of Mayo*, plate V (Nelson & Walsh 1995: 95).

BIOLOGICAL FLORA OF THE BRITISH ISLES. *J. Ecol.* **53**: 527–542 (1965).

DESCRIPTION. *Evergreen shrub* to 0.8 m tall, bushy, obligate seeder, not regenerating from a root-stock; *shoots* pale brown, densely pubescent, hairs short and simple; infrafoliar ridges not obvious. *Leaves* in whorls of 3 (or ± spirally arranged), narrow lanceolate to linear, those on non-flowering lateral shoots ("fascicles") usually shorter than 5 mm, glabrous but with spicules on ± cartilaginous edges, upper surface flattened, margins recurved and contiguous below, concealing the abaxial surface; sulcus extending from tip to petiole. *Inflorescences* aggregated towards tips of shoots and looking like a panicle, on short lateral shoots, and apical; flowers (Fig. 167) axillary, solitary in leaf axils or in whorl, 1–5 on each lateral shoot; shoot apex apparently continues to grow: *peduncle* dark purple-green, curved, terete, tomentose with some of the hairs gland-tipped, bract close to apex, and bracteoles closely pressed to calyx; bract and bracteoles ±1–1.5 mm long, leaf-like with prominent, somewhat irregular, translucent, hyaline margins stained purple, margins also with minute spicules sometimes with glandular tips, central portion swollen especially towards apex, green, with linear sulcus ±1 mm, margins parallel and contiguous. *Calyx* with 4 unequal lanceolate to triangular, ±free sepals, < 3.5 mm long, to 1 mm across, similar to bract and bracteoles, with translucent, hyaline, purple-stained (when corolla coloured), central part green, swollen (keeled) towards apex, with sulcus. *Corolla* usually rich purple (sometimes paler or white), 5–7 mm long, ovoid-urceolate, with four erect apical lobes ±1 mm long, glabrous. *Stamens* 8, ±5 mm long; *filament* curving upwards, with pronounced sigmoid bend at apex, glabrous; *anther* basifixed, with two crest-like spurs at junction with filament; *spurs* ±0.3 mm long, deeply lobed; *thecae* separate to base, ellipsoidal, ±0.8 mm long, ±0.3 mm across, pore oval, ±0.35 mm long. *Ovary* glabrous, barrel-shaped, ±2.5 mm long, emarginate, with prominent nectary ring at base; *style* c. 5 mm long, erect, terete, glabrous; *stigma* emergent, capitellate. *Seed* ±0.7–0.9(–1.0) × 0.5–0.7 mm (see Nelson 2006a). n = 12 (Fraga 1983: 536; see Nelson & Oliver 2005b).

FLOWERING PERIOD. June–September.

Fig. 167. Arthur Church's diagram of a flower of *Erica cinerea*.

Fig. 168. *Erica cinerea.* **A** part of a plant and a flowering stem × 1; **B** stem and leaves, and transverse section of a leaf × 12; **C** flower, part of corolla removed × 4; **D** anther and part of filament in side view × 16; **E** anther and part of filament in front and back views × 16; **F** fruit, the persistent calyx, corolla and stamens removed × 8; **G** seed × 40. Drawn by Stella Ross-Craig.

DISTRIBUTION. Western Atlantic Europe: Faeroe Islands, south-western Norway, Ireland, Britain, north-western Germany, Netherlands, Belgium, France, Italy (Serra 1966), Spain, Portugal. North Africa: Algeria, only at Cap Rosa. **Maps:** Hansen (1950: 30), Fægri (1960), Dupont (1962: 198, carte 41), Bannister (1965: 528), Bolòs & Vigo (1995), Preston *et alii* (2002); for detailed maps of distribution in Liguria see Serra (1966). Fig. 169.

Naturalised in North America, Australia and New Zealand (Healy 1944; Sykes 1981).

ETYMOLOGY. Latin: *cinereus* = ash-coloured, pale grey. This seems an odd name for a purple-blossomed heather, but ash-grey refers to the colour of the bark. In the work of the seventeenth-century Swiss botanist Caspar Bauhin, the descriptive Latin phrase employed for this heath was " ... *humilis cortice cinericeo, Arbuti flore*" which can be translated as low-growing, with grey bark, and a flower like the strawberry tree (*Arbutus*).

VERNACULAR NAMES. Bell heather, grey heath, Scotch heath (English); fraoch cloigíneach (Irish); Føroyskur klokkulyngur (Faeroe Islands); bruyère cendrée, bréjotte, Bucane (French); Graue Heide, Grau-Heide, Schotten-Glocken-Heide (German); bruyère cendrée (French); negrela, queiró, queiroga, torga, urze-roxa (Portuguese); argaña (Spanish); purpurljung, norsk klockljung (Swedish).

A minor curiosity is the name "cat-heather". John Jamieson's *A dictionary of the Scottish language* (1846) contained this definition: "CAT-HEATHER, *s.* A finer species of heath, low and slender, growing more in separate upright stalks than the common heath, and flowering only at the top,

Fig. 169. Distribution of *Erica cinerea*.

Aberd[een]." According to Professor George Dickie's *The botanist's guide to the counties of Aberdeen, Banff and Kincardine* (1860), "cat-heather" was applied in that region to ling, *Calluna vulgaris* (Jamieson's "common heather"). To add another twist, James Britten and Robert Holland's *A dictionary of English plant names* (1886) recorded that cross-leaved heath, *E. tetralix*, was called "cat heather" in Aberdeenshire. According to Geoffrey Grigson's masterful *The Englishman's flora* (1955: 260),"cat heather" was used in Scotland for bell heather. My enquiries a few years ago did elicit the paradoxical information that older folk in the Banchory area used "cat heather" for cross-leaved heath, whereas a retired keeper living in Braemar said that "cat heather" was "*Erica cinerea* and that alone" (Nelson 2003d, 2003e).

It may also be remarked that *Erica cinerea* is alone in having "heather" as a part of its vernacular name in English. This may have had its origins in Scotland, given that the among earliest works to make the equation were written and published there.[3] In the second volume of an edition of Allan Ramsay's *The gentle shepherd, a pastoral comedy; with illustrations of the scenary ...*, issued in Edinburgh in 1808, there is a list of "Plants found in the vicinity of the Marlfield *Loch* ..." (Ramsay 1808: 496), and *E. cinerea*, "Bell-heather, or fine leaved heath" is included as well as *E. tetralix* which is called "Rinze-heather, or cross-leaved heath." Three years later, in James Macdonald's *General view of the agriculture of the Hebrides, or Western Isles of Scotland: ...*, there is a footnote listing "the three different species [of heather] common in the Highlands and isles of Scotland": "*Erica vulgaris*, common heather; *erica cinerea*, bell heather; and *erica tetralix*, rinze heather" (Macdonald 1811: 363). (On rinze heather, see *E. tetralix*, p. 239.)

CULTIVARS. See p. 165 and Appendix I (pp. 370–373).

22. ERICA MADERENSIS. Madeira heath, Madeiran bell heather.

Madeira heath is found only at high altitudes (above 1,400 m) on Madeira. I have seen it in several localities. On Pico do Arieiro, *Erica maderensis* forms quite tight hummocks up to two metres wide and at least a half metre high (probably the result of grazing by rabbits, according to McClintock (1976, 1981), although there was little sign of grazing on any plants in December 2006) (Fig. 170). The heather is scattered across the summit and adjacent rocky slopes, usually as isolated, solitary plants (Fig. 171). On Pico Ruivo de Santana, large shrubs of *E. maderensis* hang in swathes from the precipitous slopes (Fig. 172), where grazing, even by the nimblest goats, is utterly impossible. Plants sprout from cracks in the ancient volcanic rocks that form the peaks, and evidently these shrubs live to a substantial age judging by their stout stems ("thicker than one's thumb", according to Richards (1976)).

Erica maderensis was first described as a variety of bell heather by Bentham (1839). He named it *E. cinerea* var. *maderensis*. Bentham had not seen living plants, basing his description on material collected by Dr Carl Berhard Lehmann (1811–1875). In 1904, Joseph Friedrich Nicolaus Bornmüller (1862–1948) raised Bentham's variety to specific status without any explanation.

Among plants seen on Madeira by Joseph Banks and Daniel Solander between 13 and 18 September 1768, during the early stages of the circumnavigation of HMS *Endeavour*, Captain James Cook, and recorded in a manuscript list[4], was "*Erica cinerea* ? <u>Linn.</u>"; the question mark is in the original. No specimen of this heath (nor of verifiable *E. cinerea* from Madeira) collected by Banks or Solander has been traced, so the identity cannot be checked, but if Banks and Solander reached the upper slopes of the mountains, they could well have encountered and collected *E. maderensis*.

Fig. 170. A remarkably compact plant of *Erica maderensis* on Pico do Arieiro, Madeira; the shrubs in the background are also Madeiran bell heather (E. C. Nelson).

Fig. 171. Habitat of *Erica maderensis* on Pico do Arieiro, Madeira; the vast majority of the conspicuous shrubs are Madeiran bell heather (E. C. Nelson).

McClintock (1976) ventured the opinion that *Erica maderensis* "clearly has nothing in common with *E. cinerea*, but seems to be closer to *E. terminalis*, but its pollen differs." In a later, more detailed account, he attributed that opinion to Robert (Bob) Ross (McClintock 1981), and then reversed his own view, concluding that "*E. maderensis* has much more in common with *E. cinerea* than with *E. terminalis*" (McClintock 1981: 51). Until much more work is accomplished on DNA sequences, anatomy and morphology, comparing these species must remain speculative.

In terms of gross morphology, *Erica cinerea* and *E. maderensis* are not dissimilar, and at first glance (especially of a pressed specimen) it is not surprising that the Madeiran plant was thought to be a mere variety of bell heather. They both possess included stamens and anthers with spurs. Their ovaries are glabrous. The calyx is distinctive, the lobes having hyaline margins in both species. The bracteoles are attached to the pedicel just below the calyx. The differences between the species are also relatively few. In *E. cinerea*, the spurs on the anthers are lobed and broad; in *E. maderensis* the spurs are linear and only rarely lobed. The leaves are in whorls of three in both species, but in *E. maderensis*, the foliage is not aggregated into fascicles. A further difference is in the arrangement of the flowers; in *E. cinerea*, the flowers are solitary or in whorls on short lateral shoots which in turn are aggregated into a spike that resembles (but is not) an elongated inflorescence, whereas in *E. maderensis* they are in terminal umbels. Thus the plants have an entirely different appearance at flowering time (and indeed when in fruit). The corolla shape also differs: in *E. cinerea*, the corolla tends to be urn-shaped; in *E. maderensis*, it is more tubular. McClintock (1981) did note that there was some variation in flower colour in Madeiran populations, from "deep rose pink" to almost white, although no white-flowered plant has been reported as far as I can ascertain. In general, the flowers of *E. maderensis* are mauve, not as brightly coloured nor as rich-hued as those of *E. cinerea*. A further point of difference is the capsule; in *E. maderensis*, the valves are robust, hard and woody, quite unlike the flimsy valves of *E. cinerea*. The dehisced fruits, with tattered remains of the corollas, persist on the Madeira heath for many months after they have released their seeds, and are conspicuous, like an encrustation covering the shoots (Figs 173 & 174). This is not so in *E. cinerea*; the withered corolla with the fruit inside tends to drop off the shoots before the onset of winter.

Fig. 172. A different habit: a pendulous shrub of *Erica maderensis* near Pico Ruivo de Santana, Madeira (E. C. Nelson).

Figs 173 and **174**. *Erica maderensis* in fruit, December 2006, on Pico do Arieiro, Madeira; note the persisting clusters of dehisced fruits at the tips of the shoots (E. C. Nelson).

CULTIVATION

There are few published reports about *Erica maderensis* in its natural habitats, and as few about its behaviour in cultivation (see Hall 2008). Its date of introduction in cultivation is uncertain but there are no reliable reports before 1974 when Don Richards and David McClintock returned from Madeira with propagation materials (Richards 1976; McClintock 1981). While some of the plants currently in English gardens derive from the 1974 collection, Allen Hall has also raised plants from seed obtained from Madeira by Dr Judy Rose (East Malling Research Centre, Kent) (Hall 2008). There is little variation in the flower colour of the cultivated clones, and none is distinguished by name.

Given its habitat, sharply draining, lime-free soil or compost is essential. *E. maderensis* can be grown in pots, large containers such as troughs (Fig. 175), and in raised beds and rock-gardens. Don Richards characterised the Madeira heath as a "neat and charming" plant, and suggested it

was suitable for a rock-garden with acid soil (McClintock 1981: 51). Allen Hall has several mature plants growing in large containers; they have formed mounds about half a metre across after a decade, and with their curving, pendulous stems they do not need pruning to retain this habit (Hall 2008).

In England *Erica maderensis* is not reliably hardy. Despite its habitat on the mountain peaks, where plants are subject to occasional frosts and can be covered briefly by snow in winter (Sjögren 1974), frost will damage the tips of the shoots and can occasionally be lethal to plants growing outdoors in England (Hall 1999, 2002, 2008). Thus, in frost-prone areas, it is advisable to keep plants under glass during the winter, or to give them protection during periods of intense cold.

Propagation from cuttings is possible but not as easy as by seed (Hall 1999, 2008). In cultivation, seed is ripe in late summer and early autumn (Sellers 1999).

Fig. 175. *Erica maderensis* in a large container in Joan and Allen Hall's garden, Loughborough, Leicestershire (E. C. Nelson).

HYBRIDS

No natural hybrids are known, and results from artificial cross-pollination are ambiguous, as noted above (p. 261). Dr John Griffiths reported viable seed when *Erica maderensis* was cross-pollinated with *E. cinerea* and that the resulting seedlings survived and flowered, and that "hybrid character" was confirmed. In 1984 he had used pollen from *E. cinerea* 'Alba Minor', and several seedlings were raised but they perished in a cold winter.[5]

The reciprocal cross failed for Kurt Kramer in 1994. Griffiths also tried pollinating *E. maderensis* with *E. tetralix* without obtaining seed.

Erica maderensis (Benth.) Bornm., *Bot. Jahrb. Syst.* 33: 458 (1904).

E. cinerea var. *maderensis* Benth. in DC., *Prodromus* 7: 666 (1839). Holotype: "Erica cinerea? Madeira", C. Lehmann, July 1837 (Herb. Bentham) **K**!

DESCRIPTION. *Evergreen shrub* to 0.6–0.8(–1.4) m tall, bushy or very compact; mature stems stout and woody with grey bark, to 2 cm in diameter; younger shoots whitish, densely pubescent, erect, spreading or pendulous. *Leaves* in whorls of 3, ±2.5–5(–10) mm long, linear with parallel sides, not much wider than the petiole on (?non-flowering) shoots; below flowers (on flowering shoots) variable in shape, smaller but increasing in length upwards towards the inflorescence, linear to lanceolate and then broadest below middle and tapering gradually towards very short petiole; with transparent hyaline and hirsute edges, some hairs gland-tipped; margins recurved; upper surface ±flat, sparsely hirsute becoming ±glabrous; lower surface glabrous, convex, sulcus linear extending from tip to base, sides contiguous throughout. *Flowers* in terminal umbels of 3–6 flowers; *pedicel* curving or straight, ±4 mm long, minutely and sparsely pubescent, tapering, broader at apex; *bract* at least ¾ way along, sometimes close to calyx, < 2.5 mm long, <1 mm across, ± triangular, with narrow sulcus ±1.5 mm long, margins hyaline, broad, margins fringed with minute hairs; bracteoles close to calyx, similar to bract but smaller, ±2 mm long. *Calyx* with 4 free sepals, to 3 mm long, a little unequal, lanceolate, centre green, sometimes tinged pink, with broad hyaline margins, margins minutely fringed with short hairs; sulcus conspicuous, apex acute. *Corolla* pale ("dull") pink (mauve H2) ("near white to fawnish or purplish pink": Richards (1976: 16)), ±5 mm long, narrow, tubular-ovoid, glabrous, lobes 4, semicircular, erect, to 0.5 mm long, ± 1mm wide at base, margins smooth. *Stamens* 8, included; *filament* curving and with pronounced sigmoid bend below anther, glabrous; *anthers* held above middle of corolla, basifixed, ±1.2–1.5 mm long, to 0.5 mm across at base; thecae erect, closely parallel, not fused except towards base, ±narrowly triangular in lateral view, minutely papillose-spiculose; pore 0.5 mm long, oval, towards apex of theca; spurs ±linear, to 0.7 mm long, entire or with a small linear lobe, minutely strigose, straight but may become upwardly curved. *Style* ±4 mm long, ±0.5 mm in diameter at base, straight, glabrous, tapering upwards, papillose towards apex; *stigma* at mouth of corolla or just emergent, green, ± 0.5 mm diameter, broader than apex of style, capitellate; *Ovary* ±1.5 mm tall, glabrous, green, barrel-shaped, 8-ribbed, emarginate; septa substantial; ±30 ovules per locule; nectary green, conspicuous, lobes enveloping bases of alternate filaments. *Fruit* with woody, thickened walls, very hygroscopic, opening wide on warm dry days; interior of valves glossy, pale gold when freshly opened. *Seed* ±0.5 mm long .

FLOWERING PERIOD. May–September.

DISTRIBUTION & ECOLOGY. Endemic in Madeira and confined to open, rocky areas at altitudes above 1,400 m (McClintock 1994; Hansen 1969, Sjögren 1972, both as *E. cinerea*).

Sjögren (1972), discussing the plant communities of Madeira, recorded the existence on the highest peaks of a unique community containing *Erica maderensis* (as *E. cinerea* var. *maderensis*) and *Viola paradoxa*, as well as some other endemic species. These plants inhabit "rough cliffs with many crevices and small shelves where soil can accumulate" (Sjögren 1972) (see Fig. 171). He also suggested that the Madeiran bell heather "may have been much more frequent at high altitudes" in the recent past, before a period of overgrazing which was "relatively short ... probably not more than 100 years" (Sjögren 1972: 67, 68).

Erica maderensis is classified by the IUCN as rare (Ozinga & Schaminée 2005).

ETYMOLOGY. Latin toponym: from Madeira.

VERNACULAR NAMES. Madeira heath (English; McClintock 1980b); Madeiraheide (German).

CULTIVARS. There is no named cultivar of *Erica maderensis*.

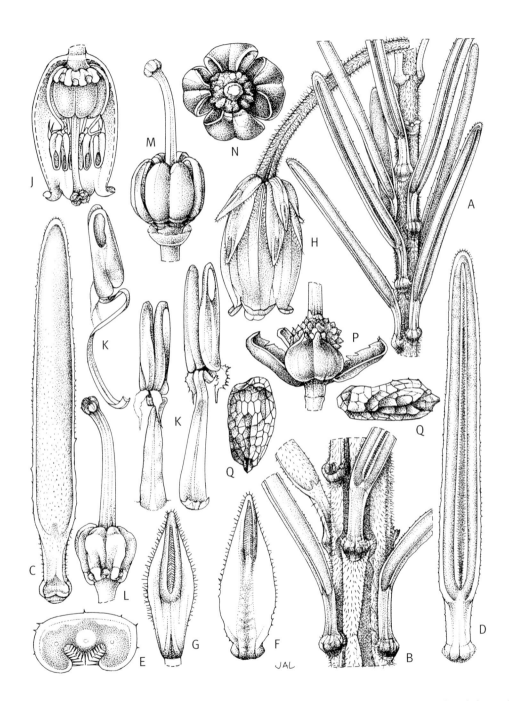

Fig. 176. *Erica maderensis*. **A** leafy shoot × 9.5; **B** leaf nodes × 16; **C** leaf from above × 14; **D** leaf from below × 14; **E** section through leaf × 46; **F** sepal × 20; **G** bracteole × 20; **H** flower from side × 7; **J** section through flower × 7; **K** stamens from side, back and front × 8; **L** gynoecium × 11; **M** capsule × 7; **N** capsule from above × 10; **P** capsule from side × 9; **Q** seed from front × 38; **R** seed from side × 38; from specimens supplied by Allen Hall. Drawn by Joanna Langhorne.

23. ERICA TERMINALIS. Corsican heath

The English name Corsican heath is ill-chosen because *Erica terminalis* occurs not just on Corsica, but also on Sardinia, in south-eastern Spain and southern Italy, and extends south into the north-western extremity of Africa, being recorded from Morocco (Fennane & Ibn Tattou 2005). Its range is fragmented but not as much as in some other species. Nowhere does *E. terminalis* overlap with *E. cinerea*, nor has it been reported on Madeira.

To add more nomenclatural confusion, this was most frequently called *Erica stricta*, which name persisted in botanical literature well into the twentieth century. However, that binomial was not validly published before 1796 — when first employed by James Donn (1758–1813), Curator of the University Botanic Garden in Cambridge, in his *Hortus cantabrigiensis* (Donn 1796), it was not accompanied by a description (Skan 1906). Thus *E. terminalis*, published by Richard Salisbury that same year, in the catalogue of his garden at Chapel Allerton (Salisbury 1796), has priority.

Erica terminalis is unusual among northern-hemisphere species in not having its pollen grains clustered into tetrads; the individual grains are separate and about 27 μm in diameter, which, not unexpectedly, is much less than the diameter of the tetrads of other species (Visset 1975; Phillips & Sparks 1975). The only other northern species to produce pollen in monads is *E. spiculifolia* (see p. 279); its grains are only about 18 μm, two-thirds the size of those of *E. terminalis* (see Phillips & Sparks 1976: 244–246, plate 1). Studies among the southern African species indicate that at least 80 of them produce monads rather than tetrads (Oliver 2000: 45).

Fig. 177. Corsican heath (*Erica terminalis*), an introduced species, at Umbra, County Derry, Northern Ireland, in September 2008. The plants, while old, all had vigorous, young shoots and were coming into bloom.

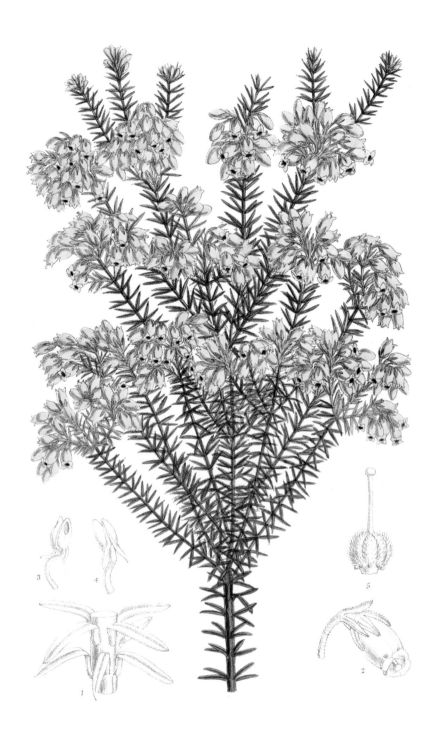

Erica terminalis   Painted by Matilda Smith, lithographed by J. N. Fitch   PLATE 21

Published in *Curtis's Botanical Magazine* tab. 8063 (1906).

In cultivation, the Corsican heath is virtually a tree heather, reaching as much as 3 metres in height and with stems as much as 0.28 m in circumference (±9 cm diameter) (McClintock 1989e). In the wild, in Corsica at least, it can be up to 2.7 metres tall with stems to 0.24 m in circumference (McClintock 1988b, 1989e).

Variants are apparently few and far between. Plants have been described with longer or shorter leaves, or with broader leaves that are pubescent underneath (and so appear grey), but such variation is to be expected in a species that inhabits widely separated regions on two continents and two large Mediterranean islands. It would be more surprising if the heather did not vary in any way.

As with all heathers, the "holy grail" of gardeners has been to find and grow a clone with white flowers. White-flowered *Erica terminalis* has been reported from Italy and Corsica on a number of occasions since 1831 but evidently never successfully propagated. In 1938, Professor René Verriet de Litardière (1888–1957), a French botanist with a particular interest in the flora of Corsica, named and described a plant with white flowers (*E. terminalis* f. *albiflora* Litard.; McClintock 1984a: 191) from that island. However, this form is elusive, and despite searches in recent decades (McClintock 1988b) no plants have been located and propagated. As an aside, it was the ambition of James Walker Porter (who raised several excellent clones of *E.* × *darleyensis*; see pp. 316–319) to produce artificially a "double white" clone of *E. terminalis* (Nelson 1984a).

CORSICAN HEATH NATURALISED IN IRELAND AND BRITAIN

There is a colony of *Erica terminalis* in Northern Ireland, on stabilised dunes at The Umbra, Magilligan, County Derry (Hackney 1992). It is said to have arisen from seed dispersed from plants in a nearby garden. Four seedlings were reported in 1922. In 1992 Hackney stated that the heather was "very plentiful over a considerable area in dune slacks within the Umbra Nature Reserve". In 2008, I found patches of Corsican heath, but would not have described the shrub as "very plentiful". No young seedlings were found, but there were numerous old shrubs with very robust trunks coming into full flower. These seem to have been heavily grazed, but were re-sprouting vigorously, and the flowers were frequented by native bumblebees (Fig. 179). Strangely, there is a record from the 1830s: Dr Lloyd collected an herbarium specimen in the "north of Ireland" in 1834 (Nelson 1979a; Reynolds 2002). The person concerned was probably Dr George N. Lloyd (1804–1889), who lectured on botany in Edinburgh and who is known to have botanised in various parts of Britain and Ireland between 1825 and 1843 (Desmond 1994).

Fig. 178. *Erica terminalis* with an orange-rumped bumblebee (*Bombus lucorum* agg.), at Umbra, County Derry, September 2008.

Fig. 179. *Erica terminalis* in cultivation in northern California (E. C. Nelson).

A bush was also found in the 1970s on a hillside at Onich, near Fort William, Scotland (Clement & Foster 1994), but no explanation has been offered for this isolated individual. Most recently, the Corsican heath has been noted on Holyhead Mountain in Anglesey, north Wales, where it evidently had been planted within an enclosure protecting telecommunication masts. Seedlings have been found outside the perimeter fence, on its south-eastern side of the enclosure, mixed with native heaths (*Erica cinerea* and *E. tetralix*, and *Calluna vulgaris*) and gorse (*Ulex europaeus* and *U. gallii*) (Bonner 2011).

CULTIVATION

Corsican heath has been cultivated in Britain since about 1765 (Aiton 1789; Aiton 1811), most probably, to start with, indoors as a "stove plant". Certainly in December 1796 when Robert Brown collected a specimen (**BM!**[6]) from a plant growing in Dickson's nursery, Leith Walk, Edinburgh, the shrub was cultivated in the stove-house (cf. Nelson 2003c). Richard Salisbury (1796) had *Erica terminalis* at Chapel Allerton in the mid-1790s; he noted it was indigenous in Corsica but did not give any other source information. In his 1802 Linnean Society monograph, Salisbury (1802), altering the name again to *E. multicaulis*, acknowledged material collected by "G. Jones", but no botanist or botanical collector named Jones with that initial (nor "W" for William, if "G" stands for Gulielmus) active in Corsica in the late 1700s is listed in standard sources.

Erica terminalis (Fig. 179) is a large shrub, flowering usually from mid-summer into late autumn and often continuing into early winter. It is tolerant of a range of soil types and is hardier than might be expected, as demonstrated by its survival in the north of Ireland. However, heavy snow can shatter branches (Canovan 2005). The russet-coloured dead flowers are an additional winter attraction, so this heather should not be pruned immediately after flowering.

Propagation by cuttings is easy, and young plants raised in this manner bloom within about 12 months. Seed can also be used; inevitably, seedlings will be variable.

HYBRIDS

As for the majority of northern *Erica* species, there are no reports of wild-occurring hybrids. Both Dr John Griffiths (Yorkshire, England) and Kurt Kramer (Edewecht-Süddorf, Germany) have used the species as a seed- and pollen-parent without much success. Failed crosses include those between *E. terminalis*, as pollen-parent, and *E. australis* (Griffiths), *E. carnea* (Griffiths), *E. ciliaris* (Griffiths), *E. manipuliflora* (Kramer) and *E. tetralix* (Griffiths), and, as seed-parent, with *E. andevalensis* (Griffiths obtained non-viable seed), *E. erigena* (Kramer), *E. sicula* subsp. *bocquetii* (Kramer) and *E. vagans* (Kramer).

Successful artificial crosses were obtained by Kurt Kramer between *Erica terminalis* and *E. cinerea* (*E. × arendsiana*, see p. 330), and also when he used *E. multiflora* pollen on *E. terminalis* (but those seedlings did not survive). He has also raised seedlings, which are extant, from *E. spiculifolia* pollinated by *E. terminalis*, but their status has yet to be assessed fully. Dr Griffiths also had some equivocal results: *E. manipuliflora* pollinated by *E. terminalis* yielded viable seed but the seedlings were identical with the seed-parent; likewise, when *E. terminalis* had been pollinated by *E. tetralix*. He obtained seedlings from *E. terminalis* after pollination by *E. vagans*, but these soon perished.

Erica terminalis Salisb., *Prodromus stirpium in horto ad Chapel Allerton vigentium*: 296 (1796). Type: "In Ins. *Corsica* indigena" **K**! (Skan 1906).
E. multicaulis Salisb., *Trans. Linn. Soc.* **6**: 369 (1802). Type: "Sponte nascentem in Ins. *Corsica*, legit G. Jones.": not traced.
E. stricta Donn, *Hortus cantabrigiensis*. (1796), nom. nud.

ILLUSTRATIONS. *Curtis's Bot. Mag.* tab. 8063 (1906); S. J. Roles, *Flora of the British Isles: illustrations* pt 2: tab. 938 (1960); J. Smythies, *Flowers of south-west Europe*: 287 (2) (Polunin & Smythies 1973); B. Keel, *Illustrations of alien plants of the British Isles* (Clement *et al.* 2005: 127).

DESCRIPTION. *Evergreen shrub* to ±2(–3) m tall; *shoots* pubescent, upright or ascending. *Leaves* in whorls of 4(–5), ± 9mm long, ±4 mm across, lanceolate to linear, very minutely pubescent; lower surface green, midrib enlarged, margins recurved, in cross-section forming two parallel channels, with sulcus that broadens to base; petiole very minutely pubescent. *Inflorescence* a single terminal umbel, or a compound inflorescence of several umbels on lateral leafy shoots; *pedicel* ±5.5 mm long, minutely pubescent; bract above the mid-point, ±2 mm long, ±0.3 mm broad, spathulate; bracteoles above ¾ way along, ±1.5 mm, narrow lanceolate, devoid of cilia. *Calyx* comprising 4 free sepals, broadly triangular, ± 2 mm long, ±2 mm across at base, very minutely pubescent or glabrous, margins devoid of cilia but with minute fringe of simple hairs, sometimes gland-tipped, sulcus green or tinged red, like leaf with linear channels. *Corolla* (5–)6–7(–8) mm, urn-shaped, elongated, glabrous, 4-lobed; lobes erect or recurving, to 1.5 mm long. *Stamens* 8, ±4.5 mm long, included; *filament* ±4.5 mm long, erect, slightly hooked at base, curved with slight sigmoid bend towards apex; *anther* 0.8 mm long, basifixed with 2 straight spurs at junction with filament, ±0.7–1.0 mm, very narrow, triangular-linear, margins ± smooth, with minute spicules on upper portion; *thecae* elliptical, broadest towards base, c. 0.4 mm across; *pollen* tricolporate, in monads, grains spherical to ovoid, 21–30 μm (polar length) × 21–31 μm (equatorial diameter) (Phillips & Sparks 1976). *Ovary* ±1.5 mm tall, >2 mm in diameter, globular, 4-locular, hirsute, ribbed, emarginate, with ±25 ovules per locule; nectary lobed; *style* ±5 mm long, ±0.3 mm diameter at base, straight, erect, glabrous, tapering towards apex; *stigma* capitate, broader than style apex, ±4 mm diameter, exserted. *Seed* 0.3–0.54 mm long (see Nelson 2006a).

FLOWERING PERIOD. May–August; often later in cultivation, with fresh blossom in September and October. In Northern Ireland, the naturalised plants at Umbra were only coming into bloom in mid-September.

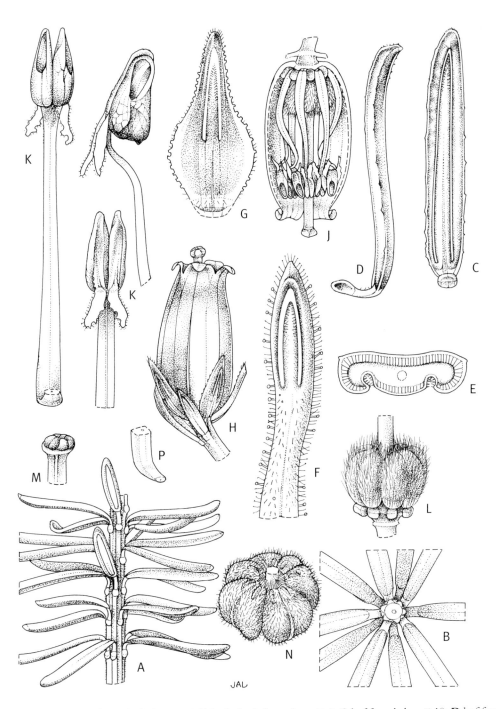

Fig. 180. *Erica terminalis.* **A** leafy shoot × 6; **B** leaf whorls from above × 6; **C** leaf from below × 10; **D** leaf from side × 10; **E** transverse section of leaf × 48; **F** bracteole × 18; **G** sepal × 10; **H** flower from side × 7; **J** longitudinal section of flower × 7; **K** stamens from front, back and side × 20; **L** ovary and nectary ring × 14; **M** style apex (stigma) × 32; **N** capsule × 7; **P** filament base × 20; from specimens grown in RHS Garden Wisley, no. W 1989 2772. Drawn by Joanna Langhorne.

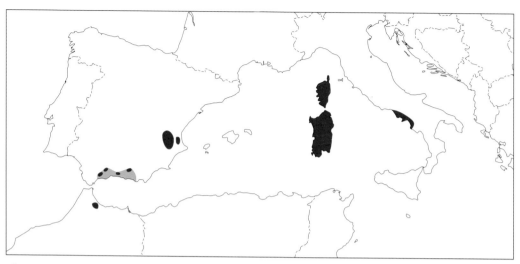

Fig. 181. Distribution of *Erica terminalis*.

DISTRIBUTION & ECOLOGY. South-western and southern Europe: Spain (southeast: Andalucía and Levante; Málaga, Granada, Valencia), France (Corsica only), Italy including Sardinia. North Africa: Morocco. Naturalised in Ireland (Hackney 1992: 247) and Great Britain (Bonner 2011). **Maps:** Benito Cebrián (1948: 58), Bolòs & Vigo (1995: 38). Fig. 181.

The Corsican heath inhabits the edges of streams, rivers and freshwater pools, usually in shade; in ravines and open woods; it is tolerant of calcareous and ultrabasic substrates, and ascends to 1,700 m altitude in Spain (Bayer 1993).

ETYMOLOGY. Latin: *terminalis* = terminal, a reference to the flowers at the tips of the shoots.

VERNACULAR NAMES. Corsican heath (English; McClintock 1980b); Steifaufreche Heide (Dippel 1887) Korsische Heide (German); brezo (Spanish).

CULTIVARS. See Appendix I (pp. 381–382). There is only one named cultivar, 'Thelma Woolner', in general circulation. Other plants with paler flowers may be seedlings of unknown origin, or clones of unrecorded provenance.

NOTES

[1] URL http://www.ngs.fo/plant/plantulisti/list_00.htm accessed 22 July 2006.

[2] David Don (1834) stated that *Erica cinerea* was the type of the genus *Erica* L. Although the present type concept had not been established in the 1830s, this has generally been accepted by succeeding botanists including Jarvis & McClintock (1990: 518) and Oliver (2000) (see p. 100).

[3] According to an internet search (using Google advanced book search on 31 May 2008), there is also one eighteenth-century work in which the equation of the "bell heather" with *Erica cinerea* is published, but I have not been able to consult the complete text of it — H. Mackenzie, *Prize essays and Transactions of the Highland Society of Scotland* (1799: 66).

[4] "Plants of Madeira"; Banks mss, State Library of New South Wales; available on-line (URL http://image.sl.nsw. gov.au/banks/60044.jpg).

[5] "Summary of hybridisation experiments"; unpublished ms by J. Griffiths, undated (c. 1989) (Major-General P. G. Turpin's papers).

[6] This specimen is labelled "Erica stricta Mr Dicksons Nursery Lieth [*sic*] Walk, Stove Dec[r] 1796 … Desc: in Mss"; however, no manuscript description has been traced (there is none in Ms B.93, cf. Nelson 2003c).

BRUCKENTHALIA, THE BALKAN HEATH

Erica spiculifolia

24. ERICA SPICULIFOLIA. Balkan heath, spike heath.

For more than one and a half centuries, until the late 1990s, the Balkan heath (or spike heath, to quote an older name) was placed in the genus *Bruckenthalia* (see, for example, Nelson 1999a) and was the only species therein. *Bruckenthalia* was stated to differ from *Erica* because it possessed partially-fused sepals but lacked bracteoles (cf. Webb & Rix 1972; Oliver 1996). The chromosome number is also different: those European species of *Erica* which had been counted have 24 chromosomes (n = 12), whereas *Bruckenthalia* possesses 36 chromosomes (n = 18). Another difference is that in *Bruckenthalia* the pollen grains are not clumped into tetrads (four grains together); they separate into individual tricolporate grains (monads), a type characteristic also of *E. terminalis* (see p. 272) (for details of the pollen of both these species, see Phillips & Sparks 1976: 244–246, plate 1).

However, as already mentioned, detailed research by Dr E. G. H. Oliver, assisted by his wife Inge, on the Cape heaths, especially those with indehiscent or partially dehiscent capsules (belonging to genera with such unfamiliar names as *Acrostemon*, *Eremia*, *Philippia* and *Thoracosperma*, dubbed the "Minor" genera of the Ericeae), led him to propose (Oliver 2000) that there was no justification for keeping these African genera separate from *Erica*, because there was so much overlap in the characters. While those characters that distinguished *Bruckenthalia* from the other European heaths apparently held true, according to published descriptions (for example, Webb & Rix 1972), they did not serve so well when *B. spiculifolia* was compared with certain African species. Although no *Erica* species has yet been detected with 36 chromosomes — but only 27 species and hybrids have been counted out of the approximately 820 species (±3%; see Nelson & Oliver 2005b) — the unique "chromosome complement [was] a rather poor character to use to uphold a genus" (Oliver 1996). In short, the characters supposed to separate *Bruckenthalia* from the enlarged genus *Erica* were worthless. Thus *Bruckenthalia* was absorbed into *Erica* (Oliver 1996, 2000), a decision quickly accepted in horticultural circles (see Nelson 1999a), and the species reverted to its original name *Erica spiculifolia*.

In fact, as pointed out here (see Table 5, p. 103), around half of the northern hemisphere *Erica* species do possess a cup-like calyx formed by partially fused sepals; and among the African species, there are some, mainly members of the disbanded genus *Philippia*, that apparently lack bracteoles. Oliver (2000) found that in the Balkan heath the two bracteoles are *not* absent but are fully recaulescent and appear as the two lateral lobes of the calyx, a situation also found in around a hundred African species. Moreover, in a survey of African species, 57 belonging to the "Minors" had pollen in monads.

To the casual observer, the most obvious characteristic of the Balkan heath is the arrangement of the small flowers, which are clustered in a terminal, racemose inflorescence.

Erica spiculifolia has a greatly fragmented distribution, from north-eastern Turkey westwards into Bosnia and Herzegovina, and from near Breaza in north-eastern Romania to Epirus in north-western Greece, a range of about 1,800 kilometres east–west and 900 kilometres north–south (for distribution details see Niedermaier 1985). The Balkan heath occurs in three separate localities in northern Turkey (Stevens 1978; Browicz 1983; Niedermaier 1985); the easternmost is on Şavval Tepe above Murgul, not far from the frontier with Georgia, where Peter Hadland Davis (see p. 292) collected it at altitudes

Erica spiculifolia   **Painted by Ferdinand Lucas Bauer**   **PLATE 22**

Published in J. Sibthorp & J. E. Smith, *Flora graeca sibthorpiana*, centuria 4, tab. 353 (1823).

between 1,600 m and 2,100 m. There is a station near Saraybaba, about 100 km to the west (Stevens 1978). *E. spiculifolia* has long been known from Uludǎg, or Bithynian Olympus (see below). In Europe, as noted, the Balkan heath ranges from northern Greece, through Bulgaria, and into the Carpathian Mountains in Romania (Browicz 1983; Niedermaier 1985). It is also reported from Albania, and states that formerly made up Yugoslavia: Macedonia, Bosnia and Herzegovina, Serbia, and Montenegro (Webb & Rix 1972; Browicz 1983; Niedermaier 1985).

POLLINATION

The entomologist Kenneth Mackinnon Guichard (1914–2002) recorded (*fide* Stevens 1978) that the flowers of *Erica spiculifolia* are visited by bumblebees (*Bombus* spp.) despite the absence of nectaries and thus lack of nectar. Guichard made this observation on Uludǎg where he gathered a specimen of Balkan heath in 1962. Knuth (1909) postulated that pollination of this species was effected by wind as well as by insects but more entomological observations are needed. The close clustering of the small flowers in terminal racemes, and the emergent styles, perhaps aid wind-pollination, although *E. spiculifolia* does not possess the very conspicuously enlarged stigma that is characteristic of the northern besom heaths (see p. 176) and those southern African species that are most certainly wind-pollinated (see Rebelo *et al.* 1985).

FOSSIL EVIDENCE

Erica spiculifolia apparently provides remarkable evidence that during the Quaternary Period the distribution patterns of heathers in the northern hemisphere were not static. Working on pollen grains extracted from peaty material found at several sites in southern England, including Beetley in Norfolk, Miss Robin Andrew (1909–1999), of the Subdepartment of Quaternary Research in Cambridge, was the first to recognise that some of the grains closely resembled the pollen of *Bruckenthalia* (Phillips & Sparks 1976; Hall 1978): "The grains [from Beetley] are very similar to *Erica terminalis*, but the latter have a much more pronounced pattern of endocracks and are larger than modern *Bruckenthalia* and the fossil grains ... ." Since Miss Andrew's work in the 1970s, other sites have yielded similar pollen, including Orford in Suffolk (Andrew & West 1977), Sel Ayre on the western coast of the Shetland Islands (Birks & Peglar 1979; Hall *et al.* 2002), the Burn of Benholme in Kincardineshire (Auton *et al.* 2000) and Camp Fauld near Peterhead in the northeast of Scotland (Whittington *et al.* 1993).

Seeds preserved in peat (for example, from Beetley; Phillips & Sparks 1976) also have been identified as coming from *Erica spiculifolia* (as *Bruckenthalia*), but this equation is flawed, and the seeds need to be re-examined. Linda Phillips reported that "Ericaceous seeds were abundant" in the Norfolk deposits, and that they "matched fossil seeds from Gort ... referred by Jessen, Andersen & Farrington (1959) to an extinct taxon, *Erica scoparia* var. *macrosperma*" (Phillips & Sparks 1976: 245). While Phillips correctly noted that

> The original identification of the seeds from Gort as a variety of *Erica scoparia* was made on the basis of the occurrence of the seeds in flowers matching *E. scoparia*, and the elongated, though thin-walled, cells of *E. scoparia* seeds from Algeciras ...

she seems to have overlooked the presence of flowers as crucial evidence in identification. "However, only small unripe seeds were found in complete flowers and the ripe seeds were found separately or associated with incomplete flowers" (Phillips & Sparks 1976). This was a serious lapse and calls into question any equation that posits the occurrence of *E. spiculifolia* in Ireland or other regions during the Quaternary interglacial periods on the basis of fossilised seeds alone.[1] On the other hand, the presence of flowering shrubs of *E. scoparia* at Gort several hundred thousand years ago is beyond question (for further comment see pp. 177–179).

The accumulated evidence from fossil pollen grains suggests that the Balkan heath was probably widespread in eastern England, north-eastern Scotland and in the Shetland Islands during interglacial periods (Whittington 1994). Its pollen has been reported also from profiles obtained in other parts of north-western Europe, including Denmark and north-western Germany (cf. Neidermaier 1985). *Erica spiculifolia* pollen has also reported from one site in Ireland, at Derrynadivva, near Castlebar, County Mayo (Coxon *et al.* 1994).

Assuming the identification of fossilised pollen is not open to question and that a pollen grain preserved in a fragment of peat is a reliable indicator of a heather's presence during previous interglacial periods, the range of the Balkan heath extended further north, when compared with its present range, by more than 13° latitude; in other words, during interglacial periods in the Late Quaternary the Balkan heath was growing at least 2,000 kilometres from its present northern limits in the Carpathian Mountains (cf. Neidermaier 1985). In common with the majority of indigenous flowering plants, *Erica spiculifolia* must have been eliminated when glaciers returned to cover northern Britain. It did not return unlike some of the other heathers including *E. mackayana* and *Daboecia cantabrica* which are also known from Scottish (Shetland Islands) and Irish interglacial deposits (see pp. 242, 244).

The assemblages of pollen preserved in the different layers of peat provide indications of the sort of vegetation that existed when the Balkan heath grew in Britain. In Shetland, for example, members of the Ericaceae (alas, not identified further), including crowberry (*Empetrum nigrum*), seem to have been dominant. Birch, pine, alder and oak probably grew nearby. The vegetation was not unlike that of present-day Shetland heaths, dominated by *Calluna*, but with *Erica spiculifolia* instead.

HISTORY: EARLY COLLECTIONS

Erica spiculifolia, described and named by Richard Salisbury (1802), was among the plants collected and preserved by John Sibthorp (1758–1796) while travelling in Asia Minor and the Aegean region during the late eighteenth century (for details of Sibthorp's travels see Lack & Mabberley 1999; Harris 2007). He gathered his specimens on the summit of Mount Olympus, called Bithynian or Mysian Olympus, to distinguish it from several other mountains with the same name. This peak is in modern Turkey and is now named Uludăg (2,543 m altitude), and *E. spiculifolia* is still present on the mountain, a component species of the dwarf shrub community that is dominated by bilberry (*Vaccinium myrtillus*) and juniper (*Juniperus communis*) (Güleryüz *et al.* 1998).

John Sibthorp visited Bithynian Olympus twice. The first occasion was in mid-August 1786 with Ferdinand Bauer (1760–1826), the botanical artist, as his companion. Bauer made several sketches of the mountain and the nearby city of Bursa (Brussa), as well as of plants found in the vicinity and on the summit. Lack & Mabberley (1999: 45) remark that very little is known of the visit to Mount Olympus, but the pair spent three nights on the mountain and Sibthorp reported that the "superior heights" of Olympus Bithynus allowed him to add a substantial number of new plants to his collection. It is certain that they saw the Balkan heath in bloom, because Bauer sketched a flowering sprig. From this sketch[2], he was to produce the most beautiful hand-coloured engraving published 37 years later in the fourth volume of Sibthorp's monumental, posthumous *Flora graeca* (Sibthorp & Smith 1823: tab. 353) (reproduced here as Plate 22, p. 280).

Sibthorp returned to Mount Olympus eight years later, when his companion was John Hawkins. They left Constantinople (Istanbul) on 20 August 1794 and spent eleven days on the trip, visiting Bursa, as well as the summit of the mountain. Whether the specimen of "Erica olympiaca" that Sibthorp gave to Richard Salisbury was collected in 1786 or 1794 cannot now be determined. However, there is every likelihood that *Erica spiculifolia* was at least first collected by Sibthorp during the 1786 ascent (Nelson 2007g).

Richard Salisbury named and described two of Sibthorp's heathers, *Erica spiculifolia* and *E. manipuliflora* (see p. 140), in his monograph on the genus *Erica*, which was published in late May 1802 within the *Transactions of the Linnean Society of London*. Of *E. spiculifolia*, Salisbury (1802) stated: "Sponte nascentem in summitate Montia *Olympus*, legit heu defunctus Sibthorpe" — "the late lamented Sibthorpe [*sic*] collected it growing naturally on the summit of Mount Olympus". He also cited a manuscript name that he attributed to Sibthorp — "E. olympiaca" — and concluded thus: "Persingularis est defunctus bractearum" — "it is very unusual [in being] devoid of bracts".

In *Flora of Turkey and the east Aegean islands* volume 6, Stevens (1978) noted that the type specimen of *Erica spiculifolia* (as *Bruckenthalia spiculifolia*) was in the Fielding-Druce Herbarium, Oxford (**OXF**), but he did not identify any particular specimen as the holotype or as isotypes. There are two sheets of *E. spiculifolia* in **OXF** (!) collected by Sibthorp, but neither bears any indications that Salisbury had seen the specimens. Like many of Sibthorp's own specimens, they were not annotated with localities or dates of collection (cf. Lack 1997a; Lack & Mabberley 1999: 102). One sheet, with three specimens attached, has a label in James Edward Smith's handwriting "E. olympica Sibth. E. spiculifolia Salisb. & F. Graec. t. 353". The inclusion of the plate number from *Flora graeca* points to this label dating after the compilation of volume 1 of *Florae graecae prodromus* (Sibthorp & Smith 1806)

Fig. 182. *Erica spiculifolia* at Tigaile Mari (1,700 m altitude) in Ciucas Mountains, Transylvania, Romania (© Michael Kesl).

in which these numbers were published. The second sheet, also with three specimens (and a fourth extraneous one which is not *E. spiculifolia*) is labelled "from Mount Olympus Herb. Sibth 257" and "879 Erica spiculifolia Salisb. Fl. Graec. t. 353 Bruckenthalia Prod. 1. 257": 879 is the number assigned to the entry for the species in *Florae graecae prodromus* (Sibthorp & Smith 1806: 257). Neither of these specimens can be the holotype.

In the herbarium of The Natural History Museum, London (**BM**!), there is a specimen annotated by Salisbury in pencil: "E. olympiaca Sibth. | — spiculifolia Salisb. It differs from all I have ever seen in having no Bractae." On the back of the sheet is the name "Sibthorp", written in ink, and the annotation "This species grows on Mt. Olympus <u>Bith.</u>" The annotation about the lack of bracts echoes the final Latin sentence of Salisbury's protologue, and he apparently is alone in attributing to Sibthorp the use as the proposed specific epithet of *olympiaca* (from the adjective *olympiacus*). In contrast, the specimens in Sibthorp's herbarium (**OXF**), as well as the text in both *Florae graecae prodromus* (Sibthorp & Smith 1806) and *Flora graeca* (Sibthorp & Smith 1823), have *olympica* (from *olympicus*) (Nelson 2007g).

Salisbury (1802) did not state explicitly that Sibthorp had sent him a specimen of this species; the wording of his acknowledgement merely states that Sibthorp had collected it on Bithynian Olympus. Sibthorp is known to have distributed duplicates to several botanists, including Joseph Banks, Peter Pallas and Joseph Franz von Jacquin (Lack & Mabberley 1999: 102). As **OXF** specimens do not bear any indications that Salisbury saw them, whereas Sibthorp's specimen (**BM**) was annotated by Salisbury, it would seem to be the one specimen used by Salisbury and, therefore, it is the holotype (Nelson 2007g).

BRUCKENTHALIA

The missing bracteoles continued to perplex botanists, and in 1831 in *Flora germanica excursoria* Heinrich Gottlieb Ludwig Reichenbach (1793–1879) decided to place the plant in a genus of its own, which he named *Bruckenthalia*. That now-abandoned generic name harks back to J. C. G. Baumgarten's *Menziesia bruckenthalia* (correctly *bruckenthalii*) published in 1816, and undoubtedly a tribute to Baron Samuel von Brukenthal (1721–1803), one-time Governor of the Grand Principality of Transylvania and advisor to Maria Theresa, Empress of Austria.

Johann Christian Gottlob Baumgarten (1765–1843), a native of Luckau (Brandenburg), studied in Dresden and Leipzig. In 1793, he arrived in Hermannstadt (Sibiu), capital of Transylvania which was then a province of the Austrian Empire. Baumgarten carried a letter of recommendation from the eminent Austrian botanist Nicolaus Joseph von Jacquin (1727–1817) addressed to Professor Joseph von Lerchenfeld (1753–1812) who made some of the earliest herbarium collections in Transylvania and enjoyed the patronage of Samuel von Brukenthal. The time was propitious: the natural sciences were flourishing in Transylvania under Brukenthal's patronage. Brukenthal made his summer residence at Freck (now Avrig), in the foothills of the Carpathian Mountains, available so that botanists from Hermannstadt, including Lerchenfeld and Baumgarten, could use it as a base for their exploration of the southern Carpathians. Thus Brukenthal also became Baumgarten's supporter.

Although he had intended to return to Leipzig, Baumgarten decided to remain in Transylvania, and so became a medical practitioner, at first in Leschkirch, Brukenthal's birth-place near to Hermannstadt. In 1801 Baumgarten moved to the city of Sighişoara (Schäßburg, in German) where he lived until his death in 1843.[3]

CULTIVATION & VARIATION

"There are few more beautiful little shrubs than this, but although quite hardy and easy to grow in any well-drained loam and leaf-mould, it is seldom seen in gardens. ... For the cool side of the rock-garden it is first-rate ..." (Johnson 1925: 84). For successful cultivation, lime-free soil is essential.

Little has changed since the mid-1920s: *Erica spiculifolia* is still most uncommon in British and Irish gardens. Why this should be is not clear, as the Balkan heath is no more difficult to grow than other heaths and is one of the hardiest species. Perhaps its rather short flowering period hinders its popularity?

The Balkan heath is a neat, low-growing shrub that, when cultivated in Britain and Ireland, blooms briefly, starting the late spring (May) when few other heathers are in flower. In the wild, the flower colour is reported to vary from deep pink to white (*E. spiculifolia* f. *albiflora* (D. C. McClint.) E. C. Nelson & D. C. McClint.) which was reported by Charles Frederick Ball as early as 1911 from the Rila Mountains in Bulgaria (Nelson 1993a, 1997). Describing the vegetation of the slopes of Musala, the highest peak in the Balkans (2,925 m altitude), Ball (1912) recorded that *Erica spiculifolia* took "the place of our Ling (*Calluna vulgaris*), on the mountain sides; it is usually pink, but deeper forms may be seen, and we found two plants with pure white flowers."

The holotype of *Erica spiculifolia* f. *albiflora* comprises some sprigs received from J. Grülich, of Sedloňov in the Czech Republic, collected at an unknown locality in the Retezat Mountains in the Southern Carpathians, Romania (McClintock 1984a; Miller & Nelson 2008). The plant proved "very difficult to propagate" vegetatively (McClintock 1984a), yet seedlings raised from the Carpathian plant are reported to have included white-flowered examples (Nelson 1997), although whether these were self- or open-pollinated is not recorded.

The only named cultivar, until 2010, was 'Balkan Rose', with deep heliotrope (H12) flowers. New selections (see p. 289) raised from seed are in cultivation with flowers ranging from white to deep red-pink. In some of these clones, originating from Kurt Kramer (Edewecht-Süddorf, Germany), the filaments were fused irregularly at bases into groups of about three, occasionally more; the reason for this is not clear, and the malformation resembles anomalies noticed in *Erica × stuartii* in Ireland.

HYBRIDS

In the wild, *Erica spiculifolia* is a rather isolated species, rarely if ever occurring in the same habitat as another species of *Erica*, as far as I can ascertain. Thus natural hybrids are probably unlikely. It is possible that in some of its Balkan montane habitats, *E. carnea* is a near neighbour because their ranges mear (cf. Neidermaier 1985).

In cultivation, however, artificial pollination has resulted in the successful raising of a number of hybrids, as yet unconfirmed by DNA studies. All of these have been raised by Kurt Kramer in Edewecht-Süddorf, Germany. Since 1987, Kramer, alone of the present-day heather breeders, has employed *Erica spiculifolia* in numerous cross-pollination attempts (Hall 1992). When used as a pollen-parent, it has so far been a failure: the attempted seed-parents were *E. carnea*, *E. cinerea*, *E. manipuliflora*, *E. tetralix* and *E. vagans*. Unsuccessful pollinators, when *E. spiculifolia* was the seed-parent, included *E. arborea*, *E. australis*, *E. erigena*, *E. lusitanica*, *E. mackayana*, *E. manipuliflora*, *E. multiflora*, *E. terminalis* and *E. umbellata*. The successful crossings were with the following pollen-parents: *E. carnea* (1986), *E. cinerea* (1986), *E. sicula* subsp. *bocquetii* (2006), *E. tetralix* (2001) and *E. vagans* (2003). Of these, the hybrid between *E. spiculifolia* and *E. carnea* is the only one that has been named: *E. × krameri* (McClintock 1998) (see p. 343).

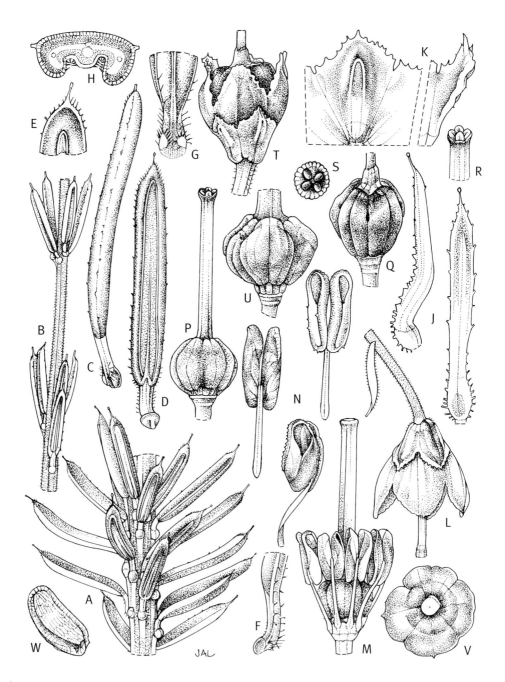

Fig. 183. *Erica spiculifolia* 'Balkan Rose'. **A** leafy shoot × 12; **B** young leafy shoot × 4.5; **C** leaf from side × 12; **D** leaf from back × 10; **E** leaf tip × 10; **F** petiole and leaf base from side × 10; **G** petiole and leaf base from front × 10; **H** transverse section of leaf × 30; **J** bract from back and side × 26; **K** calyx lobe from back and side × 22; **L** flower from side × 8; **M** gynoecium and stamens × 14; **N** stamens from back, side and front × 22; **P** gynoecium × 14; **Q** ovary × 10; **R** style apex (stigma) × 20; **S** style apex (stigma) from above × 30; **T** capsule enclosed in withered corolla and calyx × 16; **U** mature capsule × 24; **V** capsule from above × 24; **W** seed from side × 44; from specimens grown by Allen Hall. Drawn by Joanna Langhorne.

Kurt Kramer has also attempted crossing Balkan heath with some of the South African species, always using *Erica spiculifolia* as the seed-parent. Those with *E. blandfordii*, *E. curvirostris* and *E. patersonii* failed. The only successful cross to date has been *E. spiculifolia* pollinated by *E. bergiana* (see Hall 1992) — *E.* × *gaudificans* E. C. Nelson & E. M. T. Wulff (see p. 340), now quite widely cultivated in parts of the western USA (although often under the incorrect name, *E.* × *krameri*) and in a few gardens in Ireland and Britain. Two cultivars, 'Edewecht Belle' and 'Edewecht Blush', are extant (a third one is apparently lost) (Nelson 2005c). While these two cultivars depart significantly from "normal" *E. spiculifolia*, they show little sign of the influence of the other parent except in the size of the plants: shrubs of *E.* × *gaudificans* are two or three times as large as those of *E. spiculifolia*.

Erica spiculifolia Salisb., *Trans. Linn. Soc. London* **6**: 324 (1802). Type: "Sponte nascentem in summitate Montis *Olympus*, legit heu defunctus Sibthorpe": holo. **BM** (000885845)! (Nelson 2007g).
Menziesia bruckenthalia Baumg., *Enumeratio stirpium Magno Transsilvaniae principatui* 1: 333 (1816).
Erica bruckenthalii Spreng., *Neue Entdeckungen* I: 271–272 (1820).
Bruckenthalia spiculifolia Rchb., *Flora germanica excursoria* I: 414 (1831).
Erica spiculiflora G. Don, *A general history* ...III: 797 (1834).
Erica transsilvanica Willd. ex Klotsch, *Linnaea* 12: 217 (1838).
Bruckenthalia spiculiflora Benth. in A. P. de Candolle, *Prodromus systematis naturalis regni vegetabilis* VII: 694 (1839).

ILLUSTRATIONS. F. L. Bauer, *Flora graeca*, cent. 4 tab. 353 (Sibthorp & Smith 1823) (Plate 22); *Curtis's Bot. Mag.* **133**: tab. 8148 (1907).

DESCRIPTION. *Evergreen shrublet*, ±15 cm tall, stems prostrate and ascending, grey-pubescent. *Leaves* in irregular whorls of (2–)4–6, or spirally arranged, ±4–6 mm long (leaves in inflorescence usually longer, <9 mm), to 1 mm across, linear-lanceolate, flat above, glabrous except for spicules near apex, which has a long, gland-tipped hair, edges with gland-tipped hairs and minute spicules, margins revolute, sulcus not closed, midrib visible, widening towards base, to ±0.4 mm wide; petiole translucent, with a few simple hairs on edge, some gland-tipped. *Inflorescence* usually a terminal raceme with 8–40 flowers, or flowers solitary in leaf axils; *pedicels* erect, pink, straight or curving, 3–7(–10) mm long, terete, slender, puberulent with very short spreading hairs, some gland-tipped; solitary bract at base of pedicel; lowermost bract in an inflorescence ± leaf-like, uppermost pink or (in white-flowered plants) colourless, 2–3 mm long, with gland-tipped apical hair; bracteoles fully recaulescent as the two lateral lobes of the calyx. *Calyx* cyathiform, pink, creamy-pink or white, usually tightly clasping the corolla, ±1.5 mm long, with 4 triangular lobes, ±0.7 mm long (shorter than calyx tube), margins sometimes irregularly toothed, with prominent, open, greenish sulcus, ±1 mm long. *Corolla* bright pink to red–pink, very rarely white, campanulate, ±2.5–3 mm long, glabrous, 4-lobed; lobes ±1 mm long, rounded. *Stamens* 8, included, ±2 mm long; spurs not present; *thecae* ±0.8 mm long, brown; *filament* to ±1 mm long, white, glabrous, slender, straight, erect; anther attachment lateral, pore lateral in upper portion of thecae, obovate; *pollen* in monads, grains ±spherical, 17–19 μm (polar length) × 17–20 μm (equatorial diameter) (Phillips & Sparks 1976). *Style* to 4 mm long, erect, terete, glabrous, straight, emergent, tapering slightly towards tip; *stigma* only very slightly broader than style apex, capitellate; *ovary* ±1.5 mm tall, green, globose to turbinate, glabrous, with 4 locules; ±12 ovules per locule. No nectary present. *Seed* ovoid ±0.3 × 0.6 mm, epidermal cells "long and narrow ... with thick wavy walls" (see Wilson *et al.* 1973: plate 1; Phillips & Sparks 1976: plate 1, figures 14*d–f*). n = 18.

FLOWERING PERIOD. Stevens (1978) reported the flowering period in Turkey as June to September, while Denkewitz (1987: 193) has *Erica spiculifolia* in bloom from mid-July to late August, and van de Laar (1978) noted that the Balkan heath can still be in flower during September. In British gardens it blooms between May and July (Small & Small 2001).

DISTRIBUTION & ECOLOGY. South-eastern Europe: Romania, Bulgaria, Bosnia and Herzegovina, Serbia, Montenegro, Macedonia (FYR), Albania, Greece. Western Asia: northern Turkey. **Maps:** Stevens (1978: 103, map 9), Browicz (1983), Neidermaier (1985). Fig. 184.

In its native habitats, *Erica spiculifolia* is calcifuge and occurs in open woodland, scrub, and on damp moorland (Fig. 182). In Turkey it is associated with *Rhododendron caucasicum* and species of *Vaccinium* (Stevens 1978). The Balkan heath is recorded as attaining altitudes over 2,000 m in the southern part of its range: altitudes around 2,200 m are reported from Turkey and Greece (Strid 1986), 2,300 m from Bulgaria, and over 2,400 m from Macedonia (FYR) (cf. Neidermaier 1985). In alpine and subalpine habitats, *E. spiculifolia* can form a carpet.

ETYMOLOGY. Latin. Salisbury (1802: 324) gave no reason for choosing this epithet, and his description of the heath does not provide any clues. Perhaps he derived it from *spiculum*, spicule, and *folius*, leaf, and he may have been alluding to the microscopic spicules which are present on the leaves. As he evidently studied *Erica* species using a microscope — his discovery of the elaiosome on the seed of *E. australis* (see p. 221) makes that certain — he may also have observed these minute spicules. However, there is another more mundane possible explanation. *Spiculifolia*, from *spiculus*, pointed, may simply mean "with pointed or spiky leaves".

The epithet has frequently been misspelled as *spiculiflora*: both George Don (1834) and George Bentham (1839), in their respective publications, made this error.

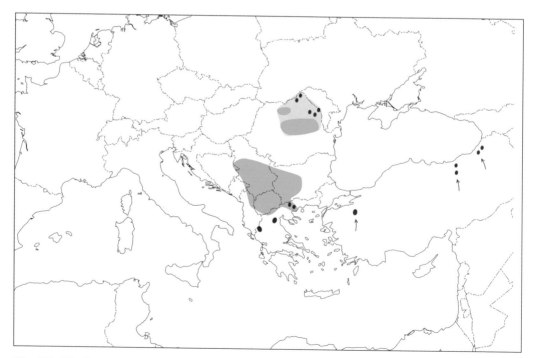

Fig. 184. Distribution of *Erica spiculifolia*.

VERNACULAR NAMES. Balkan heath, Bruckenthalia (cf. Nelson 1997), spike heath (English); bruyère de epi (French); Aehrenblütige Bruckenthalie (Dippel 1887), Siebenbürgische Heide, Siebenbürgenheide, Ahrenheide (German); coacăz, coacaza (Romanian); axljung (Dahlin 2007) (Swedish).

The name spike heath seems to derive from spike-flowered heath which is a simple translation, first used by George Don (1834), of *Erica spiculiflora*. As noted above, the epithet *spiculiflora* is a error so the name spike-flowered heath, even though it is descriptive, is based on a corruption.

CULTIVARS. As already noted, only one cultivar has been selected and named until 2010: 'Balkan Rose' is of unknown origin. Several new clones, all seedlings raised by Kurt Kramer were named and released in 2010: .'Manja' (® E.2010.02) has cerise (H6) flowers, 'Mila' (® E.2010.03) lavender (H3), and 'Raika' (® E.2010.01) is white-blossomed (*E. spiculifolia* f. *albiflora*).

Notes

[1] Reports from Kilbeg, County Waterford (Watts 1959) and Baggotstown, County Limerick (Watts 1964), rely on seeds "identical" to those found at Gort. These records may therefore indicate the presence of *Erica scoparia* var. *macrosperma* (cf. Mitchell & Watts 1979; Coxon & Waldren 1995). However, as pointed out elsewhere (p. 196, note 3), that name cannot be considered a synonym for *Bruckenthalia spiculifolia* or *E. spiculifolia*.

[2] Reproduced by Harris (2007: figure 36, p. 65). The pencil sketch of the flowering sprig of *Erica spiculifolia* is in the lower right, with an enlarged flower on its left and another on its right.

[3] Brukenthal built a palace in the finest baroque style in Sibiu (Hermannstadt), and in his will left it as a public museum which survives intact to this day. Significantly, he also promoted the study of the natural history of Transylvania (see http://www.brukenthalmuseum.ro/en/mnaturala_botanic.php (accessed 3 March 2007). My thanks to Dr Erika Schneider (University of Karlsruhe) for her assistance with information about Baumgarten and Brukenthal.

THE PENTAPERAN HEATHS

Erica sicula, including *E. sicula* subsp. *bocquetii*

The abandoned generic name *Pentapera*, published by Johann Friedrich Klotzsch (1805–1860) in 1838, provides a useful heading for this section. It signifies having parts in fives, specifically a five-celled ovary and fruit (from Greek, πέντε (*pente*), in compounds usually πέντα- (*penta-*), five; and πήρα (*pēra*), a pouch). Only two species have ever been assigned to *Pentapera*: *P. sicula* and *P. bocquetii*. The first, the Sicilian heath, was originally named *Erica sicula*, the name to which it has reverted following the "sinking" of *Pentapera* back within *Erica* by Webb & Rix (1972) in *Flora europaea*. The other species, which may conveniently be called Bocquet's heath, was reassigned to *Erica* by Stevens (1978).

These are the least familiar of the northern heathers, yet potentially among the most interesting, and they are also beautiful with large flowers in various shades of pink. They are confined to the eastern Mediterranean, where they have a very markedly disrupted range, comprising, according to current knowledge, only small populations in isolated and somewhat inaccessible localities.

The Sicilian heath and Bocquet's heath are remarkably similar. Artificially created hybrids between them are fertile (Kurt Kramer, pers. comm., 5 May 2008). Do these heathers constitute one species, perhaps split up into several subspecies, or is there more than one species, as indicated by the names in current literature? This question will only be answered definitively when plants from all the populations are submitted to detailed investigation. Meantime, our opinions can be shaped only by our observations of gross morphology and whatever patterns of variation can be detected in the wild. Garden plants can be misleading, because variation is exaggerated through deliberate (or subjective) choice of particular plants for propagation. In the same way, but to a lesser degree, herbarium specimens can be misleading, as they have also been, in reality, selected by the botanists who gathered them, and by history.

SICILIAN HEATH

The first of these five-parted heathers was found on the southern and western limestone cliffs on Monte Cofano, near Trapani (Drepani), in the northwest of Sicily. *Erica multiflora* inhabited the same place (Gussone 1821). Named after the island, *E. sicula* was described by Giovanni Gussone (1787–1866) in a catalogue of the plants in the Royal Garden at Boccadifalco near Palermo. At the behest of the Duke of Calabria, Gussone had established this experimental garden in 1817.

Unusually for the time, Gussone's description of *Erica sicula* was detailed, not just about the habitat but also about the plant. The shrubs were two or three feet in height, and in April and May they produced pale pink flowers which had either a four- or a five-part calyx and correspondingly four- or five-lobed corolla. In his observations, he made a comment that shows he had closely studied the heather, dissecting probably many flowers: the number of stamens varied, as did the number of corolla lobes and the number of sepals (calyx segments): when there were eight stamens, there were four sepals and four corolla lobes (Gussone 1821). This noticeable variability is hardly ever remarked on again until modern times.

When Johann Friedrich Klotzsch (1838) erected the new genus *Pentapera* he made no mention at all of the variability of the numbers of stamens or the lobes of the parts of the flowers that Gussone had noticed in Sicily; the floral parts were strictly in fives, the corolla five-lobed, the calyx with five sepals, and the stamens ten, hence the convenient, but somewhat inaccurate, name *Pentapera* (Klotzsch 1838: 497–498).

At first *Erica sicula* was known only from the mainland of Sicily. The next of the scattered habitats to be discovered was in the north of Cyprus: Carl Georg Theodor Kotschy (1813–1866) found *E. sicula* in the Pentadactylos Mountains during 1859 (Meikle 1985). There the heath occurs in several, separate localities, all on limestone, mostly north-facing, "usu[ally] hard of access to goats and botanists" (Viney 1994a), between about 300 m and 1,000 m altitude (Meikle 1985; Viney 1994a, 1994b). There are hummocks of *E. sicula* on the north-facing cliffs below the ruins of the Crusader castle at Buffavento, out of reach of this botanist but within range of my camera (Fig. 185).

The same heather was found as early as 1874 in the north of Cyrenaica in northern Libya (Durand & Barratte 1910), where it inhabits wadis and limestone rocks between el-Beida and Derna (Brullo & Furnari 1979).

The French naturalist, adventurer and sportsman, Louis Charles Émile Lortet[1] (1838–1909), Professor of Natural History at l'École préparatoire de Médecine in Lyon, gathered specimens of the Sicilian heath in Lebanon, between Bil'âs and Afka, at 1,100 m altitude, on 4 June 1880 (Barbey & Barbey-Boissier 1882). Lortet's interests included human physiology at altitude, and the mummification of animals in Ancient Egypt, and he is commemorated in a very handsome and endangered *Iris lortetii*, native in southern Lebanon and northern Israel.

Sometime before 1938, a collector named Stöger (about whom I can discover nothing) gathered and pressed a "scrappy" specimen of a heather, which was later identified as *Pentapera sicula*; the habitat was at Çiğlikara kasa, at an altitude between 1,800 m and 2,000 m (Stevens 1978: 97). The specimen survives in Berlin and is the earliest record of one of the Pentaperan heathers from

Fig. 185. *Erica sicula* on the cliffs at Buffavento Castle in the north of Cyprus (E. C. Nelson).

Turkey — but it is not *E. sicula sensu stricto* — I will return to this specimen shortly. No one else apparently collected the plant in Anatolia until the late 1940s (cf. Yaltirik 1967) when Peter Hadland Davis (1918–1992), then studying botany at the University of Edinburgh, saw the Sicilian heath in a limestone gorge near Kemer (Davis 1949: 155):

> Kesme Boğaz was much too thrilling to potter in. I moved in jerks, in a series of small runs. ... It was very satisfactory to find the wonderful Heath, *Pentapera sicula* (var. *libanotica*?) hanging down the shadier parts of the limestone cliffs in 3 foot bushes of Yew-green. This ... grows on the rocks of Sicily, Cyrenaica, Cyprus, Lebanon and South-West Anatolia. The large pentamerous bells are pink in the Oriental variety — at least in Cyprus — but in the Sicilian plants (which flowers more readily in cultivation) they are white.

Examining herbarium specimens from all these localities, it is at once clear that the heathers represented are all similar — they belong to the same species in my opinion — although there are indications of a substantial amount of variation within and between populations in such characters as leaf size and vesture, and flower colour and corolla shape. Meikle (1985: 1061) commented wisely, regarding the Cypriot colonies: "Until *E. sicula* has been more carefully investigated, throughout its distribution, I prefer not to subdivide the species, though it may be that future research will confirm the validity of such subdivision."

BOCQUET'S HEATH

There can be little doubt that the specimen (in Herb. J. Bornmüller, **B**!) collected by Stöger above 1,800 m at Çiğlikara "*inter Antalia et Elmali*" represents the plant that would be named *Pentapera bocquetii*: Stevens (1978) was not so certain. The heather was not recorded again for more than three decades. During April 1968, Dr Hasan Peşmen and his associates gathered it in exactly the same region. They found it in clearings in cedar of Lebanon (*Cedrus libani*) forests at an altitude of around 1,750 m. Peşmen later described the heather and, as noted, gave it the name *Pentapera bocquetii*. Stevens (1978) recognised its affinity with *Erica sicula* and reassigned it to *Erica*, as *E. bocquetii*.

Peşmen (1968) distinguished his new species from *Erica sicula* by its "rampant" habit, canescent shoots, small glabrous leaves in whorls of three (or, he noted, rarely in fours), flowers in groups of three, small bracts and flowers (including sepals, corolla, stamens and style), hairy corolla and capsule, and glabrous style. The species blossomed later (July), too — perhaps due to the effect of altitude because there were open flowers on the plants I saw in Anatolia in early May 1997 (see below). Peşmen (1968) noted that the specimens of *E. sicula* that he used for comparison were from Sicily and Cyprus as well as Turkey. Then, commenting on Yaltirik's earlier (1967) designation of *Pentapera sicula* subsp. *libanotica*, incorporating the populations from Lebanon, Cyprus and those from low altitude in Anatolia, Peşmen stated "Mais les caractères employés par Yaltirik pour distinguer ces deux taxons nous ont semblés illusoires, et les plantes de Chypre et de Sicile ne sont en vérité guère différentes."

In July 1989, Dr Gianlupo Osti and David McClintock (1989a, 1990a, 1991a; Osti 2009) went to Turkey specifically to see this plant and to bring back seed and cuttings. During this successful field-trip, they found a population at a much higher altitude, 2,450 m. The heath always grew in crevices in limestone rocks, usually in open sites within the cedar forests. In 1994, the altitudinal range was again amended when Mr & Mrs R. M. Burton discovered a population at 1,200 m, again on limestone, in a clearing within *Pinus brutia* forest (Burton 1995). These finds suggest that Bocquet's heath is probably not as uncommon as believed and that further exploration of this region will result in other colonies being located.

McClintock and Osti's expedition resulted in cuttings reaching England. Eighty-four of 130 cuttings sent to Bert Jones at Otters' Court Heathers, West Camel, Somerset, rooted. The first flowers opened there in late April 1994, but Jones (1996) noted that, even given sharp drainage, Bocquet's heath "has not grown well here". Few of the 84 young, rooted plants now survive. Subsequent attempts to root cuttings have not been very successful (Canovan 2005). The flowers tend to be "very short-lived", according to Canovan (2005): they have "the delicacy of a magnolia". Thus, Bocquet's heath remains very rare in cultivation, partly because, contrary to the expectations of a high-altitude taxon, it is decidedly capricious at more moderate altitudes — plants often will not bloom and evidently resent any disturbance (Canovan 2005).

I have seen both the Sicilian heath (*Erica sicula*) (Fig. 186) and Bocquet's heath (Figs 187 & 188) in Anatolia. In early May 1997, during a holiday at Kalkan, we went to see the colony of Bocquet's heath that Burton (1995) had reported between Sinekçibeli and Kaş. We found it without difficulty, on the limestone by the side of the road. Seeing the heather hummocks cushioning out from cracks and crevices, each bush very closely trimmed, often to the rock surface (Fig. 187), by grazing animals (probably goats), I was reminded of the way that bushes, such as juniper (*Juniperus communis*) and even yew (*Taxus baccata*), grow in The Burren in western Ireland. The heather was only just coming into bloom; Burton reported it in full flower a month later, in early June (and a month earlier, as just noted, than Peşmen). Having moved on, and had lunch, we stopped again after we crossed a small river, and went for a walk up the river bed into a shady gorge. This was not more than ten kilometres by road from where Bocquet's heath was growing. On the limestone

cliffs of the gorge, there was a heather growing in great hanging swags, just as described by Davis (1949), although not yew-green as he stated: *Erica sicula*! That the two supposedly different species can grow so close together and yet be considered distinct seems illogical. There were differences between the plants but nothing that warranted designating them as separate species.

Stevens (1978: 94–96), like Peşmen, separated *Erica bocquetii* from *E. sicula* on the basis of leaf length and the number of leaves in each whorl: according to those botanists, *E. bocquetii* has much shorter leaves, half the length of those in the other species. However, the dimensions of the leaves overlap, and leaf length is not a good indicator of separate specific status. Stevens (1978) added that in *E. sicula* there were "small spurs at the apex" of the filaments but not in *E. bocquetii*. Yet that spur character does not work either, as specimens of *E. sicula* that I have examined vary, some possessing minute, vestigial spurs and others lacking any vestiges of spurs.

The question remains whether *Erica bocquetii* is distinct enough to maintain as a separate species. My view is that the populations of Bocquet's heath are best considered as forming a subspecies of the widely

Fig. 186. *Erica sicula*: flowering shoot; between Sinekçibeli and Kaş, Anatolia, Turkey (E. C. Nelson).

Fig. 187. *Erica sicula* subsp. *bocquetii* growing on limestone in Anatolia, between Sinekçibeli and Kaş, Turkey (E. C. Nelson).

Fig. 188. Flower shoots of *Erica sicula* subsp. *bocquetii*; Sinekçibeli and Kaş, Anatolia, Turkey (E. C. Nelson).

distributed and rather variable *E. sicula*, differentiated by adaptation to habitats at high altitudes (cf. McClintock 1992a). The documented Turkish habitats (Davis 1949; Stevens 1978) of *E. sicula sensu stricto* all lie close to sea-level, whereas the plants called *E. bocquetii* are from much higher localities, between 1,200 m (Burton 1995) and 2,450 m altitude (McClintock 1991a). In all their measured characters, the high-altitude plants fall within the ranges of dimensions recorded from the other regions.

Given the fragmented distribution of *Erica sicula*, gene-flow between the widely separated populations cannot occur at present (and probably will not have taken place in recent millennia). Thus, it is to be expected that the isolated populations have developed different characteristics, although they generally retain the same morphology and ecological propensities. Keeping faith with Desmond Meikle's comment, there needs to be a lot more work done on these plants in all their habitats, and in cultivation, before any sensible classification of the several subspecies can be achieved. We need more detailed information about variation within the wild populations. Until all the populations can be sampled and, ideally, plants of known origin cultivated together under uniform conditions to eliminate environmental factors (such as any effect that altitude has on flowering time and habit), apparent variation between the populations has to be treated with great caution. What the levels of separateness are remain to be determined using the full battery of scientific methods including the relatively new techniques of DNA analyses.

Having recorded a new locality for *Erica sicula* in Anatolia without effort, I suspect that there could be many more isolated populations of this heath in clearings in the Anatolian forest and on the upper slopes of the mountains: the area is by no means well explored.

25. ERICA SICULA. Sicilian heath.

As noted above, the original locality from which Guiseppe Gussone obtained this heather was on Monte Cofano, a dolomitic limestone monadnock rising to 659 m altitude on the coast of north-western Sicily. His description is so detailed he must have collected it himself on the mountain. The late David McClintock (2002) visited Monte Cofano in April 1999, recording that the plant he saw through binoculars had white flowers, and moreover that "all the corollas collected in Sicily were white except for one lobe which was pink". His reports (McClintock 1999b, 2002) clearly indicate that he saw only one or two plants at the locality (before being seriously injured when he attempted to ascend the scree to reach them (Osti 2009)). He was later informed by friends, the Grattans, who went walking "lower down to the north", that there was plenty of *Erica sicula* "in what may have been the type locality" (McClintock 2002).

Despite reports to the contrary (for example, McClintock 1980d: 54; 2002), a small number of old plants also grow on Isola di Marettimo (Osti 2009), one of the islands of the Aegadian Archipelago, situated in the Mediterranean about 40 km to the west of Monte Cofano. There are also reports of the species on Malta (cf. Webb & Rix 1972; Pignatti 1982: 257), but there are no verified specimens and the heather certainly is not present on the island today. (*Erica multiflora* is the only heather species on Malta.)

Gussone (1821) used the phrase "pallide carnea" to describe the colour of the corolla — pale flesh-coloured — while the illustration published in *Curtis's botanical magazine* (tab. 7030; Hooker 1888) shows white flowers. I have been shown photographs of a Sicilian plant which had very pale pink blossoms. The colour of the Anatolian plants was much deeper pink, but whether corolla colour is a constant within populations must be doubted; Meikle (1985) described the Cypriot plants as having " pale to bright pink" flowers, and Viney (1994b) has "in shades of pink (even to white)". Stevens (1978) described the flowers of Anatolian *Erica sicula* as pink, and those of *E. bocquetii* as "pale purple".

CULTIVATION

"Thank you for the seeds, Lobelia Bivonæ & Erica Sicula are 2 Plants I want more than any … ."
(W. T. H. F. Strangeways to W. H. Fox Talbot, c. 1833).[2]

Erica sicula has been cultivated from time to time in Britain; the species was in the Royal Botanic Gardens, Kew, in the 1880s (as indicated by the plate published in *Curtis's botanical magazine*: Hooker 1888). It may well have been introduced into English gardens in the early 1830s if the extract from Lord Ilchester's (William Fox-Strangeways) letter to William Henry Fox Talbot (1800–1877), one of the inventors of photography and a very keen botanist, employs correct nomenclature.

A plant of *Erica sicula* was grown outdoors in a sandstone trough beside the (former) Alpine House in the Royal Botanic Gardens, Kew, during the 1980s (Fig. 189), suggesting that it is possible to keep the species outside within a sheltered space. However the plants presently known to me in cultivation are all housed under glass, including the plants presently in Kew.

Given its habitats, *Erica sicula* probably requires a very free-draining soil, and should benefit if coarse limestone chippings are incorporated to assist drainage and provide a moderately lime-rich medium.

No clone has been distinguished and named as a separate cultivar, although plants may have distinct collector's numbers. All the plants known in cultivation in Britain are of Turkish origin, as far as I can ascertain.

Fig. 189 (left). *Erica sicula* growing in a trough in the Alpine Yard, Royal Botanic Gardens, Kew (E. C. Nelson).
Fig. 190 (right). *Erica sicula*: capsules from the previous year; between Sinekçibeli and Kaş, Anatolia, Turkey (E. C. Nelson, May 1999).

Erica sicula Guss., *Catalogus plantarum hort. reg. Boccadifalco*: 74–75 (1821). Type: 'Monte Cofano presso Trapani", *Gussone*: lecto. **NAP** designated by Meikle (1985: 1060).

E. bocquetii (Peşmen) P. F. Stevens, *Flora of Turkey* **6**: 97 (1978).

E. sicula subsp. *cyrenaica* Brullo & Furnari, *Webbia* **34**: 164–166 (1979). Type: 'Uadi sopra Appolonia", *Brulla & Furnari*, 13 May 1974: holo. **CAT**.

E. sicula subsp. *libanotica* (Barb. & Barb.-Boiss.) P. F. Stevens, *Flora of Turkey & E. Aegean islands* **6**: 97 (1978); *E. sicula* subsp. *libanotica* (Barb. & Barb.-Boiss.) Brullo & Furnari, *Webbia* **34**: 166 (19 79) *comb. superfl.*

Pentapera bocquetii Peşmen, *Candollea* **23**: 271 (1968).

Pentapera sicula (Guss.) Klotzsch, *Linnaea* **12**: 498 (1838).

Pentapera sicula var. *libanotica* Barb. & Barb.-Boiss., *Herbor. Lev.*: 144 (1882). Type: "In Libano inter Billaas et Afka alt. 1100ᵐ", *J. C. E. Lortet*, 4 June 1880: not traced (? G).

Pentapera sicula subsp. *libanotica* (Barb. & Barb.-Boiss.) Yalt., *Notes Roy. Bot. Gard. Edinburgh* **28**: 13 (1967).

25a. **Erica sicula** subsp. **sicula**

ILLUSTRATIONS. *Curtis's Bot. Mag.* **114**: tab. 7030; *Flora of Libya* **54**: fig. 2; *Webbia* **34**: 165, fig. 7 (*E. sicula* subsp. *cyrenaica*).

DESCRIPTION. *Evergreen shrub* to ±0.5–0.6 m tall, bushy, often spreading to form hummocks; *shoots* erect, ascending or (when growing on cliffs) pendent, terete or grooved, densely to very sparsely pubescent with gland-tipped hairs especially when young, often glabrescent; older shoots with flaking bark. *Leaves* in whorls of 4 (–5), spreading or ascending, linear, ±(3–)8–13 × 0.8–1.3 mm, pubescent, often velvety, glabrescent, sometimes with gland-tipped hairs around edge, rarely glabrous; margins strongly revolute; sulcus linear, closed; petiole pale, usually pubescent with gland-tipped haits, erect, ±2 mm long. *Inflorescence* with 2–6(–8) flowers, rarely more, in terminal umbel on main or axillary shoots; *pedicels* 5–12 mm long, pubescent, with bract and 2(–3) bracteoles at mid-point (very rarely with additional, recaulescent bracteole); bract pale pink, lanceolate, tip acuminate, to 3.5 mm long, ±1.5 mm broad, margins glandular; bracteoles similar linear-lanceolate. *Calyx* pale pink, with 4 or 5 free sepals, ovate to lanceolate, pubescent on outside or rarely glabrous, ±4–6 mm long, ±1–2.5 mm across. *Corolla* pale to deep pink, sometimes white, urceolate, (5–)6–8.5 mm long, 4–5 mm in diameter, glabrous or pubescent outside, with 4 or 5 erect or recurved lobes, ±1–1.5 mm long and as wide, rounded. *Stamens* 8 or 10, included, to 6.5 mm long; *filament* glabrous or with hairs towards apex, flattened, erect, straight or geniculate towards apex, ±3.5 m long; *anthers* muticous, or with minute vestigal spurs; *thecae* dark, erect, ±2–3 mm long, ± 1mm across, with short hairs especially towards base or glabrous; *pore* lateral ovate-elliptical. *Ovary* 4- or 5-locular, barrel-shaped, pubescent, apex emarginate, to 3.2 mm tall; *style* erect, terete or angular, tapering, entirely glabrous or with sparse hairs on lower portion, or entirely pilose, (3.5–)5.5–6.5 mm long; *stigma* capitate, ±0.7 mm diameter, dark red–black, usually at mouth of corolla but sometimes emergent; *nectary* at base, nectar profuse. *Capsule* ± 6 × 5 mm, globose, woody and peristent, 5-lobed (Fig. 190); *seeds* 0.5–0.8 × 0.3–0.4 mm, ovoid, dark brown, glossy.

FLOWERING PERIOD. March–August.

DISTRIBUTION & ECOLOGY. Europe: Italy: Sicily only and confined to Monte Cofano (Brullo & Furnari 1979; McClintock 2002) and Isola di Marettimo. Africa: Libya: Cyrenaica only (Siddiqi 1978; Brullo & Furnari 1979; Brullo & Guglielmo 2001). Western Asia: Cyprus: in Pentadactylos Mountains (Meikle 1985; Viney 1994a, 1994b) and other scattered sites in northern Cyprus; also recorded in south-eastern Cyprus, at Aphamis, in the Larnaca region, at "3,200 ft alt [c. 1,000 m], in garigue

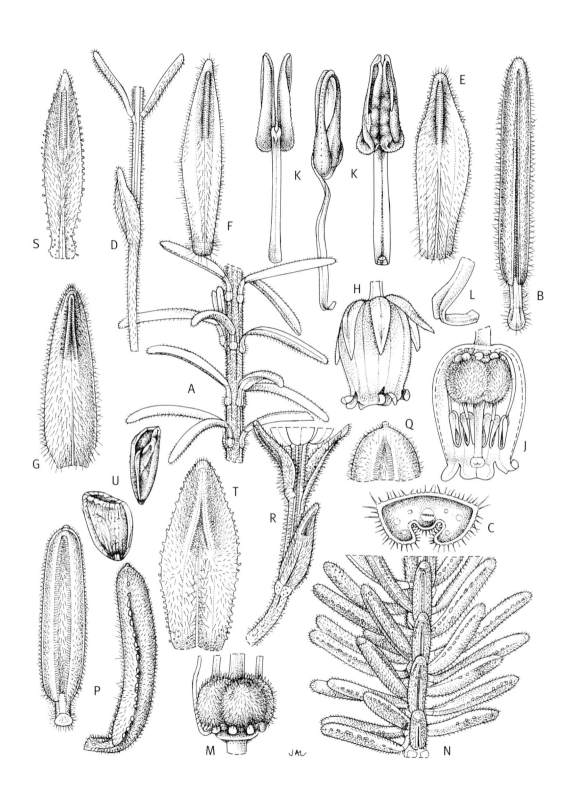

between vineyards" by Kennedy in 1937 (Meikle 1985) but it has not been seen there in recent decades (Yiannis Christophides, *in litt.* November 2007); Turkey: Anatolia only (Stevens 1978); Lebanon (Post & Dinsmore 1923). **Maps:** Browicz (1983; south-western Asia only). Fig. 192.

ETYMOLOGY. Latin toponym: *siculus* = Sicilian.

VERNACULAR NAMES. Sicilian heath (English; McClintock 1980b); Levanteheide (German).

CULTIVARS. There is no named cultivar of this subspecies, although several distinct clones are in gardens in Britain.

25b. Erica sicula subsp. **bocquetii** (Peşmen) E. C. Nelson, **comb. et stat. nov.**

Basionym: *Pentapera bocquetii* Peşmen, *Candollea* **23**: 271 (1968). Type. "Turquie, province d'Antalya, district d'Elmali, Taurus occidentalis, Çiğlikara, Dokuzgöl Mevkii, 1750 m", 15 July 1968, *S. Parlakdağ* (*Peşmen* 2158). Holo. **Izmir**: iso. **G**, **ZT**.

E. bocquetii (Peşmen) P. F. Stevens, *Flora of Turkey* **6**: 97 (1978).

ILLUSTRATIONS. *Candollea* **23**: 272 (1968); A. W. Jones, *Yearbook of The Heather Society* **1996**: 20; *The Karaca Arboretum magazine* **3** (1): 28.

DESCRIPTION. *Evergreen shrub* to 0.25 m tall, often spreading to form hummocks or closely appressed to rocks; *shoots* pubescent. *Leaves* in whorls of 3 or 4, spreading or ascending, linear, ±3–6 × 0.8–1.0 mm, pubescent on upper surface, ±glabrous underneath; margins strongly revolute; sulcus linear, closed; petiole pale, ±1 mm long. *Flowers* 2–3 (Fig. 192), rarely solitary or more, in umbel, terminal on main or axillary shoots; *pedicels* 8–15 mm long, pubescent with short, very sticky, gland-tipped hairs, with bract at base or close to mid-point, and 2 bracteoles near mid-point; *bract* pink, ovate to lanceolate, acute, to (2–)3–4 mm long, to 1.5 mm broad, margins fringed with hairs; bracteoles similar but usually shorter and broader. *Calyx* pale pink to dark red, with 5 unequal sepals, ± concave with slightly reflexed margins, free except at base, ±3–4 mm long, less than 1.5 mm across, elliptical to ovate to obovate, acute, glabrous except for a patch of hairs in middle and short, gland-tipped hairs on

Fig. 192. *Erica sicula* subsp. *bocquetii* (© A. W. Jones, reproduced by courtesy of Mrs Diane Jones).

Fig. 191 (opposite). *Erica sicula*. **A** leafy shoot × 3; **B** leaf from back × 6; **C** transverse section of leaf × 24; **D** peduncle with bract and bracteoles × 5; **E** bract × 11; **F** bracteole × 12; **G** calyx lobe × 12; **H** flower from side × 4; **J** longitudinal section of flower × 5; **K** stamens from back, side and front × 6; **L** filament base × 16; **M** ovary × 8. A–M drawn from Kew no. 1984 3706. **N** leafy shoot × 5; **P** leaf from back and side × 8; **Q** leaf tip × 17; **R** peduncle with bract and bracteoles × 7; **S** bracteole × 10; **T** calyx lobe × 10; **U** seed from front and side × 26; N–U drawn from Kew no. 1952 2107. Drawn by Joanna Langhorne.

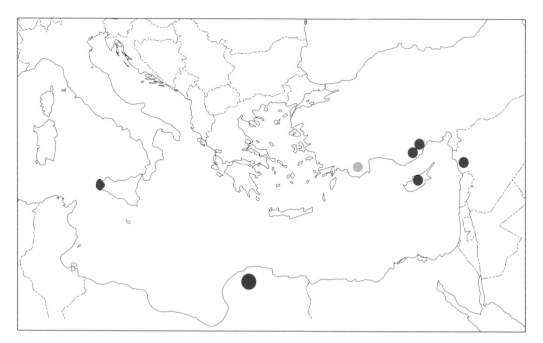

Fig. 193. Distribution of *Erica sicula* (pink) and *E. sicula* subsp. *bocquetii* (blue).

margins, tip keeled, with short (±0.5–2 mm) sulcus. *Corolla* pale to deep pink, urn-shaped, ±(5–)8 mm long, ±3.5–4 mm in diameter, 5-lobed, with sparse hairs especially on lobes; lobes ±1 mm long. *Stamens* 10, included, to 5.5 mm long; *filament* ±(2.4–)3.5(–4) mm, pink at base, white in middle and stained brown towards apex, glabrous, hooked at base, curving and geniculate towards apex; *anthers* muticous; *thecae* erect, ±2–2.5 mm long, parallel, fused towards base; *pore* lateral, ovate-elliptical. *Ovary* 5-locular, with >50 ovules per locule, barrel-shaped, very densely hirsute, to 1.5–3.5 mm tall, ±2 mm diameter, apex emarginate; *style* erect, entirely pilose, ±4–6 mm long, ±0.6 mm diameter at base, terete tapering upwards; *stigma* not exserted, capitate, ±0.6 mm diameter.

FLOWERING PERIOD. (April–)May–July.

DISTRIBUTION & ECOLOGY. Western Asia: Turkey, Anatolia only (Peşmen 1968; Stevens 1978), inhabiting clearings in pine and cedar forests, always on calcareous rocks, above 1,000 m altitude. **Map:** Browicz (1983). Fig. 193.

ETYMOLOGY. Eponym, after Professor Gilbert François Bocquet (1927–1986), Director of the Conservatoire et Jardin botaniques de la Ville de Genève, Switzerland, at the time when the species was named.

CULTIVARS. This subspecies is uncommon in cultivation; no cultivar has been selected and named.

NOTES

[1] Not Lorteto (as stated in Stevens 1978); Barbey & Barbey-Boissier (1882: 144) acknowledged that the heath was gathered "*a cl. Lorteto*", by the distinguished Lortet.

[2] The correspondence of William Henry Fox Talbot project; transcript by Professor L. J. Schaaf (2003) (URL http://foxtalbot.dmu.ac.uk/letters/ accessed 1 August 2007).

ERICA HYBRIDS

PART I. Naturally-occurring hybrids involving *E. tetralix*

Erica × *stuartii*, *E.* × *watsonii*, *E.* × *williamsii*

When northern *Erica* species grow together in the same habitat, there is always the possibility that pollen from one species will reach the stigmas of flowers of the other species, and consequently that fertilisation will take place and viable seed be produced. That seed must germinate, however, and a seedling must thrive to establish the new hybrid in a "wild" locality. Barriers to cross-fertilisation clearly occur, because the number of authenticated — even putative — natural, wild-occurring hybrids is small: just three definitely exist, although the potential number is substantially larger. The three verified hybrids are not widespread in their distribution, and they all involve cross-leaved heath, *E. tetralix*.

Erica × *stuartii* (*E. mackayana* × *tetralix*) was long believed to be endemic to western Ireland (see, for example, Stace 2005) but was discovered late last century in Asturias (Díaz González & García Rodriguez 1992; Díaz González *et al.* 1994; Mayor & Díaz 2003), and during 2005 its occurrence was detected in Galicia in north-western Spain (Fagúndez 2006). *E.* × *watsonii* (*E. ciliaris* × *tetralix*) has now also been found beyond its previously accepted limits; it is reliably recorded from southern England, where it is remarkably abundant, north-western France, and northern Spain (Díaz González & García Rodriguez 1992). There is also a very problematic record from western Ireland (Scannell 1976b). *E.* × *williamsii* (*E. vagans* × *tetralix*) is endemic to the Lizard Peninsula in Cornwall, England, and extremely uncommon there (Turpin 1984; Nelson 2007d), but the parent species co-exist in a few parts of western France, the Pyrenees and northern Spain, so its discovery elsewhere is a possibility.

There have been occasional reports of another wild-occurring hybrid, that between *Erica arborea* and *E. lusitanica* (see below, p. 322). However, no living plants of this hybrid are presently known in the wild, and unequivocal confirmation of the hybrid's existence in a natural habitat is wanting.

The botanical literature contains a handful of other supposed hybrids, phantoms of imaginative botanists, or, perhaps more fairly, imaginative parentages for unusual variants incautiously put into print without adequate research. For example, *Erica arborea* × *scoparia*, *E. arborea* × *umbellata* of Rivas Goday & Bellot (1946) which was named *E.* × *lazaroana* (Pizarro Domínguez 2007), and even *E. cinerea* × [*Calluna*] *vulgaris* which became the somewhat infamous ×*Ericalluna bealeana*, the type of which was the cultivar 'W. G. Notley'.[1] In fairness, there is a chance that such hybrids exist where the putative parents cohabit, or that they could be created in cultivation, because we know that many *Erica* species are promiscuous and that unexpected, "unnatural" hybrids can be created (see p. 326) by artificial cross-pollination of plants that have no possibility of doing so in the wild.

26. ERICA × STUARTII. Praeger's heath. [*E. mackayana* × *E. tetralix*]

This is now a familiar garden plant, popular because the new growth in late spring and early summer is brightly discoloured (Fig. 194) — not green. 'Irish Orange' and 'Irish Lemon', for example, form mounds of sometimes startling brilliance (for an explanation of the discoloured shoot tips, see Small & Alanine 1994).

For more than seven decades, this natural hybrid between *Erica tetralix* and *E. mackayana* was called *E. × praegeri*, and the English name recommended by The Heather Society, Praeger's heath, perpetuates that connection (McClintock 1980b). In the summer of 1911, the International Phytogeographical Excursion visited Connemara, and delegates saw *E. mackayana* and *E. tetralix* at Craiggamore (see p. 240), the *locus classicus* for Mackay's heath. The Danish botanist Carl Emil Hansen Ostenfeld (1873–1931) gathered specimens from plants that were intermediate between those species and subsequently decided that these represented the natural hybrid between cross-leaved and Mackay's heaths. He described the hybrid and named it *E. × praegeri* (Ostenfeld 1912) after the Irish botanist, author and librarian Robert Lloyd Praeger (1865–1953), whose wife is commemorated in *Daboecia cantabrica* 'Praegerae' (see p. 79). Specimens preserved in various herbaria indicate that the hybrid, which is abundant around Craiggamore, was collected as early as 1836. Meanwhile, a "monstrous" variant of it was described in 1890 and named *E. × stuartii*, although the connection was not recognised at that period.

Why was the Latin name changed? There were two separate reasons. Firstly, an article of the *International code of botanical nomenclature* which prohibited names that were based on monstrosities was deleted in 1975. Secondly, a sport, with flowers and foliage indistinguishable from those on plants of *Erica × praegeri*, was found on the strange clone then called 'Stuartii' (= 'Stuart's Original'; Figs 197 & 198; see also p. 230), indicating that it was merely a monstrous variant of this hybrid (McClintock 1979a). Under the principle of priority with underlies all decisions about the nomenclature of plants and animals, the earliest validly published Latin name has to be used for a species or nothospecies (primary inter-specific hybrid). Thus *E. × praegeri*, which was first published by Ostenfeld in 1912, had to be replaced by *E. × stuartii*, as that name predated it by a decade.[2]

Fig. 194. Discoloured shoots are an easy (but not always reliable) way to tell a hybrid heather; *Erica × stuartii* in Connemara (E. C. Nelson).

This hybrid (Figs 194 & 195) is especially abundant in Ireland, occurring in all five localities where *Erica mackayana* grows (see pp. 240–241; Nelson 2005a). As *E. tetralix* also occurs in all these habitats, the two species have crossed and the hybrid has arisen. The first to recognise that plants "which appear intermediate between *E. Mackayana* and *E. Tetralix*" occurred in the wild, both at Craiggamore (where they are "much more plentiful ... and ... very variable") and Carna in Connemara, was Alexander Goodman More (1874). (In August 1835, Babington (1836) had "noticed no intermediate states" at Craiggamore.)

It had long been believed that the same process did not occur in Spain, although we knew there were a few localities where these two species grew in close enough proximity for cross-pollination to occur naturally (Fraga Vila & Nelson 1984). Remarkably, as long ago as the early 1990s, Spanish botanists reported the hybrid (as *Erica × praegeri*) from several localities in central Asturias, including Sierra de la Curiscada, where *E. × watsonii* has also been found (Díaz González & García Rodriguez 1992). To our shame, those of us in Ireland and Britain who have been so long involved with heathers were utterly unaware of this. The first inkling that I had that Praeger's heath had been found in north-western Spain was when Jaimé Fagúndez told me he had found a single plant of *E. × stuartii* at Aranga, Cambás, A Coruña, in the north-western corner of Spain. The habitat, at 660 m altitude in Serra da Loba, is at a much higher altitude than any of the Irish sites (Fagúndez 2006). In Asturias, Praeger's heath is reported from even higher localities, between 660 m and 910 m above sea level (Díaz González & García Rodriguez 1992).

Erica × stuartii in Ireland has inherited from its parent *E. mackayana* the ability to proliferate from its roots, but less readily (McAllister 1996). Ecological information about the hybrid's behaviour in Spanish habitats is wanting

Fig. 195. *Erica × stuartii* in Connemara (E. C. Nelson).

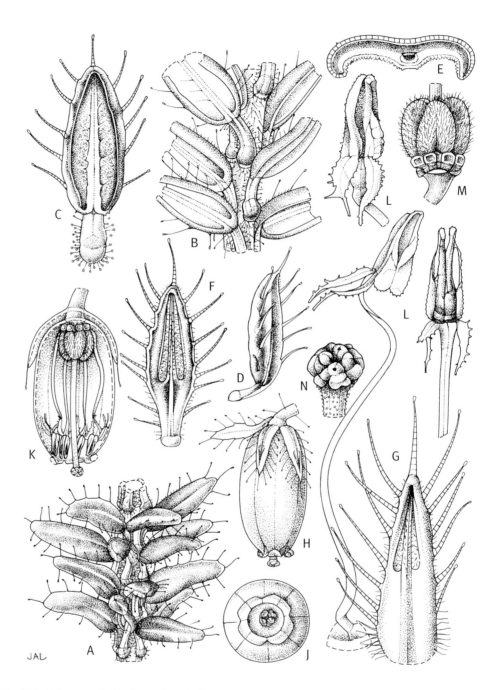

Fig. 196. *Erica × stuartii* 'Irish Lemon'. **A** leafy shoot × 10; **B** leaf nodes × 15; **C** leaf from below × 10; **D** leaf from side × 10; **E** transverse section of leaf × 26; **F** bracteole × 10; **G** calyx lobe × 18; **H** flower from side × 4; **J** flower from below × 5; **K** longitudinal section of flower × 5; **L** stamens back, side and front × 18; **M** ovary × 7; **N** style apex (stigma) × 12; from specimens grown in RHS Garden Wisley, no. W 1999 0127A. Drawn by Joanna Langhorne.

Fig. 197 (left). *Erica × stuartii*, a plant found by chance at Craiggamore, Connemara, in the summer of 2000 which has flowers indistinguishable from the original clone found by Dr Charles Stuart (see Fig. 195 and p. 230) (E. C. Nelson).

Fig. 198 (right). *Erica × stuartii* 'Stuart's Original' in cultivation, National Botanic Gardens, Glasnevin, Dublin (E. C. Nelson).

CULTIVATION & VARIATION

Erica × stuartii requires lime-free soil. Given that most cultivars are worth growing for their colourful springtime young growths, pruning immediately after the flowers have withered may be advantageous.

The clones in cultivation are all of Irish origin and all were wild-collected: no cultivars of Spanish origin have yet been introduced. These have mainly been selected for the colour of the young shoots, which can vary from bright yellow to pink ('Irish Rose' (® 120) has pink young shoot-tips). 'Stuart's Original' (Fig. 198) is grown as a curiosity, for its odd flowers.

The hybrid has been raised through artificial cross-pollination by David Wilson (British Columbia, Canada) and Dr John Griffiths (Yorkshire, England), but neither has selected and introduced new cultivars.

Erica × stuartii (Macfarl.) Mast., *Gard. Chron.*, series 3, 30: 34 (13 July 1901). Type: 'Near Roundstone Connemara', *anonymous*, 9 August 1890: neo.: **E**! (Nelson 1995a).
E. × praegeri Ostenf., *New Phytol.* 11 (4): 120–121 (1912).

DESCRIPTION. *Evergreen shrub* usually less than 0.5 m tall; shoots hirsute, with long, often gland-tipped cilia mixed with very short crispy hairs. *Leaves* lanceolate, rounded at base, in whorls of 4 (very rarely 5), ± evenly spaced but internodes increasing in length towards inflorescence, spreading horizontally or ascending, but usually not erect and parallel to stem; upper surface glabrous, or with minute erect hairs, with gland-tipped cilia around edge; margin revolute; abaxial surface with open sulcus, leaf margins not contiguous, so undersurface is visible; midrib often minutely pubescent, lateral areas white; *petiole* flattened, with gland-tipped cilia, with short straight hairs and also some longer crinkled ones. *Inflorescence* usually a terminal umbel, very occasionally several inflorescences axillary (resembling a raceme) with central shoot continuing to grow; *pedicel* ± straight, ±5 mm long, densely covered with short and long hairs, sometime gland-tipped; *bract* leaf-like, lanceolate, c. 4 mm long (including apical cilia ±1 mm), about or above mid-point, margins with long, shaggy, crinkled hairs and numerous marginal and non-marginal gland-tipped cilia; bracteoles closely pressed to calyx, spathulate, margins with long shaggy, crinkled hairs, and gland-tipped cilia (sometimes only toward top), to 3 mm long (including apical cilia ±1 mm). *Calyx* with 4 separate, often unequal sepals, to 3 mm long (including apical cilia ±1 mm), narrowly triangular, margins sparsely to densely shaggy, with gland-tipped cilia on margins (sometimes confined to apical portion), inner surface glabrous except at apex, which has shaggy hairs, outer may be sparsely hirsute, or glabrous; sulcus variable, open or closed. *Corolla* ovoid to urceolate, with 4 recurving lobes, glabrous except below and on lobes outside. *Stamens* 8, c. 5 mm long, included; *filament* hooked at base and attached under the nectary, curving, with sigmoid bend toward apex; *anther* basifixed, sometimes markedly prognathous, with spurs at attachment of filament, as long as or longer than thecae, narrowly triangular-linear and sometimes with small lobes towards tips, or outer margin deeply fimbriate; *thecae* ± triangular, ±1 mm long. *Ovary* obovate, ribbed, ±1.5 mm long, variously clothed with short hairs, emarginate, >50 ovules per locule; *style* straight, terete, erect, ±4.5–6 mm long; *stigma* capitellate, broader than style, dark, emergent; *nectary* lobed, conspicuous, encircling ovary base. *Seeds* not known.

FLOWERING PERIOD. (June–) July–September (–October)

DISTRIBUTION. Ireland: counties Donegal, Mayo, Galway and Kerry. Spain: Galicia, Asturias. **Maps:** Díaz González & García Rodriguez (1992: 137: Asturias only).

ETYMOLOGY. Eponym, named after a Scottish doctor and keen plantsman, Dr Charles Stuart (see p. 230, under 'Stuart's Original').

VERNACULAR NAMES. Praeger's heath (English; McClintock 1980b); fraoch Praeger (Irish; Nelson 2001b); Stuarts Heide (German).

CULTIVARS. See p. 230.

27. ERICA × WATSONII. Watson's heath. [*E. ciliaris* × *tetralix*]

This occurs naturally where *Erica ciliaris* and *E. tetralix* grow together, and it has been created by deliberate cross-pollination of these two species by Dr John Griffiths in England (for example, 'Claire Elise' ® no. E.2006:07) and David Wilson in Canada (for example, 'Pearly Pink' and 'Pink Pacific').

Erica × watsonii is known from the wild in England; there are records from Cornwall dating from 1831, from Dorset since 1837, and from the Blackdown Hills on the Somerset-Devon border since 1999 (Edgington 1999). The same hybrid has been recorded in western France, where it was first observed around 1860. In the early 1990s, a record was published of *E.* × *watsonii* from Sierra de la Curiscada, Asturias, northern Spain (Díaz González & García Rodriguez 1992: 135, 140). There is also an unreliable report from Ireland, based on a plant grown during the early 1970s in the "native plants" frame within the nursery at the National Botanic Gardens, Glasnevin, Dublin. This plant was reputed to have been removed from near the small population of *E. ciliaris* in Connemara (Scannell 1976b), but there is no wild-collected voucher to substantiate the claim. A mix-up of labels in the nursery is a much more likely explanation.

The hybrid was recognised, although uncertainly, and named by George Bentham in his monograph on *Erica* for Augustin-Pyramus de Candolle's *Prodromus systematis naturalis regni vegetabilis* (1839). Watson's original specimen, now in the herbarium of the Royal Botanic Gardens, Kew (**K**!), was annotated by Bentham: "It is either a hybrid or a species I have named E. watsoni." Bentham placed *E. × watsonii* under *E. ciliaris*.

Erica × watsonii is the most abundant, naturally occurring hybrid heather, at least in England; for distribution in Dorset, see Chapman (1975) and Chapman & Rose (1994). Watson's heath is distinguished from *E. ciliaris* by two reliable characters: by the presence of spurs on the anthers, and by the presence of hairs on the ovary (capsule) — Dorset heath lacks anther spurs (cf. Hull 2006) and does not have a hirsute ovary (cf. Chapman 1975: 812). From *E. tetralix* it is distinguished less reliably, by the relative length of the small spurs on the anthers, and by having a sparsely pubescent ovary — the cross-leaved heath possesses prominent spurs (cf. Hull 2006) at least half the length of the anther (Chapman 1975), and a very hirsute ovary. There also can be differences in the shape of the corolla and disposition of the flowers: in *E. ciliaris* the flowers are noticeably zygomorphic and are arranged in an elongated raceme, whereas in *E. tetralix* the flowers are in a terminal umbel and are more-or-less actinomorphic. Chapman's (1975) analysis of the hybrid swarms in Dorset relied on nine characters, including, in addition to those already described, the number of leaves in a whorl (usually three in *E. ciliaris*; four in *E. tetralix*), vesture of the abaxial midrib (glabrous in *E. ciliaris*; pubescent in *E. tetralix*), and the anther's surface texture (papillate in *E. ciliaris*; not papillate in *E. tetralix*).

CULTIVATION & VARIATION

Given that both *Erica ciliaris* and *E. tetralix* are known to exist in both "glandular" and "eglandular" forms (see pp. 169 and 234) — "eglandular" plants of *E. tetralix* are infrequent — the hybrids also exists in both states. "Eglandular" clones include 'Mary' (® no. 103), collected in Brittany, and 'H. Maxwell' (Plate 23 and p. 309).

There are also variations in flower colour, which has been exploited in selecting clones for cultivation: 'F. White' is a white with lilac tips; 'Cherry Turpin' (Plate 23 and p. 309) is extremely pale pink; the colour of the flowers of 'Dawn' (see Fig. 199 and p. 384) are rich pinkish-purple (Maxwell 1927: 69).

Erica × watsonii, like its parent species, needs lime-free soil, and plants benefit from light pruning immediately after flowering to stimulate the production of young, coloured shoots.

Erica × watsonii Benth., *Prodromus* **7**: 665 (1839). Type: 'Erica ciliaris β Downs n^r Truro', *H. C. Watson*. **K**! (McClintock 1980a).

ILLUSTRATIONS. S. J. Roles, *Flora of the British Isles: illustrations* pt 2: tab. 936 (1960); M. Blamey, *Heathers of the Lizard*, plate 2 (McClintock & Blamey 1998). Plate 23.

DESCRIPTION. *Very variable evergreen shrub* to 0.5 m tall, with gland-tipped or eglandular hairs. *Shoots* pubescent, with long, spreading cilia, and much shorter, simple hairs. *Internodes* below inflorescence equal, or unequal and lengthening upwards; infrafoliar ridges absent. *Leaves* in whorls of 4, ±3.5 mm long, >1 mm wide towards base, ovate-lanceolate, with long, gland-tipped or glandless hairs on edges, minutely pubescent above, sometimes glabrescent; margin recurved; sulcus open, abaxial midrib glabrous or minutely pubescent. *Inflorescence* an elongated raceme to 4 cm long, or contracted and umbel-like. *Pedicel* curved, pubescent with short and long, simple hairs, and longer, gland-tipped hairs; *bract* leaf-like, at or below mid-point; pair of leaf-like *bracteoles* attached close to calyx. *Calyx* green, lobes unequal, separate, ±2.5 mm long, margin with long gland-tipped hairs, and densely

fringed with short, simple hairs. *Corolla* pink, or white and pink, usually zygomorphic, ±9 mm long, ±6 mm at widest, ovate, glabrous or very sparsely pubescent; bud may have sparse, minute hairs at apex. *Stamens* 8, ±8.5 mm long, included; *filament* ±7.5 mm long, hooked at base, ±straight; *anther* ±1–1.5 mm long, with 2 vestigial spurs, 0.1–0.6 mm long, at junction with filament. *Style* ±7–8 mm long, ±0.5 mm diameter at base, straight or slightly curved, tapering towards apex, glabrous; *stigma* capitate, distinctly broader than style apex; *ovary* with very sparse hairs, especially on top, to pubescent or hirsute all over, ±1.5 mm tall, cylindrical, ribbed; 45–60 ovules per locule.

Erica × *watsonii* and *E.* × *williamsii*: hybrids of wild origin

Painted by Christabel King PLATE 23

LOWER LEFT: *E.* × *watsonii* unnamed clone: the young foliage. (Painted 11 July 1986.)

LOWER CENTRE: *E.* × *watsonii* 'H. Maxwell'. Vigorous, spreading heath (±0.3(–0.45) m tall × ±0.45 m spread); shoots with a few scattered, long, glandless hairs, young shoots orange-bronze turning yellow; mature foliage mid-green; leaves with long, glandless hairs on edges; flowers in terminal umbel-like raceme, mauve (H2 or H11; RHS 70B), ovary pubescent, green, ovules very small and numerous (<60 per locule); July–November.

Like 'Dawn' (p. 384), this cultivar came from the wild, from the heaths near Wareham, Dorset. D. F. Maxwell spotted it, and the cultivar, named after the finder's father, was introduced by Maxwell & Beale Ltd (Broadstone, Dorset) in 1925.

LOWER RIGHT: *E.* × *watsonii* 'Cherry Turpin' (® no. 20). Spreading shrub (±0.2 m tall × ±0.45 m spread); shoots with spreading, long, mainly gland-tipped hairs; foliage grey-green, leaves with gland-tipped cilia on edge, minutely pubescent above, glabrescent; inflorescence a raceme, ±4 cm long; pedicels curved, pubescent with short hairs, and very few longer, gland-tipped hairs; flowers very pale pink (paler than H8/16; RHS 56D), zygomorphic, ±9 mm long, ±6 mm at widest; calyx green, lobes ±2.5 mm long, margin with red, gland-tipped cilia, densely fringed with short, simple hairs; ovary with very sparse hairs, especially on top; June–October.

'Cherry Turpin' was collected by Mrs Turpin, wife of Major-General P. G. Turpin (West Clandon, Surrey), on Carrine Common near Truro, Cornwall, on 27 September 1978. The cultivar, originally mis-identified as *Erica ciliaris*, was introduced by J. Hall (Windlesham Nurseries, Surrey) by 1982. The flowers are described as being the "palest imaginable shade of pink" (Magor 1980: 59).

UPPER LEFT: *E.* × *williamsii* 'Lizard Downs'. Compact shrub (±0.45 m tall × ± 0.5 m spread); foliage grey-green; flowers pink (H8); stamens contorted, clustered grotesquely around ovary; filaments twisted; anthers malformed, without spurs.

This was the sixth recorded plant (see p. 313; Nelson 2007d). It resembles 'Gew Graze' in the grotesque, twisted anthers, but differs in possessing stout, gland-tipped hairs on sepal margins.

UPPER CENTRE–RIGHT: *E.* × *williamsii* 'P. D. Williams' ♀ Vigorous shrub (±0.25(–0.45) m tall × 0.5(–1) m when not pruned) spread); mature foliage green; young shoots yellow; flowers mauve (H2); ovary with sparse hairs mostly on upper half ; stamens ±3.5–4 mm long, included; anthers with small (±0.1 mm long) but prominent spurs. July–November.

This is supposed to be the clone that P. D. Williams found in the early 1900s, but if it is, it differs from published descriptions (Davey 1910, 1911; Turrill 1911; Druce 1911) and illustrations (Davey 1911) in possessing spurs on the anthers (see Nelson 2007d). Davey (1911) also stated that the stamens were "much shorter than the corolla", but this is not so; in his original account he stated that the stamens were "similar to those of [*E.*] vagans, but all included within the corolla" which is an accurate description. Davey's illustration, which is hardly believable, shows that the corolla is distinctly urceolate, with a very prominent constriction beneath the lobes, a character not exemplified by the specimen in his herbarium.

The cultivar name was suggested and published by David McClintock in 1965 for the plant then cultivated "as plain *E.* × *williamsii*" which he stated "match the type specimen" although no such specimen had been designated. He added that it had smaller, darker flowers than 'Gwavas' and posed this question: "Is it known which gathering this ... plant derives from?" (McClintock 1965: 14–15).

No matter about its history, 'P. D. Williams' is a consistently excellent heather. AGM 1993.

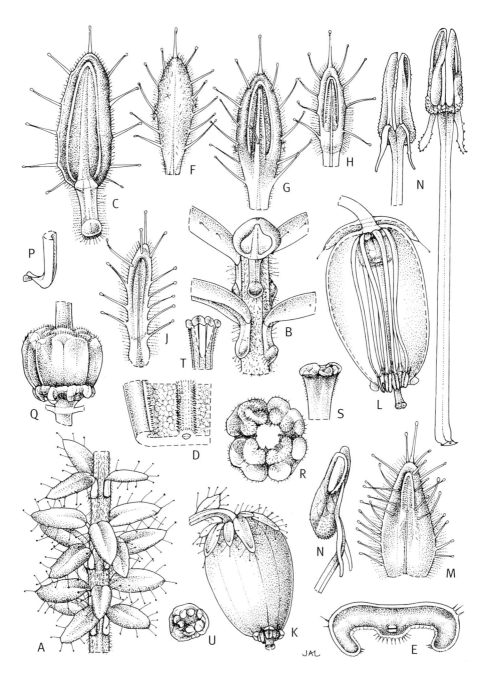

Fig. 199. *Erica × watsonii* 'Dawn'. **A** leafy shoot × 6; **B** leaf nodes × 9; **C** leaf from below × 8; **D** leaf indumentum × 35; **E** transverse section of leaf × 35; **F** bract from above × 12; **G** bract from below × 12; **H** bracteole from back × 12; **J** bracteole from front × 12; **K** flower from side × 5; **L** longitudinal section of flower × 6; **M** sepal × 15; **N** stamens from side, back and front × 20; **P** filament base × 20; **Q** ovary × 12; **R** ovary from above × 12; **S** style apex (stigma) × 12; **T** longitudinal section of style apex (stigma) × 12; **U** style apex (stigma) from above; from specimens grown by RHS Garden Wisley, no. W 1998 1331A and Allen Hall. Drawn by Joanna Langhorne.

FLOWERING PERIOD. July–September.

DISTRIBUTION. Britain, France, Spain; very rare in France and Spain, although perhaps under-recorded. **Maps:** Chapman (1975: Dorset only).

ETYMOLOGY. Eponym, after the person who apparently was the original discoverer, the Yorkshire-born botanist and plant geographer Hewett Cottrell Watson (1804–1881) (see also pp. 82, 181).

VERNACULAR NAMES. Watson's heath (English; McClintock 1980b); Watsons Heide (German); engelsk hybridljung (Swedish).

CULTIVARS. See p. 309 and Appendix I (p. 384).

28. ERICA × WILLIAMSII. Williams's heath. [*Erica tetralix × vagans*]

The history of this natural hybrid is confusing. There are many tangled references to it and several contradictory labels on herbarium specimens.

If we believe Percival Dacres Williams's account, related to Dr W. B. Turrill (1922), he found a plant on the Lizard Peninsula near St Keverne, "about ¾ of a mile to the west [of Trelanvean farm] on plot 2788", which Dr David Coombe worked out contained 4.178 acres and was "⅓ of a mile N. W. of Trelan" (see Nelson 2007d, 2007f). Williams was on the moor shooting partridges on the particular day, the same day that he found the fine cultivar of *Erica vagans* called 'St Keverne'. He took cuttings of both heathers and had a thriving plant of the hybrid in his garden at Lanarth by late August 1911, when he showed both the original wilding and the cultivated offshoot to a trio of visiting botanists, Drs Druce, Graebner and Schröter. That suggests that Williams found the plants at least a few years before 1911.

The locality pinpointed by Coombe does not equate with "Lane between Bochym & Goonhilly Downs" which was inscribed by F. H. Davey on the specimen, dated 3 November 1910, that Williams had given him. According to Davey (1910, 1911), "plants of this hybrid [had been] found ... nearly fifty years ago" by Williams's uncle, Richard Davey MP "but no record of the fact appears to have been published, and the plant remained unnoticed until ... Mr Williams had the good fortune to discover it."

After the 1911 naming of the hybrid (after P. D. Williams), accounts are much more precise. To date twelve plants have been recorded on the Lizard (see Nelson 2007d).

PLANTS FOUND ON THE LIZARD

The tally given here relies principally on the published accounts of Turpin (1982a, 1984) and McClintock & Blamey (1998), but also was checked against notes and correspondence preserved by Major-General Turpin that I have had access to (for more detailed information see Nelson 2007d).

1. c. 1860 — Richard Davey MP: "about fifty years ago" (Davey 1911), considered to be the plant from the "Lane between Bochym & Goonhilly Downs". Probably never propagated.

2. before 1910 — P. D. Williams: plot 2788 "⅓ of a mile N.W. of Trelan", grid reference SW 743193. Propagated by Williams, and in his own garden in late August 1911; the cultivar 'P. D. Williams' is believed to represent this find, but this is open to dispute. My own opinion (cf. Nelson 2007d) is that 'P. D. Williams' is not the living representation of Williams's discovery, but where it came from cannot now be ascertained.

Patrick's account (1964) that Williams "told me he actually found the plant nestling against his garden wall" is usually ignored as incorrect.

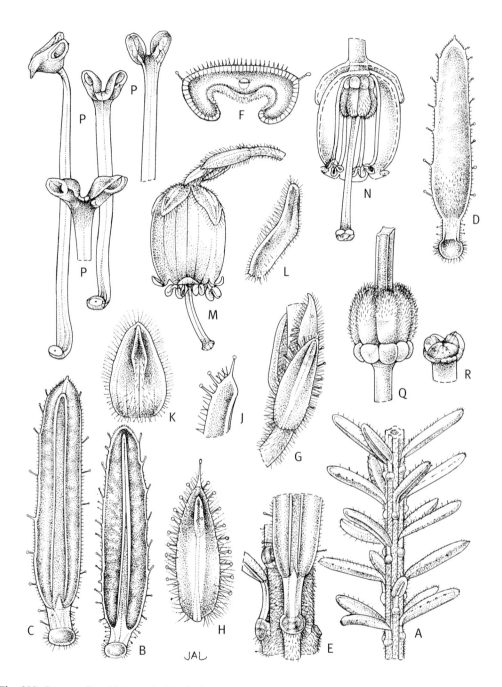

Fig. 200. *Erica × williamsii* 'Gwavas'. **A** leafy shoot × 6; **B** young leaf from back × 12; **C** mature leaf from back × 7; **D** leaf from front × 7; **E** leaf nodes × 7, **F** transverse section of leaf × 32; **G** bract and bracteoles × 21; **H** bract from back × 28; **J** tip of bract from side × 28; **K** sepal × 9; **L** sepal from side × 9; **M** flower from side × 4; **N** longitudinal section of flower × 4; **P** stamens from side, front and back × 32; **Q** ovary × 18; **R** style apex (stigma) × 26; from specimens grown by Allen Hall. Drawn by Joanna Langhorne.

3. c. 1910 "about the same time" as Williams's find — Miss M. B. Gertrude Waterer: near Cadgwith. Propagated and named 'Gwavas', which was the name of the farm where Miss Waterer was staying at the time.

4. 14 October 1924 — Miss M. E. Lavender Williams (daughter of P. D. Williams): near Zoar Chapel and the Three Brothers of Crugith (grid reference SW7519), on Crousa Downs. "It was a miserable plant and did not survive that winter" (Turpin 1982a: 8; McClintock & Blamey 1998). There are specimens in **K** (!) and **OXF** (!).

5. 8 September 1974 — Mrs Jean Paton, Mrs Barbara Garrett and Mrs J. Yorston (McClintock & Blamey 1998): Gew-graze (grid reference SW 6714). Propagated and named after the locality where it was found, but the original plant was destroyed by ploughing in April 1976.

6. 20 August 1975 — Dr Alan Leslie and R. H. W. Bradshaw: above Kynance Cove (grid reference SW6813), on Lizard Downs. Propagated and named 'Lizard Downs'. An "enormous" plant, covering about 3.25 sq metres in July 1976. This was burnt on 5 November 1989 but recovered and bloomed again in August 1990.

7. 22 June 1976 — Dr David Coombe and Peter Tinning: near Penhale in the Lizard National Nature Reserve (grid reference SW6918). Propagated and named after Dr Coombe. The original plant was burnt in 1976 but was re-established from cuttings in 1979.

8. 5 October 1981 — Mrs Cherry Turpin: near Croft Pascoe Pool (grid reference SW7319), the Lizard National Nature Reserve, beside the road to Traboe Cross. Propagated and named 'Croft Pascoe' (Nelson 2007d: 51, Fig. 3). The original survived (Fig. 201) until at least 1990 but could not be found in 1999.

Fig. 201. *Erica × williamsii* 'Croft Pascoe'; found in 1981 in the Lizard National Nature Reserve beside the road to Traboe Cross (E. C. Nelson).

9. 30 October 1983 — Dr Marion G. B. Hughes: due south of Leech Pool, near Goonhilly Earth Station (grid reference SW 718208) (Turpin 1985). Propagated and named 'Marion Hughes'.

10. 18 August 1984 — Andrew Byfield and Dr Marion Hughes: at Cow-y-jack (grid reference SW7719), on gabbro, near Polcoverack, not far from St Keverne (Turpin 1985). Propagated and named 'Cow-y-jack'.

11. 2 June 1990 — Dr David Coombe with Andrew Byfield and members of the BSBI: by a cattle track, between Kynance Farm and Kynance Cove (SW6813) on land recently added to the Lizard National Nature Reserve (Turpin 1991).

12. 9 September 1999 — Mrs Jean Julian (*olim* Mrs Jean Preston): at Kynance Cove (SW6813). Propagated and named after the finder.

OTHER PLANTS, OF GARDEN ORIGIN

There is evidence in Major-General Turpin's notes and herbarium that at least two other clones were propagated and in cultivation in the 1970s and 1980s. One was perhaps a chance seedling in the West Byfleet garden of F. B. Rice: a photograph of this exists (reproduced in Nelson 2007d: 54. Fig. 6) and is annotated "Five cuttings taken 22.7.78.", but no published reference to it has been traced. A specimen in Turpin's herbarium (now part of The Heather Society herbarium, deposited in **WSY**) is annotated "From Mrs T. Forbes Plaxtol Aug. 1986", but, again, no other information has been traced.

At present there are three other named cultivars in circulation, each the result of deliberate cross-pollination: 'Gold Button' was raised by Dr John Griffiths (Yorkshire, England) using *E. tetralix* 'Alba Mollis' with pollen from *E. vagans* 'Valerie Proudley'; 'Ken Wilson' (Fig. 202), raised by David Wilson (British Columbia, Canada), is a selected seedling from *E. tetralix* 'Hookstone Pink' pollinated by *E. vagans* 'Mrs D. F. Maxwell' (see p. 131); 'Phantom' (Nelson 2007d: 58), a white-flowered clone, was also raised in 1986 by David Wilson, from *E. tetralix* 'Alba Mollis' pollinated by *E. vagans* 'Lyonesse'.

Fig. 202. *Erica × williamsii* 'Ken Wilson', an artificial hybrid created in British Columbia, Canada, by David Wilson; in cultivation at The Bannut, Herefordshire (E. C. Nelson).

CULTIVATION

The cultivars of *Erica × williamsii* can be grown in most types of soil. They are also hardy.

Erica × williamsii Druce, *Gard. Chron.*, series 3, **50**: 388 (2 December 1911). Type: 'Erica vagans × cinerea … Lane between Bochym & Goonhilly Downs … Cornwall', 3 November 1910, *P. D. Williams*: lecto: **TRU** (Nelson 2007f).

ILLUSTRATION. M. Blamey, *Heathers of the Lizard*, plate 2 (McClintock & Blamey 1998). Plate 23.

DESCRIPTION. *Variable evergreen shrub*, each clone having distinctive characteristics; capable of regeneration from rootstock and persisting for several decades. *Shoots* erect to spreading, pubescent with very short, simple hairs and stouter, sometime gland-tipped hairs. *Leaves* in whorls of 4, ± spreading, linear-lanceolate to elliptical, ± 5 mm long, ±1 mm across at broadest part, edges with ± evenly-spaced stout cilia, often with glandular tips, and sometimes also much shorter, simple, non-glandular hairs, leaf-tip with single glandular, stout hair, upper surface flat, variably pubescent, lower surface usually glabrous, sulcus tapering, margins glabrous. *Flowers* on axillary shoots, usually in terminal umbels; *pedicels* ±4 mm long, variably hirsute with simple hairs; position of bract and bracteoles variable, to 2.5 mm long, ± 1.5 mm across, margins densely hirsute with short, simple, non-glandular hairs and longer, stouter cilia often with glandular tips, sulcus prominent, ±0.5 mm long; bracteoles usually 2, about ²/₃ way up pedicel. *Calyx* (3)4-lobed, lobes fused at base for a short distance, broadly lanceolate, to 2.5 mm long, ±1.5 mm broad, resembling bract and bracteole, margins densely hirsute with short, simple, non-glandular hairs and longer, stouter cilia often, but not always, with glandular tips (in one clone there are dendritic hairs towards the tips of the sepals), sometimes with some short, simple hairs at or below sulcus on outer surface, sulcus prominent, ±0.5 mm long. *Corolla* 4.5–5 mm long, globose–urceolate, with (3)4 erect lobes, glabrous except for occasional very sparse short, simple hairs on outer surface of lobes, lobes ± ¼ length of corolla. *Stamens* 8, erect, usually not exceeding corolla and thus included, but stamen length is very variable; *filament* straight, glabrous, sometimes fused in pairs; *anthers* very small, resembling those of *E. vagans*, often malformed and contorted, thecae dark, spreading, free, spurs absent or present; when present, very small, ±0.1 mm long, ± linear. *Style* erect, usually exceeding the corolla, to 5.5 mm long, glabrous, tapering; *stigma* broader than tip of style; *ovary* >1.5 mm long, usually sparsely pubescent with short, simple hairs scattered especially in upper part.

FLOWERING PERIOD. July–October.

DISTRIBUTION. Apparently endemic to Britain, and confined to the Lizard Peninsula, Cornwall: **Map:** Nelson 2007d: 53 (Fig. 5; updated from Turpin 1982a).

Erica × williamsii could also exist in France and Spain, where the parent species co-exist.

ETYMOLOGY. Eponym, after Percival Dacres Williams (1865–1935), of Lanarth, St Keverne, Cornwall, plantsman with a special interest in daffodils.

VERNACULAR NAMES. Williams's heath (English; McClintock 1980b); Lizardheide (German); parasolljung — "completely inexplicable" (B. Johansson, pers. comm., 28 July 2004) (Swedish).

CULTIVARS. See p. 309 and Appendix I (p. 385).

PART II. Chance hybrids from British gardens & nurseries
Erica × darleyensis, E. × griffithsii, E. × veitchii

29. ERICA × DARLEYENSIS. Darley Dale heath. [*E. carnea × E. erigena*]

As already stated, it is extremely difficult to distinguish "with the naked eye" this hybrid from its parent species. This is hardly surprising given that there is no precise, strictly morphological character that separates the parents. With the aid of a good microscope and chemicals to stain living tissue, a diligent person could examine hundreds of pollen grains to determine the percentage fertility: a large proportion of infertile, shrivelled grains seems to indicate a hybrid (cf. Jones 1979). However, there is accumulated evidence that hybrids between the northern *Erica* species have one rather particular attribute that is apparent only when there is new growth in the spring — the juvenile shoots are not pure green but are variously discoloured, tinged cream or yellow or red or pink or orange or even white (Griffiths 1987; Kramer 1991; Small & Alanine 1994). This is not a universal characteristic of known hybrids, nor is it infallible; nor is the presence of shrivelled pollen grains dependable; but these characters together are most suggestive.

This hybrid cannot arise in the wild; the geographical ranges of the parent species do not overlap. In fact, the two are separated by about 600 kilometres, the distance between the population of *Erica erigena* in the Gironde and *E. carnea* in the Alps of south-eastern France (see p. 118).

Reports that *Erica × darleyensis* is a naturalised plant — that is, one "introduced by human agency ... but now maintain[ing itself] without further intervention" (Rackham 2003: 19) — in southern England (Preston *et al.* 2002) need to be treated with considerable scepticism. Some of the records derive from places such as a military cemetery where this exotic heather was once deliberately cultivated (Nelson 2004a). Furthermore, being a hybrid, *E. × darleyensis* is usually sterile, although there is anecdotal evidence that some clones of this hybrid are capable of producing viable seeds.[3]

In the past, the hybrid was only achieved by chance when the parents grew together in a garden or a nursery, but in more recent decades, deliberate cross-pollination has been employed to produce *Erica × darleyensis*.

The first of the hybrid clones was recognised only in the 1890s in the Derbyshire nursery of James Smith & Son at Darley Dale and was marketed under such invalid and inapplicable names as *Erica* "*mediterranea hybrida*" and *E.* "*hybrida darleyensis*".[4] In 1914, the Darley Dale

Fig. 203. *Erica × darleyensis* 'J. W. Porter', raised by James Walker Porter and selected by his wife, Eileen; cultivated in Outwell, Norfolk (E. C. Nelson).

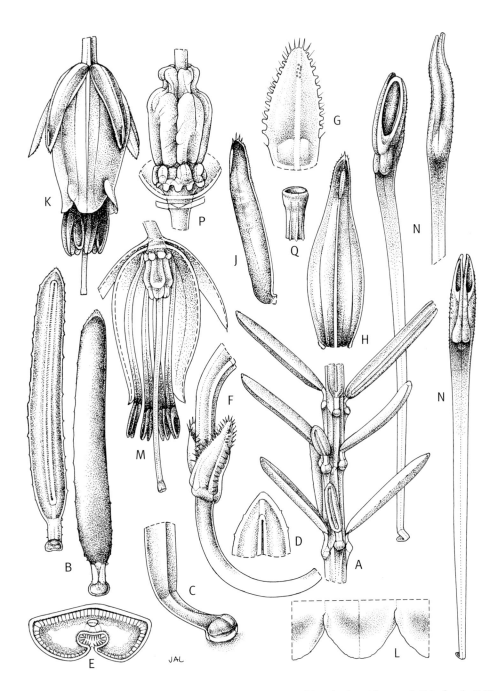

Fig. 204. *Erica* × *darleyensis* 'Silberschmelze'. **A** leafy stem × 4; **B** leaf from back and front × 8; **C** leaf node × 11; **D** leaf tip × 16; **E** transverse section of leaf × 33; **F** pedicel with bract and bracteoles × 22; **G** bracteole × 36; **H** calyx lobe × 18; **J** side view of calyx × 18; **K** flower from side × 8; **L** corolla lobes × 14; **M** longitudinal section of flower × 8; **N** stamens from side, back and front × 18; **P** ovary × 22; **Q** style apex (stigma) × 10; from specimens grown by Allen Hall. Drawn by Joanna Langhorne

hybrid was given its own binomial by William Jackson Bean (1863–1947): *E. × darleyensis*. Until the 1930s, the Derbyshire original was the only clone known but since then, many others have become available, most arising as chance seedlings in gardens and nurseries. The first to be given a cultivar ("fancy") name was 'George Rendall' in 1935. This was followed in the early 1950s by 'Arthur Johnson'. Both are still highly regarded (see below). The original clone was not named 'Darley Dale' until the mid-1960s (McClintock 1965; Brickell 1969). As of June 2008, 50 cultivar names have been accepted by The Heather Society, acting as International Cultivar Registration Authority; 16 of these are formally registered.

Among the better clones are several that were produced by James Walker Porter (1889–1963), a native of Belfast and a chemical engineer by training (Nelson 1984a, 1986b, 2007b). With a passion for growing tree and shrubs, especially heathers, from seeds, Porter raised seedlings that clearly were of hybrid origin, and from these came 'J. W. Porter', 'Jenny Porter' and 'Margaret Porter'. The parentage of these is not recorded; we do not even know whether the seeds came from *Erica carnea* or *E. erigena*.

Early in 1972, a simple experiment to try to confirm that *Erica × darleyensis* was the product of *E. carnea* crossed with *E. erigena* was conducted by Mrs Ann Parris in her garden at Usk in Monmouthshire. She attached sprigs of *E. carnea* 'Springwood Pink', in full bloom, to the flower-laden shoots of *E. erigena* 'W. T. Rackliff' and 'Brightness'. Seeds she harvested from the *E. erigena* were sown in January 1973, and Mrs Parris raised three plants from *E. erigena* 'W. T. Rackliff' and four plants from *E. erigena* 'Brightness'; all displayed discoloured young shoot tips, an indication of a hybrid (Parris 1976, 1980). None of these seedlings was worth naming, but a later attempt to cross-pollinate *E. erigena* with *E. carnea* 'Myretoun Ruby', yielded *E. × darleyensis* 'Mrs Parris's Red'.

Since then, deliberate cross-pollination of selected clones has been successfully undertaken and a handful of new cultivars selected and introduced. Kurt Kramer in Germany, using *E. carnea* 'Myretoun Ruby' and *E. erigena* 'Brightness', produced the spectacular 'Kramers Rote' (see below). In Canada, David Wilson raised the golden-foliaged 'Goldrush' (® E.2004:03) from *E. carnea* 'Golden Starlet' × *E. erigena* 'Brian Proudley', and the salmon-pink blossomed 'Irish Treasure' (® no. 215) from *E. carnea* 'Treasure Trove' × *E. erigena* 'Irish Dusk'. Dr John Griffiths's first hybrid seedlings were *E. × darleyensis* (see below, p. 333) but none was considered worthy of introduction.

CULTIVATION AND VARIATION

The Darley Dale heath is one of the easiest heathers to grow, tolerant of lime-rich as well as neutral and acidic soils, and floriferous.

Clones with white flowers have been designated *E. × darleyensis* f. *albiflora* D. C. McClint. (McClintock 1984a: 191): the type specimen came from 'Silberschmelze' (see p. 376). Plants with yellow foliage have been assigned to *E. × darleyensis* f. *aureifolia* D. C. McClint. (McClintock 1988a: 62): 'Jack H. Brummage' (see p. 374) yielded the type specimen for this *forma*.

Erica × darleyensis Bean, *Trees & shrubs hardy in the British Isles* 2: 521 (1914). Type: "Erica mediterranea hybrida (Smith, Darley Dale. 1895). Kew, 28 May 1897. [From Arboretum Herbarium] (E. carnea × E. mediterranea) Erica × darleyensis Bean Type!", *anonymous*: lecto.: **K**! (Nelson 2008d).

DESCRIPTION. *Evergreen shrub* to 1 m tall, branches ascending; shoots with some minute, simple hairs; with infrafoliar ridges. *Leaves* like *E. carnea*; ± glabrous, linear, petiole glabrous. *Flowers* pink to mauve, or white, solitary, axillary on very greatly reduced side-shoots; *pedicel* curving, ±3 mm long, glabrous, red, not markedly enlarged below calyx, <0.4 mm diameter; bract and 2 bracteoles at the mid-point together, bract ±0.7 mm long, triangular. *Calyx* truncate at base, segments fused at the base into a cup, waxy, thicker in texture than corolla, lobes ±4 mm long, ±1 mm wide, narrow, lanceolate, glabrous except for a few spicules at apex, sulcus ±0.7 mm long, obscure, apex slightly thickened. *Corolla* ±4.5–5.5 mm long, ±3 mm diameter below middle, ovoid, glabrous, lobes erect, rounded, margins ± smooth, ±1 mm long, ±1.5 mm wide at base, overlapping alternately. *Stamens* 8, ±7 mm long, anthers partly emergent; *filament* ± 5 mm long, ±0.5 mm wide, strap-shaped, not narrowing below anther, erect, straight except for hooked base, attached under nectary; *anther* ±1.5 mm long, ±0.3 mm broad, elliptical, pore ±0.8 mm long, surface with rounded protruberances visible under microscope. *Style* ±7 mm long, slender, not markedly tapered, *stigma* not broader than style apex; *ovary* to 1.5 mm tall, ±1 mm diameter, cylindrical, "stepped" above with narrow upper part, glabrous, nectary lobes prominent below; ribbed, with ±10 ovules per locule.

FLOWERING PERIOD. (November–) January–May.

ETYMOLOGY. Toponym, from Darley Dale.

VERNACULAR NAMES. Darley Dale heath (English); Bruchheide, Englische Heide (German); snöljung (Swedish).

CULTIVARS. See p. 113 and Appendix I (pp. 373–376).

30. ERICA × GRIFFITHSII. Griffiths's heath. [*E. manipuliflora × vagans*]

Griffiths's heath arose spontaneously in Britain sometime before 1949. We do not know where or when, but it may have been in the Royal Botanic Gardens, Kew. As noted elsewhere, *Erica manipuliflora* (whorled heath; see p. 140) is found around the north-eastern periphery of the Mediterranean Sea, on the coasts of the Aegean and Adriatic Seas, but does not occur west of Apulia in the "heel" of Italy or the Dalmatian coast of Croatia. *E. vagans* (Cornish heath; see p. 129) has its headquarters in western Europe, ranging from the French Pyrenees westwards into northern Spain, and north towards Brittany, with outlying colonies in Cornwall and Northern Ireland. The species do not co-exist anywhere in the wild; the distance separating them is at least 1,200 kilometres.

In 1981, John Griffiths began to experiment with heaths, cross-pollinating two he grew in his garden at Garforth, Leeds, West Yorkshire, England (Griffiths 1985, 1987, 2008; Small & Wulff 2008). His first attempts failed, but the following year he tried again and succeeded in raising several yellow-foliaged seedlings of *Erica × darleyensis*, from *E. carnea* 'Foxhollow' and *E. erigena* 'Irish Dusk'. Although none of these was worth introducing, he was spurred on to try other crosses between other hardy species, reckoning at that time that there were at least 300 possible combinations (reciprocal crosses included) between the 18 species of *Erica* then recognised as occurring in Europe. Since the early 1980s, he has, indeed, tried at least 75 different combinations of the hardy northern heathers, employing a dozen species as pollen-parents and a dozen different seed-parents.

As a seed-parent *Erica vagans* has only yielded hybrid seedlings when pollinated by *E. manipuliflora*; with four other species Griffiths's results were equivocal or the seedlings perished before they could be assessed. The reciprocal cross, applying *E. vagans* pollen to *E. manipuliflora*, also produced plants that were clearly hybrids. *E. manipuliflora* is as grudging: as a pollen-parent it produced hybrid seedlings for Griffiths only with *E. tetralix* and, as noted, *E. vagans*, and as a seed-parent it was successful only with *E. vagans*. (Kurt Kramer's and David Wilson's breeding experiments have given similar results.)

Fig. 205. *Erica × griffithsii* 'Heaven Scent'; cultivated in The Bannut, Bringsty (E. C. Nelson).

Dr Griffiths's seedlings came from *Erica manipuliflora* 'Aldeburgh', the only clone of whorled heath generally available in Britain before the late 1980s, crossed in 1983 with *E. vagans* 'Valerie Proudley'. Two cultivars came from this pioneering cross, both with yellow foliage: 'Valerie Griffiths' (® no.64), named after his wife, and 'Ashlea Gold' (see below).

It is clear that *Erica × griffithsii* has arisen spontaneously more than once. As well as the probable Kew seedling now called 'Heaven Scent' (Fig. 205), a vigorous cultivar named 'Elegant Spike' has been in cultivation since the early 1960s. This was raised by P. G. Zwijnenburg (Boskoop, Netherlands) from seed of *E. manipuliflora* received from an unspecified botanical garden. Zwijnenburg introduced this heather, which bears pale rose-pink flowers, under the name *E. vagans* 'Elegant Spike' in 1975.

In the late 1990s, a fourth cultivar of *Erica × griffithsii* was introduced: 'Jacqueline', a sport from 'Heaven Scent' of uncertain origins, is also a vigorous plant and produces long spikes of fragrant, cerise (H6) flowers. C. Kampa (Chobham, Surrey, England) sold it as an unnamed heather to a small number of landscape gardeners before the clone came to the notice of John Hewitt (Summerfield Nursery, Frensham, Surrey), who recognised its merits and named it after his daughter.

CULTIVATION

Griffiths's heath is hardy, tolerating snow and lengthy periods when the air temperature was close to 0°C (see Hall 2002). It is also tolerant of lime (Small & Small 2001).

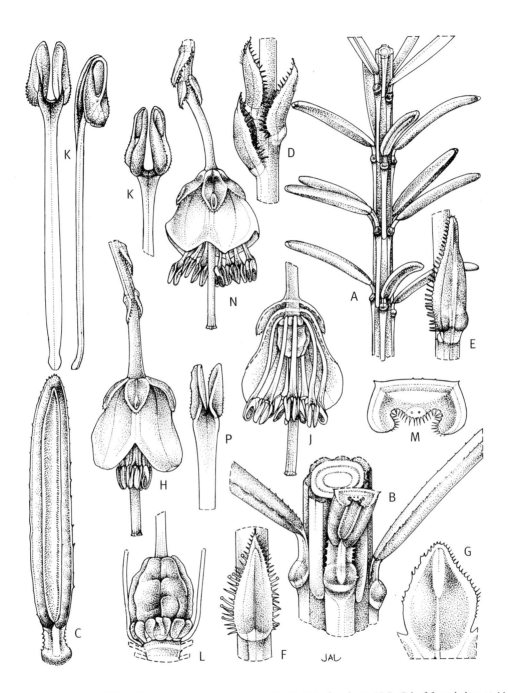

Fig. 206. *Erica × griffithsii* 'Heaven Scent'. **A** leafy shoot × 4.5; **B** leaf nodes × 17.5; **C** leaf from below × 11.5; **D** bract and bracteoles × 4.5; **E** bract × 7; **F** bracteole × 8.5; **G** calyx from back × 30; **H** flower from side × 12; **J** section through flower × 17; **K** stamens from back, side and front × 24; **L** ovary from side × 25. *E. griffithsii* 'Valerie Griffiths'. **M** section through leaf × 24; **N** flower from side × 10; **P** stamen from back × 18; from specimens grown by Allen Hall. Drawn by Joanna Langhorne.

Erica × griffithsii[5] D. C. McClint., *New Plantsman* 4 (2): 84 (June 1997). Type: 'Valerie Griffiths', "cult in horto 9 Ashley [*sic*] Close, Garforth, Leeds", 12 September 1988[6], *J. Griffiths & D. McClintock*: holo. **WSY**! (Miller & Nelson 2008).

ILLUSTRATION. 'Valerie Griffiths': *The new plantsman* 4 (2): 84 (1997). Plate 10.

DESCRIPTION. *Evergreen shrub* to 0.5 m or more tall; *shoots* erect or spreading, ribbed with pronounced infra-foliar ridges that extend at least to the next whorl of leaves, minutely and sparsely pubescent. *Leaves* in whorls of 4, variously ascending (±45°) or spreading horizontally, to 8 mm long, less than 1 mm wide, linear-lanceolate, glabrous or upper surface with forward pointing microscopic spicules at tip and in region of mid-rib, edges plain or with uneven, translucent, cartilagenous border and forward pointing spicules; margins recurved underneath, glabrous, sulcus linear and closed to elliptical and opening; petiole 0.5–0.7 mm long, flattened above, convex underneath, with minute hairs on edges. *Flowers* in axillary umbels, on short leafy shoots or ± sessile, appearing to form a long, tapering raceme, lowermost umbels usually opening first; leaflets on axillary shoots very short; umbels with 1–4 flowers; *pedicels* glabrous, slender, straight or curved, erect or ascending, ±6 mm long, ±0.3 mm diameter, green or red; with 1 bract and (1 or) 2 bracteoles near base or lower quarter of pedicel; bract ±1 mm long, linear-triangular, margins with microscopic hairs, some ±gland-tipped; bracteoles smaller, opposite when 2 present. *Calyx* 4-lobed, truncate at base; variously coloured green or pink; lobes sometimes of unequal width, ±1.3 mm long, broadest lobes ±0.7 mm across, margins fringed with microscopic hairs, sulcus ±0.5 mm long, apex usually thickened. *Corolla* campanulate-cyathiform, sometimes almost square in cross-section, ±3 mm long, lobes erect, ±1 mm long, very broad at base (to 2.5 mm), glabrous. *Stamens* 8, erect, fully emerged, 3.5–4 mm long; *filament* ±3 mm long, straight or somewhat crinkled, hooked at base and attached underneath nectary, ± even in width, glabrous; *anther* ±0.5 mm long, ovate in lateral view with pore towards apex, *thecae* dark, separate, joined only at top of filament, erect or divergent, covered with minute spicules; *pollen* present but most tetrads shrivelled and infertile (*fide* Jones 1987: 53). *Style* 4–5 mm long, straight or curved, erect or slanted, tapering slightly towards apex; *stigma* capitellate, slightly wider than style apex; *ovary* mostly glabrous but usually with some microscopic hairs on top, ±1.5 mm tall, barrel- or pear-shaped, with prominent, lobed *nectary* encircling base; ovules 22–28.

FLOWERING PERIOD. August–October.

ETYMOLOGY. Eponym (McClintock 1997a); after John Griffiths, Professor of Functional Dye Chemistry, University of Leeds.

VERNACULAR NAMES. Griffiths's heath (English); Griffiths Heide (German).

CULTIVARS. See p. 131 and Appendix I (p. 378).

31. ERICA × VEITCHII. Veitch's heath. [*E. arborea* × *E. lusitanica*]

Although the ranges of the Portuguese heath, *Erica lusitanica*, and the tree heath, *E. arborea*, overlap on the Iberian Peninsula, no incontestable reports of wild-occurring hybrids are known. The late B. E. Smythies (McClintock 1989a) suggested that places worth searching are Picoto in the Serra de Monchique, Portugal, and Huelva province in south-western Spain.

Recently, Fagúndez (2007) has identified several sheets of herbarium specimens, collected by J. J. Lastra during April 2002 at Candamo, Asturias, in northern Spain, as *Erica × veitchii*. The locality lies outside the range of *E. lusitanica*, its westernmost habitat being at Río España, near Peón, Asturias, about

40 km to the east (Lainz 1959). Fagúndez (2007) proposed that these represent a new nothosubspecies, but the name he suggested *E. × veitchii* nothosubsp. *cantabrica* was not accompanied by a Latin diagnosis or description so was not validly published. The specimens do represent an anomalous plant, according to Fagúndez (2007), lacking plumose hairs but possessing glabrous filaments, characters that distinguish *E. arborea* from *E. lusitanica*. However, all the specimens of *E. × veitchii* that I have examined from British gardens have minute hairs on the lower half of the filaments and plumose hairs scattered among the simple hairs on the shoots. Fagúndez (2007) also reported that the pollen was normal, whereas Bean (1905) noted that the original clone *E. × veitchii* produced malformed tetrads. Other quantitative characters quoted by Fagúndez (2007) are not absolutely reliable for separating *E. arborea* and *E. lusitanica*. Until fresh material of this anomalous heather from Asturias can be obtained and propagated for close, long-term comparison with the clones of *E. × veitchii* established in gardens, the evidence is too equivocal to accept that this hybrid has arisen in the wild, especially in a locality from which one of the parents seems to be absent at the present time. (In the northern hemisphere, *Erica* hybrids invariably occur, or arise in gardens, in very close proximity to *both* parents.)

Erica arborea and *E. lusitanica* have long been cultivated in Britain, Ireland and other parts of north-western Europe, so their accidental cross-pollination might be anticipated. That this has occurred more than once seems proven by the fact that different plants identified as clones of *E. × veitchii* have been introduced into cultivation in both Britain and the Netherlands. The exact parentage of these clones is in all cases unknown. The earliest example came from Veitch & Son's nursery at Coombe Wood, Exeter, Devon, and must have arisen as a seedling in the late 1800s; pressed specimens and at least one living plant were sent to the Royal Botanic Gardens, Kew, in 1902. A cultivar, originally named *E. arborea* 'Gold Tips', introduced before 1948 by Maxwell & Beale, Broadstone, Dorset, has since been determined as another example of *E. × veitchii* (Turpin 1979). It is notable in having bright green foliage, and the new growth in springtime is golden. The cultivar named 'Pink Joy' was found in a batch of seedlings at Boskoop in the Netherlands by M. Zwijnenburg, and it, too, was originally identified as *E. arborea*, because the seed has been collected from that species. Most recently, in 2006 Allen Hall identified another clone, which he had purchased under the name *E. arborea* 'Alpina' from a London nursery in the 1980s; this cultivar is named 'Westbourne Grove' (® E.2006:02) (see Registrar 2007: 71).

HISTORY

This hybrid was first brought to public notice at a meeting of the Royal Horticultural Society's Floral Committee which took place on 14 February 1905 (Anonymous 1905b[7]). It was mentioned by name within the report printed in *The gardeners' chronicle* on 18 February: "A plant nearly 4 feet high and as much in diameter ... . The flowers are white and the habit of the plant is more like that of E. arborea than E. codonodes [= *E. lusitanica*]." The heather was also briefly described within a report of the same Floral Committee meeting in *The garden* on 25 February: "... a very remarkable plant, with white bell-shaped fragrant flowers ..." (Anonymous 1905c[8]).

In the issue of *The gardeners' chronicle* published on 4 March 1905, W. J. Bean (1905) described the plant in more detail, noting characteristics that were intermediate between those of its putative parents, *Erica arborea* and *E. lusitanica* (syn. *E. codonodes*). Bean's note is deemed to constitute the protologue for *E. ×veitchii*. While no specimen was cited by Bean (1905), he did record that this plant had been "... sent to Kew a few years ago to be named", and that Messrs Veitch & Son had "presented a plant to Kew, and this has grown and flowered well." Clearly Bean had seen fresh

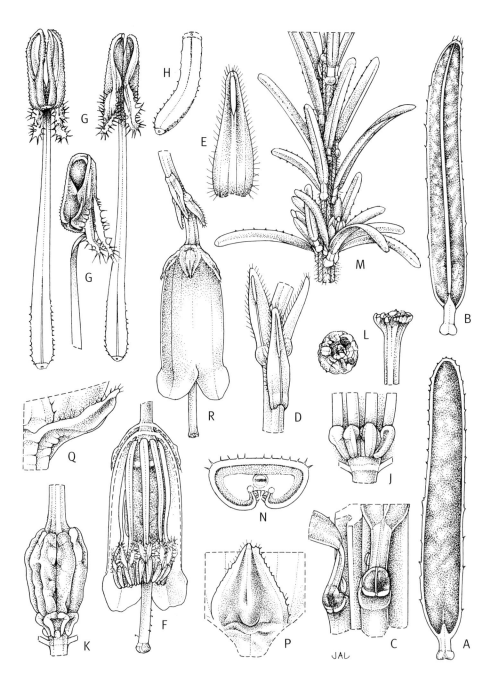

Fig. 207. *Erica × veitchii* 'Pink Joy'. **A** leaf from above × 12; **B** leaf from below × 12; **C** leaf nodes × 20; **D** bract and bracteoles × 15; **E** bracteole × 10; **F** longitudinal section of flower × 10; **G** stamens from front, back and side × 21; **H** filament base × 21; **J** nectary lobes × 12; **K** ovary × 8; **L** style apex (stigma) from above and side × 15; A–L drawn from RHS Garden Wisley, no. W 1991 1360A. *E. × veitchii* 'Exeter'. **M** leafy shoot × 6; **N** transverse section of leaf × 46; **P** calyx lobe × 20; **Q** calyx lobe from side × 20; **R** flower from side × 7; M–R from specimens grown by Allen Hall. Drawn by Joanna Langhorne.

specimens both at the RHS and at Kew, and it is likely his descriptive notes were written mainly using fresh material.

There are two specimens in the herbarium at the Royal Botanic Gardens, Kew (**K**!), that are associated with both Bean and Messrs Veitch & Son, and which could be candidates for selection as the type. A specimen collected by Bean in the Arboretum, Royal Botanic Gardens, Kew, on 27 March 1905, has to be ruled out as it would not have been available to him when he wrote the note published a fortnight earlier. The remaining specimen, from Veitch & Son's Exeter nursery, received at Kew on 9 April 1902, was available to Bean and has been designated as the lectotype.

In a letter to W. J. Bean, dated 5 January 1905 (attached to the herbarium sheet), Peter Veitch made the following comment: "The name I had thought for it is E Veitchii and if you approve I will stick to that." Bean (1905) did not claim the name as his, stating that "Messrs R. Veitch & Son, of Exeter, exhibited under this name a shrub ...". Given this, the authority may be expanded to Hort. R. T. Veitch ex Bean, in accord with *ICBN (Vienna code)* Art. 46.4 (McNeill *et al.* 2006).

CULTIVATION

Erica × veitchii can be cultivated in most soils and is relatively hardy, but the cultivars vary in their hardiness. The shrubs bloom during the spring in Britain; again, the different clones vary in their periods of maximum blossom.

Erica × veitchii Hort. R. T. Veitch ex Bean, *Gard. Chron.*, series 3, 37: 138 (4 March 1905). Type: described from cultivated specimens: "N°. 2. from the same Variety as the one sent last week for comparison ... Possibly a hybrid between E. arborea [and] E. lusitanica" [from nursery of Messrs R. Veitch & Son, Exeter], 8 April 1902, *P. C. M. Veitch*: lecto. **K**! (Nelson 2008d).

ILLUSTRATIONS. *Gard. Chron.*, series 3, 37: 228 (15 April 1905). Plate 15.

DESCRIPTION. *Variable evergreen shrub* to 4 m or more tall; represented in cultivation by at least 5 clones (cultivars). *Shoots* hirsute with both simple and plumose hairs intermixed, sometimes some hairs with glandular tips. *Leaves* linear, ±6 mm long, ±0.5 mm wide, usually sparsely hirsute; some clones have sparse, simple hairs near base and spicules on edge, while others have simple and gland-tipped hairs on the edges; margins recurved and contiguous underneath, sulcus linear, closed; *petiole* flattened, with short hairs on edges. *Inflorescence* apparently a panicle, comprising whorls of (2–)3–4 flowers in terminal clusters on long leafy shoots or on very short axillary shoots. *Flowers* white when mature, but some clones have pink buds. *Pedicel* curving, ±3 mm long, glabrous; bract about $^1/_4 - ^1/_3$ of the way along, 2 bracteoles higher, $^1/_3 - ^1/_2$ way along; *bract* ±1 mm long, margins with short, simple hairs, some with glandular tips; *bracteoles* ±0.75 mm long, similar to bract. *Calyx* green, cyathiform, base truncate, usually markedly wider than apex of pedicel, waxy, thicker in texture than corolla, glabrous, ±1.5 mm long, with 4 triangular lobes, margins with few to numerous short, simple hairs. *Corolla* tubular-campanulate to narrowly ovate, ±4 mm long, with 4 erect, rounded lobes, ±1.5 mm long. *Stamens* 8, included or tips just visible at base of corolla lobes, ±3 mm long; *filament* with spicules on lower quarter, broadest at base then narrowing and curving upwards with pronounced sigmoid bend towards apex; *anther* thecae dark crimson when immature, ellipsoid, ± 1–1.3 mm long, pore lateral, narrowly elliptical, ±0.5 mm long; 2 spurs at junction of filament, ±1 mm long, narrow, triangular, irregularly hairy, pointing downwards. *Ovary* 1–1.7 mm long, usually green, glabrous, apex emarginate and style base recessed; *style* 2–3.5 mm long, included or emergent, terete, ±0.3 mm in diameter, straight or curving, widening at tip; *stigma* cup-shaped, about twice width of style, with 4 prominent bosses.

FLOWERING PERIOD. (November–) January–April (–May).

ETYMOLOGY. Eponym: after the Veitch family which established the nursery in Exeter, Devon, England, where this hybrid was first noticed.

VERNACULAR NAMES. Veitch's heath (English; McClintock 1980b); Veitchs Heide (German).

CULTIVARS. See p. 199 and Appendix I (p. 384).

PART III. The "wilderness of uncertainty": artifical hybrid heaths

As just recounted, naturally occurring inter-specific hybrids exist within populations of both the northern hardy heaths (pp. 301–313) and the southern-hemisphere Cape heaths (see, for example, Oliver 1986). There are, or were during the nineteenth century, an uncountable number of artificial hybrids, too, the vast majority involving only Cape heaths (Nelson & Oliver 2005a).

To recapitulate briefly, the earliest of the natural hybrids of European origin to be recognised was *Erica* × *watsonii* (see p. 306): George Bentham described it in 1839 based on a gathering from Cornwall. The two other natural hybrids, *E.* × *williamsii* and *E.* × *stuartii* (as *E.* × *praegeri*), were both reported in the early 1900s. *E.* × *williamsii* (see p. 310) was described from Cornwall and has never been reported outside that county. In 1912, the Danish botanist Carl Ostenfeld described and named *E.* × *praegeri* from the west of Ireland (see p. 301). These three hybrids remain the only ones verified in natural habitats in Europe, although occasional claims for others (for example, *E.* × *veitchii*; see p. 322) are found in the botanical literature, and all three have been re-created by deliberate cross-pollination of the parents.

The other verified hardy *Erica* hybrids (cf. Nelson 1999c) all originated in cultivation, either by chance when two species that do not grow together in the wild were accidentally cross-pollinated in a garden or nursery, or artificially by deliberate selection of different species as a pollen-parent and a seed-parent.

The earliest of the chance hybrids was probably *Erica* × *darleyensis* (see p. 316). Judging by a specimen preserved in the Royal Botanic Gardens, Kew, it may have arisen in the late 1880s or early 1890s. James Smith of Darley Dale, Derbyshire, began selling the heather in 1900, so that by the time William Jackson Bean recognised that it was distinct and published the name *E.* × *darleyensis* in 1914, the Darley Dale heath was a well-known plant in English gardens. *E.* × *veitchii* (see p. 322) was next and must also have arisen in the late 1800s, this time in Messrs Veitch & Son's Nursery at Exeter in Devon. Whereas *E.* × *darleyensis* has since been produced by deliberate cross-pollination, *E.* × *veitchii* remains to be synthesised, as far as I am aware.

A third chance hybrid may have had its origins in the Royal Botanic Gardens, Kew, although it was not identified until the same cross had been deliberately raised by Dr John Griffiths in the early 1980s (Griffiths 1987). When A. W. ("Bert") Jones studied the hybrid Griffiths had produced by cross-pollinating *Erica manipuliflora* with *E. vagans*, he realised that a plant which the Maxwell & Beale nursery had sold (see p. 320), and which had been give the name 'Heaven Scent' was most probably the same (Jones 1987). In 1997, this fine, hardy hybrid was named *E.* × *griffithsii* by David McClintock (see p. 319).

All the other hybrids now in cultivation are deliberate creations — man-made heaths.

A BRIEF HISTORY OF MAN-MADE HEATHS

The history of artificial hybridisation of *Erica* is remarkable. While the southern African species, the Cape heaths, are not the subject of this monograph, they, and not the hardy northern species, were the most widely cultivated heathers in the late 1700s and early 1800s — indeed, until the late nineteenth century. Cape heaths were exceptionally fashionable, especially in Britain, and new ones commanded very high prices. Thus there was considerable potential for profit if European nurseries could produce novelties without the need to employ, at substantial expense that might not be recouped, a seed-collector at the Cape of Good Hope. The earliest attempts to produce hybrid heaths, using the Cape species, probably were carried out in the 1790s by William Rollisson (c. 1765–1842), founder of the Springfield Nursery, Upper Tooting, Surrey, but his work was not made public until after his death (see Nelson & Oliver 2005a). His eldest son, George (c. 1799–1879), published a brief note in *The gardeners' chronicle* on 8 July 1843 in which he listed the names of more than 90 hybrids raised by his father (Rollisson 1843). A few years afterwards, W. H. Story (1848) wrote that "For forty years and upwards [Rollisson] silently and successfully carried on his favourite pursuit, introducing, during that long period, most (I was going to say all) of the choicest and most favoured varieties now in cultivation." Rollisson's raising of so many hybrids of the Cape heaths was the first "extensive program of breeding new ornamental plant varieties" (Elliott 2001; Nelson & Oliver 2005a).

Whereas Rollisson kept his work secret, presumably to ensure he had few competitors, the Hon. & Very Revd William Herbert (1778–1847), an enthusiastic gardener and an authority on Amaryllidaceae and Iridaceae, had no reason for reticence. Herbert was one of the first to write about "hybridization amongst vegetables" (Herbert 1847; see also Herbert 1818, 1822, 1837) and the first explicitly to record artificial hybrids in *Erica*. He announced that he had succeeded in raising *Erica* crosses when he spoke about hybrids at a meeting of the recently-formed Horticultural Society of London on 7 July 1818. The published version of his paper concluded with this scanty comment: "and the new heaths I have already obtained, are most distinct and remarkable, the individuals of each new species [*sic*] being perfectly uniform" (Herbert 1818: 196). About a year and a half later, Herbert provided further details of his work in a "Letter to the Secretary" (who happened then to be Richard Salisbury) that was read at a meeting of the Society on 21 December 1819 (Herbert 1822). Using the terms "dust" for pollen and "mule" for a hybrid, he must have surprised his fellow plantsmen by noting that

> In an attempt to Fecundate the English Heaths with the dust of the African sorts, I was defeated by finding that the dust was shed upon the stigmas so long before the flowers expanded, that the anthers could not be taken out effectually without cutting into the bud at so early a period as to destroy its growth. The most likely cross would have been Erica cerinthoides with the dust of our E. tetralix, but E. cerinthoides does not make seed at all with me.

Remarkably, while the names given to many of Rollisson's hybrids are known — there are more than 170 entries in The Heather Society's database (see Nelson & Small 2004–2005) — not a single record of any of Herbert's hybrids has been traced, and there is no *Erica* bearing his name. The explanation clearly lies in the fact that all of Herbert's hybrid seedlings perished without maturing and blooming: "They were all lost on, or soon after, removing to Spofforth before they had flowered, though one of them was above a foot high." Not everyone approved of Herbert's work, as he related (Herbert 1837: 336):

> Soon after the publication of [my] communication to the [Horticultural] Society, I was accosted by more than one botanist with the words, 'I do not thank you for your mules,' and other expressions of like import, under an impression that the intermixture of species which had been commenced, and was earnestly recommended to cultivators, would confuse the labours of botanists, and force them to work their way through a wilderness of uncertainty.

Having so easily achieved a plethora of showy hybrids between the exotic *Erica* species from the Cape of Good Hope, it was not unnatural for European horticulturists to speculate about the possibilities of crossing these with the less showy but frost-hardy northern species. The idea was certainly current in the mid-1800s, as shown by this comment about "crossing splendid species of South Africa with the hardy natives" published in *The gardeners' chronicle* on 17 March 1849, less than two years after Herbert's death: "I need not say that a successful result would be most interesting" (Devonian 1849; see also Lancastrian 1849). In 1863, George Gordon wrote on the same topic without adding anything new.

In fact, apart from William Herbert's comment in 1819, there is no evidence that crosses between the Cape and northern hemisphere species were tried, let alone effected, during the nineteenth century. In the twentieth century, more than one attempt was made to hybridise these heaths, but without any success. Reported examples include (the African species being named first): *Erica curvirostris* × *arborea*, *E. baueri* × *australis*, and *E. subdivaricata* × *carnea* (Jones 1988). In Winnekendonck, Germany, during the 1970s, Urban Schumacher (1915–1980) attempted unsuccessfully to hybridise *E. gracilis* with *E. cinerea* (Vogel 1982). Also in the 1970s, John Crewe-Brown, in Johannesburg, tried various crosses between northern hemisphere and African species, including *E. caffra* × *australis* (Joyner 1979). Eventually, in the late twentieth century, primary hybrids between Cape and northern *Erica* species were raised. *E.* × *afroeuropaea* (p. 337), the first to be named, is the progeny of *E. arborea* and *E. baccans* (Jones 1988; McClintock 1999a). *E.* × *gaudificans* (p. 340) has *E. spiculifolia* as the seed-parent while *E. bergiana* provided the pollen (Nelson & Wulff 2007a); two clones, named 'Edewecht Belle' and 'Edewecht Blush', are presently cultivated in Europe and western North America (Nelson 2004d, 2005c).

Returning to intentional hybrids involving only the hardy northern heaths, again there is no explicit account of any being raised during the nineteenth century, although at least one anecdote suggests that attempts were made to produce them. The famous and very successful Ulster rose-breeder George Dickson I (1832–1914), of Newtownards, County Down, was said (Harkness 1985: 47) to have been "quite carried away by the new science" of plant breeding, so that he "endeavoured to cross summer and winter flowering heathers, by keeping the pollen of the one until the other was in bloom." Dickson kept the pollen sealed in goose quills. No heather was ever introduced by Dickson, so it has to be assumed that he was never successful in his heather breeding, if he ever attempted the task.

Another Ulsterman, and a person I have already mentioned, J. Walker Porter (see p. 318) was undoubtedly the pre-eminent raiser of new heather cultivars from seed during the first half of the twentieth century (Nelson 1984a, 2007b). Little is known about his methods, but he clearly stated that he had raised hybrids; whether these were accidental or deliberate is not recorded. His careful seed-collecting techniques were described by Maxwell & Patrick (1966). In a conversation recorded by Dr C. W. Musgrave (transcribed in Nelson 2007b), Porter said that he had "a great number of hybrids" between *Erica carnea* and *E. erigena* (= *E.* ×*darleyensis*) with coloured foliage: "the colour of the coloured foliage is more intense than the colour of the flowers, and most people cannot believe that it is coloured foliage". His own selections included *E.* × *darleyensis* 'Jenny Porter', named after one of his sisters, and 'W. G. Pine', and the enigmatic *E. carnea* 'Eileen Porter', named after his wife. After his death, other selections were introduced, and named, including 'Margaret Porter', after another sister, and 'J. W. Porter' (see pp. 374–375, Fig. 203 (p. 316)).

Georg Arends, of Wuppertal Ronsdorf, the renowned German plant breeder and nurseryman, recorded in his autobiography (Arends 1951: 130) that he had raised seedlings from *Erica terminalis* crossed with *E. cinerea* (see below).

Although Arends apparently succeeded in raising a hybrid involving *Erica cinerea* (bell heather), the species is generally neither a good seed-parent nor a useful pollen-parent (cf. Griffiths 1985). Several attempts have been made to produce hybrids from bell heather and the spring-flowering *E. carnea*. In the late 1960s, Dr Ans Heyting, at Boskoop in The Netherlands, obtained seeds; the progeny was weak and perished before flowering, so the cross could not be verified.[9] In the 1980s, the German plantsman and heather connoisseur ("Pflanzenliebhaber und Heidekenner") Lothar Denkewitz, who is otherwise known principally as a breeder of irises, placed pollen from *E. cinerea* 'Pink Ice' on *E. tetralix* 'Tina'. Two of the resulting seedlings were selected and named 'Samtpfötchen' and 'Riko', but both are indistinguishable from *E. tetralix*, implying that the expected hybrid was not formed. Denkewitz also tried cross-pollinating *E. tetralix* and *E. carnea* without success, while the dozen seedlings from *E. ciliaris* 'Corfe Castle' pollinated by *E. tetralix* were all like *E. ciliaris*.

For scientific research purposes, rather than for horticultural reasons, Professor David Allardyce Webb (1912–1994) of Trinity College, Dublin, succeeded in fertilising *Erica tetralix* with pollen from *E. mackayana*, the most probable way in which *E. × stuartii* arises in wild localities in Ireland (Webb 1955, 1975: 339). Once again the seedlings were not "raised to maturity" (Webb 1975), perishing before they flowered, so their hybrid status could not be established (Griffiths 1985). The date of this experiment is not recorded but must have been between 1955 and 1975.

As already mentioned (p. 318), in 1972, Mrs Ann Parris "tried a simple experiment" to confirm that *Erica × darleyensis* was the product of *E. carnea* crossed with *E. erigena*. She tied sprigs of *E. carnea* 'Springwood Pink' to the flowering branches of *E. erigena* 'W. T. Rackliff' and 'Brightness'. From the few seeds she harvested, Mrs Parris raised seven plants. The young shoots were all discoloured, an indication of a hybrid (Parris 1976; Griffiths 1985; Small & Alanine 1994).

More direct pollination methods, placing the pollen directly on the stigma of the seed-parent, were to be employed subsequently. Whereas Ann Parris's method was haphazard, the work of Kurt Kramer, Dr John Griffiths and David Wilson, the principal hybridists active since the 1980s (Small & Wulff 2008), was not.

In the following section, the hybrids are arranged alphabetically according to the person who is believed to have first created the cross.

Georg Arends's hybrid

In his autobiography, Georg Adalbert Arends (1863–1952), who raised many excellent garden plants, reported that he had succeeded in crossing *Erica cinerea* and *E. terminalis* but that the seedlings had been killed during a harsh winter (Arends 1951). *E. cinerea* evidently was a favourite plant and Arends is credited with introducing several extant cultivars, including 'Atropurpurea' and 'Splendens'. He also produced *E. ×darleyensis* 'Erecta' and 'Silberschmelze' (see p. 376).

There is really no reason to disbelieve Arends's claim, especially since the hybrid has been re-created by his compatriot Kurt Kramer. To mark Arends's achievement, as well as his other contributions to heather cultivation, *E. terminalis × cinerea* is now named in his honour (Nelson 2007e).

32. ERICA × ARENDSIANA. Arends's heath. [*E. terminalis* × *E. cinerea*]

Bell heather (*Erica cinerea*) and Corsican heath (*E. terminalis*) do not coexist in the wild, although their areas of distribution almost meet in the Mediterranean basin: the populations of *E. cinerea* in Liguria are not far removed from those of *E. terminalis* in Corsica. However, a natural hybrid between these heathers is improbable.

The plants presently in cultivation were raised by Kurt Kramer by fertilising *Erica terminalis* (Corsican heath) with pollen from *E. cinerea* (bell heather). Seedlings were raised before 2000, and at least two clones are extant; these are named 'Ronsdorf' and 'Charnwood Pink' (see below). The reciprocal cross failed for Kramer in 1986. While the hybrid has not been confirmed by DNA analysis, the gross morphology of now-mature shrubs and the anomalous flowering behaviour, the presence of discoloured young shoots (a characteristic of other hybrids, as just noted), contorted styles and stamens, and malformed pollen suggest that it is, indeed, a hybrid (Nelson 2007e).

Allen Hall (Nanpantan, Loughborough, Leicestershire) has cultivated *Erica × arendsiana* 'Charnwood Pink' (Figs 208 & 209) for more than four years. Both parents are summer-flowering species, yet *E. × arendsiana* blossoms in the Halls' garden in central England throughout the autumn, winter and early spring sporadically. Allen Hall (pers. comm., 2006) examined the pollen of the clone that he grows and found it to be very irregularly formed, neither in perfect tetrads like that of *E.*

Figs 208 and **209**. *Erica × arendsiana* 'Charnwood Pink'; cultivated in Joan and Allen Hall's garden, Loughborough, Leicestershire (E. C. Nelson).

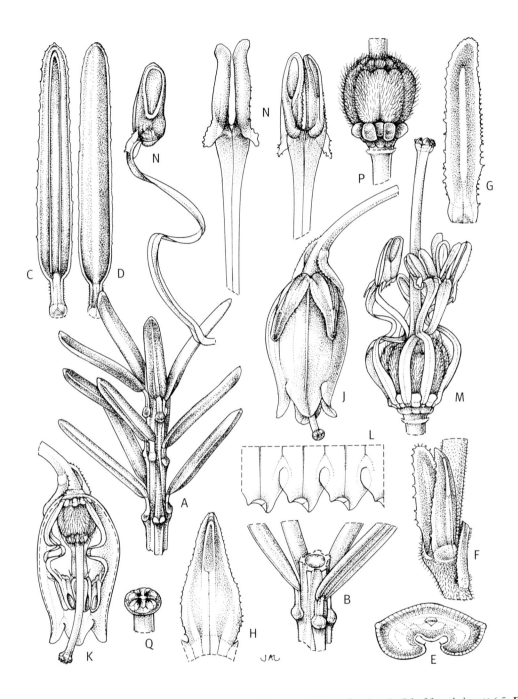

Fig. 210. *Erica × arendsiana* 'Charnwood Pink'. **A** leafy shoot × 4.5; **B** leaf node × 9; **C** leaf from below × 6.5; **D** leaf from above × 6.5; **E** section through leaf × 18; **F** bract and bracteole × 15.5; **G** bracteole × 25; **H** calyx lobe × 6.5; **J** flower × 7; **K** section through flower × 7; **L** lobes of corolla × 12; **M** gynoecium and stamens × 11; **N** stamens from side, back and front × 22; **P** ovary × 11, **Q** style apex (stigma) × 6.5; from specimens supplied by Allen Hall and E. C. Nelson. Drawn by Joanna Langhorne.

cinerea, nor as separate grains (monads) like *E. terminalis*. Many grains were misshapen, and they were irregularly clustered; while most were aggregated in tetrads, there were also monads, pairs and triads, and some clusters had more than four grains. Furthermore the pollen he observed would not take up acetocarmine stain and, thus, are apparently infertile.

Erica × arendsiana is hardy, surviving outdoors, unprotected, temperatures as low as -5°C. It is a curiosity rather than a highly ornamental heather.

Erica × arendsiana E. C. Nelson, *Heathers* **4**: 56–57 (2007). Type: cultivated 10 Upper Green, Loughborough LE11 3SG, United Kingdom, 19 February 2006, *A. Hall, s. n.*: holo.: **WSY**!.

DESCRIPTION (based on 'Charnwood Pink'). *Variable, evergreen shrub* to 1 m tall after 5 years; habit upright; shoots sparsely pubescent, with prominent infrafoliar ridges. Mature *foliage* mid-green, but young shoots discoloured; *leaves* in whorls of 4, spreading, glabrous, to 8 mm long, broadest towards base (to 1.4 mm wide) and tapering towards apex; apex with microscopic spicules; margins revolute, sulcus green, to 0.4 mm wide and opening at base. *Inflorescences* of 1–4 (or more) flowers in terminal umbels on short, leafy axillary shoots, or in a large, many-flowered terminal umbel; usually several inflorescences ranged towards end of a shoot and so appearing racemose. *Flowers* dull pink, palest at base, erect or pendulous; *pedicel* 5–8 mm long, red, sparsely pubescent, straight or curving; bract positioned somewhat above the mid-point, ±1 mm long, narrow-lanceolate, greenish, margins fringed; two bracteoles together in a whorl or separated, about 2/3 way along the pedicel, ±1.5 mm long, leaf-like, narrow, linear, with fringed margins and, at apex, a prominent sulcus, green, about 1/3 length of bracteole. *Calyx* 4-lobed; lobes green, unequal, to 3 mm long, to 1 mm broad, obovate-triangular, thickened towards apex, free or fused for a very short length at base; margins scarious, with short, simple hairs, sulcus green, open, widest at base, about half the length of the lobe. *Corolla* ovoid, 6 mm long, to 3 mm broad, glabrous, 4-lobed; lobes ±1 mm long, erect. *Stamens* 8; *filament* white, ±0.3 mm across at base, tapering gradually upwards and then much expanded (to 0.5 mm across) immediately below thecae, sigmoidally curved, sometimes malformed and twisted or contorted; *anther* dark brown after anthesis; *thecae* ±1 mm long, narrow-elliptical, surface papillate and minutely hirsute, ±0.4 mm wide, fused only at base, erect and parallel; pores lateral, situated towards top of theca, ±0.5 mm long; *spurs* narrow, pointed, minutely pubescent, 0.2 mm long, projecting downwards; *pollen* present at least in some anthers, infertile and malformed. *Style* frequently malformed and contorted, with a prominent bend near base; if *not* malformed, ±5 mm long, white shading to deep pink at apex, straight, tapering slightly towards apex ±0.2 mm diameter, emergent; *stigma* capitate, dark pink, ±0.3 mm diameter; *ovary* green, barrel-shaped, ±1.5 mm diameter, to 1 mm tall, conspicuously ribbed, densely tomentose, 4-locular, with ±15 ovules per locule; lobed *nectary* encircling the base, nectar copious. *Fruit* not formed.

FLOWERING PERIOD. September–April.

ETYMOLOGY. Eponym; a tribute to the German plantsman Georg Arends, who was the first to raise this hybrid.

VERNACULAR NAMES. Arends's heath (English); Arends Heide (German).

CULTIVARS. See Appendix I (p. 363). One clone, with pale flowers, is in cultivation in Great Britain, and there are at least two growing in Germany.

John Griffiths's hybrids

Dr John Griffiths, by profession a research scientist interested in the chemistry of dyes, is a purely amateur, yet remarkably successful, heather breeder. For him, producing hybrids between the hardy species of *Erica* is a "fascinating, challenging, and exasperating hobby" (Griffiths 1985).

John and his wife, Valerie, lived for almost four decades in Garforth, a town in the West Riding of Yorkshire situated close to the city of Leeds, and they had a modest garden of about a third of an acre (Griffiths 2008). In 1981, without any experience of plant-breeding, he began his attempts to raise hybrids from hardy heathers. As he related in 1985, only four years after commencing, whereas

> hybridisation seems attractively simple ... with flowers the size of an old halfpenny piece or larger, no one seems to have any advice to offer for flowers as small as those of our heathers, and if anyone cares to remove the anthers from a heather flower with a pair of scissors without all but annihilating the flower they will soon realise why such advice is rather scarce.

It is interesting to compare that comment by Dr Griffiths with what the Revd William Herbert wrote a century and a half earlier (see p. 327).

Realising that the techniques which worked for *Rhododendron* and *Fuchsia* were of little use, Griffiths began to experiment, and developed his own working methods. While collecting pollen was "relatively simple", there were unexpected discoveries: for example, that some species (including *Erica carnea*, *E. erigena*, *E. vagans* and *E. ciliaris*) produced copious quantities of pollen, while others such as *E. cinerea* and *E. tetralix*, produced relatively little, and the white-flowered cultivar of the latter named 'Alba Mollis' never yielded any (Griffiths 1985). Once collected, the pollen could be transferred to the stigma of a chosen flower by an elegantly, simple tool:

> To apply the pollen one should fashion a small pointed stick out of some plastic material, preferably black, and rub this on a piece of cloth. On touching the pollen with this, one will find that the static charge will cause the pollen to adhere firmly to the point. The pollen is then readily transferred to the sticky tip of the stigma, the dark colour of the stick enabling the pollen to be seen clearly. After a little practice one will find the whole operation very simple ... .

John Griffiths's earliest attempts at crossing heathers yielded large amounts of seed that germinated, but the vast majority of the seedlings were "pure species" and usually inferior to the cultivars that had been the seed-parents (Griffiths 1985). While many of his heather seedlings were worthless, John was learning which cultivars were good seed-parents, just as he learned which species had the best pollen yields.

By 1985, John Griffiths was able to report some success. He had essentially re-created a number of well-known hybrids by artificial means, although some of these had yet to be proven: *Erica carnea* × *erigena* (*E.* × *darleyensis*); *E. ciliaris* × *tetralix* (*E.* × *watsonii*); *E. tetralix* × *vagans* (*E.* × *williamsii*); and *E.* × *stuartii* by employing pollen from Spanish *E. mackayana* on *E. tetralix*. More significant were several novel combinations. *E. manipuliflora* pollinated by *E. vagans* had resulted in excellent seed, and from this came vigorous seedlings, although "hybrid character cannot be claimed with any certainty. However, it is interesting," Griffiths wrote (1985: 29–30), "that those seedlings from the cross with [*E. vagans*] 'Valerie Proudley' as parent show a fair proportion with golden foliage." Also, two cultivars of *E. tetralix*, 'Con Underwood' and 'Bartinney', pollinated by *E. manipuliflora*, yielded seedlings that "look quite distinct from *E. tetralix* of the same age, and resemble a vigorous *E.* × *williamsii*" (Griffiths 1985: 32).

Subsequent research and observations led to the descriptions and naming of two new hybrids. *Erica* × *griffithsii* (see p. 320) is the name for the progeny of *E. manipuliflora* crossed with *E. vagans*, which, as noted above, had been produced by chance, perhaps in the Royal Botanic Gardens, Kew, several decades earlier. The hybrid which John Griffiths had effected between *E. manipuliflora* and *E. tetralix*, however, was entirely new; it has been named *E.* × *garforthensis* (see below).

Since 1981, John Griffiths has used 15 species as possible seed-parents and 14 as sources of pollen; and he has attempted 75 different combinations. He has had numerous failures; for example, *Erica australis* has never yielded seed when pollinated by other species. On the other hand, *E.* × *griffithsii* 'Valerie Griffiths' is now a highly regarded heather and readily available from nurseries in Europe and North America (see p. 378).

33. ERICA × GARFORTHENSIS. Garforth heath. [*E. manipuliflora* × *E. tetralix*]

The parent species, cross-leaved heath (*Erica tetralix*) and whorled heath (*E. manipuliflora*), do not grow together anywhere in the wild. In fact, their ranges are separated by several hundred kilometres, so there is no chance of this hybrid forming except in cultivation.

The original seedlings of this artificial hybrid were produced from *Erica manipuliflora* 'Aldeburgh', a hardy clone of the whorled heath which was the only cultivar widely available in the 1980s (see pp. 147, 380). Early in September 1983, Dr John Griffiths applied its pollen to seven flowers of the white-blossomed *E. tetralix* 'Bartinney' (Griffiths 1987: 49), and four of these flowers formed seed. This was collected and sown in the following November. By June 1984, 15 seeds had germinated, and three years later the tallest seedling was 15 cm in height. About half the plants were "tall, fastigiate, grey green" while the others were "short, fastigiate, light green, [with] gold tips."[10] By 1988, none had yet produced flowers. Eventually they were assessed to be "pretty much the same" (McClintock 2000), so all, but one, were discarded. That survivor is now named 'Craig' (® E.2007:04) (see p. 377). After 15 years, and annual pruning, *E.* × *garforthensis* 'Craig' was 80 cm tall, but it had never flowered profusely and was deemed "to be of scientific interest only" (McClintock 2000).

In 1996, David Wilson, of Chilliwack, British Columbia, Canada, independently made the same cross using different cultivars as parents. His chose *Erica tetralix* 'Melbury White' as the seed-parent, and used pollen from *E. manipuliflora* 'Korčula', a cultivar that had been collected on the Adriatic island of that name in 1978 (see p. 147). While the parent clones of the cross-leaved heath used by Wilson and Griffiths both have white flowers, the clones of whorled heath are markedly different; 'Korčula' has fragrant, pale shell-pink (H16) flowers, whereas 'Aldeburgh' has lilac pink (H11) flowers. Nine of the Canadian seedlings were grown on for comparison, and were propagated. Compared with the sole surviving Garforth plant, the seedlings raised by David Wilson produced a more profuse display of larger flowers although, according to McClintock (2000), these are paler and not as fragrant. Of the nine plants that were evaluated, only one so far has been selected for naming; clone no. 4 was named 'Tracy Wilson' after Irene and David Wilson's younger daughter (Wilson 2000).

Erica × garforthensis D. C. McClint., *Yearbook of The Heather Society* 2000: 17 (2000). Type: "Garden at 9 Ashlea Close, Garforth, Leeds", September 1998, *J. Griffiths*: holo.: **WSY!**

ILLUSTRATION. *Yearbook of The Heather Society* 2000: 18 (2000).

DESCRIPTION. *Evergreen shrub*, variable in habit depending on parentage; upright, erect, to ±1 m

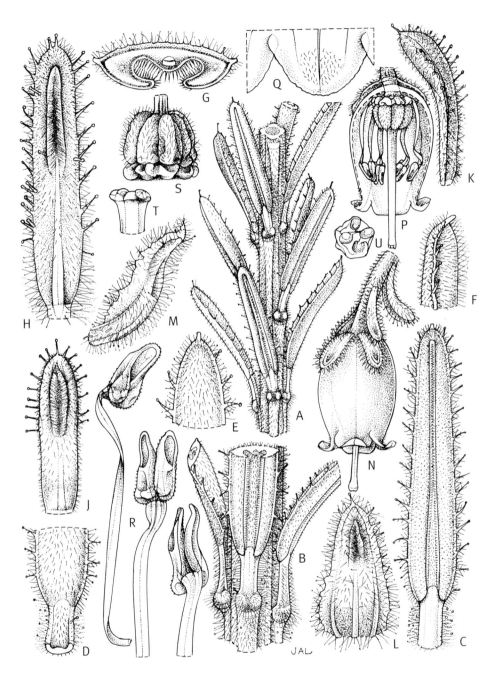

Fig. 211. *Erica × garforthensis* 'Craig'. **A** leafy shoot × 7.5; **B** leaf nodes × 16; **C** leaf from back × 16; **D** base of leaf from above × 16; **E** leaf tip from front × 16; **F** leaf tip from side × 16; **G** section through leaf × 53; **H** bract from back × 25; **J** bracteole from back × 7; **K** bracteole from side × 7; **L** calyx lobe from back × 22; **M** calyx lobe from side × 22; **N** flower from side × 8; **P** section through flower × 8, **Q** corolla lobes × 20; **R** stamens from side, front and back × 13; **S** ovary × 20; **T** style apex from side × 28; **U** style apex from above showing the 4-lobed stigma × 28; from specimens grown by John Griffiths. Drawn by Joanna Langhorne.

tall (after 25 years), or compact and spreading reaching ± 0.25 m tall, to ±0.35 m broad after 4 years; shoots densely hirsute with short, simple, eglandular and glandular hairs. *Leaves* in whorls of 4, ±5 mm long, less than 1 mm across, elliptical-lanceolate to linear, densely pubescent (*not* glabrous as stated in the protologue) on upper surface, edges with longer, gland-tipped hairs, margins recurved underneath but not contiguous, sulcus open with densely hirsute margins, exposed midrib pubescent; petiole pubescent, ±1 mm long, flattened above, convex underneath, sometimes sharply curving. *Flowers* in umbels on short, leafy, lateral shoots, or terminal; ±5 flowers per umbel; *pedicel* ±4 mm long, straight or curving, hirsute with long, crinkled, simple hairs; position of *bract & bracteoles* variable; in 'Craig' *bract* at base of pedicel, in 'Tracy Wilson' about mid-point; ±2.5 mm long, ±linear-lyrate (broadening at base and towards apex, narrowest in middle), green, hirsute with long, eglandular and glandular hairs, thickened at apex, sulcus open, ±1 mm long; 2 *bracteoles* below middle of pedicel in 'Craig' or above in 'Tracy Wilson', above the bract, similar, linear, ±1.7 mm long, sulcus ±0.4 mm. *Calyx* 4-lobed, ±2 mm long, lobes unequal, fused at base, lanceolate, to 1 mm broad, fringed with simple, eglandular hairs, outer surface pubescent with simple hairs, apex green and inner surface hirsute, sometime tinged pink on margin, sulcus open, hirsute. *Corolla* pink or white, ±5.5 mm long, ±3.5 mm diameter, urceolate, with sparse, short, simple hairs on outer surface of lobes, otherwise glabrous; lobes rounded, spreading, ±1 mm tall, ±1.7 mm across at base, overlapping alternately. *Stamens* 8, ±4 mm long, erect, included; *filament* glabrous, straight except for hooked base, attached beneath nectary; *anthers* erect at first but often contorted later, with spurs of different and varying lengths, *thecae* ±0.5 mm long, often somewhat misshapen; no pollen detected during microscopic examination. *Style* ±5 mm long, ±straight, terete, ±0.3 mm diameter at base, tapering slightly upwards, glabrous; *stigma* capitellate, ±0.4 mm diameter, dark; *ovary* ± 1 mm tall, ±1.2 mm diameter, cylindrical, faintly ribbed, entirely pubescent, 20–30 ovules per locule; *nectary* distinct, dark, lobed; nectar present. *Seed* unknown, plant reported to be sterile.

FLOWERING PERIOD. July–October.

ETYMOLOGY. Toponym: Garforth in the West Riding of Yorkshire is where Dr and Mrs Griffiths lived (Griffiths 2008) and where this hybrid was first raised in the early 1980s.

VERNACULAR NAME. Garforth heath (English).

CULTIVARS. See Appendix I (p. 377). Two clones have been named.

Kurt Kramer's hybrids

A native of north-western Germany, Kurt Kramer gained practical experience as a gardener when he was younger and later worked in several large nurseries (Small 2001). Using his parents' farm, he began to grow heathers, aiming to produce plants of higher quality than other nurseries. He was watchful, too, spotting a sport on *Calluna vulgaris* 'Peter Sparkes' which he propagated and introduced — 'Annemarie' (see p. 55) was his first new heather. To date, 35 cultivar names have been registered by Kramer for his *Calluna* cultivars, but he has raised and introduced around twice that number of named hardy heathers.

While Kurt Kramer has established an unequalled reputation for exceptional *Calluna* cultivars, especially the so-called "bud-bloomers", he has also been active in cross-breeding *Erica* species. He began hybridising heathers in a casual way in 1974 (Kramer 1991), but became much more active after 1980. Among his earliest introductions resulting from deliberate cross-pollination were cultivars of *E. carnea*: 'Golden Starlet' (see p. 366), 'Rotes Juwel' and 'Schneekuppe'. Another of his early achievements was the outstanding, and deservedly ubiquitous, *E.* × *darleyensis* 'Kramers Rote' (Fig. 222; see p. 375), which came from a 1981 cross between *E. carnea* 'Myretoun Ruby' and *E. erigena* 'Brightness'. Curious to find out what species will produce hybrids, Kramer has employed 17 hardy species as pollen-parents, and 14 as seed-parents, making more than 70 different crosses, including reciprocal ones. Fourteen of these yielded seeds that germinated and seedlings that survived — remarkable results. Further, he has used nine different Cape heaths, trying to cross these, mostly without success, with a limited number of European heaths — *E. arborea*, *E. australis*, *E. cinerea*, *E. spiculifolia* and *E. tetralix*.

Erica spiculifolia has been his most surprising (Kramer 1991) and, eventually, prolific seed-parent, producing seedlings when pollinated by *E. carnea*, *E. cinerea*, *E. sicula* subsp. *bocquetii*, *E. tetralix* and *E. vagans*, as well as by *E. bergiana*, one of the Cape heaths. *E.* × *krameri* (p. 343) arose from the cross with *E. carnea*, and *E.* × *gaudificans* (p. 340) has the Cape heath, *E. bergiana*, as its other parent. Kramer also has raised two note-worthy hybrids from *E. arborea*: *E.* × *oldenburgensis* (p. 346) had *E. carnea* as the pollen-parent, while *E.* × *afroeuropaea* (see below) came from the cross with *E. baccans* from South Africa.

34. ERICA × AFROEUROPAEA. [*E. arborea* × *E. baccans*]

This artificial hybrid was synthesised by Kurt Kramer in 1986 when he deliberately cross-pollinated *Erica arborea* var. *alpina* and *E. baccans*. A single, apparently sterile clone was introduced into British gardens in 1992 by Denbeigh Heather Nurseries. The hybrid is not widely distributed — it is not reliably hardy in Britain (see Hall 2002) and needs to be given greenhouse protection during winter to preform well. However, a plant, now at least 1.5 m tall, has survived outdoors in a protected spot in the Halls' garden at Loughborough, for nine years and blossoms into the late summer. This clone, named

'Vossbarg' (see below), is also in cultivation in the western USA. In Britain, grown under glass, this hybrid comes into flower in spring. I saw this in bloom at Glenmar Heather Nursery, Bayside, in northern California, during August 2004; there, the plants were in pots in a plastic tunnel (Fig. 212).

Erica × *afroeuropaea* was the first hybrid between a European heather and a Cape heath to have been reported and the first to be named. *E. baccans* (berry heath, bessieheide), an endemic South African species restricted to mountain slopes in the Cape Peninsula, is an erect shrub with pink flowers. The other parent was a European clone of the tree heather, *E. arborea*, often treated as a cultivar ('Alpina'), collected in the mountains of central Spain by Georg Dieck from Zöschen, Germany, in 1892 (see p. 199).

It may be worth repeating that *Erica arborea* is not confined to Europe, and that it is also native in north and east Africa and in western Asia (see pp. 200–209).

Fig. 212. *Erica* × *afroeuropaea* 'Vossbarg'; Glenmar Heather Nursery, Bayside, northern California (E. C. Nelson).

Erica × afroeuropaea D. C. McClint., *New Plantsman* 6: 207 (1999). Type: Edewecht-Süddorf, Germany, nursery at Edammerstrasse 26, 10 April 1989, *K. Kramer*: holo.: **WSY**0100110!

ILLUSTRATION. *New Plantsman* 6: 207 (1999).

DESCRIPTION. *Evergreen, upright shrub* to 1.5 m or more tall; branches ascending, shoots hirsute with unequal hairs, some long, some short, some gland-tipped. *Leaves* linear, to 2.5–6 mm long, ascending, petiole ±0.7 mm, green, upper surface flat with forward pointing microscopic hairs scattered throughout, edges with spicules and a few gland-tipped hairs. *Flowers* in terminal umbels of 2–4; *pedicel* ±3 mm long, straight or curving, glabrous, tapering towards the apex where it is broadest (±0.5 mm diameter); bract and 2 bracteoles towards the base, similar in shape, colour and texture; bract ±2 mm long, narrow lanceolate, colourless with red tip, sulcus visible (but not conspicuous), linear, ±1 mm long, margins with a few gland-tipped short hairs especially in basal half; bracteoles similar, broader at base (±0.5 mm) but gland-tipped hairs more equally dispersed over margins from tip to base. *Calyx* composed of 4 free sepals, lateral sepals overlapping dorsal and ventral sepals at base and clearly a separate whorl, ovate-triangular, ±2.5 mm long, ±2.5 mm across, pink or green, glabrous except for a few short gland-tipped hairs on margin, tips thickened, sulcus linear, ±1 mm long. *Corolla* white changing to pink (H8–H11), 4–5 mm long, urn-shaped, four-cornered at base, with four erect, rounded lobes to 1 mm long. *Stamens* 8, ±3 mm long, included; *filament* curving, with sigmoid bend at apex, with a few hairs (some gland-tipped) scattered at base, tapering gradually upwards; *anther* pale red when young, apparently without pollen; *thecae* erect, ±1 mm long, not fused except at base near junction with filament, with microscopic hairs, with 2 large, hirsute, deltoid *spurs* almost as long as theca (±0.9 mm). *Ovary* glabrous, ±1.5 mm diameter and height, squat, barrel-shaped, ribbed, green throughout or tinged red on one side, ±20 ovules per locule; *nectary* encircling base; *style* glabrous, ±straight, stout, ±2–2.5 mm long, ±0.5 mm in diameter at base, tapering towards tip; *stigma* obconical, ±0.6 mm diameter, dark reddish-purple or dark grey, emergent.

FLOWERING PERIOD. Spring and summer.

ETYMOLOGY. Latin toponyms: *afro–* = prefix indicating African; *europaeus* = European.

VERNACULAR NAME. Eineweltheide (German); no English name has been coined.

NOTES. The protologue (McClintock 1999a) is confused and contradicts the characteristics of the type material in the following respects. McClintock stated "*frutex … glaber*" ("glabrous shrub"), "*bracteola media*" ("bracteoles half way above [*sic*] petiole"), "*stylus inclusus, stigma capitatum*" ("style included; stigma capitate"). In fact the plant is not glabrous; the stems have scattered, simple hairs, and there are minute hairs on the leaves and anthers. His description is also contradictory, as he seems to have mistakenly used "petiole" (leaf stalk) when he should have employed "pedicel" (flower stalk); the bracteoles are on the pedicel and are attached to the lower half (between the mid-point and the attachment to the plant stem). The total length of the gynoecium is approximately the same as the length of the corolla; so the stigma, which is slightly wider than the style apex (but hardly capitate), sometimes protrudes a little at the mouth of the flowers, especially the most mature ones.

A plastic nursery tag attached by wire to the holotype is inscribed "E. arborea × baccans HC63D". McClintock's annotation reads, in full: "Erica arborea × E. baccans Edewecht-Süddorf, Germany Nursery greenhouse at Edammerstr 26 c. 10 April 1989 Kurt Kramer overall fertility <0.14% Fl[ower]s Red Purple 63D. Type specimen."

CULTIVARS. As noted, there is only one clone, 'Vossbarg', in cultivation, and the description above was based on living material of it.

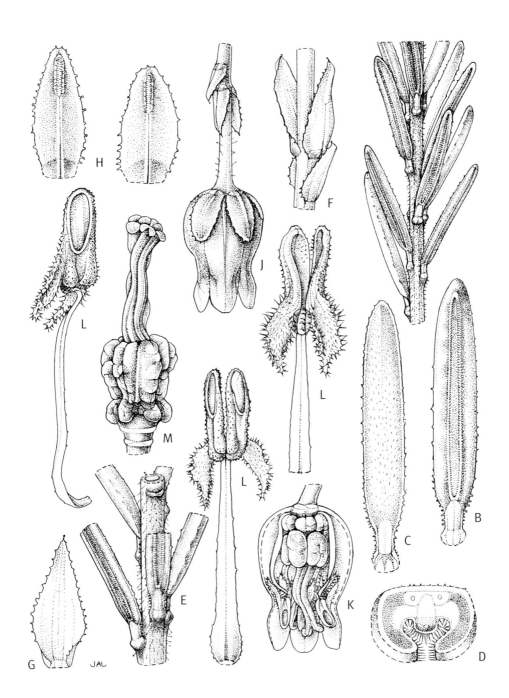

Fig. 213. *Erica* × *afroeuropaea* 'Vossbarg'. **A** leafy stem × 5.5; **B** leaf from below × 12; **C** leaf from above × 12; **D** section through leaf × 18; **E** leaf nodes × 8; **F** bract and bracteoles × 12.5; **G** bracteole × 15; **H** sepals from back × 14; **J** flower from side × 6; **K** section through flower × 7; **L** stamens from side, front and back × 27; **M** gynoecium × 14; from specimens grown by Allen Hall. Drawn by Joanna Langhorne.

35. ERICA × GAUDIFICANS. Edewecht heath. [*E. spiculifolia* × *E. bergiana*]

Among the many novel heathers produced by Kurt Kramer is a second hybrid between a hardy, northern-hemisphere heath and one of the Cape heaths. Three seedlings were produced from *Erica spiculifolia* (Balkan heath) that had been artificially fertilised during the spring of 1987 with pollen from the African (Cape) species *E. bergiana* (K. Kramer, pers. comm., 23 March 2004).

Erica bergiana is an unmistakable heather notable for its urn-shaped flowers (arranged in clusters of four) with sharply reflexed sepals, and hirsute leaves and stems; these characters are clearly demonstrated in Fay Anderson's painting in Baker & Oliver (1967: plate 88) and the photograph in Schumann & Kirsten (1992: 143). *E. spiculifolia* (see p. 279) is also most distinctive, unlike any of the other northern hemisphere species. It has tiny, bell-shaped flowers clustered in a leafless, terminal inflorescence (raceme), and leaves that are hairless (apart from some microscopic spicules at the tips).

Superficially, Kramer's seedlings resemble *Erica spiculifolia*, although mature plants are much more vigorous. They display no morphological evidence that *E. bergiana* was the pollen-parent, so chromosome counts or DNA analysis will be essential to confirm beyond doubt that they are hybrids.

In March 1992, members of The Heather Society who visited Kramer's nursery at Edewecht-Süddorf saw "rows of *Bruckenthalia* × *E. bergiana* hybrids" (Hall 1992). Another group of members saw the hybrid during April 1994 (Sellers 1994: 4). Cuttings of this new heather were donated to The Heather Society by Kurt Kramer and were successfully propagated by David Small (Denbeigh Heathers). Subsequently, Denbeigh Heathers provided further cuttings and young rooted plants to heather enthusiasts in various countries; for example, cuttings were made available to members of The Heather Society at the annual conference held at Penrith in 1997 (see Everett & Jones 1998: 63). While neither Kurt Kramer Heidezüchtung nor Denbeigh Heathers put this plant on the market, as far as I can ascertain, it has now been widely dispersed. During July and August 2004, I saw mature plants in Ireland and western USA (Fig. 215), and I am aware of other examples in cultivation in England and France.

Careful examination of these plants indicates that two clones are in cultivation; these have been named 'Edewecht Belle' and 'Edewecht Blush' (Nelson 2004d, 2005c). 'Edewecht Blush' is distinguished (see below) by hairless flower stalks (pedicels) and white (colourless) flowers which appear pink due to the colour of the immature anthers. The second clone, 'Edewecht Belle', is easily distinguished by its pubescent pedicels and pink flowers. Both clones have malformed stamens, but the extent of malformation is not consistent from flower to flower. The filaments are usually broadened and fused, and the anthers are degenerate. Similar malformations have been reported in *Erica mackayana* (Nelson 1995b); the cultivar 'Maura' demonstrates the characteristic well.

My observations in Ireland and North America suggest that 'Edewecht Blush' flowers best when grown in the open; shading greatly reduces the quantity of blossom. It can be in bloom for many months, beginning as early as April and continuing into winter. Pruning delays the onset of the flowering season, because the immature inflorescences are removed. In the mildest areas, where plants are not affected by frost, it may bear flowers for twelve months. 'Edewecht Belle' is apparently much less common in cultivation; Allen Hall (Loughborough, Leicestershire) has informed me that his plant reaches peak blooming in mid-June, and he has noted that it is hardy and "flowers freely to make a good garden display" (Hall 2002).

A close-up photograph of flowers of one of the original seedlings (Nelson 2005c), taken about 1990 by Kurt Kramer, clearly shows "double" flowers. In these, the malformed stamens are obvious

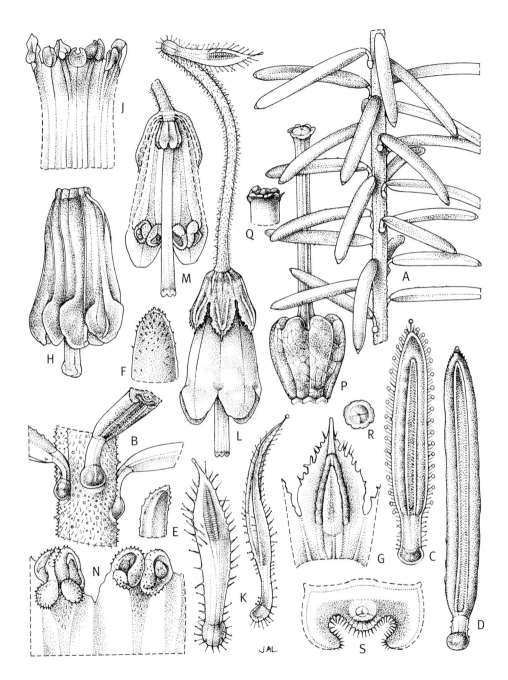

Fig. 214. *Erica × gaudificans* 'Edewecht Blush'. **A** leafy shoot × 6.5; **B** leaf nodes × 27; **C** young leaf from below × 10; **D** adult leaf from below × 15; **E** leaf tip from side × 12; **F** leaf tip from front × 20 ; **G** calyx lobe from back × 22; **H** fused stamens × 14; **J** fused stamens from front × 26; *E. × gaudificans* 'Edewecht Bell'. **K** leaf from below and side × 22; **L** flower from front × 17; **M** section through flower × 10; **N** stamens displayed from inside × 14; **P** gynoecium × 16; **Q** style apex from side × 19; **R** style apex (stigma) from above × 16; **S** longitudinal section through leaf × 22; from specimens grown by Allen Hall. Drawn by Joanna Langhorne.

at the mouth of the corolla, appearing as an inner cluster of closely-packed "petals"; thus the internal "pseudocorolla" is as long as, or a little longer than, the corolla, not shorter as in 'Edewecht Blush' and 'Edewecht Belle'. None of the flowers I have examined resemble this state, and so I conclude that the third seedling was distinct, too, but does not seem to have survived in cultivation.

CULTIVATION

The Edewecht hybrid heath is hardy in central England, tolerating snow and air temperature close to 0°C (see Hall 2002).

Erica × gaudificans E. C. Nelson & E. M. T. Wulff, *Heathers* 4: 57–58 (2007). Type: 'Edewecht Belle'; cultivated 10 Upper Green, Nanpantan, Loughborough LE11 3SG, United Kingdom, 30 June 2004, *Allen Hall, s. n.*: holo.: **WSY** 0077089!

ILLUSTRATION. *Heathers* **2**: 21, 22 (2005).

DESCRIPTION. *Bushy, evergreen shrub*, vigorous to 0.5 m tall. *Shoots* densely puberulous. *Leaves* bright green, scattered, almost spirally arranged or in pairs or in "disarticulated" whorls of 3 or 4; linear, with parallel sides, to 6.5 mm long, ±0.5 mm wide; sulcate; with hairs along the margins and sometimes on the under-rolled sides; apex pointed, with spicules (microscopic hairs), terminal hair usually with gland at tip. *Flowers* in contracted, terminal racemes, 8–14(–20) per cluster, central peduncle to ±20 mm. *Pedicels* straight or curving, 3–5 mm long, red or tinged red, hairless or pubescent, with 1 bract at base (leaf-like, <3 mm long, <0.4 mm wide), or occasionally with the

Fig. 215. *Erica × gaudificans* 'Edewecht Blush'; cultivated by Edith Davis in Fort Bragg, California (E. C. Nelson).

bract ¼ way from base; very rarely with 1 or 2 proximate bracteoles. *Calyx* cup-shaped, clasping base of corolla, hairless, ±1.25 mm long, 4-lobed; lobes ±0.75 mm long, pale green with colourless, fringed margins, fused irregularly for up to 0.5 mm (but sometimes one sulcus extending almost to base). *Corolla* campanulate or broad funnel-shaped, or conical, white or pink, glabrous, 3.5–4.5 mm long, ±1.5 mm in diameter, 4-lobed; lobes erect, rounded, ±1 mm wide, ±0.8 mm long (¼ length of tube); buds can be pink. *Stamens* 8, not manifest (included), variously malformed, some simple, petaloid staminodes; others have broadened filaments; *filaments* may coalesce to form a "pseudocorolla"; *anthers* degenerate, often remaining fused in a collar, when normal, without awns; pollen observed. *Ovary* indistinctly 4-celled (4-locular) or 8-celled, glabrous, inversely pear-shaped, obovoid or barrel-shaped; *stigma* simple, truncate, with 4 dark-red ovoid carpels. *Ovules* numerous.

FLOWERING PERIOD. (April–) June–August (–October)

ETYMOLOGY. Latin: *gaudifico*, I make joyful.

The explanation of this name is as follows. *Gaudifico* was suggested by the observation that when Kurt Kramer visited the Wulffs' garden in Oregon during the summer of 2006, he was delighted to see vigorous, floriferous plants of 'Edewecht Blush'. Several visitors commented on his reaction: they noticed that he had a big smile on his face (he is naturally of rather solemn mien) (Nelson & Wulff 2007a).

VERNACULAR NAME. Edewecht heath; Edewecht Heide (German).

CULTIVARS. See Appendix I (pp. 377–378). Two clones are in cultivation and have been named (see Nelson 2005c).

36. ERICA × KRAMERI. Kramer's heath. [*E. carnea* × *E. spiculifolia*]

This hybrid arose from the successful cross-pollination by Kurt Kramer of *Erica spiculifolia* (Balkan heath) and *E. carnea* (winter heath) in 1987 (McClintock 1998). These two species grow in central Europe, but they have different habitats. They may grow close to each other in a few places, so there is a slight possibility that this hybrid exists in the wild. However, as well as occupying different ecological niches, their flowering times do not overlap much. *E. carnea* blooms as soon as the snow that has been covering the shrubs throughout the winter melts in spring or early summer. *E. spiculifolia* tends to bloom in the wild during late summer and early autumn. In western European gardens, which never have long-lying snow, such as in Ireland or Britain, these flowering times shift, with *E. carnea* blooming through winter into early spring and *E. spiculifolia* in mid- or late spring.

Kramer raised a large number of seedlings of this hybrid but eventually retained only a few. I am aware of five numbered clones in cultivation outside Germany. Clones no. 4 ('Rudi') and no. 7 ('Otto') (see p. 378) are the only ones that have been formally named and are probably the only ones commercially available at present. Mrs Brita Johansson, an expert gardener and heather enthusiast who lives in Vargön, Sweden, has experience of cultivating seven or eight clones of *Erica ×krameri*, and she considers clone no. 6 the best for flowers, at least in south-western Sweden. "*E. ×krameri* has <u>one</u> big advantage. The foliage is fresh green all the year, never dark and drab … ." In southern Sweden, clones nos 2 and 5 display the most vivid springtime foliage — the small, short, new shoots are brightly discoloured in early June but quickly turn green (Brita Johansson, *in litt.*, 27 August 2006; Johansson 2007). According to Mrs Eileen Petterssen (2006), Kramer's heath is a "wandering smotherer" in her garden on the west coast of Norway where it is hardy (Petterssen 2004).

Kramer's heath is not glabrous as stated by McClintock (1997b); the hairs may not be readily

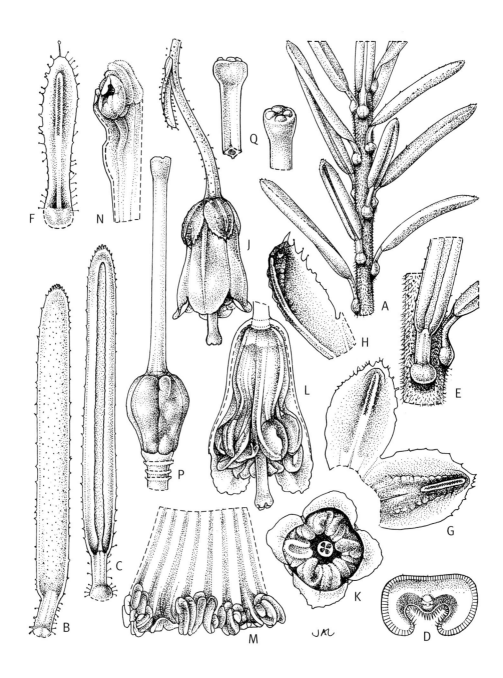

Fig. 216. *Erica × krameri*. **A** leafy shoot × 8.5; **B** leaf from above × 22; **C** leaf from below × 22; **D** section through leaf × 42; **E** leaf nodes × 32; **F** bract × 30; **G** calyx lobes from back × 12; **H** calyx lobe from side × 12; **J** flower × 10; **K** flower from below × 14; **L** section through flower × 18; **M** fused stamens × 14; **N** malformed stamens from side × 16; **P** gynoecium × 20; **Q** style apex × 32; from specimens supplied by E. C. Nelson. Drawn by Joanna Langhorne.

seen by the unaided eye but are clearly visible with a hand lens or microscope. It has tiny flowers resembling those of *Erica spiculifolia*; these tend to be sparse and insignificant, according to Brita Johansson (*in litt.*, 27 August 2006). Some flowers have malformed styles. McClintock (1997b) stated that the ovary was abortive. In the flowers of 'Rudi' and 'Otto' which I examined, the four- or eight-locular ovaries contained numerous ovules in each locule; evidently these do not develop and so no fruits are known. The hybrid is proven by the diploid chromosome complement of 30 (McClintock 1997b), comprising 12 chromosomes from *E. carnea* and 18 from *E. spiculifolia*.

Erica × krameri D. C. McClint., *Der Heidegarten* heft **41**: 19 (1997): Type: "Cult. Edewecht, NW Germany, 27. Juli 1990."[11] *D. McClintock & K. Kramer.* holo.: **WSY** (0028344)!

ILLUSTRATIONS. K. Kramer (McClintock 1997b, 1998).

DESCRIPTION. *Evergreen sub-shrub*, 0.2–0.5 m tall, spreading (vigorously in some conditions) to at least 0.5 m across; shoots erect, sparsely pubescent, sometimes hairs gland-tipped; new shoots discoloured (orange-red) briefly in spring; mature foliage bright mid-green. *Leaves* spreading or erect, in whorls of 3 or 4, small, linear, 2.5–6 mm long, 0.6–0.8 mm broad, glabrous except for sparse, short microscopic hairs near tip, but some leaves with gland-tipped hairs on edge and near tip; margins recurved; sulcus with parallel sides, about ⅓ width of leaf; tip acute, translucent, sometime glandular; *petiole* amber-coloured, ±0.6–0.8(–1.1) mm long, with few, scattered hairs, some with glandular tips. *Flowers* small, >20 clustered in a compact, cylindrical, terminal raceme to 1 cm long; peduncle straight, erect; *pedicel* curved or straight ±2–3.5 mm long (sometime articulated at base), glabrous or sparsely pubescent with short, dark gland-tipped hairs; *bract* recaulescent and bracteoles absent; or lanceolate *bract* attached about the middle of pedicel and 1 *bracteole* present below the calyx. *Calyx* with 4 fused lobes, to 2 mm long, translucent, colourless, lobes about ⅔ length of calyx, acuminate, irregularly toothed near tips, appressed to corolla. *Corolla* ±(2–)4 mm long, ±2 mm in diameter, white to pink (H1 (Amethyst) to H12 (heliotrope)), urnshaped with slight constriction below lobes to campanulate or cupshaped with sides parallel and without noticeable constriction below lobes; lobes ±1 mm long, rounded. *Stamens* 8, not exserted, muticous; *filaments* (1.5–)2–3 mm long, geniculate towards apex; *anthers* ±1 mm long, held just below corolla lobes, *thecae* not fused except towards base, smooth, dark brown; *pollen* present, 99% sterile (*fide* McClintock 1997b). *Style* (when not malformed) to 5–6 mm long, longexserted, sometimes pink; *stigma* not broader than end of style to capitate; *ovary* 8- or 4-locular, glabrous, topshaped to obovoid, ±1.3–1.6 mm tall, ±1 mm in diameter; *ovules* numerous. 2n = 30 (*fide* McClintock 1997b, 1998).

FLOWERING PERIOD. June–November.

ETYMOLOGY. Eponym: after Kurt Kramer, nurseryman and heather enthusiast, of Edewecht-Süddorf, Germany, who created this artificial hybrid and several others, including *Erica × afroeuropaea* (p. 337), *E. × gaudificans* (p. 340) and *E. × oldenburgensis* (p. 346). "In recognition of his pioneering work on hybridising heathers", Herr Kramer was given The Heather Society's Award of Honour, a singular distinction, the award being presented during the First International Heather Conference, held at Elmshorn, Schleswig-Holstein, Germany, in August 2000 (Small 2001).

VERNACULAR NAME. Kramer's heath (English); Kramers Heide (German).

CULTIVATION. Kramer's heath requires an acid soil. It is hardy (USDA zones 5–9) and has a prolonged flowering period, from June to November, in the northern hemisphere.

CULTIVARS. See Appendix I (pp. 378–379). Only two clones have been selected for propagation and naming (see Flecken 2002).

37. ERICA × OLDENBURGENSIS. Oldenburg heath. [*E. arborea* × *E. carnea*]

This is hybrid between a pair of heathers that do not grow side by side in the same habitats, although they surely come close in the Balkan Peninsula. In some ways, this seems an incongruous coupling: the lofty tree heath (*Erica arborea*) and the diminutive, shrubby winter heath (*E. carnea*).

Kurt Kramer fertilised *Erica arborea* with pollen from *E. carnea* for the first time in 1986. He raised several hundred seedlings, some of which, by 1992, were as much as 0.75 m tall. Both white- and pink-flowered plants were produced (Hall 1992). In 1992, twelve clones were selected for propagation, and cuttings were acquired by David Small, Denbeigh Heather Nursery, Creeting St Mary, Suffolk. The clones were not named at this time but were designated with nursery code numbers (DHN12/92–21/92, 30/92–31/92). In April 1994, during a visit by English members of The Heather Society to Kramer's nursery in Edewecht, Germany, two of the seedlings were selected for naming, and the hybrid was subsequently given the name *E. × oldenburgensis*. The two selections have been named 'Ammerland' (DHN17/92; Kramer no. 6) (Fig. 217) and 'Oldenburg' (DHN15/92; Kramer no. 4).

Erica × oldenburgensis D. C. McClint., *Der Heidegarten* heft 39: 35 (June 1996). Type: 'Oldenburg', Baumschule Kurt Kramer, Edammer Str. 26, 26188 Edewecht, Deutschland, 6 April 1994, *D. McClintock, K. Kramer, D. Small et al., s. n.*: holo. "Hb. Heather Society" **WSY**0100341!

ILLUSTRATIONS. *Der Heidegarten* heft 37 (1995); Small & Small (1998: 85).

DESCRIPTION. *Variable, evergreen shrub* to 1 m tall; stems sparsely pubescent with very short, simple hairs and gland-tipped hairs; infrafoliar ridges visible. *Leaves* in whorls of 3, linear, to 1 cm long, ±0.8 mm broad, flattened, light to dark green, edges with a very few gland-tipped hairs or microscopic spicules; margins recurved and parallel underneath, sulcus very narrow, linear, undersurface sometimes visible, margins densely lined with microscopic hairs; petiole flattened with a few gland-tipped hairs on edges. *Inflorescence* appearing to be a spike, clustered along upper parts of shoots; flowers on long, leafy axillary shoots, or on very short shoots (seeming to be in leaf axils), in terminal whorls of (1–)3(–5); peduncle ±5–7 mm long, curved, glabrous, translucent; bracts and bracteoles similar, about ¼–⅓ away from base, ±1–1.5 mm long, triangular, translucent, tip thickened, sulcus visible, margins with simple and gland-tipped hairs. *Calyx* ±3 mm long (but can be less than 2 mm long), with 4 unequal lobes, fused for ±0.5 mm at the base, waxy, thicker in texture than corolla, usually

Fig. 217. *Erica × oldenburgensis* 'Ammerland'; cultivated in Outwell, Norfolk (E. C. Nelson).

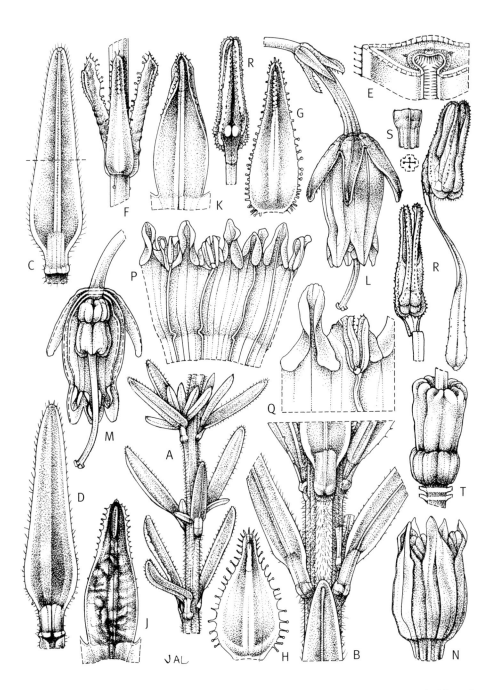

Fig. 218. *Erica × oldenburgensis*. **A** leaf shoot × 6; **B** leaf nodes × 14; **C** leaf from below × 12; **D** leaf from above × 12; **E** section through leaf × 40; **F** bract and bracteoles × 21; **G** bract × 32; **H** bracteole × 32; **J** calyx from back × 14; **K** calyx from inside × 14; **L** flower from side × 8; **M** section through flower × 8; **N** inner whorl of fused stamens × 10; **P** fused stamens displayed from inside × 10; **Q** detail of fused stamens × 21; **R** stamen from back, front and side × 18; **S** stigma apex with section of style × 32; **T** ovary and nectaries × 30; supplied by E. C. Nelson. Drawn by Joanna Langhorne.

white with keeled, green or yellow-green tips, with sulcus ±0.5 mm long, margins with sparse, simple and gland-tipped hairs. *Corolla* ovoid, ±4–5(–6) mm long, with 4 erect or patent lobes, ±1 mm long, glabrous. *Stamens* often malformed, rarely perfect, usually included but sometimes emergent; when malformed, filaments are petaloid, variously fused and thecae absent or grotesquely misshapen; if perfect, *filament* curving with prominent sigmoid bend towards apex, c. 4 mm long; *anther* brown, basifixed, without spurs; *thecae* ±1 mm long, elongated lanceolate-triangular, slightly prognacious, with microscopic spicules or papillae; pollen observed. *Ovary* tub- or bollard-shaped, green or stained with red, 4-locular, with ±15 ovules per locule, depressed at top, apex emarginate; *style* erect, emergent, ±5 mm long, ±0.25 mm wide, tapering very slightly but not expanded at apex, tinged green or red; *stigma* truncate. Ovules abortive; seeds not formed.

FLOWERING PERIOD. (December–) January– May.

ETYMOLOGY. Oldenburg is the name of a district (Landkreis), and of its capital, an historic, university city, in Lower Saxony, north-western Germany. (Ammerland is an adjoining district.)

VERNACULAR NAMES. Oldenburg heath (English); Oldenburger Heide (German).

NOTES. This binomial had first appeared in print two years before its valid publication. In the following sources there was no accompanying Latin diagnosis nor designation of a holotype: *Der Heidegarten* heft **35**: 14 (June 1994); *Ericultura* **93**: 8 (June 1994); *Bulletin of The Heather Society* **5** (2): 5 (Summer [i.e. June] 1994); *Deutsche Baumschule* 9/1994: 413 (September 1994); *Ericultura* **96**: 30 (March 1995); *Ericultura* **98**: 14 (September 1995). While a Latin diagnosis was published in the May 1995 issue of *Deutsche Baumschule*, McClintock (1995) omitted to cite a holotype and so, again, the binomial was not validly published there. E. ×*oldenburgensis* was eventually validated when a Latin diagnosis, accompanied by the citation of a holotype, was published in the June 1996 issue of *Der Heidegarten* (McClintock 1996).

CULTIVARS. See Appendix I (p. 381). From the seedlings raised by Kramer, only two cultivars have been selected for release and naming.

"WILDERNESS OF UNCERTAINTY"
Other unnamed hybrids

In this final section, I wish to place on record some heathers of hybrid origin that have been raised in the last three decades but which as yet have not been given a Latin name. In one instance, a single clone has been selected and named as a cultivar. These plants are all rare, grown only by the keenest heather enthusiasts, and they will perhaps always be so; but their existence needs to be documented. In some instances, hybrid formulae have been used to designate the plants, although whether the plants are veritable hybrids has yet to be proven.

Bearing in mind that both Dr John Griffiths and Kurt Kramer have each attempted around 70 different combinations of species, and that there are others who take an interest in trying to raise novel hybrids, there is a good chance that other hybrids will be produced, propagated and distributed.

Erica andevalensis × *E. tetralix*

About 1987, Kurt Kramer cross-pollinated *Erica tetralix* 'Foxhome' and *E. andevalensis*, and viable seed was produced. Of the seedlings raised, one was selected, after it had flowered, for naming.

'Pat Turpin' (® no. 100): Twiggy, broad, spreading shrub (±0.35 m tall × ±0.55 m spread); foliage greygreen; flowers magenta (H14) above and mauve (H2) on underside; July–November.

This cultivar, which blooms continuously through the summer and autumn, was named in honour of Major-General Patrick G. Turpin (1911–1996) CB, OBE, Chairman of The Heather Society between 1977 and 1992. At the time of introduction, the name *Erica* × *stuartii* was considered to be applicable, because McClintock (1989a, 1992) had reduced *E. andevalensis* to a subspecies within *E. mackayana*. Consequently, a proposal by Small (1992) to publish a new binomial (*E.* × "turpiniana") was abandoned. Should the separateness of the three cross-leaved species be established beyond doubt, this artificial hybrid would represent a distinct hybrid and thus would merit a separate binomial. It is currently being sold as a cultivar of *E.* ×*stuartii*.

This same hybrid has also been raised by Dr John Griffiths.

Erica lusitanica × *E. carnea*

There is no possibility of this hybrid arising in the wild; the species' ranges do not overlap. I have seen a photograph of a seedling raised from this artificial cross by Kurt Kramer, and it is clearly different from the parent species. However, I have not yet been able to examine specimens.

Kramer raised nine seedlings, of which five bloomed in the spring of 2008.

Erica lusitanica × *E. erigena*

Barry Sellers (Norbury, London) has carried out a few deliberate crossings; like others, he failed to get any seed from *Erica australis* pollinated by *E. arborea*, whereas he has raised, as others have, *E.* × *darleyensis*. His singular achievement is an apparent hybrid from the Portuguese heath, *E. lusitanica*, pollinated by the Irish heath, *E. erigena* (see p.125): as noted the surviving clone is named 'Lucy Gena' (® E.2009: 03). This combination could occur in the wild, because the parents do grow in close proximity in Portugal and perhaps in north-eastern Spain.

The reverse cross, *Erica erigena* pollinated by *E. lusitanica*, was attempted by Kurt Kramer but failed. However, Dr John Griffiths also tried this combination and obtained viable seed; but the seedlings were lost before they could be assessed.

E. sicula × *E. spiculifolia* & *E. sicula* subsp. *bocquetii* × *E. spiculifolia*

These are among Kurt Kramer's hybrids, as yet unproven, involving the Balkan heath, *Erica spiculifolia*. They need to be investigated using DNA analysis. One difficulty with the known hybrid progeny of *E. spiculifolia* is that they tend to resemble that parent, perhaps because of the fact that there are likely to be six unpaired chromosomes in the cells of the hybrid. These chromosomes will have come from *E. spiculifolia*, and they seem very influential in terms of morphology.

According to Kurt Kramer's response (e-mail 13 July 2007) to my request for information about hybrids that he has created, living plants of the following hybrids exist or existed:

E. spiculifolia × *E. manipuliflora* (Kramer 2008; Kramer & Schröder 2009)
E. spiculifolia × *E. terminalis*
E. spiculifolia × *E. vagans*
E. spiculifolia × *E. tetralix* (successful but lost).
E. terminalis × *E. multiflora* (successful but lost).

Notes

[1] The idea of a hybrid between *Erica cinerea* and *Calluna vulgaris* has been around for a century or so; McClintock (1980c) has reviewed the subject in his paper on bell heathers with split corollas. It was supported by N. E. Brown (who had described some of the "minor" genera of Cape heaths for *Flora capensis*). Horticulturists also espoused the idea, notably A. T. Johnson (1956: 85) who mentioned two cultivars 'W. G. Notley' and 'Winifred Whitley' as "a brace of bi-generic hybrids (*E. cinerea* + *Calluna vulgaris*). Interesting botanically, they have no much garden value", he added. Chapple (1962: 152) concurred with the following nonsense: "Rarely does an alliance take place in which Heath and Heather become one and, in this instance ['W. G. Notley'], the marriage is highly successful."

"The mischief was widespread and the culmination ... was to mislead that fine dendrologist the late Dr G. Krüssmann" (McClintock 1980c). In 1960 Krüssman published × *Ericalluna bealeana*, based on 'W. G. Notley' (see also Krüssmann 1965), and a skirmish of words ensued as McClintock sought (1965: 15–17) to convince Krüssmann of his error. "The end of this was at Cologne Airport in 1976, when Dr Krüssmann greeted me with a recantation, conceding that I ... was right" (McClintock 1980c).

McClintock's *coup de grace* was to name the variant of bell heather possessing "deeply divided corollas with depauperate generative parts" — 'W. G. Notley' again provided the "holotype" — after Krüssmann: "From my respect and affection for him and in oblique reference to his non-existent bigeneric hybrid genus ..." (McClintock 1980c).

Ironically, *Erica cinerea* var. *kreussmaniana* [sic] D. C. McClint. was not validly published because the holotype was not composed of a single gathering as defined under the *International code of botanical nomenclature* (see Miller & Nelson 2008). Krüssmann's nothospecies × *Ericalluna bealeana* is invalid because he failed to cite a holotype. Ironically too, despite being based on an erroneous notion, the nothogeneric name × *Ericalluna* Krüssm. remains available "if and when known or postulated plants of *Calluna* Salisb. × *Erica* L. should appear" (McNeill *et al.* 2006: 124. Art. H.9 Ex. 2.).

[2] As noted by Webb (1967), he once held the view that "*Erica stuartii*" was "a very peculiar mutant" of *E. mackayana*. However, he revised that opinion and told the English botanist James Edgar Dandy (1903–1976) that the strange heather was "more like a mutant of *E.* × *praegeri*", whereupon Dandy "took this rather casual *obiter dictum* as a firm taxonomic opinion, with the result that this unique plant and the common and familiar hybrid are now regarded as two nothomorphs which must share the name *E.* × *stuartii*" (cf. Dandy 1958b). In retrospect, Professor David Webb's *obiter dictum* was quite correct, as was Dandy's decision to make *E.* × *praegeri* a synonym of *E.* × *stuartii*.

[3] David Edge (Forest Edge Nurseries, Wimborne, Dorset), pers. comm., September 2007; Peter Bingham (Kingfisher Nurseries, Gedney Hill, Lincolnshire), pers. comm., September 2007.

[4] The binomial *Erica hybrida* was in use as early as 1802 (Nelson & Small 2005) for a Cape heath, or a hybrid from a Cape heath; while subsequently the name may have been validly published, its application is most uncertain, contradictory and confused, rendering it valueless. Thus, to employ this *omnium gatherum* binomial for any plant originating as a cross between heathers indigenous to the northern hemisphere is incorrect. Alas, the practice persists in some nurseries.

[5] The binomial was in print as early as 1993 in *Der Heidegarten* heft **34**: 6 (1993), but as no Latin diagnosis accompanied the name, it was not validly published there (see Miller & Nelson 2008).

[6] McClintock's annotation reads: "Garden at 9 Ashlea Close, Garforth, Leeds, 22 Sept 1988 DMᶜC J Griffiths".

[7] The accompanying text cannot be construed as constituting a diagnosis under *ICBN (Vienna code),* Art 32.2 because it does not "distinguish the taxon from others".

[8] This can also be dismissed as validating the binomial under *ICBN (Vienna code)*, Art 32.2.

[9] Nothing about Dr Heyting's work is available in publications, as far as I can ascertain. Margareth Hop, Dr Heyting's successor as plant-breeder in Applied Plant Research, Lisse, Netherlands, informed me (e-mail, 30 March 2006) that she had "inherited the crossing records of Dr Heyting. At the end of the 1960s she has tried crossing E[rica] *cinerea* and *E. carnea* many times. As far as I can tell, all of these crossings resulted in no seeds, or seeds that didn't germinate."

[10] "Summary of hybridisation experiments"; unpublished ms by J. Griffiths, undated (c. 1989) (Major-General P. G. Turpin's papers).

[11] McClintock's annotation reads: "Bruckenthalia spiculifolia × Erica carnea Nursery at Edammerstr 26, Edewecht-Süddorf, 27 July 1990 DMcC K Kramer 3 of the clones Type."

APPENDIX I
Recommended & interesting cultivars

The cultivars that are described here are those that currently (2007) hold the Royal Horticultural Society's Award of Garden merit (indicated by ♛) or are recommended by The Heather Society (indicated by ⬡) (Small & Cleevely 1999; see also Davis 2006; Everett 2006): in September 2011, The Heather Society issued a revised list of "100 recommended heathers" (see *Bulletin of The Heather Society* 7 (14) (Autumn 2011)). The AGM heathers should be readily available in Great Britain, at least.

Also included here are a small number of cultivars of botanical or horticultural interest, including some selections of the most recently raised artificial hybrids.

Many other named cultivars are mentioned in the preceding pages, and some descriptions are provided for recently named clones of the artificial hybrids. Cultivar names that are formally registered with The Heather Society are indicated by the symbol ® followed by the registration number (see Nelson & Small 2000, and subsequent Supplements to *International register of heather names* published annually in the *Yearbook of The Heather Society* (2001–2003) and *Heathers* **1–8** (2004–2011)).

The cultivars are arranged alphabetically according to their established names, and in the sections concerning *Daboecia* and *Erica*, also alphabetically under the Latin name of the taxon to which they belong.

CALLUNA: LING

Calluna vulgaris: ling, heather.

'Alexandra' PBR ♛ Upright, compact, bushy heather (±0.3 m tall × ±0.4 m spread); foliage dark green; young flower buds bicoloured (white and pale crimson), darkening to deep crimson (H13) and finally ageing a uniform dark ruby (H5), never opening; August–December (Small 1996b).

A "startling success" (Small 1996b), introduced in 1993. One of the many so-called "bud-bloomers" raised by Kurt Kramer, Edewecht-Süddorf, Germany. These "bud-bloom" cultivars were raised as hardy substitutes for the Cape heath *Erica gracilis* L., and millions are now sold principally for winter "window-box" planting and decorating graves on All Saints' Day (Small 1996b).

'Alexandra' was named by Kramer after the late Doris Nefedov (née Treitz; 1942–1969), a popular Lithuanian-born singer and songwriter, who used the stage-name Alexandra. AGM 2002.

'Alicia' PBR ⬡ ♛ (*Calluna vulgaris* f. *alba*). Floriferous, bushy, compact, upright shrub (±0.3 m tall × ±0.4 m spread); foliage bright green; flower buds white, larger than those of 'Melanie' (see below), never opening; August–December.

Another of the so-called "bud-bloomers" raised and introduced about 1993 by Kurt Kramer, Edewecht-Süddorf, Germany. While similar to 'Melanie', 'Alicia' has a stiffer habit, larger buds, and a longer "flowering" period. AGM 2002.

'Allegro' ♛ Vigorous, neat, tall shrub (±0.5 m tall × ±0.6 m spread); foliage dark green; flowers ruby (H5), in long spikes; August–October.

This distinctive clone (Small & Cleevely 1999) was a seedling from 'Alportii Praecox', raised by P. Bakhuyzen & Zonen, Boskoop, Netherlands; introduced by Bakhuyzen & Zonen in 1977. AGM 1993.

'Amilto' 🏵 Broad, upright heather (±0.35 m tall × ±0.5 m spread) notable for its coloured foliage which in winter is orange-red and bronze; the young shoots are bronzy yellow in spring, maturing to yellow in summer; flowers deep magenta (H4); August–September.

This is a very hardy plant and it needs harsh conditions to show its brightest foliage colours, so it is likely to perform best in more northern and cooler places. 'Amilto' was a chance seedling found by J. J. M. C. van Steen, Etten-Leur, Netherlands, and was introduced by P. G. Zwijnenburg, Boskoop, Netherlands, in 1982. The name is an acronym of Algemene Middenstands Industrie- en Landbouwtentoonstelling.

'Anette' PBR ♛ Upright, bushy plant (±0.35 m tall × ±0.4 m spread); foliage green throughout the year; flower buds magenta (H14), never opening; August–November.

Paul Wolf, Seligenstadt, Germany, found 'Anette' as a sport on 'Melanie' in 1993; he named it after his daughter-in-law. Plant breeders' rights were granted in 1997 to Kurt Kramer, Edewecht-Süddorf, Germany, who released this fine cultivar in 1996. Small & Small (2001) regard this "bud-bloomer" as outstanding, and recommend 'Anette' for planting in containers such as window-boxes, while Wulff (2007) commended it for retaining "its true green foliage color in winter". AGM 2002.

'Annemarie' 🏵 ♛ — see plate 2 (centre) (pp. 54–55).

'Anthony Davis' 🏵 ♛ (*Calluna vulgaris* f. *alba*). Erect, compact shrub (±0.45 m tall × ±0.5 m spread); foliage grey-green; flowers white, in long spikes; August–September.

'Anthony Davis' was found by Peter G. Davis, Timber Tops, Haslemere, Surrey, who introduced it in 1970. He named this fine cultivar after his son. Soon after its introduction, 'Anthony Davis' was praised as outstanding, "a flower-arranger's joy" (*The Heather Society bulletin* **2** (3): 6 (Autumn 1974)). AGM 1993.

'Arabella' PBR ♛ Floriferous heath; open habit, erect (±0.3 m tall × ±0.4 m spread); foliage dark green; flowers "luminous pure red"; August–September.

Bred and selected by Kurt Kramer, Edewecht-Süddorf, Germany, before 1995; Kramer also introduced 'Arabella', which is protected by plant breeders' rights (granted 15 October 1996). Small & Small (2001) reckoned that this showed great promise.

'Athene' PBR 🏵 Upright heather with stiffly erect shoots (±0.4 m tall × ±0.4 m spread); foliage dark green throughout the year; flower buds glowing ruby red to cerise (H5–H6), never opening and their colour lasting from September into November.

'Athene' (® C.2004:01), named after the Greek goddess, is one of Kurt Kramer's "Gardengirls"™ series of introductions protected by plant breeders' rights. This cultivar was found by Wilfried Holzwart, Hoisdorf, Germany, as a sport on 'Amethyst' on 23 September 2002. Like most of the "Gardengirls" ™ plants it is intended for temporary use in tubs and planters to provide a colourful display in autumn.

'Beoley Gold' ♀ (*Calluna vulgaris* f. *alba*). Upright shrub (±0.35 m tall × ±0.5 m spread); foliage yellow throughout the year; flowers white, not abundant; August–September.

Underhill (1990) noted that 'Beoley Gold' did not perform as well as expected in trials during the late 1960s at the Royal Horticultural Society's Gardens, Wisley, in Surrey but elsewhere, as in Worcestershire, it had brilliant foliage. Now regarded as one of the best heathers for foliage colour, the colour does depend on aspect, being more golden on the side exposed to the sun (south-facing in northern hemisphere) (Rogers 2006). 'Beoley Gold' is thought to have been a spontaneous seedling from 'Gold Haze'; like 'Beoley Crimson' it was found at Beechwood Nursery by J. W. Sparkes and was introduced by Sparkes by 1963. AGM 1993.

'Con Brio' 🏵 Neat, bushy heather (±0.35 m tall × ±0.45 m spread) with a fine, unusual fragrance; foliage in winter is reddish bronze turning dark yellow-green in summer; flowers ruby (H5); August–September.

This arose as a sport on 'Allegro' (see above) from which it is distinguished by the yellow and bronze colours of the foliage in summer and winter. Introduced by P. Bakhuyzen & Zonen, Boskoop, Netherlands, in 1981, the same nursery that raised 'Allegro'. The flowers are distinctly fragrant.

'County Wicklow' 🏵 ♀ Floriferous, spreading, compact ling (±0.25 (– 0.35) m tall × ±0.35 (– 0.5) m spread; foliage bright green; flowers "double", shell pink (H16); August–October.

'County Wicklow' needs to be carefully pruned, otherwise it is rather straggly. As its name suggests, this splendid heather was found in County Wicklow, Ireland. Miss Meta Archer (Nelson 1981b, 2000e) noticed it growing at Lough Dan in the Wicklow Mountains, and the clone was propagated and introduced by Maxwell & Beale Ltd, Broadstone, Dorset, in 1933. A water-colour portrait of 'County Wicklow', signed "OJB" (who has not been identified), was published in Maxwell & Beale's 1935 catalogue (Cleevely 1997). Leslie Slinger (1907–1974), proprietor of the Slieve Donard Nursery, Newcastle, County Down, Northern Ireland, praised it for its "mat-like growth": "... 'County Wicklow' appeals to me so much that it must come in any short-list of heathers which I favour" (Slinger 2007). AGM 1993.

'Dark Beauty' ᴾᴮᴿ ♀ Compact heath (±0.25 m tall × ±0.35 m spread); foliage dark green; flowers "semi-double", cerise (H6) ageing to rich ruby (H5); August–October.

'Dark Beauty' was a sport produced by 'Darkness' (see below) after X-ray treatment in 1986; the flowers are darker than those of its parent. It was propagated and introduced in 1989 by H. Hoekert, Oldebroek, Netherlands, and is protected by a grant of plant breeders' rights to F. Maassen, Straelen, Germany. Platt (1992) described 'Dark Beauty' as "very choice": "the semi-double flowers have a wonderful colour, opening blood red and later turning purple red." Small & Cleevely (1999) regarded 'Dark Beauty' as outstanding, while Small & Small (2001) praised it as one of the best cultivars introduced in recent years.

However, like many cultivars that arose as sports, this can revert. In North America these paler-blossomed reversions have apparently been propagated and are sometimes sold under the name 'Dark Beauty'. Thus anyone seeking this clone needs to see it in flower. AGM 2002.

'Darkness' 🐾 ♔ Erect, compact, floriferous heath (±0.25 m tall × ±0.35 m spread); foliage dark green; flowers crimson (H13) in very dense spikes; August–October.

This is an outstanding ling, raised — or found — by M. C. ("Denny") Pratt before 1962. According to Underhill (1990), he gave propagation material to the nearby University of Liverpool Botanic Gardens, Ness, The Wirral, Cheshire, so that this was introduced by Ness Botanic Garden. The name is a pun on Ness. AGM 1993.

A sport from 'Darkness', having green foliage flecked with red and yellow, is named 'Denny Pratt'. 'Dark Beauty' (see above) and 'Dark Star' (see below) are also its progeny.

'Dark Star' ♔ Compact, neat ling (±0.2 m tall × ±0.35 m spread); foliage dark green; flowers variously described as a mixture of single and "double", or as "semi-double", crimson (H13); August–October.

Ernst Stührenberg, Wiesmoor, Germany, found this as a sport on 'Darkness' in 1981, and it was introduced in 1984. It is more showy than 'Darkness' but not as red as 'Dark Beauty' (Small & Small 2001). AGM 1993.

'Elsie Purnell' 🐾 ♔ Open, upright shrub (±0.4 m tall × ±0.75 m spread); young foliage dark green turning grey-green and brown-green in winter; flowers "double", lavender (H3) in very long spikes; August–October.

'Elsie Purnell' was a sport on 'H. E. Beale', and was found by J. W. Sparkes, Beechwood Nursery, Beoley, Redditch, Worcestershire, in 1951. While he first exhibited the clone in 1954, 'Elsie Purnell' was not introduced commercially until the early 1960s. The winter foliage is drab but the long flower-spikes are outstanding. AGM 1993.

Elsie was the wife of Ralph Purnell, of Edinburgh and Solihull, Warwickshire, and, like her husband, was a keen member of the Alpine Garden Society (Underhill 1990). There is also a cultivar of *Calluna* named after Ralph Purnell

'Firefly' 🐾 ♔ — see plate 2 (upper right) (pp. 54-55).

'Gold Haze' ♔ (*Calluna vulgaris* f. *alba*). Upright, sprawling plant (±0.3 (– 0.45) m tall × ±0.5 m spread); foliage yellow-green in spring with yellow tips to the shoots, turning yellow in summer and fading to yellow-green in winter; flowers white in long spikes; August–October.

'Gold Haze' is yet another of Sparkes's introductions; it was a sport on 'Hayesensis' found and introduced by J. W. Sparkes, Beechwood Nursery, Beoley, Redditch, Worcestershire, before 1961. Slinger (2007) was disappointed: "'Gold Haze' came into cultivation with a great reputation but it disappoints me; maybe this is because with me the plant has always a hard look and will quite quickly grow into a plant which only maintains its golden colour at the ends of the branches." AGM 1993.

'J. H. Hamilton' ♔ Dwarf, spreading, neat ling (±0.1 m tall × ±0.25 m spread); foliage dark green; flowers "double", bright pink (H8); July–September.

This cultivar was found near Moughton, north of Settle, Yorkshire. When introduced by Maxwell & Beale Ltd, Broadstone, Dorset, in 1935, it was named 'Mrs J. H. Hamilton'; mysteriously, the next year the name was changed by the nursery. An anonymous water-colour portrait of 'J. H. Hamilton' was included in the nursery's 1936 catalogue (Cleevely 1997). AGM 1993.

J. H. Hamilton was a director of Maxwell & Beale Ltd.

'Joy Vanstone' ♇ Upright shrub (±0.3 (– 0.5) m tall × ±0.5 (– 0.75) m spread); foliage golden green ("straw-coloured") in summer, orange in winter; flowers lavender (H3); August–September.

J. W. Sparkes, Beechwood Nursey, Beoley, Redditch, Worcestershire, introduced this cultivar before 1963, and named it after Mrs Joy Vanstone (née Waldron), Lowson Ford, Warwickshire, wife of Jack Vanstone, then director of J. V. White of Birmingham. "A beautiful, much under-rated plant", according to Small & Small (2001). AGM 1993.

'Kerstin' 🏴 ♇ Hardy, erect, vigorous bush (±0.4 m tall × ±0.6 m spread); young shoot tips pale yellow and red in spring, mature foliage deep lilac-grey; flowers mauve (H2); August–September.

A seedling, deliberately raised and selected by Brita Johansson, Vargön, Sweden, in 1983, and named (® no. 62) by Mrs Johansson in 1988 after her daughter. The cultivar was introduced by Mrs Tessa Forbes, Plaxtol Nurseries, Plaxtol, Kent, in 1988. AGM 2002.

'Kinlochruel' (*Calluna vulgaris* f. *alba*) ♇ — see plate 2 (lower right) (pp. 54–55).

'Mair's Variety' ♇ (*Calluna vulgaris* f. *alba*). Large, strong-growing, upright shrub (±0.5 m tall × ±0.6 m spread); foliage bright green; flowers white, in long, tapering spikes (to 0.25 m long); August–September.

An ideal heather for using as a cut-flower, 'Mair's Variety' has been in cultivation since the early 1930s but its history is not recorded. Johnson (1932) called it a "stranger ... a peculiarly robust [ling] notable for the great length of its flower-spike." Slinger (2007) noted that 'Mair's Variety' "... has the longest stem of flowers but to me it is untidy but I daresay useful for cutting." AGM 1993.

'Melanie' ᴾᴮᴿ (*Calluna vulgaris* f. *alba*). Erect heath (±0.35 m tall × ±0.4 m spread); hairy foliage green; flower buds white, never opening, in long spikes; August–November.

This ling is one of the so-called "bud-bloomers" and is another outstanding cultivar, recommended for flower-arrangers. 'Melanie' was raised in 1984 and named by Kurt Kramer, Edewecht-Süddorf, Germany. It is a selected seedling from the deliberate crossing of 'Marleen' and 'Hammondii'. The cultivar is protected by plant breeders' rights (granted in 1991). 'Melanie' is a fantasy name (® no. 85).

'Mullion' ♇ Neat, spreading, low-growing ling (±0.25 m tall × ±0.5 m spread) with copiously branched shoots; foliage dark green; flowers lilac pink (H11) in branching spikes; August–September.

While the Maxwells were on their honeymoon in 1923, they found several superb heathers: this is one of these post-nuptial cultivars. 'Mullion' came from Mullion Cove on the Lizard Peninsula, Cornwall. The cultivar was named and introduced by Maxwell & Beale Ltd, Broadstone, Dorset, in 1928. Johnson (1932) described 'Mullion' as "among the best ... its peculiar attractions being a close, compact, prostrate habit, a wealth of much-branched and copiously flowered spikes of a bright bluish pink, a long season and bright moss-green foliage." AGM 1993.

'My Dream' ♇ (*Calluna vulgaris* f. *alba*). Erect shrub (±0.45 m tall × ±0.5 m spread); foliage dark green; flowers "double", pure white; August–October.

This is a showy ling, a sport from 'H. E. Beale' (to which it has a tendency to revert). The sport was noticed on 15 September 1974 during a visit by the Midland Group of The Heather Society to Graham Cookes' garden, Little Froome, Drayton Lane, Fenny Drayton, Warwickshire. 'My Dream' was named by Mr Cookes in 1976 (® no. 2), and introduced by Blooms of Bressingham, Diss, Norfolk, in 1978. AGM 2002.

'Peter Sparkes' 🌸 ♛ Vigorous, branching, spreading shrub (±0.35 m tall × ±0.5 (−1) m spread); foliage dark grey-green; producing long spikes covered densely with "double", rose-pink (H7) flowers; August −November.

'H. E. Beale' has produced a clutch of sports, some of which are now highly regarded: 'Elsie Purnell' (see above), 'My Dream' (see above) and 'Peter Sparkes', all AGM recipients, are the best. The others are 'Hatje's Herbstfeuer', 'Heike', 'Schurig's Sensation', 'Snowball' and 'Sonja'.

'Peter Sparkes' was found and introduced before 1957 by J. W. Sparkes, Beechwood Nursery, Beoley, Redditch, Worcestershire. AGM 2002.

'Radnor' ♛ Compact yet vigorous, spreading plant (±0.3 m tall × ±0.5 m spread); foliage bright green; flowers "double", pale pink (H16) on short, curving spikes; August–October.

Found in Radnor Forest near Presteigne, Wales, by Miss H. M. Appleby in 1954 and introduced by Dingle Hollow Nursery, Stockport, Cheshire, shortly afterwards. 'Radnor' blossoms earlier than 'H. E. Beale' and is probably the best "double" pink-flowered ling for a restricted space. AGM 1993.

'Red Beauty' 🌸 Upright heather with erect shoots; foliage dark green; flowers "double" in dense long "spikes".

This name has been used for several clones of *Calluna* that were submitted for plant breeders' rights by Franz Maassen and Gerd Canders, Straelen, Germany; this one is CLL302. It is very similar to 'Dark Beauty' but has longer flowering shoots and redder flowers.

'Red Favorit' 🌸 Low-spreading, bushy heather (±0.2 m tall × ±0.7 m spread); foliage dark green with bronze tints in winter; flowers "double", deep crimson (H13), buds red; August–September.

'Red Favorit' was a sport on 'J. H. Hamilton' found in Germany and introduced by Reinhold & Hinrikus Barth, Bad Zwischenahn, in 1983.

'Red Star'. Erect heath (±0.4 m tall × ±0.6 m spread); foliage dark green, shading to bronze-green in winter; flowers "double", cerise (H6); September–October.

The colour of the flowers of this outstanding clone is redder than 'Annemarie', from which it was a sport deliberately induced by irradiation, selected and introduced by Kurt Kramer, Edewecht-Süddorf, Germany, in 1984.

'Red Star' itself has sported several times, but the resulting named cultivars are not exceptional. As with 'Dark Beauty', plants sold under this name in North America need to be checked when in bloom to ensure that they are true to type, and not pale-flowered reversions.

'Robert Chapman' 🌸 ♛ — see plate 1 (lower centre middle) (pp. 52–53).

'Roland Haagen' ♛ Low-growing, compact shrub (±0.15 m tall × ±0.35 m spread); foliage golden-yellow in summer, darkening to orange, with darker shoot tips, in winter; flowers mauve (H2); August–September.

Recommended (Small & Small 2001): an Irish cultivar derived from a sport on a "normal" wild plant that was spotted near Carrick-on-Suir, County Waterford, by Rinus Zwijnenburg, Boskoop, Netherlands, whilst on holiday in 1970. 'Roland Haagen' was named after a friend who was on holiday with Zwijnenburg when the sport was found. The cultivar was introduced by P. G. Zwijnenburg in 1972. AGM 1993.

'Serlei Aurea' ♔ (*Calluna vulgaris* f. *alba*). Upright, low-growing, "tidy" shrub (±0.3 m tall × ±0.4 m spread); foliage golden-yellow, with more pronounced yellow tips to the shoots in summer and autumn; flowers white, single; September–October.

This arose as a sport on the old cultivar 'Serlei' and differs from it in having golden foliage. 'Serlei Aurea' was introduced by W. G. Slocock, Goldsworth Nursery, Woking, Surrey, in 1927. The origins of 'Serlei' are unknown, and the derivation of the name is also lost. AGM 1993.

'Silver Knight' — see plate 1 (lower centre right) (pp. 52–53).

'Silver Queen' 🎖️ ♔ Broad, spreading heather (±0.4 m tall × ±0.55 m spread); foliage very downy with long silvery hairs and so appearing silver-grey; flowers single, lavender (H3); August–September.

There is a fair certainty that more than one plant called 'Silver Queen' has existed in nurseries in Ireland and Britain. The earliest introduction was from Daisy Hill Nursery, Newry, County Down, Northern Ireland, about 1937. This was described in the nursery's catalogues as "An interesting and charming find. Its growths are quite prostrate and are covered with a silvery tomentum, its flowers are pale pink'" (Nelson & Grills 1998). However, it is known that in later years Daisy Hill Nursery also marketed a more upright clone as 'Silver Queen', and probably it is this one that has persisted. Unravelling the skeins of its history is now impossible, as the nursery no longer exists (Nelson & Grills 1998). The first 'Silver Queen' was collected in the wild, somewhere near Newry.

To complicate matters further, another 'Silver Queen' of English origin was introduced by Primrose Hill Nursery, Haslemere, Surrey, before 1952; this was said to have been found by Mrs Olive Cowan in her garden at Farnham, Surrey.

'Silver Queen', in its current incarnation, was given an AGM in 1993.

'Silver Rose' 🎖️ ♔ Upright, bushy heather (±0.4 m tall × ±0.5 m spread) with silver-grey, hairy foliage; flowers single, lilac-pink (H11); August–October.

John Letts found this before 1968 growing wild on Sunningdale Golf Course, Surrey. Evidently the golf course was a good place for hunting for heathers, as the same place yielded *Calluna vulgaris* 'Finale' and a clutch of *Erica cinerea* cultivars: 'Apricot Charm', 'Rock Pool', White Dale', 'Windlebrooke' and probably also 'Golden Tee'. 'Silver Rose' and most of those other heathers were also introduced by J. F. Letts, Foxhollow, Windlesham, Surrey. AGM 1993.

'Sir John Charrington' ♔ Low-growing, broad, upright heath (±0.2 (– 0.4) m tall × ±0.4 (– 0.6) m spread); foliage light bronze in spring, turning yellow-green in summer, after which the shoot tips go mahogany and scarlet before all the foliage becomes red, staying like that throughout the winter; flowers single, deep lilac-pink (H11); August–October.

Sir John Charrington (1886–1977) was the founder of The Heather Society and its first chairman and later its President, and this ling was named after him and presented to him at a birthday-celebration lunch held in his honour at the Royal Horticultural Society's Gardens, Wisley, on 30 July 1966. He was "a doyen of the coal trade and a gentleman of great charm" (McClintock 1987).

The cultivar was raised, and was introduced in 1966, by J. W. Sparkes, Beechwood Nursery, Beoley, Redditch, Worcestershire. AGM 1993.

'Sirsson'. More spreading than 'Sir John Charrington' (±0.3 m tall $\times$ ±0.5 m spread); foliage golden in summer, changing to bright orange-red in winter; flowers pink (H8); August–September.

Don Richards found this chance seedling in his garden at Rydal Mount, Eskdale, Cumbria, about 1978. The odd and witty name arose because 'Sirsson' is thought to be a seedling from 'Sir John Charrington'— thus "Sir's son". Geoffrey Yates, Ash Landing Gardens, Far Sawrey, Ambleside, Cumbria, introduced 'Sirsson' before autumn 1980.

'Sister Anne' 🌼 ♀   Spreading, low-growing, compact heath (±0.1 m tall $\times$ ±0.25 m spread); foliage grey-green in summer, acquiring a dull bronze tone in winter; flowers single, mauve (H2); August–October.

Miss Anne Moseley collected 'Sister Anne' about 1929 on the Lizard Peninsula, Cornwall, and it was propagated and introduced before 1940 by W. E. Th. Ingwerson, Birch Farm Nursery, Gravetye, East Grinstead, Surrey. Miss Moseley was a nurse, but it was because her sister called her "Sister Anne" that the heather got its name. AGM 1993.

'Spring Cream' 🌼 ♀ (*Calluna vulgaris* f. *alba*). Compact shrub (±0.35 m tall $\times$ ±0.45–0.6 m spread); new shoot-tips in spring cream, turning green, and then yellow in autumn and winter; flowers single, white; August–November.

Found, propagated, and introduced about 1966, by J. W. Sparkes, Beechwood Nursery, Beoley, Redditch, Worcestershire. AGM 1993.

'Sunset' ♀   Spreading, low-growing plant (±0.2 m tall $\times$ ±0.45 m spread); foliage red in autumn and winter, acquiring bronze tints in spring, and changing to golden yellow in summer; flowers single, lilac-pink (H11); August–October. AGM 1992.

Another cultivar that was found, propagated and introduced (in 1963) by J. W. Sparkes, Beechwood Nursery, Beoley, Redditch, Worcestershire. AGM 1993.

'Tib' 🌼 ♀   Spreading, open heath (±0.3 m tall $\times$ ±0.4 m spread); foliage dark green; flowers "double", heliotrope (H11/H12); July–October.

Collected in 1934 in the Pentland Hills, Scotland, by Miss Isobel Young, from Currie, Edinburgh, 'Tib' was introduced by Maxwell & Beale Ltd, Broadstone, Dorset, in 1938. Its name was Miss Young's nickname.

'Tib' is one of the earliest of the "double"-blossomed lings to bloom, and a redoubtable prize-winner. As well as holding the Royal Horticultural Society's AGM (1993), 'Tib' gained more prizes than any other heather on the show tables at RHS shows between 1972 and 1996: 16 firsts, 13 seconds, 8 thirds and 1 fourth (Nelson 2005d).

'Velvet Fascination' 🌼 ♀ (*Calluna vulgaris* f. *alba*). Tall, erect shrub (±0.5 m tall $\times$ ±0.7 m spread); foliage silvery green, not changing markedly in winter; flowers single, white; August–September.

This arose as a sport on 'Silver Knight' in H. Hoekert's nursery, Oldebroek, Netherlands, in 1979; introduced in 1983. Small & Small (2001) described it as a "first-class plant which does not become drab in winter". AGM 2002.

'White Coral' 🏅 (*Calluna vulgaris* f. *alba*). Spreading, low-growing heath (±0.2 m tall × ±0.4 m spread); foliage bright green all year; flowers "double", white; September–October.

Kurt Kramer, Edewecht-Süddorf, Germany, found this as a sport on 'Kinlochruel' about 1984, and he introduced the cultivar in 1990.

Similar sports, differing from the parent cultivar in having plain green foliage that does not acquire bronze tint, have occurred elsewhere. The late Herman Blum found one in the heather garden at De Voorzienigheid, Steenwijkerwold, Holland, also in 1983. John H. Bell noticed a small shoot on a plant of 'Kinlochruel' growing at Whinfell Nursery, Craketrees Farm, Whinfell, Cumbria, in 1994; he propagated this and named the cultivar 'Whinfell White', but the plant was indistinguishable from 'White Coral'.

'White Lawn' 🏆 (*Calluna vulgaris* f. *alba*). Completely prostrate heather (±0.05 m tall × ±0.4 m spread); foliage green; flowers single, white, in long spikes; August–September.

This distinctive cultivar was a wilding found during the 1930s near Loch Fyne, Argyll, Scotland, by a Mr Graham, a forester. It circulated as a unnamed heather for several decades: Mrs Olive Cowan grew the clone in her garden at Farnham, Surrey, about 1967, having received it from Gilbert Barratt of Mitchett, near Camberley, also in Surrey, who in turn had obtained it from a Scottish nursery. John Hall, Windlesham Court Nursery, Surrey, introduced 'White Lawn' after it was named by David McClintock in 1977. AGM 1993.

'Wickwar Flame' 🏅 🏆 Vigorous, low-growing, spreading heather (±0.5 m tall × ±0.65 m spread); winter foliage bronze to bright red changing to golden yellow in summer, and then to orange and golden bronze in autumn; flowers single, mauve (H2/H3); August–September.

Found, and introduced in 1970, by G. Osmond, Archfield Nursery, Wickwar, Wooton under Edge, Gloucestershire. This plant needs exposure to sun and cold to produce its striking colours, and, as with all coloured-foliage *Calluna* cultivars, the colours vary with aspect: plants will look quite a different colour when viewed from the north than when viewed from the south. AGM 1993.

'Yvette's Gold' 🏅 Spreading, bushy heather (±0.3 m tall × ±0.5 m spread); foliage golden; flowers white; August–October.

This Canadian heather was a chance seedling found by Yvette Knutson at Chilliwack, Sardis, British Columbia, before 1983 when she was working in David Wilson's nursery. Wilson propagated and introduced 'Yvette's Gold' about 1992.

DABOECIA: SAINT DABEOC'S HEATHS

Daboecia azorica: Queiró, Azores heath.

'Arthur P. Dome'. Almost prostrate sub-shrub (±0.15 m tall × ±0.4 m spread), slow-growing; foliage green; flowers ruby; June–July.

Art Dome, Seattle, Washington, USA, obtained this clone from Mick Jamieson, Oak Bay Nursery, Victoria, British Columbia, Canada, on 27 March 1940 (Dome 2000; Dome *et al.* 2001). This is the only clone that has been named.

Daboecia cantabrica: **Saint Dabeoc's heath.**

'Arielle' ♀ Robust, compact, upright shrub (±0.3 m tall × ±0.5 m spread); foliage dark green; flowers glowing magenta (H14); July–October.

A deliberately raised seedling, selected, named, and introduced in 1995, by Kurt Kramer, Edewecht-Süddorf, Germany. This is a better "red" St Dabeoc's heath than earlier introductions, with larger flowers than 'Waley's Red' (see below), and "promises to be better than all the other" (Small & Small 2001) clones with similar flower colour. AGM 2003.

'Bicolor' ♀ Spreading, vigorous shrub (±0.35 m tall × ±0.75 m spread); foliage green; flowers variable, purple or white, or intermediate shades of pale mauve, and often striped white and beetroot (H9); June–November.

Douglas Fyfe Maxwell's description cannot be bettered: "A delicious piece of nonsense, the comic of the group. With complete inconsequence flowers of white, pale pink and ... rosy-purple, are slap-dashed about. A purple flower may have a white flower on the same stem, or a quarter, or a half of it may be white; every combination of the three is possible" (Maxwell & Patrick 1966). An earlier, anonymous (1874a, 1874b) writer remarked that

> its chief peculiarity ... is that its flowers are sometimes purple, sometimes white, sometimes of various intermediate shades of blush-white, pallid purple, or pink. In the majority of cases the spikes bear flowers uniform in colour, all purple, all blush, or all white, but these are so mixed up that it appears as though two or three varieties were accidentally associated in the tuft ... and not unfrequently there appears a spike on which both pure purple and pure white flowers are associated, or even sometimes may be seen flowers in which one-half of the tube is white and the other half purple.

Plants should be seen in flower before they are acquired to ensure that they are as variously multicoloured as described.

This "comic's" history is not known with any certainty, and it has had a number of other Latin names including "*striata*" and "*variicolor*". All that can be concluded is that the cultivar now called 'Bicolor' was being marketed by the third quarter of the nineteenth century. A plant resembling it was mentioned in *The garden*, William Robinson's recently established weekly paper, on 14 November 1872. Waterer's Knap Hill nursery had it as early as 1874 (Anonymous 1874a, 1874b), while the Irish nursery, Rodger, McClelland & Co., Newry, County Down, listed it (as *Menziesia polifolia*) in 1882 (cf. Nelson & Small 2000: **1**: 82).

Underhill (1990) recorded that he had "heard of seedlings with the bicolour blooms arising from the seed of 'Bicolor'." AGM 1993.

'Blueless' 🏵 Compact, bushy evergreen shrub (±0.35 m tall × ±0.5 m spread); foliage dark green; flowers rose pink (H7) flowers; June–October.

This seedling was one of several that came from the late Don Richards (1915–2009), Rydal Mount, Eskdale, Cumbria, who, incidentally, invented and developed the Dewpoint Cabinet. It was introduced by Geoff Yates, Far Sawrey, Ambleside, before 1980. Comparing the red/pink-blossomed cultivars of St Dabeoc's heath, Yates (1985) recorded that the flowers of 'Blueless' are an even clearer pink than those of 'Praegerae', 'Waley's Red' and 'Cupido': "the colour of a plant in full bloom being less hard, but not quite so bright." The name is a pun, of course.

'Cinderella' 🏆 Neat, bushy evergreen shrub (±0.3m tall × ±0.35m spread); foliage dark green; flowers large, essentially white with a pale pink (H16) flush at the base of the corolla and sometimes also at its "mouth", and brownish-purple sepals; June–October,

This is described as a sport of 'Bicolor', the bizarre plant which has multi-coloured, even variegated, flowers. 'Cinderella' was introduced in 1970 by P. G. Zwijnenburg, Boskoop, Netherlands. Its fault, however, is that it can revert to 'Bicolor'.

'Charles Nelson' 🏆 — see plate 5 (p. 71), plate 6 (pp. 78–79), and Fig. 219.

'David Moss' 🏆 (*Daboecia cantabrica* f. *alba*). Floriferous, vigorous, compact shrub (±0.35 m tall × ±0.45 m spread); glossy foliage dark green; flowers white in long (0.15–0.3 m), erect racemes; June–October.

This presumably arose spontaneously as a seedling before 1968, when it performed well at a trial of white-blossomed clones of St Dabeoc's heath carried out at the Royal Horticultural Society's Gardens, Wisley. Subsequently, this selection was named — after his son "because he is a cricketer — white flannels" (McClintock 1984a) – and introduced by W. Moss, Afonwen, Mold, Flintshire, Wales. AGM 1993.

'Hookstone Purple' 🏆 Tall, vigorous, spreading shrub (±0.45 m tall × ±0.85 m spread); foliage green; flowers large, amethyst (H1); May–November, but not blooming well in dry summers.

This long-flowering cultivar was introduced about 1960 by G. Underwood & Son, Hookstone Green Nursery, West End, Woking, Surrey. Maxwell & Patrick (1966) dismissed it as "similar colour but not so good as the type", yet its prolonged blooming season makes it a useful heather.

Fig. 219. *Daboecia cantabrica* 'Charles Nelson' in cultivation (E. C. Nelson).

'Praegerae' — see plate 6 (pp. 78-79).

'Rodeo' ♀ Upright heather (±0.45 m tall × ±0.6 m spread); foliage bright green; flowers pale shell-pink (H16), almost white; June–October.

'Rodeo' arose as a sport on 'Alba Globosa' in 1981, and was named and introduced in 1983 by P. Bakhuyzen & Zonen, Boskoop, Netherlands. AGM 2003.

'Waley's Red' 🏵 ♀ Vigorous, spreading shrub (±0.35 m tall × ±0.5 m spread); foliage green; flowers magenta (H14); June–October.

Frank Raphael Waley (1893–1987), Sevenoaks, Kent, collected this clone in northern Spain about 1970. The cultivar was named and introduced by B. & V. Proudley, St Briavels, Gloucestershire, in 1976. There is a trace of blue in the corolla, so it is not as pure a red as 'Praegerae'; otherwise, it is a better plant (Small & Small 2001). AGM 1993.

Daboecia × *scotica*: **Hybrid Saint Dabeoc's heath.**

'Barbara Phillips' ♀ Floriferous, spreading bush (±0.4 m tall × ±0.65 m spread); foliage dark green; flowers amethyst (H1); June–November.

'Barbara Phillips', named after the wife of Brigadier C. F. Lucas-Phillips, was selected by Peter G. Davis, Marley Common, Haslemere, Surrey, from a clutch of about 600 seedlings that had been raised from open-pollinated *Daboecia* × *scotica* 'William Buchanan'. Thus its correct name is also *D.* × *scotica* and not *D. cantabrica* (Nelson & Small 2000: 1: 72). AGM 2003.

'Jack Drake' ♀ Neat, compact heather (±0.15 m tall × ±0.3 m spread); foliage dark green; flowers ruby (H5) ; May–July.

One of the seedlings that arose in the garden of William Buchanan, Bearsden, Glasgow, before 1960; introduced (as "Seedling no. 3") by Inshriach Nursery, Aviemore, Inverness-shire, and subsequently named after the nursery's owner, Jack Drake. AGM 1993.

'Katherine's Choice' 🏵 Erect, evergreen, dwarf shrub (±0.2 m tall × ±0.25 m spread); foliage green; flowers cerise; May–July.

A seedling, raised and selected by David Edge, Wimborne, Dorset, and named after his daughter.

'Silverwells' ♀ (*D.* × *scotica* f. *albiflora*). Floriferous, neat, spreading shrub (±0.15 m tall × ±0.4 m spread); foliage green; flowers very profuse, white, May–September.

This seedling was raised by Alex Duguid at Edrom Nurseries, Coldingham, Berwickshire, in 1968, and was subsequently introduced by Edrom Nurseries. Silverwells was the name of the house of the proprietors of Edrom Nurseries, the Misses Logan Home. AGM 1993.

'William Buchanan' 🏵 ♀ Compact, floriferous shrub (±0.35 m tall × ±0.55 m spread); foliage dark green; flowers deep crimson (H13); May–October.

One of the batch of chance seedlings that appeared before 1960 in William Buchanan's garden, Bearsden, Glasgow; selected (as "Seedling no. 1") and introduced by Inshriach Nursery, Aviemore, Inverness-shire. This cultivar yielded the holotype (**E**) for *Daboecia* × *scotica* D. C. McClint. AGM 1993.

ERICA: HEATHS & HEATHERS

Erica andevalensis

'Blanco de Odiel'(® no. E.2005:04) (*Erica andevalensis* f. *albiflora*). Upright, twiggy shrub (±1 m tall × ±0.35 m spread); foliage green; flowers white with amber-coloured anthers; July–September.

Collected by the Río Odiel, Huelva, Spain, on 18 July 1982 by D. McClintock, E. C. Nelson and D. J. Small (see Nelson & McClintock 1984).

Erica arborea: tree heath.

'Albert's Gold' 🌼 ♈ (*Erica arborea* f. *aureifolia*) — see plate 15 (pp. 198–199).

'Alpina' ♈ (*Erica arborea* var. *alpina*) — see plate 15 (pp. 198–199).

'Estrella Gold' 🌼 ♈ (*Erica arborea* f. *aureifolia*). Floriferous, broad, erect shrub (±1.2–1.5 m tall × ±0.75 m spread); young shoots bright yellow, mature foliage yellow-green, turning more golden in winter; flowers white; April–May (Small 1996a).

This clone came from Serra da Estrela, east of Coimbra, in Portugal. Rinus Zwijnenburg, Boskoop, Netherlands, collected it, and it was named and introduced by P. G. Zwijnenburg, Boskoop, Netherlands, by 1974. 'Estrella Gold' is relatively hardy. AGM 1993,

Erica × *arendsiana*: Arends's heath

'Charnwood Pink' (® E.2007:11. Erect, sparsely branched shrub (±1 m tall after 5 years); foliage green, young shoots discoloured; leaves in whorls of 4; pale pink flowers (H4 (RHS75D) lilac to H16 (RHS65C) shell pink; mainly in Autumn.

A seedling raised in 2000 by Kurt Kramer, Edewecht-Süddorf, Germany, and distributed in Britain through The Heather Society. This clone provided the holotype of *Erica* × *arendsiana*. 'Ronsdorf' (® E.2007:04) has darker, lavender (RHS75A, H3) flowers, and does not appear to be cultivated outside Germany at present.

Erica australis: southern heath.

'Mr Robert' ♈ (*Erica australis* f. *albiflora*) — see plate 17 (lower left) (pp. 224–225).

'Riverslea' 🌼 ♈ Floriferous, upright shrub (±1.2 m tall × 0.85 m spread); foliage dark green; flowers lilac-pink (H11); April–June.

A chance seedling (Maxwell & Patrick 1966), spotted by Aubrey Pritchard, and introduced by Maurice Pritchard & Sons, Riverslea Nursery, Christchurch, Hampshire, before 1946. The flowers are more numerous and deeper in colour than in other cultivated clones. 'Riverslea' can become rather ungainly and sprawling so it needs to be well pruned. AGM 1993.

'Trisha'. Bushy, erect shrub; foliage sage green; bract and bracteoles stained dark red; calyx with 4–9 unequal, free sepals; corolla deep lilac, with 4, 5 or 6 lobes, and may have a "flap" forming an extra lobe or a petaloid sepal fused to the outside; stamens and gynoecium appear to be normal; March–May.

A seedling raised by Kurt Kramer, Edwecht-Süddorf, Germany, and selected and named (® E.2007.01) by David Edge, Forest Edge Nursery, Wimborne, Dorset.

Erica carnea: winter heath.

'Adrienne Duncan' 🌣 ♀ Spreading, low-growing heather (±0.15 m tall × 0.35 m spread); foliage dark green with bronze tints; flowers dark heliotrope (H12); January–April.

This "superb" (Small & Cleevely 1999) cultivar was a chance seedling found by Lieutenant-Colonel J. H. Stitt in his garden, Drumcairn, Blairgowrie, Perthshire, Scotland, about 1955, and was introduced by Tabramhill Gardens, Newstead Abbey, Nottinghamshire, before 1973. Stitt named the clone after Lady Duncan of Jordanstone because she admired the plant. It is more vigorous and has a longer flowering period than 'Vivellii', which has similar bronze-tinged foliage (Small & Small 2001: 94). AGM 1993.

'Ann Sparkes' 🌣 ♀ (*Erica carnea* f. *aureifolia*) — see plate 8 (lower middle) (pp. 112–113); Fig. 220.

'Aurea' (*Erica carnea* f. *aureifolia*). Spreading, yet compact and low-growing (±0.15 m tall × 0.35 m spread); golden foliage is best in spring and early summer, when the shoot tips are orange; flowers opening pink (H8), changing with age to heliotrope (H12); January–May.

This is an old cultivar, introduced by C. Verboom of Boskoop, Netherlands, before 1928. Chapple (1964) described it as a very good plant, but two decades and more later, McClintock (1988a) and Underhill (1990) had reservations. All the same, 'Aurea' was still recommended (Small & Cleevely 1999), although more than one clone may be marketed under this name. The Heather Society rescinded its recommendation in 2011.

'Challenger' ♀ Spreading, low-growing subshrub (±0.15 m tall × 0.45 m spread); foliage dark green, tinted bronze; flowers with "bold" magenta (H14) corolla and crimson (H13) calyx (Fig. 220); January–April.

A cultivar introduced in 1986 by H. van Gemeren and S. C. van der Wilt, Boskoop, Netherlands. AGM 1993.

'Eva' 🌣 Neat, low, bushy heather (±0.2 m tall × ±0.3 m spread); foliage dark green acquiring bronze tints in winter; flowers "light red", abundant; January–April.

'Eva' was raised by D. Lohse, Bullenkuhlen, Germany, and introduced in 1995. It begins to flower earlier than the other "red" winter heaths; Small & Small (2001) suggested it would "become a firm favourite in the winter garden".

'Foxhollow' 🌣 ♀ (*Erica carnea* f. *aureifolia*). Vigorous, spreading heather (±0.15 m tall × 0.4 m spread); foliage yellow with bronze tips of the shoots, darker and richer in winter; flowers sparse, starting heliotrope (H12) turning to shell pink (H16) (Fig. 220); January–April.

This is one of John Letts' cultivars and was named after his house. The flowers may be sparse but the foliage is "outstanding" (Small & Small 2001); "superb" (Small & Cleevely 1999). AGM 1993.

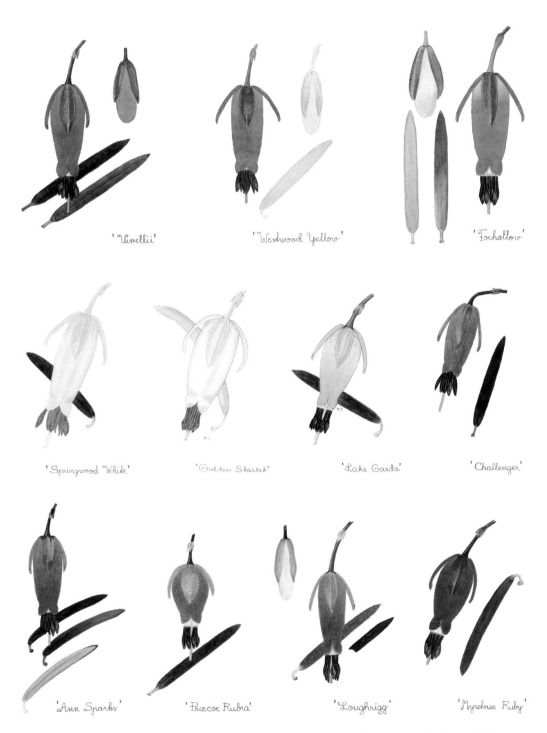

Fig. 220. Studies of the flowers and leaves of some of the award-winning and recommended cultivars of *Erica carnea*; a collage of original water-colours by Brita Johansson © (reproduced by permission). The individual florets are enlarged approximately 4 times.

'Golden Starlet' ⬡ ♀ (*Erica carnea* f. *aureifolia*; *E. carnea* f. *alba*). Compact, low, spreading plant (±0.15 m tall × 0.4 m spread); foliage dazzling yellow in early summer, but leaves becoming bicoloured (yellow and pale green), and so turning lime-green in winter; flowers white (Fig. 220); December–March.

A seedling from 'Foxhollow' deliberately crossed with 'Snow Queen' by Kurt Kramer, Edewecht-Süddorf, Germany, who selected and introduced this "strongly recommended" (Small & Small 2001) cultivar in 1984. AGM 1993.

'Hilletje' ⬡ Spreading, low, bushy heather (±0.15 m tall × ±0.3 m spread); golden-green foliage in summer darkening and becoming orange-red in winter; flowers dark lilac-pink (H11) darkening to heliotrope (H12); December–February.

This Dutch cultivar was found as a sport on 'Praecox Rubra', one of the famous Backhouse heathers selected by Richard Potter (see p. 367). The sport was found by M. Verwey, Boskoop, Netherlands, in 1976. 'Hilletje' was introduced in 1980.

'Ice Princess' ♀ (*Erica carnea* f. *alba*). Spreading, low-growing, with ascending flowering shoots (±0.15 m tall × 0.35 m spread); foliage bright green; flowers white, in long sprays; February–April.

Raised in 1984 by Kurt Kramer, Edewecht-Süddorf, Germany, this cultivar was named (® no. 117) and introduced by the British Heather Growers Association in 1993. Kramer deliberately cross-pollinated 'Snow Queen' and 'Springwood White' (see below) to produce this seedling (no. W230). AGM 2002.

'Isabell' ⬡ ♀ (*Erica carnea* f. *alba*). Floriferous, spreading heather with erect flowering shoots (±0.15 m tall × 0.35 m spread); foliage bright green; flowers white; February–April.

'Isabell' is from the same deliberately made cross as 'Ice Princess', and was raised, selected, named (® no. 97) and introduced (in 1987) by Kurt Kramer, Edewecht-Süddorf, Germany. AGM 2002.

'King George' ⬡ Spreading, compact heather (±0.15 m tall × ±0.25 m spread); foliage dark green; flowers pink (H8) becoming darker with age; December–March.

'King George', collected in the wild in Switzerland by Richard Potter of Backhouse Nurseries of York, can start to bloom in October. It was originally described as having "Madder carmine to purple suffused carmine" flowers with dark brown anthers.

This was listed as 'King George V' in Backhouse's special 1911 catalogue marking the coronation of King George V.

'Lake Garda' ♀ Vigorous, spreading heather (±0.15 m tall × 0.4 m spread); foliage dark green; sepals two-thirds length of corolla, fused at base for 0.3 mm; corolla pale pink (H8) to shell-pink (H16) (Fig. 220); flowers arranged evenly around shoots; January–March.

A wild-collected clone, found during March 1977 on a hill overlooking Lago di Garda in northern Italy by David McClintock; on the same occasion, he also found some white-flowered plants (McClintock 1984a: 185). This cultivar was selected for propagation and subsequently introduced by Mrs Tessa Forbes, Plaxtol Nurseries, Plaxtol, Kent, in 1984 (® no. 37).

'Loughrigg' ⬡ ♀ Vigorous heather, good at carpeting (±0.15 m tall × 0.55 m spread); foliage dark glaucous green, tinged bronze in winter; flowers opening pink (H8), and getting progressively darker until rose-pink (H7) or heliotrope (H12) (Fig. 220); January–May.

'Loughrigg', named after the local fell (peak), was a seedling from 'Vivellii' introduced by Geoffrey Hayes Ltd, Grasmere, Westmorland, England, in the 1940s. Chapple (1966) reported that it came about in the mid-1930s as the result of a deliberate (but evidently unsuccessful) attempt by Leonard Hotson to fertilise 'Vivellii' using pollen from a white-flowered *Erica erigena*. Slinger (2007) noted that 'Loughrigg' has the "nearest to purple flowers ... and its pleasing foliage contrasts attractively with the flowers." AGM 1993.

'March Seedling' 🏵 ♀ Floriferous, vigorous and spreading (±0.2 m tall × 0.5 m spread); foliage deep green; flowers pale heliotrope (H12); February–May.
    'March Seedling' probably arose as a chance, self-sown seedling, and was introduced by C. J. Marchant, Stapehill, Wimborne, Dorset, by 1954. Its principal merit is its relatively late (for *Erica carnea*) flowering period.

'Myretoun Ruby' 🏵 ♀ — see plate 8 (lower left) (pp. 112–113), Fig. 220.

'Nathalie' PBR ♀ Compact, upright plant (±0.15 m tall × 0.4 m spread); foliage dark green; flowers purple (H10); January–April.
    Described as having the "deepest and brightest" flowers (Small & Small 2001), this cultivar is a selected seedling (no. 271) from open-pollinated 'Myretoun Ruby' raised by Kurt Kramer, Edewecht-Süddorf, Germany, in 1981. Selected, named (® no. 105) and introduced by Kramer in 1992, 'Nathalie' is protected by plant breeders' rights (granted on 27 October 1993). AGM 2002.

'Pink Spangles' 🏵 ♀ Compact, vigorous heather (±0.15 m tall × 0.45 m spread); foliage green; flowers with pale shell-pink (H16) calyx, and darker pink (H8) corolla; January–May.
    A chance, self-sown seedling found near a plant of *Erica erigena* by Neil Treseder in Mrs P. H. Davey's garden at Saffron Meadow, Devoran, Cornwall; introduced (as *E. × darleyensis*) by Treseder's Nursery, Truro, Cornwall, in 1966, but subsequently confirmed as *E. carnea* (Jones 1979). Underhill (1990) described it as "one of the most exciting" of the winter heath clones, and an ideal groundcover plant. Slinger (2007) praised it: "big pink flowers appear early in the New Year and the colour deepens as the flowers age — an excellent plant." AGM 1993.

'Praecox Rubra' ♀ Vigorous, with semi-prostrate shoots, spreading (±0.15 m tall × 0.4 m spread); foliage dark green sometimes tinged brown; flowers "rich rose carmine" (heliotrope, H11) with dark brown anthers (Fig. 220); November–May.
    Another of a dozen *Erica carnea* clones collected during the late 1800s and early 1900s in Switzerland by Richard Potter (1844–1922) of Backhouse Nurseries (York) Ltd, and released by Backhouse Nurseries in the autumn of 1912 (Backhouse 1911). "I always have had a great liking for 'Praecox Rubra'," writes Leslie Slinger (2007), "which is easily recognised by its comparatively small flower of deep rich red." AGM 1993.

'R. B. Cooke' ♀ Floriferous, spreading heather (±0.2 m tall × 0.45 m spread); foliage mid-green; flowers plentiful, corolla pink (H8) ageing to mauve (H2), with shell-pink (H16) calyx; December–May.
    A seedling (of uncertain origin) introduced in 1971, and named after Randle Blair Cooke (1890–1973), of Corbridge, Northumberland, a timber-broker and keen plantsman, a prominent member of the Scottish Rock Garden Club, who specialised in rhododendrons (Buchan 2007). AGM 1993.

'Rosalie' PBR 🏵️ 🏆 Low-growing with upright shoots (±0.15 m tall × 0.5 m spread); foliage green, tinted bronze; flowers bright pink; January–April.

Described as being suitable for growing in containers (Small & Small 2001), this cultivar is another of the seedlings from open-pollinated 'Myretoun Ruby' that were raised by Kurt Kramer, Edewecht-Süddorf, Germany, in 1981. Selected, named (® no. 99) and introduced by Kramer in 1990, 'Rosalie' is protected by plant breeders' rights (granted on 18 December 1991). AGM 2002.

'Rotes Juwel' (® 98) 🏵️ Slow-growing, spreading heather (±0.15 m tall × ±0.3 m spread); foliage dark green acquiring red tints in winter; flowers beetroot (H9); November–April.

'Rotes Juwel' has purer red flowers than all other cultivars of *Erica carnea*. The cultivar was a seedling (no. 264) of 'Myretoun Ruby' that had been open-pollinated, and was selected and introduced by Kurt Kramer, Edewecht-Suddorf, Germany, in 1990.

'Springwood White' 🏵️ 🏆 (*Erica carnea* f. *alba*) — see plate 8 (middle left) (pp. 112–113), Fig. 220..

'Sunshine Rambler' 🏆 (*Erica carnea* f. *aureifolia*). Spreading, moderately vigorous heather (±0.2 m tall × 0.4 m spread); foliage bright yellow all year, tinged red in winter; flowers open pink (H8) and darken to heliotrope (H12), in spikes to 9 cm long; February–April.

A spontaneous seedling in Mrs Olive Cowan's garden, Petherton, Temple Close, Moor Park, Farnham, Surrey, which was introduced by G. B. Rawinsky, Primrose Hill Nursery, Haslemere, Surrey, in 1971. AGM 1993.

'Treasure Trove'. Slow-growing, compact but rather twiggy shrub (±0.15 m tall × 0.4 m spread); foliage green; flowers salmon (H15); February–April.

The flower colour is a very distinctive one within this species (Small & Cleevely 1999). A spontaneous seedling, found by Amy Doncaster, Chandlers Ford, Hampshire, before 1978; introduced by Macpenny's Nurseries, Christchurch, Dorset, by 1978. Mrs Doncaster (née Baring) (1894–1995) was a plantsman of great reputation (Buchan 2007).

'Vivellii' 🏆 — see plate 8 (upper left) (pp. 112–113), Fig. 220.

'Westwood Yellow' 🏵️ 🏆 (*Erica carnea* f. *aureifolia*) — see plate 8 (lower right) (pp. 112–113), Fig. 220.

'Winter Snow' (® 118) 🏵️ Low, spreading heather (±0.15 m tall × ±0.35 m spread); foliage bright green; flowers white, profuse; February–May.

Kurt Kramer, Edewecht-Suddorf, Germany, cross-pollinated 'Springwood White' and 'Snow Queen' in 1984, and seedling W205 was selected by Kramer for the British Heather Growers Association to name and introduce. 'Winter Snow' was released commercially in 1993.

'Wintersonne' PBR 🏵️ Spreading, bushy heather (±0.2 m tall × 0.65 m spread); foliage striking bronzy red in winter, but paler in summer; flower buds distinctively tinged orange, opening lilac-pink (H11), flower darkening to magenta (H14), late-blooming; February–May.

Another seedling raised in 1984 by Kurt Kramer, Edewecht-Süddorf, Germany, from open-pollinated 'Myretoun Ruby'; selected and named (® no. 96) by Kramer (plant breeders' rights granted 27 October 1993).

Erica ciliaris: **Dorset heath.**

'Corfe Castle' 🏵 ♆ Compact, bushy heather (±0.2 m tall × 0.35 m spread); foliage green; buds cerise (H6), mature flowers rose-pink (H7); August–October.

George Osmond, Archfield Nursery, Wickwar, Wooton under Edge, Gloucestershire, collected this cultivar near Corfe Castle, Dorset, before 1965 when he introduced it. Its flowers have a distinctive colour, and Underhill (1990) noted that the foliage turns bronze in winter. AGM 1993.

'David McClintock' 🏵 Vigorous, spreading bush with loose, open habit (±0.25 m tall × 0.5 m spread); foliage grey-green; flowers bicoloured, white at base shading to beetroot (H9) but "changing to pink late in season"; August–September.

This is another wild-collected clone, found north of Carnac in Brittany, France, by David McClintock (Platt, Kent) in 1962. The original plant was "inextricably mixed with a dwarf gorse bush". The cultivar was named after McClintock by accident; he had intended 'Carnac' as its name (Nelson & Small 2000: **1**: 108). 'David McClintock' was introduced by B. & V. Proudley, Aldenham Heather Nurseries, Hertfordshire, in 1968. An AGM awarded by the Royal Horticultural Society in 1993 was recently rescinded.

'Mrs C. H. Gill' ♆ — see plate 13 (upper left centre) (pp. 164–165).

'Ram' 🏵 Compact, bushy heather (±0.15 m tall × ±0.4 m spread); foliage is pale, yellowish green throughout the year; flowers magenta (H14); August–October.

'Ram' came from P. Bakhuyzen & Zonen (Boskoop, Netherlands) and was introduced in 1984.

Fig. 221. *E. ciliaris* f. *alba* 'Stoborough'; cultivated in northern California, with St Dabeoc's heaths (E. C. Nelson).

'Stoborough' ♀ (*Erica ciliaris* f. *alba*). Vigorous, erect heather (±0.6 m tall), possessing cilia without glandular tips; foliage green; flowers large, described as "globular" or "pitcher shaped", carried on upright stems, white; July–October. Fig. 221.

Maxwell & Patrick (1966) commended this as "the tallest grower, ... strong, stout and virile"; they gave two feet (60 cm) as its height.

Maxwell himself found 'Stoborough' "on a bank between the hamlets of Stoborough and Arne, on the Great Heath" (Maxwell & Patrick 1966); it is likely to be the "good" white-flowered Dorset heath mentioned by Maxwell (1927) as "in the hands of the Dorset Nursery, Broadstone". The original plant survived on the bank, "no more than 25 yds from the road", for several years after Maxwell found it, but was then dug up. Maxwell & Beale Ltd, Broadstone, Dorset, introduced 'Stoborough' in 1929.

Claims that 'Stoborough' was an "old variety that had already been introduced" were firmly dismissed by Maxwell (see Maxwell & Patrick 1966: 124). AGM 1993.

Erica cinerea: bell heather.

'Alba Minor' 🌼 ♀ (*Erica cinerea* f. *alba*). Floriferous, neat, carpeting heather (±0.2 m tall × ±0.55 m spread); foliage dense, bright green; flowers white; June–October.

A "variety" with this name was noted in *The garden* in 1888. If this is the same plant, 'Alba Minor' has been grown for not less than a century, and is "still one of the best" (Small & Small 2001). AGM 1993.

'C. D. Eason' 🌼 ♀ — see plate 13 (upper right) (pp. 164–165).

'C. G. Best' ♀ Compact heather, forming a dome-shaped bush with long, upright flowering shoots (±0.25 (–0.4) m tall × ±0.5 m spread); foliage green, noticeably tufted; flowers rose-pink (H7); June–October.

Like 'C. D. Eason', this was found in the wild near Maxwell & Beale's nursery by another member of the staff, C. G. Best, and like Eason's cultivar, this was introduced in the 1929 catalogue. Maxwell & Patrick (1966) described the flower colour as "milky pink with no blue in it ... neither cerise or crushed strawberry, yet both colours spring to the tongue." One of the outstanding cultivars in trials during the early 1990s at the Royal Boskoop Horticultural Society (van de Laar 1998); the others were 'C. D. Eason' (see plate 13), and 'Glasnevin Red' (see below). AGM 1993.

'Celebration' 🌼 (*Erica cinerea* f. *alba*; *E. cinerea* f. *aureifolia*). Low-growing heather with long, spreading and curving shoots (±0.2 m tall × ±0.4 m spread); summer foliage "intense" bright gold, turning lime-green in winter; flowers white; July–September.

This chance seedling, found by John L. Jones in his own nursery, Glynwern Nursery, Cilcennin, Lampeter, Dyfed, Wales, possesses an unusual combination of foliage and flower colours; white blossom with yellow or gold foliage is uncommon in *Erica cinerea*. Jones introduced 'Celebration' in 1995 under the name 'White Star', but under rules then applying, that name had to be replaced because it was already in use in the denomination class (since dismantled, see Registrar 2004). Lortz (2002) reported that this cultivar has "the best golden foliage" among heathers growing in her nursery at Shelton in the State of Washington, USA.

'Cevennes'. Compact, spreading shrub, good at carpeting ground (±0.25 (– 0.3) m tall × ±0.6 m spread); foliage green; flowers mauve (H2), in upright, stiff spikes; July–October.

Sir Oscar Warburg (1876–1937), Epsom, Surrey, and his son, Dr Edmund F. Warburg (1908–1966), found this clone in the Cévennes, France, and donated it, with other plants, to the Royal Horticultural Society's Gardens, Wisley. Mulligan (1963) recalled that it was growing at Wisley when he was assistant to the Director in the 1930s so it must have been found in the late 1920s.

'Champs Hill' ♀ Floriferous, upright bush (±0.35 m tall × ±0.45 m spread); foliage green; flowers in long (to 15 cm) spikes, dusky cerise (H6) fading to rose-pink (H7); July–October.

A spontaneous seedling in the Bowermans' garden, Champs Hill, Coldwaltham, Sussex, spotted by Mrs Margaret Bowerman in 1976, and named (® no. 22) in 1982. AGM 2003.

'Cindy' ♀ Low, spreading heather (±0.25 m tall × ±0.5 m spread); foliage very dark green; flowers large, purple (H10); June–November.

The plant grown under this name is recommended and gained an AGM in 1993, but, as noted by Underhill (1990), the clone does not really match Letts's original description, which was of a heather with a "strong upright habit" with dark foliage of "an unusual bronze-green" (Letts 1966: 49). 'Cindy', found during "one of our rambles in Cornwall" (Letts 1966), was named after one of the Letts' Sheltie dogs. AGM 1993.

'Eden Valley' 🆂🆂 ♀ Prostrate, spreading, tidy heather (±(0.15 –) 0.2 (– 0.3) m tall × ±0.55 m spread); foliage deep green; flowers bicoloured, corolla white at base shading to lavender (H3) at tip; July–October.

This unusual and attractive cultivar was found on the bank of a track at Trink Hill, south-west of St Ives, Cornwall, by Miss Martha Betha Gertrude Waterer (1822–1974) of Ludgvan near Penzance, Cornwall, in 1926. 'Eden Valley' — the name of Miss Waterer's house (Garrett & McClintock 1985) — was introduced by Knap Hill Nursery, Woking, Surrey, in 1933. AGM 1993.

'Fiddler's Gold' ♀ (*Erica cinerea* f. *aureifolia*). Free-flowering, spreading heather (±0.25 m tall × ±0.45 m spread); young shoots bright yellow with red, maturing to golden yellow; flowers lilac-pink (H11); July–August.

A spontaneous seedling in Edward Plummer's garden, Fiddlestone Lodge, Burton, Wirral, Cheshire, which was propagated and named at the University of Liverpool Botanic Gardens, Ness, before 1959. The parent could have been either 'Golden Drop' or 'Golden Hue' (see below), or both of those clones, as the seedling came up in a gravel path near them. 'Fiddler's Gold' is more floriferous than 'Golden Hue' and more vigorous than 'Golden Drop' (Hulme 1994). AGM 1993.

Erica vagans 'Fiddlestone' (see p. 383) is another outstanding heather from Plummer's Wirral garden.

'Golden Hue' ♀ (*Erica cinerea* f. *aureifolia*). Changeable, strong, tall plant with an elegant habit (±0.4 m tall × ±0.8 m spread); foliage golden yellow in summer and autumn; in winter "copper-gold", the shoots having red tips which turn orange shaded with red in spring; flowers amethyst (H1); July–October.

"The original plant was collected in Dorset some years ago from a railway bank. The curious part about it was that only half — that is, one side — of the plant had golden foliage; the rest of it was the ordinary shade of green" (Maxwell 1927). Maxwell & Beale Ltd introduced 'Golden Hue' in 1925. Winifred Walker's water-colour portrait was published first in D. F. Maxwell's book *The low road* (1927: opp. p. 54) and later in the nursery's 1930 catalogue (Cleevely 1997).

Maxwell (1927) described the winter foliage as "golden-red, a welcome touch of brightness in a drab world". AGM 1993.

'Glasnevin Red'. Sturdy heather, forming a dome-shaped bush (±0.2–0.3 m tall × ±0.25 m spread); foliage dark green; flowers "striking ruby" (Small & Small 2001) or "dark wine-red" (van de Laar 1998) (H5); June–October.

This was a seedling found in a batch of 'C. D. Eason' at the National Botanic Gardens, Glasnevin, Dublin, Ireland, before 1961, when propagations were distributed to the Daisy Hill Nursery, Newry, and the Slieve Donard Nursery, Newcastle, County Down, Northern Ireland (Nelson 2000e: 73–74). It may have arisen from 'C. D. Eason'. This cultivar was granted an award of merit by the Royal Horticultural Society in 1968, yet has a reputation (at least in Ireland) for being short-lived and sensitive to frost (Nelson & Walsh 1984: 105). However, 'Glasnevin Red' was one of the three outstanding cultivars in trials during the early 1990s at the Royal Boskoop Horticultural Society (van de Laar 1998), the others being 'C. D. Eason' (see plate 13) and 'C. G. Best' (see above).

'Hookstone White' ♀ (*Erica cinerea* f. *alba*). Large, upright heather (±0.45 m tall × ±0.8 m spread); foliage greyish green in spring, bright green later; flowers in dense spikes (usually ±0.15–0.2 m long), pure white; July–October.

This exceptional heather was collected from the side of Greyspot Hill, Chobham Ridges, Surrey, by George E. Underwood in 1936, and was introduced by G. Underwood & Son, Hookstone Green Nursery, West End, Woking, Surrey, by 1960. AGM 1993.

'Knap Hill Pink' ♀ Neat, spreading shrub (±0.3 m tall × ±0.6 m spread); foliage dark olive-green; flowers magenta (H14), in long spikes; June–September.

A very confused plant, like several other reputedly old cultivars; this one was introduced perhaps as early as 1925 and then was named 'Rosea Knap Hill Variety'. Early descriptions all suggest that the clone had "pink" flowers rather like 'Rosea'. Plants of this at trial in the Netherlands are recorded as having clear pink flowers (heliotrope H12) (van de Laar 1998), whereas those in The Heather Society's trials in the early 1970s had magenta flowers (Vickers 1976), as described here. I suspect there is more than one plant being grown under this name. The magenta-blossomed clone gained the AGM in 1993.

'Lime Soda' 🅙🅢 ♀ Floriferous, spreading heather (±0.3 m tall × ±0.55 m spread); foliage lime-green; flowers lavender (H3); June–October.

A spontaneous seedling found by Brian G. ("Jack") London in his garden at Taverham, Norwich, Norfolk, which was introduced by Neil Brummage, Heathwoods Nursery, Taverham, in 1978. AGM 2002.

'Pentreath' 🅙🅢 ♀ Tidy, spreading, low-growing bush (±0.25 m tall × ±0.6 m spread); foliage blue-green in spring, changing to dark green; flowers rich beetroot (H9); June–October.

An outstanding (Small & Small 2001) heather with dark flowers found in the wild near Pentreath, Cornwall, by Mr and Mrs Donald Waterer, Bagshot, Surrey, and introduced by Knap Hill Nursery, Woking, Surrey, in 1951. AGM 1993.

'Pink Ice' ♈ Dwarf, twiggy bush (±0.15 m tall × ±0.35 m spread); foliage bronze-green in spring, turning to silvery dark green; flowers large, rose-pink (H7); July–November.

An "unusually compact" seedling found by John Letts in the heather garden at Foxhollow, Windlesham, Surrey, and named and introduced by him before 1966. Letts (1966) described it as having "soft pink flowers over a long period", and van de Laar (1998) noted the colour as pure pink (heliotrope H12). AGM 1993.

'P. S. Patrick' ♈ Upright, strong, hardy bush (±0.2 (–0.4) m tall × ±0.45 m spread); foliage grey-green in spring, turning dark green, glossy; flowers large, vivid purple (H10) to amethyst (H1) in upright spikes; June–September.

Paul Spurgeon Patrick (c. 1892–1974) was, at the time of the discovery of this heather, employed in Maxwell & Beale's nursery, Broadstone, Dorset. He spotted the heather about 1928 on the nearby Corfe Mullen Hills, and Maxwell & Beale Ltd introduced the cultivar in 1930. Douglas Fyfe Maxwell, a close friend, described it as "the finest of the deep purple-flowered" heathers; "the bush is deep green, strong, spiky, and virile" (Maxwell & Patrick 1966). Patrick was a founder member and Vice-President of The Heather Society and Honorary Editor of its *Yearbook* (1963–1965, 1969–1974). AGM 1993.

'Stephen Davis' 🏵 ♈ Neat, spreading heather (±0.2 (–0.3) m tall × ±0.45 m spread); glossy foliage dark green; flowers bright magenta (H14); June–September.

A wild-collected clone found on Marley Common, Haslemere, Surrey, by P. G. Davis, and introduced by him by 1969. Davis named the heather after his son. AGM 1993.

'Velvet Night' 🏵 ♈ Distinctive, spiky, spreading plant (±0.2 m tall × ±0.6 m spread); foliage dark green; flowers dark beetroot (H9) tinged with crimson; June–September.

This is one of the darkest of the bell heathers in cultivation. Maxwell & Patrick (1966) described the colour of the flowers as "deep velvety purple-maroon ... which recalls that of the "black" tulips", and suggested planting 'Velvet Night' with 'Domino' or 'G. Osmond'. F. J. Stevens, one of the directors of Maxwell & Beale Ltd, found this on the moors near the company's Naked Cross Nursery, Corfe Mullen, Dorset, in 1957. AGM 1993.

'Windlebrooke' ♈ (*Erica cinerea* f. *aureifolia*). Spreading heather (±0.15 m tall × ±0.45 m spread); summer foliage golden-yellow, changing to orange-red in winter; flowers mauve (H2); July–September.

The shoots of 'Windlebrooke' spread out like a star-fish (Underhill 1990). John Letts collected this cultivar on Sunningdale Golf Course, Surrey, and introduced it in 1972. Windle Brook is a local stream. AGM 1993.

Erica × *darleyensis*: **Darley Dale heath.**

'Arthur Johnson' 🏵 ♈ Vigorous, spreading heather, forming a dome-shaped bush (±0.6 m tall × 0.8 (–1.5) m spread); young shoots cream, mature foliage green ("apple-green"); flowers pink (H8) maturing to heliotrope (H12) closely arrayed in long (to 0.2–0.3m) spikes, slightly fragrant; late November–April.

A spontaneous seedling in Arthur Tysilio Johnson's garden at Conwy, north Wales, before 1952; perhaps from *Erica erigena* 'Hibernica' and *E. carnea* 'Ruby Glow' (Johnson 1956: 83). Johnson, a one-time schoolmaster, later a prolific author and expert plantsman, himself opined that this cultivar was "undoubtedly one of the most important additions made to the heath family [*sic*] during the last fifty years" (Johnson 1956). AGM 1993.

'Eva Gold' 🌼 Bushy, spreading heather (±0.3m tall × ±0.5m spread); foliage golden throughout the year; flowers magenta (H14); January–April.

'Furzey' 🌼 ♀ Dense, spreading shrub (±0.4 m tall × 0.85 m spread); young shoots creamy pink and red, mature foliage dark green; flowers lilac-pink (H11) maturing to heliotrope (H12); December–May.

A chance seedling found by Captain Hugh Dalrymple at Furzey Gardens, Minstead, Lyndhurst, Hampshire, which was introduced by John F. Letts, Windlesham, Surrey, before 1963. Originally identified as *Erica carnea*. Leslie Slinger (2007) wrote of 'Furzey': "a plant of great lasting qualities and in bloom for most of the winter" (Slinger 2007). AGM 1993

'Ghost Hills' ♀ Floriferous, spreading heather (±0.4 m tall × 0.9 m spread); young shoots cream, mature foliage light green; flowers pink (H8), aging to heliotrope (H12) with prominent protruding anthers; November–May.

This may have been a sport on 'Darley Dale', but its exact origin is uncertain. Jack Brummage (see below) found and introduced the cultivar about 1960. Named after Brummage's house: Ghosthill Plantation, near Taverham, may have been the origin of the name. AGM 1993

'Jack H. Brummage' (*Erica × darleyensis* f. *aureifolia*). Neat shrub forming a compact dome (±0.3 m tall × 0.65 m spread); summer foliage bright yellow turning more orange or gold in winter; flowers heliotrope (H12); January–June.

A spontaneous seedling growing close to *Erica carnea* 'Aurea', found by Jack Brummage, Heathwoods Nursery, Taverham, Norwich, Norfolk, and introduced by Heathwoods Nursery before 1966.

'Jenny Porter' 🌼 ♀ Vigorous, neat heather (±0.45 m tall × 0.6 m spread); young shoots pale cream with pink tints, turning green; flowers very pale lilac (H4), almost white, with dark "almost chocolate" anthers; December–May.

This "charming" heather (Maxwell & Patrick 1966) was raised, and probably also named, by J. W. Porter (see below), Carryduff, Northern Ireland, one of a series of hybrids he produced (Nelson 1984a: 32; 2007b, 2009d). Jenny, a keen gardener, was one of Porter's unmarried sisters. AGM 1993.

Erica × darleyensis 'Margaret Porter', which produces shell-pink flowers in profusion, was another of Porter's seedlings and was named by him after another of his sisters (Nelson 2007b).

'J. W. Porter' 🌼 ♀ Neat shrub (±0.25 m tall × 0.4 m spread); young shoots red and cream, turning dark green; flowers often sparse, pink (H8) to heliotrope (H12); January–May (Fig. 203, p. 316).

James Walker Porter was an expert plantsman, raising many plants, including heathers, from seed (Nelson 1984a, 2007b, 2009d). This was one of his seedlings, selected and named by his wife, Eileen, and introduced by John F. Letts, Foxhollow, Windlesham, Surrey, in 1970. Its principal virtue is the spectacular (Small & Cleevely 1999) display of the young shoots, because flowers are sparse and flowering unreliable (Small & Small 2001). AGM 1993.

'Kramers Rote' 🏆 (KRAMER'S RED). Vigorous, spreading heather (±0.35 m tall × 0.6 m spread); foliage dark green tinged bronze; flowers "glowing" magenta (H14); January–April. Fig. 222.

Kurt Kramer, Edewecht-Süddorf, Germany, produced this heather by deliberately cross-pollinating *Erica carnea* 'Myretoun Ruby' (see p. 113) and *E. erigena* 'Brightness' in 1981. Selected and named by Kramer, 'Kramers Rote' has been praised as "the most outstanding [heather] of recent times" (Small & Small 2001). AGM 1993.

Under the *International code of nomenclature for cultivated plants* (Brickell *et al.* 2004) translating a cultivar name is not permitted. Thus the English version of 'Kramers Rote' (which does not have an apostrophe) is deemed to be a trade designation by The Heather Society, acting as International Cultivar Registration Authority for the genus (Nelson & Small 2000).

'Lucie' 🏵. Floriferous, vigorous (±0.35 m tall × 0.6 m spread) and resistant to disease; foliage dark green; flowers large, in dense, broad (to 2.5 cm across) spikes; corolla to 7 mm long, to 4 mm across, magenta (H14) shading to amethyst (H10); calyx lobes to 4 mm long, the same colour as the corolla; December–April.

This outstanding heather, the first to receive the Premier Award of The Heather Society, was a sport on 'Kramers Rote' from which it differs by having longer flowers in broader "spikes'), 'Lucie' was discovered at les Pépinières Renault, Gorron, France, in 1997, and is protected by plant breeders's rights.

Fig. 222 (left). The very popular *Erica* × *darleyensis* 'Kramers Rote'; a tray of young flowering plants ready for sale and labelled quite typically with the trade designation KRAMER'S RED (E. C. Nelson).

Fig. 223 (right). *Erica erigena* 'W. T. Rackliff; cultivated in Outwell, Norfolk (E. C. Nelson).

'Mary Helen' ♀ (*Erica × darleyensis* f. *aureifolia*). Moderately vigorous, low, spreading plant (±0.25 m tall × 0.45 m spread); foliage yellow-green or golden in summer and turning bronze in winter; flowers pink (H8), February–April.

A spontaneous seedling found about 1980, and introduced in 1984, by Peter Foley, Holden Clough Nursery, Bolton-by-Bowland, Lancashire. This is slower growing than 'Jack H. Brummage' (see above) and has darker foliage.

'Moonshine' (® 175) 🏆 Bushy heather (±0.3 m tall × ±0.5 m spread); foliage bright yellow throughout the year; flowers shell-pink (H16) becoming pink (H8) as they age; November–April.

'Moonshine' was spotted as a sport on 'Darley Dale', the original clones of this hybrid heather, in 1996 by Mike Ayres who was production manager at John Hall's nursery in Windlesham. The foliage does not darken much in winter.

'Silberschmelze' 🏆 (MED WHITE, MOLTEN SILVER, WHITE BEADS) (*Erica × darleyensis* f. *alba*). Floriferous, vigorous yet compact, spreading heather (±0.4 m tall × 0.8 m spread); young shoots cream, mature foliage dark green; flowers white (often described as ashen white or silvery white); December–May.

Georg Arends, Wuppertal Ronsdorf, Germany, found this as a sport on a plant which he called *Erica mediterranea hybrida* — this is generally believed to have been *E. × darleyensis* 'Darley Dale'. He introduced 'Silberschmelze' in 1937. 'White Perfection' (see below) was a sport of 'Silberschmelze'.

As noted above (see 'Kramers Rote'), a cultivar name should not be translated. 'Silberschmelze' holds the dubious record for the largest number of misspellings (at least 13 are recorded on The Heather Society database) and replacement names in English (some of which are translations); all such names, including MED WHITE (or MEDITERRANEAN WHITE), MOLTEN SILVER and WHITE BEADS (perhaps the most commonly encountered trade designations), are rejected and their use is discouraged.

'Spring Surprise' 🏆 Bushy, upright heather (±0.45 m tall × ±0.6 m spread); foliage dark green without the usual "coloured" young shoots in spring; flowers dark rose-pink; March–May.

This seedling (KK 31) was deliberately raised and selected by Kurt Kramer, Edewecht-Suddorf, Germany, before 1992, and was introduced by Kingfisher Nursery, Gedney Hill, Lincolnshire, in 1994.

'White Perfection' 🏆 ♀ (*Erica × darleyensis* f. *alba*) — see plate 8 (upper right) (pp. 112–113).

Erica erigena: Irish heath.

'Golden Lady' ♀ (*Erica erigena* f. *aureifolia*; *E. erigena* f. *alba*). Outstanding, upright, slow-growing shrub (±0.75 m tall × 0.55 m spread); foliage golden-yellow throughout the year; flowers sparse, white; February–April.

A sport on 'W. T. Rackliff' (see below), introduced, and shown at the Royal Horticultural Society's Chelsea Flower Show, by Bressingham Gardens, Diss, Norfolk, in 1977. AGM 1993.

Erica erigena f. *aureifolia* D. C. McClint. was described and named to accommodate this yellow-foliaged variant (McClintock 1988a: 62).

'Irish Dusk' 🌸 ♛ — see plate 8 (centre) (pp. 112–113).

'Thing Nee' (® 143) 🌸 Upright, bushy heather attaining ±0.8 m tall × ±0.8 m spread in 12 years; mature foliage mid-green throughout the year, but the young shoots are yellow in spring and summer, turning light green in autumn; flower with lilac (H4) sepals and pink (H8) corolla; February–May.

The foliage is described as "much more gold" (Small & Small 2001) than 'Golden Lady' of which this was a seedling. 'Thing Nee' (a Thai name pronounced "ting nay") was raised by Barry Sellers at Chandler's Ford, Hampshire, in 1984.

'W. T. Rackliff' 🌸 ♛ (*Erica erigena* f. *alba*). Very compact, upright, neat shrub, usually dome-shaped, ultimately as tall as wide (±0.75 (–1.5) m tall × 0.75 m spread); young foliage green, turning rich, deep green; flowers pure white; February–June. Fig. 223.

Maxwell & Beale Ltd, Broadstone, Dorset, introduced this excellent heather in 1935 but "have no record or memory of where it was obtained" (Maxwell & Patrick 1966: 101). Other sources suggest, without any new evidence, that it was collected in Ireland. The person after whom it was named is equally elusive and unknown.

McClintock (1987) praised this as "the shapeliest" of all hardy heathers. 'W. T. Rackliff' "needs no pruning and will grow slowly to 1½ m, always neat, its flowers decorating it in their green stage before opening to white stars. It is best grown as a solitary specimen, to show off its comeliness", he wrote (McClintock 1987: 63). AGM 1993.

Erica × *garforthensis*: **Garforth heath.**

'Craig' (® E.2007:04). Compact shrub when pruned (±1 m tall after 25 years); mature foliage grey-green, young shoots bright green; flowers fragrant, pink (H8).

The only surviving plant of Dr Griffiths's original batch of seedlings from *Erica tetralix* 'Bartinney' × *E. manipuliflora* 'Aldeburgh'. It is named after Dr and Mrs Griffiths's first grandson, Craig Simpson.

'Tracy Wilson' (® no. 160). Floriferous, compact, spreading heather (±0.25 m tall × ±0.45m spread); foliage light green, slightly variegated; young shoots tipped creamy yellow; flowers in compact, cylindrical spikes, appearing white with a hint of pale pink as they age, slightly fragrant; July–October.

Clone no 4, from *Erica tetralix* 'Melbury White' × *E. manipuliflora* 'Korčula', raised, selected and named by David Wilson (Chilliwack, British Columbia, Canada).

Erica × *gaudificans*: **Edewecht heath.**

'Edewecht Belle' (® E.2004:02). Bushy, vigorous shrub heather (±0.5 m tall); flowers pink (H9), bell-shaped to conical, ± 4.5 mm long; peduncles pubescent, dark red ; ovary 8-celled, not distinctly stipitate, but broadly obovoid to barrel-shaped with emarginate apex; style to 3.7 mm long, only slightly broader at base, not exserted.

Raised by Kurt Kramer, Edewecht-Süddorf, Germany.

'Edewecht Blush' (® E.2004:01). Bushy, vigorous, spreading heather (±0.5 m tall); flowers colourless but appearing pink ("blushing") when young; peduncles usually glabrous, tinged red; ovary indistinctly 4-celled, inversely pear-shaped, stipitate; style to c. 4.5mm long, usually prominently exserted.

Like 'Edewecht Belle', this was raised by Kurt Kramer, Edewecht-Süddorf, Germany (see Fig. 215, p. 342).

Erica × *griffithsii*: **Griffiths's heath.**

'Ashlea Gold'. Compact heath (±0.4 m tall × 0.6 m spread); foliage bright yellow eventually turning green, but appearing "deep gold"; flowers in a narrow spike; pedicels dark red (H5), corolla pink (H8), anthers dark, ovary with short hairs on top; July–October.

This was one of the original batch of seedlings raised by Dr John Griffiths, Garforth, Leeds, West Yorkshire, in 1983, from *Erica manipuliflora* 'Aldeburgh' deliberately crossed with *E. vagans* 'Valerie Proudley'.

'Heaven Scent' — see plate 10 (upper mid-left) (pp. 130–131) (see also Fig. 205, p. 320).

'Jacqueline' 🪙 Vigorous, erect, bushy heather (±1 m tall × ±0.6 m spread); foliage dark grey-green with grey; flowers in long "sprays", cerise (H6); July–December.

The flowers of most hardy heathers are scented (to attract insects) and the fragrance of 'Jacqueline' is very noticeable, as is that of its "parent" 'Heaven Scent'. 'Jacqueline' arose as a sport sometime before 1998, and has a tendency to revert to 'Heaven Scent': reverted shoots need to be removed when they are detected.

'Valerie Griffiths' (® 64) 🪙 Bushy, spreading heather (±0.4 m tall × ±0.55 m spread); summer foliage yellow, turning golden in winter; flowers pale pink in "spikes" to 10 cm long; July–October.

Dr John Griffiths, Garforth, Leeds, West Yorkshire, raised this seedling in 1983 from *Erica manipuliflora* 'Aldeburgh' deliberately pollinated with *E. vagans* 'Valerie Proudley'. Named after Dr Griffiths's wife, this heather was introduced by Denbeigh Heather Nurseries: "... one of the best introductions of the 1990s" (Small & Small 2001).

Erica × *krameri*: **Kramer's heath.**

'Otto' (® no. 176). Leaves (2.5)–3.8–(5.7) mm long; flowers bell- or cup-shaped, ±4 mm long, to 2 mm across, without noticeable constriction below lobes; bract recaulescent; bracteoles absent; stamens included; anthers dark; style exserted, glabrous, ±6 mm long; stigma peltate; June–November.

This seedling, raised in 1990, was selected on 28 September 2002 by Kurt Kramer, David Small and Jos Flecken and is named after the German actor and comedian, Otto Waalkes (b. 1948).

'Rudi' (® 177). Leaves (4)–5.5–(6) mm long; flowers bright pink (Amethyst (H1) to heliotrope (H12)), urn-shaped and slightly constriction below lobes; bract about mid-point on pedicel, 1 bracteole below calyx; stamens visible at mouth of corolla; filaments 2–3 mm long; anthers dark brown; style exserted, glabrous; stigma peltate; August–October.

Like 'Otto' this was raised in 1990 and selected for naming on 28 September 2002 by Kurt Kramer, David Small and Jos Flecken. Its name comes from the stage-name of the Dutch-born entertainer, singer and comedian, Rudi Carrell (*olim* Rudolf Wijbrand Kesselaar; 1934–2006).

Erica lusitanica: **Portuguese heath.**

'George Hunt' (*Erica lusitanica* f. *aureifolia*). Tall, upright shrub (±0.7 m tall × ±0.45 m spread); foliage yellow throughout the year; flowers white; March–April.

Like other yellow-foliaged heathers, 'George Hunt' is frost- and wind-sensitive but an excellent plant in mild, sheltered sites. This was found by Mr Hunt, Lymington, Hampshire, in 1959, and was very briefly protected by plant breeders' rights.

'Sheffield Park'. Tall, upright heather (±0.7 m tall × ±0.45 m spread); foliage green; flower-buds deep pink buds, opening pure white; March–April (Fig. 135, p. 215).

A chance seedling at Sheffield Park, Sussex, introduced in 1996 by Liss Forest Nursery, Hampshire. A shrub of this, in the heather garden at the Royal Horticultural Society's Gardens, Wisley, was in bloom in late September 2006.

Erica mackayana: **Mackay's heath.**

'Galicia' 🏅 Vigorous, bushy, spreading heather (±0.3 m tall × ±0.55 m spread), flowering shoots branching so each stem has more than one inflorescence; foliage dark green with prominent gland-tipped hairs on leaf margins; flowers large, to 7 mm long, "rich purple" (heliotrope H12; RHSCC 70A–72C) above shading almost to white underneath; July–October.

Although hundreds of cuttings were collected and successfully rooted during our expedition to Spain in 1982, only a handful of clones were regarded as worth naming and introducing. 'Galicia' came from Monte Castello, Galicia, Spain (DHN 84/82) and was collected there on 22 July 1982 (Nelson 1989b; McClintock 1983a); it was introduced by Denbeigh Heather Nurseries, Creeting St Mary, Suffolk.

'Shining Light' 🏅 ♛ (*Erica mackayana* f. *eburnea*). Bushy, vigorous and floriferous heather (±0.25 (– 0.6) m tall × ±0.5 (– 1) m spread); foliage rich dark green; leaves with conspicuous gland-tipped hairs on edges and tips; flowers large, pure white, in terminal umbels; June–September. Fig. 224.

Several white-flowered clones were collected in north-western Spain by David McClintock, Charles Nelson and David Small during an expedition to study Spanish species of *Erica* in 1982. 'Shining Light' came from Monte Castello, Galicia, and was gathered on 23 July (Nelson 2000e). Subsequently, these Spanish heathers were grown in a trial plot at the National Botanic Gardens, Glasnevin, Dublin, and a few, like this one, were selected for naming and introduction. Mrs Tessa Forbes, Plaxtol Nurseries, Plaxtol, Kent, introduced 'Shining Light' (as 'Shining White', a rejected name) in 1989 (Nelson 1989b).

Because it is not readily available 'Shining Light' has recently been stripped of the AGM awarded to it in 1993. No matter, it remains one of the finest white heathers and succeeds as well in California as it does in Ireland and Britain. It is frost-tolerant (Hall 2002).

A remarkable characteristic is that it continues to produce inflorescences on lateral shoots below the first terminal umbels, thus prolonging the period of bloom. 'Shining Light' produces viable seeds, and Allen Hall (2002) has reported raising seedlings that are indistinguishable from the parent.

Fig. 224. *Erica mackayana* 'Shining Light'; cultivated in the National Botanic Gardens, Glasnevin, Dublin (E. C. Nelson).

Erica manipuliflora: **whorled heath.**

'Aldeburgh'. Erect, neat shrub (±1 m tall × ±0.8 (–1) m spread) with pale grey stems; foliage green; flowers lilac pink (H11); August–October.

David Small discovered this heather growing as a hedge at a house named Talltrees in Aldeburgh, Suffolk, and introduced it before 1976 through his nursery, Denbeigh Heather Nursery, Creeting St Mary, Ipswich. The origin of 'Aldeburgh' cannot be determined. It was the first named clone in this species to be made available commercially.

'Olivia Hall' (® no. 142). Erect heather (±1 m tall); foliage green (RHS CC 137B); flowers in long spikes; sepals pale green; corolla broad-campanulate, white in warm climates and indoors, otherwise very pale lilac (H4); anthers exserted, dark beetroot (H9); August–October.

Mrs Hall collected this at Lara Bay, Antalya, Turkey, on 1 October 1978, when it appeared to have pure white flowers. However the flowers are actually very pale pink and have dark, not tan, anthers (see Joyner 2007).

Erica multiflora: **many-flowered heath.**

'John Tucker' (® no. E.2006.06). Upright shrub (±0.5 (– 1) m tall); foliage glossy, dark green; flowers opening white, in 3s in leaf axils; corolla narrow-urceolate, to 4.5 mm long, 2.5 mm diameter (1.5 mm at throat); anthers dark brown, exserted slightly; calyx and pedicel white. February–May.

A somewhat mysterious clone, originally labelled *Erica* "longifolium". This heather was raised from seed in the late 1980s or early 1990s by the late John Tucker, Worthing, Sussex, and subsequently propagated vegetatively. The origin of the original seed is uncertain: it was said to have come from South Africa, which is most improbable. There are no other named clones available.

Erica × *oldenburgensis*: **Oldenburg hybrid heath.**

'Ammerland' ⬤⬤ Compact, upright shrub forming a rounded bush (±0.7 m tall × ±0.7 m spread); foliage green; young shoot tips briefly orange; flowers opening white, maturing to lilac-pink (H11), corolla ±5 mm long, with erect lobes; calyx lobes ±3 mm long, white with green tips; peduncle ruby (H5); ovary tinged red; March–May (Fig. 217; p. 346).

Selected from a group of seedlings raised from *Erica arborea* deliberately crossed with *E. carnea* by Kurt Kramer, Edewecht-Süddorf, Germany, in 1986. This cultivar was introduced by the raiser in 1994. In central England this is frost-hardy (see Hall 2002).

'Oldenburg'. Broad, rounded shrub (±0.7 m tall × ±0.7 m spread); foliage green; flowers white, corolla ±4.5 mm long, lobes curling outwards; calyx lobes ±3 mm, white with green-yellow tips; peduncle translucent, white; ovary green with brown-green nectary ringing it at base, apex emarginate; stamens with malformed filaments, often fused and sometimes forming a secondary corolla; March–May.

Like 'Ammerland', 'Oldenburg' was selected from a group of seedlings raised from *Erica arborea* deliberately crossed with *E. carnea* by Kurt Kramer, Edewecht-Süddorf, Germany, in 1986.

Erica platycodon subsp. *maderincola*: **Madeira besom heath.**

'Madeira Gold' (® no. 36). Dense, upright shrub (±1 m tall × ±0.8 m spread); foliage plain green; flowers small, pink-tinged, whitish; summer

A misleading name. This clone arose as a sport on a plant that had been collected during 1974 in Madeira, and was originally described as having yellowish green foliage. Consequently it provided the holotype for *Erica scoparia* f. *aureifolia* D. C. McClint. (McClintock 1988a). The sport was noticed in the McClintocks' garden, Bracken Hill, Platt, Kent, in 1982, and the cultivar was propagated and introduced in 1984 by Mrs Tessa Forbes, Plaxtol Nursery, Kent. However the foliage is at best yellowish green.

Erica × *stuartii*: **Praeger's heath.**

'Irish Lemon' ⬤⬤ ♈ — see plate 18 (centre left) (p. 230).

Erica terminalis: **Corsican heath.**

'Golden Oriole' (® no. 75) — Upright shrub (±1 m tall × ±1 m spread); foliage greenish-yellow, the shoot tips turning red-bronze in winter (Platt 1991); flowers lilac pink..

'Golden Oriole' was a seedling, found in 1989 by Mr and Mrs D. I. Lockie in their garden at Sherford, Taunton, Somerset. It was originally described variously as having "yellow foliage" (Registrar 1990) or "yellow new growth" (Platt 1991), but it seems to have "reverted".

'Thelma Woolner' 🕷️ Upright, bushy heather (±0.65 m tall × ±0.5 m spread); foliage dark green; flowers dark lilac-pink (H11).

This is not a reliable clone, being frost sensitive. It came from Monte Arqueri in the Gennargentu Mountains of central Sardinia, whence it was collected by Mr and Mrs Lionel Woolner in 1967.

Erica tetralix: cross-leaved heath.

'Alba Mollis' 🕷️ ♀ (*Erica tetralix* f. *alba*) — see plate 18 (upper centre) (pp. 230–231).

'Con Underwood' ♀ Spreading shrub (±0.25 m tall × ±0.5 m spread); foliage appearing grey-green because upper surface of each leaf is hirsute; leaf edges and lower sides with long, gland-tipped cilia; flowers large, magenta (H14); sepals with long, red, gland-tipped hairs; July–November.

Noticed in Aldershot area, Surrey, by Mrs Constance Underwood in August 1938. She drew her husband's attention to the plant, and while it was not impressive, the "small crimson flowers" were evidently sufficiently distinctive for Mr Underwood to consider it worth propagating. "We cut the plant out with a knife and took it home and planted it very carefully." The heather survived throughout the Second World War, after which it was "rediscovered", propagated, and introduced in 1948 by G. Underwood & Son, Hookstone Green Nursery, West End, Woking, Surrey. Described variously as "one of the best" (Underhill 1971) and "probably the best" (Maxwell & Patrick 1966) cultivar of the cross-leaved heath. AGM 1993.

'Pink Star' 🕷️ ♀ (*Erica tetralix* f. *stellata*). Compact heath (±0.2 m tall × ±0.35 m spread); foliage grey; flowers arrayed horizontally and evenly spaced in a star-like umbel, bright pink (H11); July–October.

Mr and Mrs John Letts found this "most attractive" cultivar in a "roadside field whilst motoring through Cornwall". AGM 1993.

'Riko' 🕷️ Compact, bushy heather (±0.2 m tall × ±0.35 m spread); foliage grey-green; flowers ruby (H5), June–September.

Lothar Denkewitz, Hamburg, Germany, raised this somewhat mysterious cultivar in 1991. He had tried to fertilise *E. tetralix* 'Tina' with pollen from *E. cinerea* 'Pink Ice' but the resulting seedlings all turned out to be indistinguishable from *E. tetralix*: the attempted hybrid was evidently not produced. Riko was named after Denkewitz's grandson, "... ein kräftiger, fröhlicher Kerl ist": a strong, happy guy. 'Samtpfötchen' is a sister-seedling.

Erica umbellata: dwarf Spanish heath.

'Anne Small' (® no. 158). Bushy, spreading shrub (±0.35 m tall × ±0.45 m spread); foliage bright green; flowers with white corolla and golden-brown anthers; May–June.

This was raised from seed collected by Dr Maria Isabel Fraga Vila in Galicia, Spain. One seedling had white flowers and was selected and named after Mrs Anne Small who was Administrator of The Heather Society (1986 to 2006).

'David Small' (® no. 159). Bushy heather with upright shoots (±0.45 m tall × ±0.55 m spread); foliage dark green; flowers vivid amethyst (H1) to heliotrope (H12) with dark anthers (see Tyler 2001).

During a field-trip in Spain undertaken by David McClintock, David Small and Dr Charles Nelson in July 1982, cuttings of selected variants of *Erica umbellata* were collected at Cabo Villano, La Coruña. From these, in 1999, this clone was selected and named after David Small, then the Chairman of The Heather Society. 'David Small' grew outdoors, unprotected, in a ceramic sink at Outwell, Norfolk, for more than 10 years and flowered profusely (Nelson 1999b).

Erica vagans: **Cornish heath.**

'Birch Glow' 🏵 🏆 Floriferous, compact, dome-shaped shrub (±0.25 m tall × ±0.5 m spread); foliage dark green; flowers bright, deep rose-pink (H7); August–November (about two weeks later than clones of similar colour).

This was a chance seedling, found by Walter Edward Theodore Ingwersen (1883–1960) on his Birch Farm Nursery, Gravetye, East Grinstead, Sussex. The foundling was among some plants of 'St Keverne', while 'Mrs D. F. Maxwell' grew in the next row. An outstanding plant which was in circulation by the early 1960s. AGM 1993.

'Cornish Cream' 🏆 (*Erica vagans* f. *alba*). Vigorous, floriferous shrub forming a large dome (±0.35 m tall × ±0.6 m spread); foliage bright green; flowers "off-white" in very long (±0.2 m), tapering spikes; August–November.

An "exciting" (Underhill 1990) and outstanding clone that was found on Goonhilly Downs on the Lizard Peninsula, Cornwall, and introduced by Treseder & Sons, Truro, Cornwall, before 1966. AGM 1993. It is very hardy, like 'Birch Glow'.

'Fiddlestone' 🏵 Dome-shaped heather (±0.3 m tall × ±0.6 m spread); foliage green; flowers cerise (H6); August–October.

Found as a seedling at Fiddlestone Lodge, Burton-in-the-Wirral, Cheshire, the home of Edward Plummer (d. 1959), probably in the 1950s. This is described as "a brighter edition" (Hulme 1994) of 'Mrs D. F. Maxwell' (and the two may be mixed in the nursery-trade today). The University of Liverpool Botanic Gardens, Ness, introduced and named 'Fiddlestone'.

'Golden Triumph' 🏵 (*Erica vagans* f. *alba*). Vigorous, spreading, floriferous shrub (±0.4 m tall × ±0.65 m spread); young shoots bright golden; mature foliage green; flowers white; August–October.

Alan Wayne Newsham found this as a sport on 'Lyonesse' (see below) in 1982 at Twin Acre Nursery, Knutsford, Cheshire. It has all the virtues of its parent with the bonus of colourful shoot-tips in springtime.

'Kevernensis Alba' 🏵 🏆 (*Erica vagans* f. *alba*). Compact, bushy shrub (±0.25 m tall × ±0.45 m spread); foliage green; flowers small, pure white, in narrow, compact spikes; anthers pale golden brown; August–October.

P. D. Williams (after whom *Erica × williamsii* is named: see p. 311), St Keverne, Cornwall, is reputed to have collected 'Kevernensis Alba' on the Lizard Peninsula, Cornwall; it was in cultivation before 1927. Letts (1966) applauded this as a most attractive plant: "honeybees' paradise". AGM 1993.

'Lyonesse' 🏆 (*Erica vagans* f. *alba*) — see plate 10 (lower left) (pp. 130–131).

'Mrs D. F. Maxwell' �₿ 🏆 — see plate 10 (lower far right) (pp. 130–131).

'Valerie Proudley' 🏆 (*Erica vagans* f. *alba*, *E. vagans* f. *aureifolia*) — see plate 10 (lower middle) (pp. 130–131).

'Yellow John' 🌿 Vigorous, bushy heather (±0.35 m tall × ±0.45 m spread); young foliage bright yellow becoming pink-tinged with age; flowers lilac (H4); August–October.
  'Yellow John' is more vigorous than 'Valerie Proudley' and also is distinguished from that older cultivar by its pink, not white, flowers. It was a seedling found by Jan B. A. Dekker, Mijdrecht, Netherlands, in 1982; introduced by P. G. Zwijnenburg, Boskoop, Netherlands, in 1986.

Erica × *veitchii*: Veitch's heath

'Exeter' 🌿 🏆 Very floriferous, vigorous shrub (±1.25 m tall × 0.65 m spread); foliage light green; flowers sweetly scented, corolla white; stigma flattened, pink; March–May.
  The least hardy of the cultivars, this is claimed to be the original clone from Veitch's nursery in Exeter introduced by Veitch & Sons in 1905. If it is the original clone, 'Exeter' is entitled to the praise given by Gertrude Jekyll (1913) — "splendid"! The name 'Exeter' was bestowed in 1969 (Brickell 1969). AGM 1995.

'Gold Tips' 🌿 🏆 — see plate 15 (pp. 198–199).

Erica × *watsonii*: Watson's heath.

'Claire Elise' (® E.2006:07) 🌿 Spreading, compact, bushy heather (±0.15 m tall × ±0.4 m spread); young shoots with golden tips in spring, turning dark green, foliage with gland-tipped hairs; leaves in whorls of 3 or 4; flowers deep magenta-pink (H1) in short spikes; July–October.
  Named after his daughter, this is one of Dr John Griffiths's deliberately created hybrids. He used *Erica ciliaris* 'Corfe Castle' (see p. 365) and *E. tetralix* 'Con Underwood' (p. 377) as the parents. The cross was made in August 1982 and the seedling first blossomed five years later.

'Dawn' 🏆 Low-growing heath (±0.15 m tall × ±0.35 m spread); shoots with spreading, long, gland-tipped hairs; mature foliage greyish-green; tips of new shoots red, turning golden later; leaves with gland-tipped cilia on edge, glabrescent but retaining simple, minute hairs on tip; flowers rich pinkish-purple (H1/H8) in a compact raceme, ±2 cm long; corolla zygomorphic, ±9 mm long, ±6 mm at widest; calyx green, lobes ±2 mm long, margin with gland-tipped cilia, densely fringed with short, simple hairs; ovary pubescent all over. July–October. Fig. 199 (p. 310).
  Douglas Fyfe Maxwell found 'Dawn' on moors near Wareham, Dorset, about 1922–1923, at the same time as a clone subsequently called 'Gwen'. 'Dawn' was introduced by Maxwell & Beale Ltd (Broadstone, Dorset) in 1925. Maxwell himself described the flowers as 'the colour diffused across the eastern heavens "when dawn's left hand was in the sky" — hence 'Dawn' (Maxwell 1927). Another source (McClintock 1971b: 27) indicated that 'Dawn' was named after a niece of H. E. Beale. AGM 1993.

Erica × *williamsii*: **Williams's heath.**

'Cow-y-Jack'. Spreading, open shrub (to 0.25 m tall × 0.45 m spread); mature foliage green; young shoots brilliant yellow; flowers sparse, pink (H8 above, H16 underneath). July–October.

'Cow-y-Jack', the tenth plant reported from the Lizard Peninsula, Cornwall (see p. 314; Nelson 2007d), is "by far the best" cultivar for springtime colour (Small & Small 2001).

'Ken Wilson'. Spreading but compact heath (±0.3 m tall × ± 0.5 m spread); foliage green, including young shoots, pubescent; flowers magenta (H14) "without a trace of blue" (Small & Small 2001); corolla often with more than 4 lobes; stamens sometimes petaloid with pale pink filaments that may be expanded or fused; anthers may be grotesquely malformed or almost absent; July–November (Fig. 202, p. 314).

A man-made cultivar, raised by David Wilson, Chilliwack, British Columbia, Canada, and named after his father, the late Ken Wilson (1920–2006) who was one-time Supervisor of the University of British Columbia's Botanic Gardens (Knight 2006; Wilson 1998). It is a selected seedling from *Erica tetralix* 'Hookstone Pink' pollinated by *E. vagans* 'Mrs D. F. Maxwell'. Small & Small (2001) note that "the flower colour needs to be seen to be believed."

'P. D. Williams' — see plate 23 (upper centre-right) (pp. 308–309).

'Phantom' (® E.2007:09; see Nelson 2007d: 58). Low-growing, compact heather (±0.3 m tall × ± 0.15 m spread after 4 years); foliage light green; young shoots yellow; calyx green, corolla white; stamens 8, free or variously and irregularly fused in groups of 2–3; anthers pale red when young turning pale tan. July–September.

Raised in 1986 by David Wilson, 'Phantom' is in cultivation in Britain and Canada. The parents were *Erica tetralix* 'Alba Mollis' pollinated by *E. vagans* 'Lyonesse'. (McClintock & Blamey (1998: 8) referred to this vaguely).

APPENDIX II
Richard Salisbury's letter

Richard Salisbury's study of *Erica* species, undertaken mostly (we must assume) before he spoke to the Linnean Society of London on 6 October 1801, was far ahead of its time. In his published monograph (Salisbury 1802), he employed a simple notation system which included the following as prime characters:

> leaves: number per whorl
> anthers: whether muticous or spurred
> stamens: whether included or exserted
> inflorescence: whether axillary, pseudo-axillary or terminal

However, apart from *Calluna* and *Salaxis* he stopped short of designating genera or subgeneric taxa, perhaps expecting to return to the subject in a subsequent publication which never materialised.

That he continued to think about the subdivision of this huge genus is evident from a letter, perhaps never sent, that was written (from internal evidence, namely the reference to Herbert's paper to be "published in the next part of the Hort. Tr.") late in 1819. We do not know who the intended recipient was, but there is ample evidence that the contents of this letter, or perhaps of another, now-lost manuscript very similar to it, were known to both George Don and George Bentham, because Don, first, and Bentham, later, attributed certain generic and section names to Salisbury, and those names (or allied names) occur in the letter.

The manuscript is partly transcribed here — the original which is in The Botany Library, The Natural History Museum, London (Salisbury mss, vol. 3 p. 280). The name of the recipient and the date have been removed deliberately from the manuscript; consequently also part of the second page of the letter is absent.

> The old Genus Erica containing 400 species which I have wild specimens of about 150 unexamined, I now regard as an Order, distinguished from all other Bicornes by their Anthers coalescing (ante anthesin) round the pores where the Pollen afterwards escapes, & this so forcibly that the 2 lobes of difft anthers will often separate at this rachis (connectif of the French) if pulled asunder, than at their Pores. I have divided these 250 species which I know into about 60 genera, 20 consisting of only 1 species each; 20 or thereabouts of from 2 to 4 species, the rest of more, & Wm Herbert who is studying heathers, has in a paper which will be published in the next part of the Hort. Tr. mentioned the propriety of dividing more. What can be more distinct than Tubiflora & my Anagallidiflora — I divided the Order into Sections & these into subdivisions.

> Sect 1: Per. polyspermum, medio-loculorum dehiscens Div. ?m. Corolla 1–4 lineas longa, 4–8-andria, nunc irregularis. Folia 3–4-na.
>   ★ Australes
> Here I insert the Cape of Good Hope genera
>   ★★ Boreales (confined to our Hemisphere)

> Gypsocallis.   Pseudo-ax. exss. m. E. vagans &
> Tesquaria   Pseudo-ax exs. m. E. multiflora &c
> Peronium   term. exs. m E. umbellata L. sui generis

Eremocallis   lat. incl.c. E. cinerea L.

(I believe there are 2 species under this; & Dryander thought so too; he never could believe the Swedish plants the same as ours which is the Portugal plant certainly, from Sir Thomas Gage's specimens & seeds I raised.

Tylosporum ... E. australis L. this Portugal shrub grows 10 feet height & differs from all others in its seeds which have a Carancule as large as a Mimosa seed.

Tetralix ... E. tetralix. the Corsican Stricta is another species

[cut]

Achlamys ... E. Olympiaca Sibth.

This differs from all, besides, other strong characters in having no Bractes whatever ...

I cannot rummage in my papers to send you my full generic characters now ... I remember however that Gypsocallis differs from all the other Genera of the section in the anthers being erect & forming a ring so open & the lobes of different anthers cohering so closely, that till they split the filaments appear to belong to neither & so be placed between the anthers thus [diagram] There is only another of Gypsocallis besides vagans, and they are both confined to Gypsaceous soil – It was found by Forskhol near the Dardanelles & I think he calls it verticillata ...

E. carnea, herbacea & Purpurascens L. are all 1 plant & form a 2d Species of Tesquaria with Erica multiflora, which I regard as the type – I found it not far from Avignon – E. mediterranea, which fills the sombre solitudes of Estremadura is the 3d & only species of Tesquaria besides –

APPENDIX III
White heather

There are no references before 1855 to the superstition that white heather is lucky (McClintock 1970b; Nelson 2006c). To exemplify the silence about white heather being a token of good luck, Ann Pratt's *Flowering plants of Great Britain* (1855) contained substantial material about heather, including the statement that *Calluna vulgaris* (ling) "... is an exceedingly beautiful plant ... occasionally bearing white flowers." She quoted George Luxford (1807–1854), editor of *The phytologist* and lecturer in botany at St Thomas' Hospital, London, who opined that it was "a very elegant plant. The red and the white-flowered varieties, with their smooth, deep green, closely imbricated leaves, are pretty and delicate ...". There is no mention in Pratt's work of any heather being a lucky plant. The earliest source is Queen Victoria's book *Leaves from the journal of our life in the Highlands from 1848 to 1861*, which was originally issued in an edition for private circulation only but was subsequently published by Smith, Elder & Co. and, as Hibbert (2000: 329–330) related, sold 100,000 copies within three months. The date of publication of the public edition was 1868, so the earliest reference to lucky white heather really should be so dated. In the sequel, *More leaves from the journal of a life in the Highlands from 1862 to 1882* (Victoria 1884), there was at least one more reference to lucky white heather.

As for actually using white heather during the Victorian era, sprigs were included in the bouquets of the bride or bridesmaids at many royal weddings including significantly that of the Princess Royal (Princess Victoria) and Crown Prince Friedrich Wilhelm of Prussia in the Chapel Royal, St James's Palace, London on 25 January 1858. The following day, *The Times* reported that the bridesmaids' dresses, designed "by the illustrious bride herself" were garnished with bouquets of pink roses and white heather, and that a "bouquet of the same flowers [was] worn in the girdle and upon each shoulder" (Nelson 2008b). Given the date of the wedding, the heather is very unlikely to have been *Calluna vulgaris* (ling) which is presumed to have been the plant plucked by the Crown Prince when he proposed to Princess Victoria (see Nelson 2006c).

Given that *Leaves from the journal of our life in the Highlands from 1848 to 1861* (Victoria 1868) was a sensational best-seller, there is no reason not to agree with David McClintock (1970b) that the tradition attributing good luck to white heather achieved popularity during the second half of the reign of Queen Victoria. The evidence from other publications supports this suggestion. In the numerous books dealing with the language of flowers that were published before about 1870, and which I have managed to examine, there are no entries for white heather, only for "heath" and its meaning is always "solitude"! However, in at least one post-1868 publication — Tyas's *Speaking flowers* (1875: 98–99) — there is an entry for white heather and its meaning is given as "good luck": "it is so regarded in Scotland, as we read in our beloved Queen's "Journal in the Highlands"...", Tyas explained! Under "Good luck", Tyas (1875: 145) noted that it was signified by "white Heather (Calluna vulgaris, β alba)".

"White heather was used by the British royal family as a sort of charm, considered lucky for brides", according to *Curious chapbooks and hysterical histories* (quoted in Nelson 2006c), but was it a peculiarly *Scottish* tradition, ancient, and so untraceable, yet revealed and then popularised by Queen Victoria?

Leaves from the journal of our life in the Highlands from 1848 to 1861 is a heavily edited version of the Queen's personal journal of her life at Balmoral. The first reference to white heather and its lucky association is contained in the account of the day, 29 September 1855, on which Crown Prince Friedrich Wilhelm of Prussia asked the 14-year old Princess Royal, Princess Victoria, the eldest daughter of Queen Victoria and Prince Albert, to be his future wife. The story goes that when riding on Creag nam Ban, not far from Balmoral, "Fritz" (the Crown Prince's nickname) picked a piece of white heather and gave it to Vicky, because, so "Fritz" said, white heather was an the emblem of "good luck".

Was this really an old Scottish tradition? If the act of picking the sprig of white-flowered heather was a spontaneous one on the part of the Crown Prince, and if it was his affirmation that this was good luck, then we may consider it an imported, European superstition. To suppose that on this, his first visit to Scotland, he was prompted by, or even deliberately instructed by, the Queen or Prince Albert or a member of the royal household to fulfil the requirements of an old Scottish belief (as suggested by McClintock 1970b) requiring him to find, collect and present the white heather as a means of contriving to declare his love seems rather far-fetched. White-flowered heathers do occur in the Scottish Highlands but are not that easily found: in fact, we don't actually know which heather it was that the Crown Prince plucked and presented to the Princess Royal, because the accounts do not record its exact identity.

It is clear, however, that the action of the Prussian Crown Prince lingered in the memory of the British Royal Family and became a motif. Perhaps the most telling incident occurred seven years later, on 3 September 1862, when the recently widowed Queen was at Laeken, outside Brussels, to meet Prince and Princess Christian of Denmark and their daughter Alexandra, the fiancée of the Prince of Wales, "Bertie". "Her whole appearance", wrote Queen Victoria, "was one of the greatest charm, combined with simplicity and perfect dignity. I gave her a little piece of white heather, which Bertie gave me at Balmoral, and I told her I hoped it would bring her luck" (Hibbert 1984: 167). Thus the Princess Royal's younger brother employed the same token to convey a message to his bride-to-be.

In an interesting reversal of the Victorian, royal tradition, nowadays at weddings in Scotland (and elsewhere), white heather is more often worn by the groom and his party than by the bride. Yet there is a constant in these actions: white heather is a token between couples, a love-token boding good fortune or good luck. There is no element of chance — of chancing upon a plant of white-blossomed heather in the wild — in the same way as a lucky four-leaved clover. In any case, the element of chance, if it ever was a part of the lore, has been extirpated by the ready availability at every season of a white-blossomed heather, whether *Calluna* or *Erica*, or even dried inflorescences of *Limonium sinuatum*, sea lavender (Richards 1990).

There is another twist to this tale, and it is the odder because neither Wallace (1903) nor McClintock (1970b) made any mention of it. According to Internet sources (see Nelson 2005b, 2006c), there is a "Celtic" legend, "an old, old tale and a sad one", about Malvina and her lover, Oscar. Oscar, a gallant hero, was away doing battle. One day, a messenger brought Malvina a bunch of purple heather, Oscar's last token of love before he was slain. She burst into tears, and her tears fell on some heather, the flowers of which turned white. Ever afterwards, as she sorrowfully wandered the moors, crying for her dead lover, those of her tears that dripped on to heather instantly turned the flowers white. "Although this is a symbol of my sorrow", she declared, "may the white heather bring good fortune to all who find it." That fable has clear links with the notorious Ossian forgeries of the Revd John Macpherson (1713–1765), Minister of Sleat on the Isle of Skye, and his namesake (they were not related), the writer James Macpherson (1736–1796). Yet, there is no trace of the story about Malvina, Oscar and the transformation of purple heather into white in James Macpherson's original Ossian text (Nelson 2006c).

BIBLIOGRAPHY

Aiton, W. (1789). *Hortus kewensis*. Volume **2**. G. Nicol, London.

Aiton, W. T. (1811). *Hortus kewensis*. Volume **3**. Second edition. Longman, Hurst, Rees, Orme, & Brown, London.

Alcock, R. H. (1876). *Botanical names for English readers*. L. Reeve, London.

Alemon (1881). Joldwynds, Surrey. *Gard. Chron.*, n.s., **16**: 206.

Allaby, M. (1998). *A dictionary of plant sciences*. Oxford University Press, Oxford.

Allan, M. (1982). *William Robinson 1838–1935. Father of the English flower garden*. Faber & Faber, London.

Allen, D. E. (1978). *The naturalist in Britain. A social history*. Penguin Books, Harmondsworth.

Allen, D. E. & Hatfield, G. (2004). *Medicinal plants in folk tradition*. Timber Press, Portland & Cambridge.

Alm, T. (1999). Heather (*Calluna vulgaris*) in Norwegian folk tradition. *Yearb. Heather Soc.* **1999**: 49–52.

Anderson, E. B. (1965). *Gardening on chalk and limestone*. W. H. & L. Collingridge, London.

Andrew, R. & West, R. G. (1977). Pollen spectra from Pliocene Crag at Orford, Suffolk. New Phytol. 78: 709–714.

Andrews, H. C. (1794–1830). *Coloured engravings of heaths*. 4 volumes. The author, London.

Andrews, H. C. (1805–1828). *The heathery; or a monograph on the genus Erica*. 6 volumes. The author, London.

Anonymous (1845). "In der Sitzung der Gesellsch. naturforsch. Freunde zu Berlin ...". *Flora* **23** (1): 190.

Anonymous (1846). Royal Dublin Society. Discovery of new planets [*sic*]. *The medical times* **19**: 164 (16 December).

Anonymous (1874a). Garden gossip. *Florist. & Pomol.* (February 1874): 41.

Anonymous (1874b). Various-flowered variety of St. Dabeoc's heath. *Garden* (London 1871–1927) **5**: 138–139 (14 February).

Anonymous (1877). Drummond Castle. *Gard. Chron.*, n.s., **7**: 688–689.

Anonymous (1905a). *Erica lusitanica* in Dorset. *J. Bot.* **43**: 220–221.

Anonymous (1905b). The Royal Horticultural [Society] Floral Committee ... *Erica × Veitchi* ... *Gard. Chron.*, ser. 3, **37**: 106 (18 February).

Anonymous (1905c). Royal Horticultural Society. Floral Committee (February 14). *Garden* (London 1871–1927) **67**: ix (25 February).

Anonymous (1909). Some Connemara plants — resume of a lecture delivered by W. J. C. Tomlinson. *Rep. (Annual) & Proc. Belfast Naturalists' Field Club*, ser. 2, **6**: 294–296.

Anonymous (1983). *Heather trials 1976–81*. [The Heather Society.] (See also Vickers (1976).)

Anonymous (1985). *Daboecia cantabrica* (Huds.) C. Koch dans le sud-Armoricain. *Index seminum 1986*: 17–20. Jardin botanique, Ville de Nantes.

Aparicio, A. (1994). Números cromosomáticos de plantas occidentales, 708. *Anales Jard. Bot. Madrid* **51** (2): 280.

Aparicio, A. (1995). Seed germination of *Erica andevalensis* Cabezudo & Rivera (Ericaceae), an endangered edaphic endemic in southwestern Spain. *Seed Sci. Technol.* **23**: 705–713.

Aparicio, A. (1999). *Erica andevalensis*. In: G. Blanca, B. Cabezudo, J. E. Hernández-Bermejo, C. M. Herrera, J. Molero Mesa, J. Muñoz & B. Valdés (1999). *Libro rojo de la flora silvestra amenazada de Andalucía*. Tomo **1**: *Especies de peligro de extinción* pp. 119–122. Junta de Andalucia.

Aparicio, A. & García-Martín, F. (1996). The reproductive biology and breeding system of *Erica andevalensis* Cabezudo & Rivera (Ericaceae), an endangered edaphic endemic in southwestern Spain. implications for its conservation. *Flora* **191**: 345–351.

Arber, A. (1986). *Herbals their origin and evolution. A chapter in the history of botany 1470–1670*. Third edition (with an introduction and annotations by William T. Stearn). Cambridge University Press, Cambridge.

Arends, G. A. (1951). *Mein Leben als Gärtner und Züchter*. Ulmer, Stuttgart.

Asensi, A., Bennett, F., Brooks, R., Robinson, B. & Stewart, R. (1999). Copper uptake studies on *Erica andevalensis*, a metal-tolerant plant from southwestern Spain. *Communications in soil science & plant analysis* **30** (11–12): 1615–1624.

Ashworth, S. (2010). 12th June. In: O. Mountford & J. Akeroyd, BSBI [overseas field meeting]: Romania, p. 50. *BSBI news* **114**: 49–55.

Aubert, G. (1967). La répartition d'*Erica multiflora* L. dans le bassin Méditerranéen et en Provence. *Ann. Fac. Sci. Marseille* **39**: 5–12.

Auton, C. A., Gordon, J. E., Merritt, J. W. & Walker, M. J. C. (2000). The glacial and interstadial sediments at the Burn of Benholm, Kincardineshire: evidence for onshore pre-Devensian ice movement in northeast Scotland. *J. Quatern. Sci.* **15**: 141–156.

Babington, C. C. (1836). On several new or imperfectly understood British and European plants. *Proc. Linn. Soc. London* **17**: 451–464 (published between 18 July and 8 August).

[Backhouse, J.] (1882). Erica maweana. *Florist & Pomol.*: 75.

Backhouse, James & Son Ltd (1911). *Erica carnea winter flowering*. The Nurseries, York.

Baker, H. A. & Oliver, E. G. H. [1967]. *Ericas in southern Africa with paintings by Irma von Below, Fay Anderson and others*. Purnell, Cape Town & Johannesburg.

Baker, M. (2005). *Erica scoparia* L. Besom heath (Ericaceae). *Tasweeds* no. 28: 10–11.

Ball, C. F. (1912). Botanizing in Bulgaria. *Gard. Chron.*, ser. 3, **51**: 252–253, 274–275.

Ball, J. (1877). Spicilegium florae Maroccanae. *J. Linn. Soc., Bot.* **16**: 281–742.

Ballester, A., Vieitez, A. M. & Vieitez, E. (1979). The allelopathic potential of *Erica australis* L. and *E. arborea* L. *Bot. Gaz.* 140: 433–436.

Bannister, P. (1965). Biological flora of the British Isles. *Erica cinerea* L. *J. Ecol.* **53**: 527–542.

Bannister, P. (1966). Biological flora of the British Isles. *Erica tetralix* L. *J. Ecol.* **54**: 795–813.

Bannister, P. (1996). The frost resistance of heaths and heathers. *Yearb. Heather Soc.* **1996**: 39–42.

Bannister, P. (2007). A touch of frost? Cold hardiness of plants in the southern hemisphere. *New Zealand J. Bot.* **45**: 1–33.

Bannister, P. & Polwart, A. (2001). The frost resistance of ericoid heath plants in the British Isles in relation to their biogeography. *J. Biogeogr.* **28**: 589–596.

Barclay-Estrup, P. (1974). The distribution of *Calluna vulgaris* (L.) Hull in western Canada. *Syesis* **7**: 129–137.

Barclay-Estrup, P. (1988). The distribution and some ecological aspects of *Calluna vulgaris* in Canada. Unpublished typescript (presented at the International conference on lowland heaths, University of Masscahusetts, March 1988, Nantucket Island).

Barclay-Estrup, P. (1991). Scottish heather, *Calluna vulgaris* (L.) Hull, in eastern Canada. *Naturaliste Canad. (Rev. Écol. Syst.)* **118**: 47–55.

Barbey, C. & Barbey-Boissier, W. (1882). *Herborisations au Levant: Egypte, Syrie et Méditerranée*. Georges Bridel, Lausanne.

Baron, J. (2006) [Photograph of] *Erica tetralix* 'Meerstal'. *Heathers* **3**: 71.

Barrie, F. R. (2006). Report of the General Committee: 9. Report of Special Committee 3C of the Berlin Congress (*Taxon* **41**: 552–583. 1992). *Taxon* **55** (3): 795–796.

Bauer, F. A. (1796[–1803]). *Delineations of exotick plants cultivated in the Royal garden at Kew*. W. T. Aiton, London.

Bauhin, C. (1623). ΠΙΝΑΞ *theatri botanici*. Ludovici Regis, Basle.

Bauhin, C. (1671). ΠΙΝΑΞ *theatri botanici*. Reprint. Impensis Johannis Regis, Basle.

Bauhin, J. & Cherler, J. H. (1650). *Historia plantarum universalis ....* **I**. Typographia Caldoriana, Yverdon.

Baumgarten, J. C. G. (1816). *Enumeratio stirpium Magno Transsilvaniae principatui*. Volume **1**. Libraria Camesinae, Vienna.

Bayer, E. (1993). Erica, in *Flora iberica*. Vol. **4**: 485–506. CSIC, Madrid.

Bayer, E. & López González, G. (1989). Los brezos españoles. *Quercus* **35**: 21–36.

Bean, W. J. (1905). *Erica veitchii* ×. *Gard. Chron.*, ser. 3, **37**: 138 (4 March).

Bean, W. J. (1914). *Trees and shrubs hardy in the British Isles*. John Murray, London.

Bean, W. J. (1933). New and interesting plants. *New Fl. & Silva* **5**: 197.

Bean, W. J. (1950). *Trees and shrubs hardy in the British Isles*. Seventh edition. John Murray, London.

Beentje, H. J. (2006). Ericaceae. *Flora of tropical East Africa*. Royal Botanic Gardens, Kew.

Beijerinck, W. (1936). Die Geographische Verbreitung von *Calluna vulgaris* (L.) Salisb. [*sic*]. (*Mitteilung no.* **11** *der Biologische Station Wijster (Drente), Holland.*) *Recueil Trav. Bot. Néerl.* **33**: 341–350.

Beijerinck, W. (1940). *Calluna* a monograph on the

Scotch heather. *Verh. Kon. Ned. Akad.Wetensch., Afd. Natuurk.* **38** (4).

Beith, M. (2004). *Healing threads. Traditional medicines of the Highlands and Islands.* Birlinn Ltd, Edinburgh.

Benito Cebrián, N. de (1948). *Brezales y brezos. Síntesis geobotánica de las formaciones de Ericoideas y resumen monográfico de las especies españolas.* Instituto Forestal de Investigaciones y Experiencias, Madrid.

Bentham, G. (1839). Tribus III *Ericeæ* (*Calluna, Pentapera, Erica, Bruckenthalia*). In: A.-P. de Candolle (ed.), *Prodromus systematis naturalis regni vegetabilis.* Vol. **7**: 612–694. Treuttel & Würtz, Paris.

Besançon, H. (1978a). Les bruyères rares du littoral aquitain. *Bull. Soc. Linn. Bordeaux, sect. Mycol.,* 1978: 27–19.

Besançon, H. (1978b). Les bruyères rares du littoral aquitain (2ᶜ partie). *Bull. Soc. Linn. Bordeaux, sect. Mycol.,* 1978: 8–12.

Birks, H. J. B. & Ransom, M. E. (1969). An interglacial peat at Fugla Ness, Shetland. *New Phytol.* **68**: 777–796.

Birks, H. J. B. & Peglar, S. M. (1979). Interglacial pollen spectra from Sel Ayre, Shetland. *New Phytol.* **83**: 559–575.

Bisgrove, R. (1990). *The National Trust book of the English garden.* Viking, London.

Blamey, M. & Grey-Wilson, C. (1993). *Mediterranean wild flowers.* HarperCollins, [London].

Blanca, G., Cabezudo, B., Hernández-Bermejo, J. E., Herrera, C. M., Molero Mesa, J., Muñoz, J. & Valdés, B. (1999). *Libro rojo de la flora silvestra amenazada de Andalucía.* Tomo **1**: *Especies de peligro de extinción.* Junta de Andalucia.

Blanca, G., Onieva, M. R. L., Lorite, J., Lirola, M. J. M., Mesa, J. M., Quintas, S., Girela, M. R., Ángeles Varo, M. de los & Vidal, S. (2001). *Flora amenazada y endémica de Sierra Nevada.* Universidad de Granada & Consejería de Medio Ambiente, Junta de Andalucía.

Boivin, B. (1966). Énumération des plantes du Canada II—Lignidées. *Naturaliste Canad.* **93**: 371–437.

Bolòs, O. de & Vigo, J. (1995). *Flora dels països Catalans. Vol. III (Pirolàcies – Compostes).* Editorial Barcino, Barcelona.

Bonner, I. (2011). Heathers on Holyhead Mountain, Anglesey (v. c. 52). *BSBI Welsh Bull.* **88**: 14.

Bornmüller, J. F. N. (1904). Ergebnisse zweier botanischer Reisen nach Madeira und den Canarischen Inseln. *Bot. Jahrb. Syst.* **33** heft 3: 387–492.

Bostock, J. & Riley, H. T. (eds) (1855). *The natural history of Pliny … with copious notes and illustrations.* H. G. Bohn, London.

Bowen, H. (2000). *The flora of Dorset.* Pisces Publications, Newbury.

Bradley, R., Burt, A. J. & Read, D. J. (1981). Mycorrhizal infection and resistance to heavy metal toxicity in *Calluna vulgaris. Nature* **292**: 335–337.

Bradlow, F. R. (1994). *Francis Masson's account of three journeys at the Cape of Good Hope 1772–1775.* Tablecloth Press, Cape Town.

Bramwell, D. & Bramwell, Z. (1974). *Wild flowers of the Canary Islands.* Stanley Thornes Ltd, London & Burford.

Brian, M. V. (1977). *Ants.* (New naturalist no. 59.) Collins, London.

Brickell, C. D. (1969). Preliminary notes on the nomenclature of some cultivated heathers. *J. Roy. Hort. Soc.* **94** (9). 373–377.

Brickell, C. D. & McClintock, D. C. (1987). Proposal to conserve *Erica carnea* against *E. herbacea. Taxon* **36**: 477–481. (See also Brummitt, R K. (1990). Report of the Committee for Spermatophyta 37. *Taxon* **39**: 294.)

Brickell, C. D., Baum, B. R., Hetterscheid, W. L. A., Leslie, A. C., McNeill, J., Trehane, P., Vrugtman, F. & Wiersema, J. H. (eds). (2004). International code of nomenclature for cultivated plants. *Acta Hort.* **647.**

Britten, J. & Holland, R. (1886). *A dictionary of English plant names.* English Dialect Society, London.

Brodie, J. & Watts, D. (2001). *Erica tetralix* L. (cross-leaved heath) in The Burren (Co. Clare, Ireland). *Yearb. Heather Soc.* **2001**: 7–8.

Brough, P. R. & McClintock, D. C. (1978). Heaths with parts in fives and sixes. *Watsonia* **12**: 156–157.

Broughton, D. A. & McAdam, F. H. (2002). *The vascular flora of the Falkland Islands: an annotated checklist and atlas.* Falklands Conservation, Stanley.

Browicz, K. (1983). *Chorology of trees and shrubs in south-west Asia and adjacent regions.* Vol. **2**. Polish Scientific Publishers, Warsaw & Poznan.

Brown, N. E. (1905–1906): Ericaceae. In: W. T. Thistelton-Dyer (ed.), *Flora capensis* **4** section 1: 315– 418. Lovell Reeve & Co., London.

Browne, J. (2004). Gray, Samuel Frederick (1766–1828). *Oxford dictionary of national biography.* Oxford University Press, Oxford. (URL http://www.oxforddnb.com/view/article/11353 (accessed July 29, 2007.)

Brullo, S. & Furnari, F. (1979). Taxonomic and nomenclatural notes on the flora of Cyrenaica (Libya). *Webbia* **34** (1): 155–174.

Brullo, S. & Guglielmo, A. (2001). Considérations phytogéographiques sur la Cyrénaïque septentrionale. *Bocconea* **13**: 209–222.

Brummitt, R. K. (1993). Report of the Committee for Spermatophyta: 38. (961) Conserve *Erica vagans* L. *(Ericaceae)* with a new type ... . *Taxon* **42**: 696.

Brummit, R. K. (2009). Report of the Nomenclature Committee for Vascular Plants: 60. *Taxon* **58** (1): 280–292

Bruneau de Miré, P. & Quézel, P. (1959). Sur la présence de la bruyère en arbre (*Erica arborea* L.) sur les sommets de l'Emi Koussi (Massif du Tibesti). *Compt. Rend. Sommaire Séances Soc. Biogéogr.* **34**: 66–70.

Buchan, U. (2007). *Garden people. Valerie Finnis and the golden age of gardening* Thames & Hudson, London.

Bullar, J. & Bullar, H. (1841). *A winter in the Azores: and a summer at the baths of the Furnas.* J. Van Voorst, London. 2 vols.

Bullock, J. M. & Clarke, R. T. (2000). Long distance dispersal by wind: measuring and modelling the tail of the curve. *Oecologica* **124**: 506–521.

Bullock, J. M. & Moy, I. L. (2004). Plants as seed traps: inter-specific interference with dispersal. *Acta oecologica* **25**: 35–41.

Burton, R. M. (1995). A new locality for *Erica bocquetii. Karaca Arbor. Mag.* **3** (1): 27–29.

Butcher, R. W. (1961). *A new illustrated British Flora.* Leonard Hill, London.

Cabezudo, B. & Rivera, J. (1980). Notas taxonomicas y corologicas sobre la flora de Andalucia occidental. 2. *Erica andevalensis* Cabezudo & Rivera, sp. nov. *Lagascalia* **9**: 223–226.

Calmann, G. (1977). *Ehret flower painter extraordinary. A biography.* Phaidon, London.

Candolle, A.-P. de (1839). *Prodromus systematis naturalis regni vegetabilis.* Volume **7**. Treuttel & Würtz, Paris

Canovan, R. (2005). Heathers: coping with the weather and alkaline clay. *Heathers* **2**: 1–12.

Canovan, R., Petterssen, E., Osti, G., Kay, S. & Nelson, E. C. (ed.) (2008). Northern Spain, July 2007. *Heathers* **5**: 44–52.

Capelo, J. H., Bingre, P., Arsénio, P. & Espírito-Santo, M. D. (1998). Notes do Herbário da Estação

Florestal Nacional (LISFA) fasc. VII. XVIII: Uma ericácea nova para a flora portuguesa. *Silva Lusit.* 6 (1): 119–120.

Carballeira, A. (1980). Phenolic inhibitors in *Erica australis* L. and in associated soil. *J. Chem. Ecol.* **6**: 593–596.

Castroviejo, S. (1982). Sobre la flora Gallega, IV. *Anales Jard. Bot. Madrid* **39** (1):157–165.

Cazalbou, R. (2000). Sur l'étymologie du mot *Bruja*: interlude botanico-phonétique. *Communication présentée au IXe Colloque de Linguistique Hispanique, Lille, 16–18 mars 2000* (URL http://www.mshs. univ-poitiers.fr/libero/cazalbou.htm).

Chapman, E. F. (1949). *Cyprus trees and shrubs.* Cyprus Government Printing Office, Nicosia.

Chapman, S. B. (1975). The distribution and composition of hybrid populations of *Erica ciliaris* L. and *E. tetralix* L. in Dorset. *J. Ecol.* **63**: 809–823.

Chapman, S. B. & Rose, R. J. (1994). Change in the distribution of *Erica ciliaris* L. and *E.* × *watsonii* Benth. in Dorset, 1963–1987. *Watsonia* **20**: 89–95.

Chapple, F. J. (1964). *The heather garden.* Revised edition. W. H. & L. Collingridge Ltd, London.

Chapple, F. J. (1965). *Erica umbellata* in Lancashire. *Yearb. Heather Soc.* **1** (3): 52.

Chapple, F. J. (1966). Carnea heath. *Gard. Chron.* **159**: 203–204.

Charrington, J. (1962). A heather society. *J. Roy. Hort. Soc.* **87** (8): 369.

Chilton, L. (1999). *Plant list for Lanzarote.* Second edition revised. Marengo Publications, Hunstanton.

Chittenden, F. J. (1931). [*Daboecia*]. *Gard. Ill.* **53** (14 March): 164.

Church, A. (1908). *Types of floral mechanism.* Clarendon Press, Oxford.

Clapham, A. R., Tutin, T. G. & Warburg, E. F. (1962). *Flora of the British Isles.* Second edition. Cambridge University Press, Cambridge.

Cleevely, R. J. (1997). The heather illustrations in the catalogues of Maxwell & Beale with an account of the botanical artist Winifred Walker (1882–1966). *Yearb. Heather Soc.* **1997**: 19–30.

Cleevely, R. J. (2004). Gray, John Edward (1800–1875). *Oxford dictionary of national biography.* Oxford University Press, Oxford. (URL http://www. oxforddnb.com/view/article/11344, accessed 29 July 2007.)

Cleevely, R. J. & Oliver, E. G. H. (2002). A preliminary

note on the publications dates of H. C. Andrews' *Coloured engraving of heaths* (179–1830). *Arch. Nat. Hist.* **29**: 245–264.

Cleevely, R. J., Nelson, E. C. & Oliver, E. G. H. (2003). More accurate publication dates for H. C. Andrews' *The Heathery*, particularly volumes 5 and 6. *Bothalia* **33** (2): 195–198.

Clement, E. J. & Foster, M. C. (1994). *Alien plants of the British Isles*. Botanical Society of the British Isles, London.

Clement, E. J. & Smith, D. P. J. & Thirlwell, I. R. (2005). *Illustrations of alien plants of the British Isles*. Botanical Society of the British Isles, London.

Coats, P. (1970). *Great gardens of Britain*. Spring Books, London.

Coe, M. (ed.) (1985). *Oxford illustrated encyclopedia. Vol. 2 The natural world*. Oxford University Press, Oxford.

Coker, P. D. (1968). A preliminary report; the site of *Erica vagans* near Belcoo, Northern Ireland. Unpublished.

Colgan, N. & Scully, R. (1898). *Cybele hibernica*. Second edition. Edward Ponsonby, Dublin.

Cook, E. T. (1908). *Trees and shrubs for English gardens*. London.

Costa, J. C., Aguiar, C., Capelo, J. H., Lousã, m. & Neto, C. (1998). Biogeografia de Portugal continental. *Quercetea* **1**: 5–56.

Coutinho, A. X. P. (1913). *A flora de Portugal (plantas vasculares)*. Aillaud Alves, Paris & Lisbon.

Coxon, P., Hanon, G. & Foss, P. (1994). Climatic deterioration and the end of the Gortian Interglacial in sediments from Derrynadivva and Burren Townland, near Castlebar, County Mayo, Ireland. *J. Quatern. Sci.* **9**: 33–46.

Coxon, P. & Waldren, S. (1995). The floristic record of Ireland's Pleistocene temperate stages. In: R. C. Preece (ed.), *Island Britain, a Quaternary perspective*. (Geological Society special publication no. 96: 243–267.)

Cruz, A. & Moreno, J. M. (2001). Lignotuber size of *Erica australis* and its relationship with soil resources *J. Veg. Sci.* **12**: 378–384.

Cruz, A., Pérez, B.,Velasco, A. & Moreno, J. M. (2003). Variability in seed germination at the inter-population, intra-population and intra-individual levels of the shrub *Erica australis* in response to fire-related cues. *Pl. Ecol.* **169**: 93–103.

Curtis, J. (1824–1839). *British entomology, being illustrations and descriptions of the genera of insects found in Great Britain and Ireland*. [J. Curtis], London. 16 vols.

Curtis, T. G. F. (2000). The change in the legal status of *Erica ciliaris* L. Dorset heath in Ireland. *Yearb. Heather Soc.* **2000**: 89–92.

Curtis, W. (1800). *Erica ciliaris*. Ciliated heath. *Bot. Mag.* **14**: tab. 484.

Dahlin, M. (2007). Because of a heather show ... . *Heathers* **4**: 11–14.

Dahm, J. (2004). In memoriam Hermann Blum. *Ericultura* no. 132: 13–15.

Dallimore, W. (1910). The heath garden. In: F. W. Meyer (ed. E. T. Cook), *Rock and water gardens: their making and planting: with chapters on wall and heath gardening* pp. 196–222 (Chapter 32). Country Life Ltd., London.

Dandy, J. E. (1958a). *The Sloane herbarium*. British Museum (Natural History), London.

Dandy, J. E. (1958b). *List of British vascular plants, incorporating The London catalogue of British plants*. British Museum (Natural History), London.

Darwin, T. (1996). *The Scots herbal. The plant lore of Scotland*. Mercat Press, Edinburgh.

Dauphin, P. & Aniotsbehere, J. C. (1993). Les galles de France. *Mem. Soc. Linn. Bordeaux* **2**: 1–316.

Davey, F. H. (1910). *Erica vagans* × *cinerea*. *J. Bot.* **48**: 333–334.

Davey, F. H. (1911). A botanical report for 1909, 1910 ... *Erica vagans* × *cinerea*. *J. Roy. Inst. Cornwall* **18**: 383.

Davies, M. S. (1984). The response of contrasting populations of *Erica cinerea* and *E. tetralix* to soil types and waterlogging. *J. Ecol.* **72**: 197–208.

Davis, E. (2006) A demonstration bed of The Heather Society's recommended heathers. *Heathers* **3**: 1–5.

Davis, P. H. (1949). A journey in south-west Anatolia. *J. Roy. Hort. Soc.* **74**: 154–164.

Dawson, W. R. (1934). *Catalogue of the manuscripts in the library of the Linnean Society of London. Part I.—The Smith papers*. The Linnean Society, London.

Day, R. (1995). Dr. Paul Barclay: 1977 Newfoundland plant collection. *Sarracenia* **5** (2): 7–8.

Denkewitz, L. (1987). *Heidegärten*. Ulmer, Stuttgart.

Désamoré, A., Laenen, B., Devos, N., Popp, M., González-Mancebo, J. M., Carine, M. A. &

Vanderpoorten, A. (2011). Out of Africa: north-westwards Pleistocene expansions of the heather *Erica arborea*. *J. Biogeogr.* **38**: 164–176.

Desmond, R. G. C. (1977). *Dictionary of British and Irish botanists and horticulturists including plant collectors and botanical artists*. Third edition. Taylor & Francis, London.

Desmond, R. G. C. (1994). *Dictionary of British and Irish botanists and horticulturists*. Revised edition. Taylor & Francis, & The Natural History Museum, London.

Devonian, (A) (1849). Hybrid heaths. *Gard. Chron.* 17 March: 166.

Dias, E., Elias, R. B. & Nunes, V. (2004). Vegetation mapping and nature conservation: a case study in Terceira Island (Azores). *Biodivers. & Conserv.* **13**: 1519–1539

Díaz González, T. E. & García Rodriguez, A. (1992). Comportamiento ecológico y distribución en Asturias de *Erica* × *praegeri* Ostenf. (*Erica mackaiana* × *Erica tetralix*). **XII** *Jornadas de fitosociologia*, panel 097. Libro de resúmenes Oviedo, 23 al 25 de Septiembre de 1992.

Díaz González, T. E., Fernández Prieto, J. A., Nava Fernández, H. S. & Fernández Casado, M. de los A. (1994). Catálogo de la flora vascular de Asturias. *Itin. Geobot.* **8**: 529–600.

Díaz-Villa, M. D, Marañón, T., Arroyo, J., Garrido, B. (2003). Soil seed bank and floristic diversity in a forest-grassland mosaic in southern Spain. *J. Veg. Sci.* **14**: 701–709.

Dickie, G. (1860). *The botanist's guide to the counties of Aberdeen, Banff and Kincardine*. A. Brown & Co., Aberdeen.

Dickson, C. & Dickson, J. H. (2000) *Plants and people in ancient Scotland*. Tempus, Stroud.

Dieck, G. (1899). *Die Moor- und Alpenpflanzen (vorzugsweise Eiszeitflora) des Alpengartens Zöschen bei Merseburg und ihre Cultur*. G. Dieck, Zöschen.

Dillenius, J. J. (1724). *Johannis Raii synopsis methodica stirpium britannicarum ... Editio tertia*. W. & J. Innys, London (Facsimile edition 1973. The Ray Society, London.)

Dippel, L. (1887). *Handbuch der Laubholzkunde*. Bd 1. Paul Parey, Berlin.

Dizerbo, A. H. (1974). *La vegetation et la flora de la Presqu'île de Crozon (Finistere)*. Société pour l'Étude et la Protection de la Nature en Bretagne, Brest.

Dodoens, R. (1583). *Stirpium historiae pemptades sex, sive libri XXX*. C. Plantin, Antwerp.

Domac, R. (1963). Flora otoka Molata. *Acta Bot. Croat.* **22**: 83–98 (not seen).

Dome, A. P. (2000). *Daboecia azorica* [photograph]. *Heather news* **23** (4): [ii].

Dome, A. P., Lortz, K. & Mackay, D. A. M. (ed.). (2001). *Daboecia azorica* 'Arthur P. Dome'. *Heather news* **24** (2): 2–4.

Don, D. (1834). An attempt at a new arrangement of the Ericaceæ. *Edinburgh New Philos. J.* **17**: 150–160.

Don, G. (1834). *A general system of gardening and botany*. Volume **3**. C. J. G & F. Rivington, London.

Donn, J. (1796). *Hortus cantabrigiensis, or, a catalogue of plant, indigenous and foreign cultivated in the Walkerian botanic garden, Cambridge*. The author, Cambridge.

Dony, J. G., Jury, S. L. & Perring, F. H. (1986). *English names for wild flowers*. Second edition. The Botanical Society of the British Isles, London.

Druce, G. C. (1911). New or noteworthy plants. *Erica tetralix* × *vagans*. *Gard. Chron.*, ser. 3, **50**: 388.

Dulfer, H. (1963). Bemerkungen über einige Linnésche *Erica*-Arten. *Bot. Jahrb. Syst.* **82** (4): 429–434.

Dunk, K. von der, Lotto, R., Lübenau, R. & Philippi, G. (1987). *A guide to bryologically interesting regions in Germany ... edited & translated by J.-P. Frahm, prepared for a bryological fieldtrip during the XIV Botanical Congress, Berlin*.

Dupont, P. (1962). *La flore atlantique européenne. Introduction à l'étude phytogéographique du secteur ibéro-atlantique*. (Thèses présentées à la faculté des sciences de l'université de Toulouse). Edouard Provat, Toulouse.

Durand, E. & Barratte, G. (1910). *Florae libycae prodromus*. Romet, Geneva.

Eager, A. R., Nelson, E. C. & Scannell, M. J. P. (1978). *Erica ciliaris* in Connemara 1846–1854. *Irish Naturalists' J.* **19**: 244–245.

Eason, J. (1993). C. D. Eason, heather propagator. *Yearb. Heather Soc.* **4** (1): 32–34.

Edgington, M. J. (1999). *Erica ciliaris* L. (Ericaceae) discovered in the Blackdown Hills, on the Somerset-Devon Border (v.c. 3). *Watsonia* **22**: 426–428.

Edwards, M. (1996). Some bees, wasps and other insects associated with British heathlands. *Yearb. Heather Soc.* **1996**: 31–35.

Eighme, L. E. (1982). Heathers for California. *Pacific Coast nurseryman and garden supply dealer* **41** (1): 44–46, 48.

Elliott, W. B. (2001). *Flora and illustrated history of the garden flower*. Scriptum-Cartago, London.

Elphinstone, M. (2002). *Hy Brasil. A novel*. Canongate, Edinburgh.

Elphinstone, M. (2003). *Erica hybrasiliensis*. *Yearb. Heather Soc.* **2003**: 59–61.

Evans, E. (1977). *Irish heritage. The landscape, the people and their work*. W. Tempest, Dundalk.

Everett, D. M. (1983). Famous heather nurseries — famous heather names. Part 2. *Yearb. Heather Soc.* **3** (1): 52–61.

Everett, D. M. (2000a). *Erica arborea* — the pipe smoker's dream. *Yearb. Heather Soc.* **2000**: 11–14.

Everett, D. M. (2000b). *The Heather Society's guide to Everyone can grow heathers!* The world of heather booklet series, no. 1. The Heather Society, Creeting St Mary, Suffolk.

Everett, D. M. (2006). One hundred (and one) recommended heathers. *Heathers* **3**: 6–8.

Everett, D. M. (2007). Abbotswood gardens — on hundred years of heathers. *Heathers* **4**: 35–40.

Everett, D. M. & Jones, A. W. (1998). Heathers, rather than daffodils, in Lakeland. *Yearb. Heather Soc.* **1998**: 59–63.

Fægri, K. (1960). Coast plants. In: K. Fægri, Lid J. Gjærevoll & R. Nordhagen (eds), *Maps of distribution of Norwegian vascular plants*. Vol. 1. Oslo University Press, Oslo.

Fagúndez, J. (2006). Two wild hybrids of *Erica* L. (Ericaceae) from northwest Spain. *Bot. Complutensis* **30**: 131–135.

Fagúndez, J. & Izco, J. (2003). Seed morphology of *Erica* L. sect. *Chlorocodon* Benth. *Acta Bot. Gall.* **150** (4): 401–410.

Fagúndez, J. & Izco, J. (2004a). Seed morphology of *Erica* L. sect. *Tylospora* Salisb. ex I. Hansen. *Israel J. Pl. Sci.* **52**: 341–346.

Fagúndez, J. & Izco, J. (2004b). Seed morphology of *Daboecia* (Ericaceae). *Belgian J. Bot.* **137** (2): 188–192.

Fagúndez, J. & Izco, J. (2007). A new European heather: *Erica lusitanica* subsp. *cantabrica* subsp. nova (Ericaceae). *Nordic J. Bot.* **24** (4): 389–394.

Fagúndez, J. & Izco, J. (2009). Seed morphology of *Erica* L. sect. *Loxomeria* Salisb. ex Benth., sect. *Eremocallis* Salisb. ex Benth. and sect. *Brachycallis* I. Hansen, and its systematic implications. *Pl. Biosyst.* **143**: 328–336.

Fagúndez, J. & Izco, J. (2010). Seed morphology of the European species of *Erica* L. sect. *Arsace* Salisb. ex Benth. (Ericaceae). *Acta Bot. Gall.* **154**: 45–54.

Fagúndez, J., Juan, R., Fernández, I. & Izco, J. (2010). Systematic relevance of seed coat anatomy in the European heathers (Ericeae, Ericaceae). *Pl. Syst. Evol.* **284**: 65–76.

Farmer, D. (1997). *Oxford dictionary of saints*. University Press, Oxford.

Fennane, M. & Ibn Tattou, M. (2005). *Flora vasculaire du Maroc inventaire et chorologie*. Vol. **1**. Travaux de l'Institut Scientifique, Rabat.

Fennane, M., Ibn Tattou, M., Mathez, J., Ouyahya, A. & Oualidi, J. (eds) (1999). *Flore pratique du Maroc*. Vol. **1**. Travaux de l'Institut Scientifique, Rabat.

Ferguson, H. (1991). Cornish heath naturalized in California. *Heather news* no. 53: 8.

Fernández-Palacios, J. M. & Arévalo, J. R. (1998). Regeneration strategies of tree species in the laurel forest of Tenerife (The Canary Islands). *Pl. Ecol.* **137**: 21–29.

Field (1888). Hardy heaths. *Garden* (London 1871–1927) **34** (15 December): 565–566.

Fitzherbert, S. W. (1903). *The book of the wild garden*. London.

Flecken, J. G. (2002). Otto en Rudi. *Ericultura* 127: 23–24.

Focke, W. O. (1881). *Die Pflanzen-mischlinge: ein beitrag zur biologie der gewächse*. Gebrüder Borntrager, Berlin.

Forsskål, P. (1775). *Flora ægyptiaco-arabica ... post mortem auctoris edidit Carsten Niebuhr*. Möller, Copenhagen.

Forster, G. (1787). Plantae Atlanticae ex insulis Madeirae, St Jacobi, Adscensionis, St Helenae, et Fayal reportae. *Commentat. Soc. Reg. Sci. Gott.* **9**: 13–74 (extracted in Guppy 1917: 491–492).

Forsyth, A. (1873). Hardy heaths. *Florist & Pomol.* **12** (August): 175–176.

Foss, P. J. & Doyle, G. (1988). Why has *Erica erigena* (the Irish heather) such a markedly disjunct European distribution. *Pl. Today* **1** (5): 161–168.

Foss, P. J. & Doyle, G. (1990). The history of *Erica erigena* R. Ross, an Irish plant with a disjunct European distribution. *J. Quatern. Sci.* **5** (1): 1–16.

Foss, P. J., Nelson, E. C. & Doyle, G. (1987). The distribution of *Erica erigena* Ross in Ireland. *Watsonia* 16: 311–327.

Fraga Vila, M. I. (1982). *Aportacion al estudio taxonomico de las especies de los generos Erica y Calluna presentes en Galicia*. Unpublished thesis. Facultad de Biologia, Universidad de Santiago de Compostela.

Fraga [Vila], M. I. (1983). Números cromosomáticos de plantas occidentales, 223–233. *Anales Jard. Bot. Madrid* **39** (2): 533–539.

Fraga Vila, M. A. (1984a). Valor taxonomico de la morfologia de las semillas en las especies del genero *Erica* presentes en el no de España. *Acta Bot. Malacit.* **9**: 147–152.

Fraga [Vila], M. I. (1984b). Notes on the morphology and distribution of *Erica* and *Calluna* in Galicia, north-western Spain. *Glasra* **7**: 11–23.

Fraga [Vila], M. I. & Nelson, E. C. (1984). Studies in *Erica mackaiana* Bab. II. Distribution in northern Spain. *Glasra* 7: 25–33.

Francisco-Ortega, J. & Santos-Guerra, A. (1999). Early evidence of plant hunting in the Canary Island from 1694. *Arch. Nat. Hist.* **26**: 239–267.

Frenot, Y., Chown, S. L., Whinam, J., Selkirk, P. M., Convey, P., Skotnicki, M. & Bergstrom, D. A. (2005). Biological invasions in the Antarctic: extent, impact and implications. *Biol. Rev.* **80**: 45–72.

Frenot, Y., Gloaguen, J. C., Massé, L. & Lebouvier, M. (2001). Human activities, ecosystem dsturbance and plant invasions in subantarctic Crozet, Kerguelen and Amsterdam islands. *Biol. Conserv.* **101**: 33–50.

Frutos, M. (1986). *Tumores vegetales de España: fitomasagallas*. CSIC, Madrid. (not seen.)

Fuchs, L. (1542). *De historia stirpium commentarii insignes*. Isingrin, Basle.

G. (1848). Trentham Hall, one of the seats of the Duke of Sutherland. *Gard. Chron.* 14 & 21 October: 687, 703.

Gallet, S. & Roze, F. (2002). Long-term effects of trampling on Atlantic heathland in Brittany (France): resilience and tolerance in relation to season and meteorological conditions. *Biol. Conserv.* **103**: 267–275.

Gallet, S., Lemauviel, S. & Roze, F. (2004). Responses of three heathland shrubs to single or repeated experimental trampling. *Environm. Managem.* **33**: 821–829.

Gamst, S. B. & Alm, T. (2003). Klokkelyng (*Erica tetralix*) i Tromsø (Troms) — ny nordgrense, eller bare et gjenfunn? [*Erica tetralix* in Tromsø (Troms) — new northern limit, or just a rediscovery?] *Polarflokken* 27 (1): 3–8.

Garrett, B. G. & McClintock, D. C. (1985). Miss Waterer of Eden Valley 1882–1974. *Yearb. Heather Soc.* **3** (3): 46–53.

Gay, J. (1836). Duriaei iter asturicum botanicum, anno 1835 susceptum. *Ann. Sci. Nat. Bot.*, ser. 2, **6**: 113–137, [213]–225, 340–355 [NB an error in pagination occurs on pp. 213–224, these being erroneously numbered 113–124].

Gay, P. A. (1955a). Distribution and variation of *Erica mackaiana*. In: J. E. Lousley (ed.), *Species studies in the British flora* pp. 119–122. Arbroath.

Gay, P. A. (1955b). Problems of hybridization and species limits in some *Erica* species. In: J. E. Lousley (ed.), *Species studies in the British flora* pp. 126–127. Arbroath.

Gerard, J. (1597). *The Herball or generall Historie of Plantes*. J. Norton, London.

Giardina, G., Raimondo, F. M. & Spadaro, V. (2007). A catalogue of plants growing in Sicily. *Bocconea* **20**.

Gibson, B. R. & Mitchell, D. T. (2006). Sensitivity of ericoid mycorrhizal fungi and mycorrhizal *Calluna vulgaris* to copper mine spoil from Avoca, County Wicklow. *Biol. & Environm.: Proc. Roy. Irish Acad.* **106 B**: 9–18.

Gilbert-Carter, H. (1950). *Glossary of the British flora*. Cambridge University Press, Cambridge

Gimingham, C. H. (1960). Biological flora of the British Isles. *Calluna vulgaris. J. Ecol.* **45**: 455–483.

Gimingham, C. H. (1972). *Ecology of heathlands*. Chapman & Hall, London.

Gimingham, C. H. (1995). Heaths and moorland: an overview of ecological change. In: D. B. A. Thompson, A. J. Hester & M. D. Usher (eds), *Heaths and moorland: cultural landscapes* pp. 9–19. HMSO, Edinburgh.

Gimingham, C. H., Chapman, S. B. & Webb, N. R. (1979). European heathlands. In: R. L. Specht (ed.), *Ecosystems of the World*. **9A** *Heathlands and related shrublands descriptive studied* pp. 365–413 (chapter 14). Elsevier, Amsterdam.

Giroux, J. (1936). *Erica multiflora* L. ou "Bruyère à fleurs nombreuses". Imprimerie Mari-Lavit. Montpellier.

Godman, F. du C. (1870). *Natural history of The Azores, or Western Islands*. John van Voorst, London.

Godwin, H. (1975). *History of the British flora. A factual basis for phytogeography*. Second edition. Cambridge University Press.

González-Rabanal, F. & Casal, M. (1995). Effect of high temperatures and ash on germination of ten species from gorse shrubland. *Vegetatio* **116**: 123–131.

Gordon, G. (1863). Hints for raising a hardy race of cross-bred heaths. *J. Hort. Cottage Gard.* **4** (n.s.): 43–44.

Gray, S. F. (1821). *A natural arrangement of British plants according to their relations to each other as pointed out by Jussieu, de Candolle, Brown, &c. including those cultivated for use.* 2 volumes. Baldwin, Cradock & Joy, London.

Gray, V. (1970) *Erica multiflora* L. *Yearb. Heather Soc.* **1** (8): 16–18.

Greuter, W. (1975). Quisquiliae floristicae graecae, 1–3. *Candollea* **30**: 323–330.

Griffiths, J. (1985). Hybridisation of the hardy Ericas. *Yearb. Heather Soc.* **3** (3): 17–34.

Griffiths, J. (1987). Hybridisation of the hardy Ericas. Part 2. *E.* × *williamsii*, *E.* × *watsonii* and new hybrids from *E. manipuliflora*. *Yearb. Heather Soc.* **3** (5): 42–50.

Griffiths, J. (1999). A hardy form of *Daboecia azorica*. *Yearb. Heather Soc.* **1999**: 34–35.

Griffiths, J. (2008). A heather hybridizer's Yorkshire garden. *Heathers* **5**: 9–16.

Grigson, G. (1955). *The Englishman's flora.* Phoenix House, London.

Güleryüz, G., Arslan, A., Gökçeoglu, M. & Rehder, H. (1998). Vegetation mosaic around the first center of tourism development in the Uludag Mountain, Bursa-Turkey. *Turkish J. Bot.* **22**: 317–326.

Gunther, A. E. (1974). A note on the autobiographical manuscripts of John Edward Gray (1800–1875). *J. Soc. Bibliogr. Nat. Hist.* **7**: 35–76.

Gunther, A. E. (1975). *A century of zoology at the British Museum through the lives of two Keepers 1815–1914.* Dawsons of Pall Mall, London.

Gunther, R. W. T. (1928). *Further correspondence of John Ray.* The Ray Society, London.

Gunther, R. W. T. (1945). *Early science in Oxford. Vol. XIV Life and letters of Edward Lhwyd.* Oxford.

Guppy, H. B. (1917). *Plants, seeds, and currents in the West Indies and Azores. The results of investigations carried out in those regions between 1906 and 1914.* William & Norgate, London.

Gussone, G. (1821). *Adnotationes ad catalogum plantarum quae asservantur in regio horto serenissimi Francisci Borbonii Principis Juventutis in Boccadifalco prope Panormum.* Angeli Trani, Naples.

Guthrie, F. & Bolus, H. (1905). *Erica*. In: W. T. Thistelton-Dyer (ed.), *Flora capensis* **4** sect. 1: 4–315. Lovell Reeve & Co., London.

Hackney, P. (1992). *Stewart & Corry's Flora of the north-east of Ireland vascular plant and charophyte sections.* Third edition. Institute of Irish Studies, The Queen's University of Belfast.

Hadfield, M. (1985). *A history of British gardening.* Revised edition. Penguin Books, Harmondsworth.

Hagerup, E. & Hagerup, O. (1953). Thrips pollination in *Erica tetralix*. *New Phytol.* **52**: 1–7.

Hall, A. (1978). Some new palaeobotanical records for the British Ipswichian Interglacial. *New Phytol.* **81**: 805–812.

Hall, A. (1992). Visit to Germany and Holland, March 1991. *Yearb. Heather Soc.* **3** (10): 45–51.

Hall, A. (1997). Some observations on *Erica scoparia* subspecies cultivated in Surrey. *Yearb. Heather Soc.* **1997**: 5–10.

Hall, A. (1999). Heathers in the glasshouse. *Yearb. Heather Soc.* **1999**: 11–21.

Hall, A. (2002). The hardiness of some of the less familiar heathers in the English Midlands. *Yearb. Heather Soc.* **2002**: 37–42.

Hall, A. (2006). Notes on heather seeds I: photographing seeds. *Heathers* **3**: 53-54.

Hall, A. (2008). *Erica maderensis* in England. *Heathers* **5**: 17–25.

Hall, A. M., Gordon, J. E., Whittington, G., Duller, G. A. T. & Heijnis, H. (2002). Sedimentology, palaeoecology and geochronology of Marine Isotope Stage 5 deposits on the Shetland Islands, Scotland. *J. Quatern. Sci.* **17** : 51–67.

Hannon, G. E., Wastegård, S., Bradshaw E. & Bradshaw, R. H. W. (2001). Human impact and landscape degradation on the Faroe Islands. *Biol. & Environm.: Proc. Roy. Irish Acad.* **101** B (1–2): 129–139.

Hansen, A. (1969). Checklist of the vascular plants of the archipelago of Madeira. *Bol. Mus. Munic. Funchal* **29**: 1–62.

Hansen, I. (1950). Die europäischen Arten der Gattung *Erica* L. *Bot. Jahrb. Syst.* **75**: 1–81.

Harkness, J. (1985). *The makers of heavenly roses.* London, Souvenir Press.

Harris, K. M. (1966). Midge galls on *Erica* carnea. *Yearb. Heather Soc.* **1** (4): 46–48.

Harris, S. (2007). *The magnificent Flora graeca. How the Mediterranean came to the English garden.* Bodleian

Library, Univeristy of Oxford, Oxford.

Harvey, J. H. (ed.) (1983). *The Georgian garden. An eighteenth-century nurseryman's catalogue. John Kingston Galpine*. The Dovecote Press, Stanbridge, Wimborne, Dorset.

Harvey, J. H. (1988). *The availability of hardy plants of the late eighteenth century*. Garden History Society.

Haslerud, H.-D. (1974). Pollination of some Ericaceae in Norway. *Norweg. J. Bot.* **21**: 211–216.

Hayden, R. (1976). Mary Delany 1700–1788 and her paper mosaicks. *Yearb. Heather Soc.* **2** (5): 11–14.

Hayden, R. (1980). *Mrs Delany her life and her flowers*. Colonnade Books, London.

Haytor, G. (1825). Appendix [to accompany ... Diagram of Colours]. In: [G. Sinclair], *Hortus ericaeus woburnensis: or, a catalogue of heaths, in the collection of the Duke of Bedford at Woburn Abbey*. [London.]

Healy, A. J. (1944). Some additions to the naturalised flora of New Zealand. *Trans. Roy. Soc. New Zealand* **74**: 221–231.

Hedberg, I. & Hedberg, O. (2003). *Erica* L. In: I. Hedberg, S. Edwards & Sileshi N. (eds), *Flora of Ethiopia and Eritrea* **4** (1): 46–47. National Herbarium, Biology Department, Addis Ababa University and Department of Systematic Botany, Uppsala University.

Hedberg, O. (1957). Afroalpine vascular plants. A taxonomic revision. *Symb. Bot. Upsal.* **15**: 1.

Hegi, G. (1927). *Illustrierte Flora von Mittel-Europa, mit besonderer Berucksichtigung von Deutschland, Oesterreich und der Schweiz*. Vol. **5** (3). J. F. Lehmann, München.

Hemsley, W. B. (1905) *Erica lusitanica. Curtis's Bot. Mag.* **131**: tab. 8018.

Henrey, B. (1975). *British botanical and horticultural literature before 1800 ...*. London, Oxford University Press. 3 vols.

Hepper, F. N. & Friis, I. B. (1994). *The plants of Pehr Forsskål's "Flora aegyptiaco-arabica" collected on the Royal Danish expedition to Egypt and the Yemen 1761–63*. Royal Botanic Gardens, Kew.

Herbert, W. (1818). Instructions on the treatment of the *Amaryllis longifolia* ... with some observations on the production of hybrid plants. *Trans. Hort. Soc. London* **3**: 187–196.

Herbert, W. (1822). On the production of hybrid vegetables: with the results of many experiments in the investigation of the subject. *Trans. Hort. Soc. London* **4**: 14–47, 48–50.

Herbert, W. (1837). On crosses and hybrid intermixture in vegetables. In: *Amaryllidaceæ* pp. 335–380. J. Ridgway, London.

Herbert, W. (1847). On hybridization amongst vegetables. *J. Hort. Soc. London* **2**: 1–28, 81–107.

Hereman, S. (1868). *Paxton's botanical dictionary*. Second edition. Bradbury, Evans & Co., London.

Hermann, P. M. & Palser, B. F. (2000). Stamen development in the Ericaceae. I. Anther wall, microsporogenesis, inversion and appendages. *Amer. J. Bot.* **87**: 934–957.

Herrera, J. (1987). Flower and fruit biology in southern Spanish Mediterranean shrublands. *Ann. Missouri Bot. Gard.* **74**: 69–78.

Herrera, J. (1988). Pollination relationships in southern Spanish Mediterranean shrublands. *J. Ecol.* **76**: 274–287.

Hibbert, C. (1984). *Queen Victoria in her letters and journals*. London.

Hibbert, C. (2000). *Queen Victoria, a personal history*. London.

Hitchcock, A. (2003). *Erica verticillata* is brought back from the brink of extinction. *Yearb. Heather Soc.* **2003**: 45–50.

Hitchcock, A. (2006). Restoration conservation at Kirstenbosch. *Veld & Flora* **92** (1): 40–44.

Hitchcock, A. (2007). The return of *Erica verticillata*. *Veld & Flora* **93** (1): 14–17.

Hofmann, U. (1968). Untersuchungen an Flora und Vegetation der Ionischen Insel Levkas. *Vierteljahrsschr. Naturf. Ges. Zürich* **113** (3): 209–256.

Hogan, F. E., Hogan, J. & MacErlean, J. C. (1900). *Luibleabran: Irish and Scottish Gaelic names of herbs, plants, trees, etc.* M. H. Gill & Son, Dublin.

Hohenester, A. & Welss, W. (1993). *Exkursionsflora für die Kanarischen Inseln mit Ausblicken auf ganz Makaronesien*. Verlag Eugen Ulmer, Stuttgart.

Holland, P. (translator) (1601). *The historie of the world: commonly called, The naturall historie of C. Plinius Secundus. Translated into English ...*. A. Islip, London.

Holmboe, J. (1914). Studies on the vegetation of Cyprus based upon researches during the spring and summer 1905. *Bergens Mus. Skr.*, n.s., Bd1, no. 2.

Hood, R. (2006). Urze durazia — the heather of Porto Santo. *Heathers* **3**: 50–52.

Hooker, J. D. (1888). *Erica sicula. Curtis's Bot. Mag.* **114**: tab. 7030.

Hooker, W. J. (1833). *Erica mediterranea* var. β. In: J. Sowerby, *English botany*, supplement, **2**: tab. 2774. Richard Taylor, London.

Hooker, W. J. (1835). Another heath found in Ireland. *Compan. Bot. Mag.* **1**: 158.

Hooker, W. J. (1840). *Flora boreali-americana*. Part 12. H. G. Bohn, London.

Hooker, W. J. & Arnott, G. A. W. (1855). *The British flora*. Seventh edition. Longman, Brown, Green & Longmans, London.

Hort, A. F. (1916). *Theophrastus Enquiry into plants with minor works on odours and weather signs*. 2 vols. (Reprint 1961.) (The Loeb Classical Library.) W. Heinemann, London & Harvard University Press, Cambridge, Massachusetts.

Hosking, J. R., Conn, B. J. & Lepschi, B. J. (2003). Plant species recognised as naturalised for New South Wales over the period 2000–2001. *Cunninghamia* **8**: 175–187.

Howkins, C. (1997). *Heathland harvest. The uses of heathland plants through the ages*. Chris Howkins, Addlestone.

Huckerby, E., Marchant, R. & Oldfield, F. (1972). Identification of fossil seeds of *Erica* and *Calluna* by scanning electron microscopy. *New Phytol.* **71**: 387–392.

Hudson, W. (1762). *Flora anglica*. J. Nourse & G. Moran, London.

Hudson, W. (1778). *Flora anglica*. Second edition. J. Nourse, London

Hull, D. (2006). The mysteries of *Erica ciliaris* and *E. tetralix*. *Heathers* **3**: 47–49.

Hull, J. (1808). *The British flora; or, a systematic arrangement of British plants*. **I**. Second edition. Manchester.

Hulme, J. K. (1994). Edward Plummer of Fiddlestone Lodge. *Yearb. Heather Soc.* **1994**: 11–13.

Iglesias-Díaz, M. I. & González-Abuín, F. (2004). Investigating the rooting potential by cutting propagation in wild populations of heathers in NW Spain (Galicia). *Acta Hort.* **630**: 279–285.

Ingram, C. (1970). *A garden of memories*. H. F. & G. Witherby, London.

Jackson, B. D. (1877). A life of William Turner. Reprinted (1965) in *William Turner … Facsimiles with introductory matter by …* . pp. [11]– 26. The Ray Society, London.

Jackson (Sparkes), L. (2009). My granddad, Joe Sparkes (1894–1981). *Heathers* **6**: 19–24.

Jamieson, J. (1808). *An etymological dictionary of the Scottish language*. Edinburgh. [Not seen.]

Jamieson, J. (1846). *A dictionary of the Scottish language … abridged … by J. Johnstone*. W. Tait, Edinburgh.

Jansen, J. (1935). Over eenige in ons land aangetroffen vormen van *Calluna vulgaris*. *Ned. Kruidk. Arch.* **45**: 126–128.

Jardim, R., Fontinha, S. & Fernandez, F. (1998). Pico Branco: a peculiar floristic locality on Porto Santo. *Bol. Mus. Munic. Funchal (Hist. Nat.)* **50** (285): 43–57.

Jardim, R. & Francisco, D. (2000). *Endemic flora of Madeira. Flora Endémica da Madeira*. Múchia Publicações, Funchal.

Jarvis, C. E. (1992). *Erica* Linnaeus … *nom. cons. prop. Taxon* **41**: 563.

Jarvis, C. E. (2007). *Order out of chaos. Linnaean plant names and their types*. The Linnean Society in association with The Natural History Museum, London

Jarvis, C. E. & McClintock, D. C. (1990). Notes on the typification of fourteen Linnaean names for European species of *Erica*, *Calluna*, and *Andromeda* (Ericaceae). *Taxon* **39**: 517–520.

Jekyll, G. (1913). *Wall and water gardens with chapters on the rock-garden and the heath-garden*. Fifth edition. Country Life, London.

Jessen, K. (1948). *Rhododendron ponticum* L. in the Irish inter-glacial flora. *Irish Naturalists' J.* **9**: 175–175.

Jessen, K., Andersen, S. & Farington, A. (1959). The interglacial deposit near Gort, Co. Galway, Ireland. *Proc. Roy. Irish Acad.* **60** B 3.

Johansson, B. (2007). Further observations on green foliage. *Heather News Quart.* **30** (3): 13.

Johnson, A. T. (1928). *The hardy heaths and some of their nearer allies*. The Gardeners' Chronicle, London.

Johnson, A. T. (1932). Some of the newer heaths *New Fl. & Silva* **5**: 156–160.

[Johnson, A. T.] (1935). *Erica carnea* Springwood. *My garden* **5**: 15 (reprinted in *Heathers* **4**: 34).

Johnson, A. T. (1942). Ericaceae and Vacciniaceae. Shrubs of garden merit and interest. *J. Roy. Hort. Soc.* **67**: 363.

Johnson, A. T. (1950). *The mill garden*. Collingridge, London.

Johnson, A. T. (1951). *Erica carnea* 'Springwood'. *J. Roy. Hort. Soc.* **86**: 92.

Johnson, A. T. (1956). *Hardy heaths and some of their nearer allies*. Revised edition. London.

Johnson, S. (1755). *A dictionary of the English language*. London.

Johnson, T. (1633). *The herball or generall historie of plantes. Gathered by John Gerarde of London … very much enlarged and amended …* . London. (Facsimile edition 1975. Dover Publications Inc, New York.)

Johnston, I. (translator) (2007). *Aeschylus The Oresteia*. Richer Resources Publications, Arlington (Virginia).

Jones, A. W. (1977). Observations on lime tolerance. *Yearb. Heather Soc.* **2** (6): 38–40.

Jones, A. W. (1978). A famous nursery — Maxwell and Beale. *Yearb. Heather Soc.* **2** (7): 13–17.

Jones, A. W. (1979). The classification of hardy winter-flowering heaths with notes on *Erica* × *darleyensis*. *Yearb. Heather Soc.* **2** (8): 38–46.

Jones, A. W. (1987). Notes on *Erica manipuliflora*, *E. vagans* and their hybrids. *Yearb. Heather Soc.* **3** (5): 51–57.

Jones, A. W. (1988). A trip to Holland and Germany. *Bull. Heather Soc.* **4** (4): 6–8.

Jones, A. W. (1996). *Erica bocquetii*. *Yearb. Heather Soc.* **1996**: 20.

Jones, A. W. (1997a). *Erica* 'Heaven Scent'. *Yearb. Heather Soc.* **1997**: 38.

Jones, A. W. (1997b). Yet another "nomenclatural change"? *Bull. Heather Soc.* **5** (10, Spring): 11.

[Jones, A. W. & Jones, D. H.] (eds) (1981). [Galls on *Erica*.] *Yearb. Heather Soc.* **2** (10): 22.

Jones, W. H. S. (1961). *Pliny Natural history with an English translation*. Vol. **6** Libri XX–XXIII. (The Loeb Classical Library.) William Heinemann, London & Harvard University Press, Cambridge, Massachusetts.

Joyner, P. (1979). [South West Group news]. *Bull. Heather Soc.* **2** (17): 5.

Joyner, P. (2007). *Erica manipuliflora* 'Olivia Hall'. *Heathers* **4**: 4, 32.

Julian, J. & Newton, P. (2004). The *Calluna* collection that never was. *Heathers* **1**: 18–25.

Jussieu, A. de (1802). Mémoire sur la plante nommée par les botanistes *Erica daboecia*, et sur la nécessité de la rapporter à un autre genre et à une autre famille. *Ann. Mus. Natl. Hist. Nat.* **1**: 52–56. (An English summary was published during 1806, in *Ann. Bot.* **2**: 167–169.)

Jussieu, A. de (1828). Tetralix. In: F. Cuvier (ed.), *Dict. Sci. Nat.* **43**: 505.

Kaffke, A., Matchutadze, I., Couwenberg, J. & Joosten, H. (2002). Early 20th century Russian peat scientists as possible vectors for the establishment of *Calluna vulgaris* in Georgian *Sphagnum* bogs. *Suo* **53**: 61–66.

Kawano, M., Han, J., Kchouk, M. E. & Isoda, H. (2009). Hair growth regulation by the extract of aromatic plant *Erica multiflora*. *J. Nat. Med.* **63**: 335–339.

Kazanis, D. & Arianoutsou, M. (2004). Long-term post-fire vegetation dynamics in *Pinus halepensis* forests of Central Greece: a functional group approach. *Plant Ecol.* **171**: 101–121.

Keogh, J. (1735). *Botanalogia universalis hiberniae*. Cork.

Kingston, N. & Waldren, S. (2006). Biogeography of the Irish 'Lusitanian' heathers. In: S. J. Leach, C. N. Page, Y. Peytoureau & M. N. Sanford, *Botanical links in the Atlantic arc.* (BSBI Conference report no. 24: 147–156). Botanical Society of the British Isles & English Nature.

Klotzsch, J. F. (1838). *Pentapera*, in Additamenta et emendata in Ericearum genera et species. *Linnaea* **12**: 497–498.

Knapp, S. (2003). *Potted histories. An artistic voyage through plant exploration*. Scriptum Editions, London.

Knight, A. (2006). In memoriam [Ken Wilson]. *Heather News Quart.* **29** (4): 2.

Knoche, H. (1923). *Vagandi Mos: Reiseskizzen eines Botanikers*. **I**. Die Kanarischen Inseln 1923. Librairie Istra, Strasbourg. [not seen; see Page 1979.]

Knuth, P. (1909). *Handbook of flower pollination … translated by J. R. Ainsworth Davis*. Clarendon Press, Oxford.

Koch, K. H. E. (1872). *Dendrologie*. Vol. **2**. Erlangen.

Kotze, D. (1996). Improved seed germination of Cape *Erica* species by plant-derived smoke. *Yearb. Heather Soc.* **1996**: 37–38.

Kramer, K. (1991). Züchtung von Heidepflanzen. *Der Heidegarten* **29**: 6–9.

Kramer, K. (2008). Die Arthybriden der Eriken. *Der Heidegarten* **63**: 16–17.

Kramer, K. & Schröder, J. (2009). Kreuzungen mit der Siebenbürgen Heide. *Der Heidegarten* **66**: 19–21.

Kron, K. A., Judd, W. S., Stevens, P. F., Crayn, D. M., Anderberg, A. A., Gadek, P. A., Quinn, C. J. & Luteyn, J. L. (2002). A phylogenetic classification of Ericaceae: molecular and morphological evidence. *Bot. Rev. (Lancaster)* **68**: 335–423.

Krüssmann, G. (1960). ×*Ericalluna bealeana*. *Deutsche Baumschule* **12** (6): 154–156.

Krüssmann, G. (1965). Weitere Beobachtungen an Ericalluna. *Deutsche Baumschule* **1X** (10): 302.

Krüssmann, G. (1984). *Manual of cultivated broadleaved trees and shrubs … .* (Translated by M. E. Epp.) Batsford, London. (Original edition, 1978. *Handbuch de Laubgeholze*. Parey, Berlin.)

Kubitzki, K. (ed.) (2004). *Families and genera of vascular plants*. Vol. **6** (Flowering plants: Dicotyledons: Celastrales, Oxalidales, Rosales, Cornales, Ericales). Springer, Berlin.

Kunkel, G. (1976). *Biogeography and ecology in the Canary Islands*. (*Monogr. Biol*. **30**.) Junk, The Hague.

Kunkel, G. (1977). Inventario florístico de la laurisilva de La Gomera, Islas Canarias. *Naturalia hispanica* **7**: 1–137.

Lacaita, C. C. (1929). [*Erica mackaii* Hook.], in Duriæi iter asturicum botanicum. *J. Bot.* **67**: 256–258.

Lack, H. W. (1996). Die frühe botanische Erforschung der Insel Kreta. *Ann. Naturhist. Mus. Wien, B* **98** suppl.: 183–236.

Lack, H. W. (1997a). The Sibthorpian Herbarium at Oxford — guidelines for its use. *Taxon* **46**: 253–263.

Lack, H. W. (1997b). Die Frontispize von John Sibthorps "Flora Graeca". *Ann. Naturhist. Mus. Wien, B* **99**: 615–654.

Lack, H. W. & Mabberley, D. J. (1999). *The Flora Graeca story. Sibthorp, Bauer, and Hawkins in the Levant*. Oxford University Press.

Ladero, M. (1970). Nuevos táxones para la flora de Extremadura (España). *Anales Inst. Bot. Cavanilles* **27**: 85–104.

Lahondère, C. (2006). *Erica erigena* R. Ross (Irish heath) in the Medóc region of S. W. France. In: S. J. Leach, C. N. Page, Y. Peytoureau & M. N. Sanford, *Botanical links in the Atlantic arc*. (BSBI Conference report no. 24: 177–180). Botanical Society of the British Isles & English Nature.

Laínz, M. (1959). Aportaciones al conocimiento de la flora cántabro-astur III. *Collect. Bot.* (Barcelona) **5**: 671–696.

Laird, M. & Weisberg-Roberts, A. (eds) (2009). *Mrs Delany and her circle*. Yale University Press, New Haven & London.

Lamarck, J. B. A. P.M. de & Candolle, A.-P. de (1805). *Flore française, ou descriptions succinctes de toutes les plantes qui croissent naturellement en France*. Third edition. H. Agasse, Paris.

Lamb, J. G. D. (1964). On the possible occurrence of *Erica mackaiana* Bab. in Co. Mayo. *Irish Naturalists' J.* **14**: 213–214.

Lamb, J. G. D. (1994). Quicker propagation of heathers on a large scale. *Yearb. Heather Soc.* **1994**: 9–10.

Lamb, [J. G. D.] (1999). *Erica umbellata* in central Ireland. *Yearb. Heather Soc.* **1999**: 30.

Lambie, D. (1994). *Introducing heather Scotland's most remarkable plant*. The Scottish Collection, Fort William.

Lancastrian, (A) (1849). Hardy heaths &c. *Gard. Chron*. 31 March: 197–198.

Lange, J. (1999). Lyngen fortæller: the lore and uses of heather in Denmark, and the origins of the word ling. *Yearb. Heather Soc.* **1999**: 46–48.

Lange, J. M. C. (1863). *Pugillus plantarum imprimis hispanicarum*. Part 3. Copenhagen.

Lawson, G. (1866). Notice of the occurrence of heather (*Calluna vulgaris*) at Saint Ann's Bay, Cape Breton Island … read 5 December 1864. *Proc. Trans. Nova Scotian Inst. Nat. Sci.* **1** (3): 30–35.

L'Écluse, C. de (1576). *Rariorum aliquot stirpium per Hispanias observatarum historia*. Plantin, Antwerp.

L'Écluse, C. de (1583). *Rariorum aliquot stirpium per Pannoniam, Austriam, & vicinas quasdam prouinvias observatarum historia*. Plantin, Antwerp.

L'Écluse, C. (1601). *Rariorum plantarum historia*. Plantin, Antwerp.

Legg, C. J., Maltby, E. & Proctor, M. C. F. (1992). The ecology of severe moorland fire on the North York Moors: seed distribution and seedling establishment of *Callua vulgaris*. *J. Ecol.* **80**: 737–752.

Letts, J. F. (1966). *Handbook of hardy heaths and heathers … .* John F. Letts, The Heather Specialist.

Lewis, C. T. (1915). *An elementary Latin dictionary*. Clarendon Press, Oxford.

Lewis, C. T. & Short, C. (eds) (1890). *A Latin dictionary*. Clarendon Press, Oxford.

Lhwyd, E. (1712). Some father observations relating to the antiquities and natural history of Ireland. In a letter from the late Mr. Edw. Lhwyd … to Dr Tancred Robinson, F. R. S. *Philos. Trans.* **27**: 524–526.

Liddell, H. G. & Scott, R. (1890). *A Greek-English lexicon*. Clarendon Press, Oxford.

Lindley, J. (1829). *A synopsis of the British flora*. Longman, Rees, Orme, Brown, & Green.

Lindley, J. (1834). *Erica codonodes* Bell bearing heath. *Edwards's Bot. Reg.* **20**: tab. 1698.

Lindsell, A. (1937). Was Theocritus a botanist? *Greece & Rome* **6**: 78–93 (reprinted, with expanded footnotes, in Raven (2000): 63–75).

Linnaeus, C. (1738). *Hortus cliffortianus*. Amsterdam.

Linnaeus, C. (1753). *Species plantarum*. L. Savius, Stockholm.

Linnaeus, C. (1754a). *Flora anglica*. L. M. Hojer, Uppsala. (Facsimile edition, 1973. The Ray Society, London).

Linnaeus, C. (1754b). *Genera plantarum*. Fifth edition. Stockholm.

Linnaeus, C. (1759). Flora anglica. *Amoenitates academicae* **4**: 88–111. (Facsimile edition, 1973. The Ray Society, London).

Linnaeus, C. (1762). *Species plantarum*. Second edition. Stockholm.

Linnaeus, C. (1770). *Dissertationem botanicam de Erica*. J. Edmann, Uppsala.

Linnaeus, C. (1771). *Mantissa plantarum altera*. L. Salvius, Stockholm.

Litardière, R. V. de (1938). *Erica*. In: J. I. Briquet & R. V. de Litardière (eds), *Prodr. Fl. Corse* **III**: 179.

Lloret, F. & Lopez-Soria, L. (1993). Resprouting of *Erica multiflora* after experimental fire treatments. *J. Veg. Sci.* **4**: 367–374.

Loddiges, C. & Sons. (1817–1833). *The botanical cabinet*. John & Arthur Arch, John Hatchard, C. Loddiges & Sons, C. Cooke, London. 20 vols.

Lortz, K. (2002). *Heather and heathers. Colors for all seasons*. Heaths & Heathers, Shelton, WA.

Lothian, J. [1845]. *Practical hints on the culture and general management of alpine or rock plants*. W. H. Lizars, Edinburgh.

Loudon, J. C. (1838). *Arboretum et fruticetum britannicum*. London.

Mabey, R. (1996). *Flora britannica*. Chatto & Windus, London.

Macdonald, J. (1811). *General view of the agriculture of the Hebrides, or Western Isles of Scotland: ...* . S. Doig & A. Stirling, Edinburgh.

Machado, A. (1976). Introduction to a faunal study of the Canary Islands' laurisilva, with special reference to the ground-beetles. In: G. Kunkel (ed.), *Biogeography and ecology in the Canary Islands*. (*Monogr. Biol.* **30**: 347–411) Junk, The Hague.

Mackay, J. T. (1831a). Notice of a new indigenous heath, found in Cunnamara. *Trans. Roy. Irish Acad.* **16**: 127–128.

Mackay, J. T. (1831b). Natural history in Ireland. ... *Erica mediterranea ...* . *Mag. Nat. Hist.* **4**: 167.

Mackay, J. T. (1836). *Flora hibernica comprising the flowering plants ferns ... of Ireland*. William Curry Jun & Co., Dublin.

Mackay, J. T. (1997). Irish heathers, 1846: an unpublished address to the Royal Irish Academy ... edited by E. C. Nelson. *Yearb. Heather Soc.* **1997**: 31–37.

MacLeod, C. I. (1963). Report of the first general meeting. The Secretary's report. *Yearb. Heather Soc.* [**1** (1)]: 6–7.

Maginess, D. (1981). The gall midge *Wachtliella ericina*. *Yearb. Heather Soc.* **2** (10): 21–22.

Magor, E. W. M. (1980). A quest for the white Dorset heath. *Yearb. Heather Soc.* **2** (9): 55–59.

Mahon, B. [1982]. Traditional dyestuffs in Ireland. In: A. Gailey & D. Ó hÓgáin (eds), *Gold under the furze. Studies in folk tradition presented to Caoimhín Ó Danachair*. The Glendale Press, Dublin.

Maire, R. (1931). *Erica cinerea* var. *numidica*. *Contributions à l'étude de la flore de l'Afrique du Nord* no. 1070.

Mäkinen, Y. & Tiikkainen, J. (1966). *Erica tetralix* in Finland. *Ann. Bot. Fenn.* 3:410–417.

Malcolm, W. (1771). *A catalogue of hot-house and green-house plants, fruit and forest trees*. London.

Mallik, A. U., Hobbs, R. J. & Legg, C. J. (1984). Seed dynamics in *Calluna–Arctostaphylos* heath in north-eastern Scotland. *J. Ecol.* **72**: 855–871.

Malyschev, L. I. (1997). *Calluna*. In: L. I. Malyshev (ed.), *Flora Sibiri*. Tom **11**: 24–25, 231 (Pyrolaceae–Lamiaceae (Labiatae)) [in Russian]. Novosibirsk.

Mansanet, J., Alcober, J. A., Boira, H., Peris, J. B. & Currás, R. (1980). Contribución al estudio ecológico de la *Erica erigena* R. Ross en el Reino de Valencia. *Anales Inst. Bot. Cavanilles* **37**: 117–124.

Márquez, B., Hidalgo, P. J., Ángeles Heras, M., Velasco, R. & Córdoba, F. [2004]. *Erica andevalensis*: un brezo endémico y en peligro de extinción de la zona mineral de Huelva. *Journadas técnicas de ciencias ambiamentales*. (http://www.uhu.es/francisco.cordoba/investigacion/divulgacion/pdf/Erica%20andevalensis.pdf accessed 17 December 2006.)

Márquez García, B., Hidalgo Fernández, P. J. & Córdoba García, F. (2006). How does the plant *Erica andevalensis* survive despite highly elevated soil metal contamination? *SETAC globe* **7** (1): 37–38. (http://communities.setac.net/download/cat-TheGlobe/TheGlobe-0701-Full.pdf accessed 2 December 2007).

Marotta, S. & Franco, J. C. (2001). Is the genus *Lusitanococcus* Neves a junior synonym of *Cucullococcus* Ferris (Hemiptera: Coccoidea: Pseudococcidae)? *Entomologica* **33**: 127–131.

Martín-Cordero C., Reyes, M., Ayuso, M. J. & Toro, M. V. (2001) Cytotoxic triterpenoids from *Erica andevalensis*. *Z. Naturf. C.* **56**: 45–48.

Marrs, R. H. & Bannister, P. (1978) The adaptation of *Calluna vulgaris* (L.) Hull to contrasting soil types. *New Phytol.* **81**: 753–761.

Marrs, R. H. & Proctor, J. (1978). Chemical and ecological studies of heath plants and soil of the Lizard Peninsula, Cornwall. *J. Ecol.* **66**: 417–432.

Martyn, T. (ed.) (*c.* 1795–1807). *The gardener's & botanist's dictionary*. Vol. **1** part 2. F. C. & J. Rivington, *et al.*, London. 2 vols [the title page of vol. **1** part 2 is dated 1807: but see Nelson 2004c.]

Marzol-Jaén, M. V. (2011). Historical background of fog water collection studies in the Canary Islands. In: L. A. Bruijnzeel, F. N. Scatena & L. S. Hamilton (eds), *Tropical montane cloud forests,* pp. 353–358. Cambridge University Press, Cambridge.

Masson, F. (1779). An account of the island of St. Miguel. *Philos. Trans.* **68**: 601–610.

Mather, L. J. & Williams, P. A. (1990). Phenology, seed ecology, and age structure of Spanish heath (*Erica lusitanica*) in Canterbury, New Zealand. *New Zealand J. Bot.* **28**: 207–215.

Mayor, M. & Díaz, T. E. (2003). *La flora asturiana*. Edición actualizada. Real Instituto de Estudios Asturíanos, Oviedo.

Maxwell, D. F. (1927). *The low road. Hardy heathers and the heather garden*. Sweet & Maxwell Ltd, London.

Maxwell, D. F. & Patrick, P. S. (1966). *The English heather garden*. Macdonald & Co.

McAllister, H. (1996). Reproduction in *Erica mackaiana*. *Yearb. Heather Soc.* **1996**: 43–46.

McCalla, W. (1834). Botany — the wild mountains of Roundstone, Cunnemara. *Dublin Penny J.* **2**: 217.

McClintock, D. C. (1965). Notes on British heaths 2. Hybrids in Britain. *Yearb. Heather Soc.* [**1** (3)]: 9–17.

McClintock, D. C. (1966a). *Companion to flowers*. G. Bell & Sons.

McClintock, D. C. (1966b). Notes on British heaths 3. Variation. *Yearb. Heather Soc.* [**1** (4)]: 32–43.

McClintock, D. C. (1966c). Double-flowered hardy heaths. *J. Roy. Hort. Soc.* **91**: 438–443.

McClintock, D. C. (1968). Further notes on *Erica ciliaris* in Ireland. *Proc. Bot. Soc. Brit. Isles* **7**: 177–178.

McClintock, D. C. (1969). *Daboecia azorica* and its hybrids with *D. cantabrica*. *J. Roy. Hort. Soc.* **94**: 449–453.

McClintock, D. C. (1970a). White-flowered *Daboecia cantabrica*. *Irish Naturalists' J.* **16** (12): 391–392.

McClintock, D. C. (1970b). Why is white heather lucky? *Country life* 15 January: 159 (reprinted, without illustrations, in *Heather news* **17** (4): 11–12 (Fall 1994); *Bull. Heather Soc.* **5** (no. 7): 10–11 (1996)).

McClintock, D. C. (1971a). Recent developments in the knowledge of European Ericas. *Bot. Jahrb. Syst.* **90**: 509–523.

McClintock, D. C. (1971b). Personal names used for our hardy heathers. *Yearb. Heather Soc.* **1** (8): 25–29.

McClintock, D. C. (1972). The wild heathers of Ireland. In: *Report of the Recorders' Conference Dublin September 1972:* pp. 24–35. BSBI Irish Regional Committee.

McClintock, D. C. (1974a). Anti-clipping. *Bull. Heather Soc.* **2** (no. "21"): 5–6.

McClintock, D. C. (1974b). Birds' nests in heather gardens. *Bull. Heather Soc.* **2** (3): 7.

McClintock, D. C. (1976). Postscript to Richards, D. A. (1976). Mostly *Erica maderensis* and *Daboecia azorica*. *Yearb. Heather Soc.* **2** (5): 19–20.

McClintock, D. C. (1977). The double-flowered heathers. *Yearb. Heather Soc.* **2** (6): 25–27, 29–32, 34–37.

McClintock, D. C. (1978). St Dabeoc's heaths and their hybrids. *The Garden* **103**: 114–116.

McClintock, D. C. (1979a). The status, and correct name for, *Erica* 'Stuartii'. *Watsonia* **12**: 249–252.

McClintock, D. C. (1979b). The chromosome numbers of heathers. *Plantsman* **1**: 63–64.

McClintock, D. C. (1980a). The typification of *Erica ciliaris* L., of *E. tetralix* L. and of their hybrid, *E.* × *watsonii* Bentham. *Bot. J. Linn. Soc.* **80**: 207–211.

McClintock, D. C. (1980b). *A guide to the naming of plants*. Second edition. The Heather Society.

McClintock, D. C. (1980c). Bell heathers with split corollas. *Plantsman* **2**: 182–191.

McClintock, D. C. (1980d) *Erica sicula*. *Yearb. Heather Soc.* **2** (9): 45–55.

McClintock, D. C. (1981). The bell heather in Madeira. *Yearb. Heather Soc.* **2** (10): 48–51.

McClintock, D. C. (1982a). The stories of some Irish heather cultivars. *Moorea* **1**: 37–41.

McClintock, D. C. (1982b). *Erica tetralix* met trosvormige Bloiewijze. *Levende Natuur* **84** (5–6): 189.

McClintock, D. C. (1982c). Dorset or ciliate heath. *BSBI news* no. **30**: 23.

McClintock, D. C. (1983a). *Iter hispanicum ericaceum*. *Yearb. Heather Soc.* **3** (1): 33–40.

McClintock, D. C. (1983b). The Carna colony of *Erica mackaiana*, a new variety. *Irish Naturalists' J.* **21**: 85–86.

McClintock, D. C. (1984a). The white heathers. *Plantsman* **6** (3): 181–191.

McClintock, D. C. (1984b). Daboecias with erect flowers. *Yearb. Heather Soc.* **3** (2): 46–47.

McClintock, D. C. (1984c). Sports, reversion witches broom and the like. *Yearb. Heather Soc.* **3** (2): 57–68.

McClintock, D. C. (1986). Harmonising botanical and cultivar classification with special reference to hardy heathers. *Acta Hort.* **182**: 277–283.

McClintock, D. C. (1987). A history of hardy heathers. *Hortus* **3**: 59–63.

McClintock, D. C. (1988a). Hardy Ericas with coloured foliage, and other newer cultivars. *Hortus* **7**: 53–62.

McClintock, D. C. (1988b). *Erica terminalis* var. *albiflora*. *Bull. Heather Soc.* **4** (5): 6.

McClintock, D. C. (1988c). A name for *Erica manipuliflora* with white flowers. *Yearb. Heather Soc.* **3** (10): 43–44.

McClintock, D. C. (1989a). The heathers of Europe and adjacent areas. *Bot. J. Linn. Soc.* **101** (3): 279–289.

McClintock, D. C. (1989b). (961) Proposal to conserve *Erica vagans* L. (Ericaceae) with a conserved type. *Taxon* **38** (3): 526–527.

McClintock, D. C. (1989c). The "Viridiflora" variants of Cornish heath. *Yearb. Heather Soc.* **3** (7): 51–54.

McClintock, D. C. (1989d). The *Erica scoparia* in Madeira. *Yearb. Heather Soc.* **3** (7): 32–35.

McClintock, D. C. (1989e). Heather records. *Yearb. Heather Soc.* **3** (7): 54–60.

McClintock, D. C. (1990a). *Erica bocquetii* and other Turkish ericas. *Yearb. Heather Soc.* **3** (8): 48–53.

McClintock, D. C. (1990b). *Erica scoparia* and its variants. *Yearb. Heather Soc.* **3** (8): 53–61.

McClintock, D. C. (1990c). Gall midge damage on Ericas. *Bull. Heather Soc.* **4** (11): 4.

McClintock, D. C. (1991a). The ericas of Turkey. *Karaca Arbor. Mag.* **1**: 6–9.

McClintock, D. C. (1991b). Who was "Our Mr Richard Potter"? *Yearb. Heather Soc.* **3** (9): 17–18.

McClintock, D. C. (1991c). *Erica scoparia* var. *macrosperma*, an updating. *Yearb. Heather Soc.* **3** (9): 33.

McClintock, D. C. (1992a). The heathers of Europe and neighbouring regions — an up-date. *Yearb. Heather Soc.* **3** (10): 12–16.

McClintock, D. C. (1992b). David Wilson's hybridisations. *Yearb. Heather Soc.* **3** (10): 51–53.

McClintock, D. C. (1994). *Erica*. In: J. R. Press & M. Short (eds), *Flora of Madeira*, pp. 248–249. The Natural History Museum, London.

McClintock, D. C. (1995). Erstbeschreibung auf Lateinish *Erica* × *oldenburgensis*. *Deutsche Baumschule* 5/1995: 198.

McClintock, D. C. (1996). *Erica* × *oldenburgensis* jetzt nomenklatorisch gültig beschrieben. *Der Heidegarten* **39**: 35.

McClintock, D. C. (1997a). *Erica* × *griffithsii*, a new hybrid heather. *New Plantsman* **4** (2): 83–84.

McClintock, D. C. (1997b). *Erica spiculifolia* × *Erica carnea* = *Erica* × *krameri*. *Der Heidegarten* **41**: 18, 19–20.

McClintock, D. C. (1998). *Erica* × *krameri*: the hybrid between *E. carnea* and *E. spiculifolia*. *Yearb. Heather Soc.* **1998**: 26.

McClintock, D. C. (1999a). Cape heaths crossed with European species. *New Plantsman* **6** (4): 207.

McClintock, D. C. (1999b). A presidential predicament Mount Cofano — Friday 16th April 1999. *Bull. Heather Soc.* **5** (17): 3–5.

McClintock, D. C. (2000). Yet another *Erica* hybrid: *E.* × *garforthensis*. *Yearb. Heather Soc.* **2000**: 17.

McClintock, D. C. (2002). News of *Erica sicula* in Sicily. *Yearb. Heather Soc.* **2002**: 15–16.

McClintock, D. C. & Blamey, M. (1998). *Heather of the Lizard*. The Cornwall Garden Society.

McClintock, D. C. & Rose, F. (1970). Cornish heath in Ireland. *Irish Naturalists' J.* **16** (12): 387–399.

McClintock, D. C. & Wijnands, O. (1982). *Erica carnea* or *E. herbacea*. *Taxon* **31**: 319–320.

McGuire, A. F. & Kron, K. A. (2005). Phylogenetic relationships of European and African Ericas. *International journal of plant sciences* **166**: 311–318.

McNab, W. (1832). *A treatise on the propagation, cultivation, and general treatment of Cape heaths, in a climate where they require protection during the winter months*. Thomas Clark, Edinburgh.

McNeill, J., Barrie, F. R., Burdet, H. M., Demoulin, V., Hawksworth, D. L., Marhold, K., Nicolson, D. H., Prado, J., Silva, P. C., Skog, J. E., Wiersma, J. H. & Turland, N. J. (2006). *International code of botanical nomenclature (Vienna Code)*. (*Regnum Veg.* **146**.) A. R. G. Gantner Verlag KG, Ruggell.

Meades, S. J., Hay, S. G. & Brouillet, L. (2000). *Annotated checklist of the vascular plants of Newfoundland and Labrador*. [URL http://www.digitalnaturalhistory.com/meades.htm accessed 8 June 2007.]

Medagli, P. & Ruggiero, L. (2002). Wild heathers of Apulia (southern Italy). *Yearb. Heather Soc.* **2002**: 43–46.

Meikle, R. D. (1985). *Flora of Cyprus*. Vol. **2**. The Bentham-Moxom Trust, Royal Botanic Gardens, Kew.

Mesléard, F. & Lepart, J. (1989). Continuous basal sprouting from a lignotuber: *Arbutus unedo* L. and *Erica arborea* L., as woody Mediterranean examples. *Oecologia* **80**: 127–131.

Mesléard, F. & Lepart, J. (1991). Germination and seedling dynamics of *Arbutus unedo* and *Erica arborea* on Corsica. *J. Veg. Sci.* **2**: 155–164.

Messel, L. (1918). *A garden flora trees and flowers grown in the gardens at Nymans … 1890–1915*. Country Life & George Newnes, London; Charles Scribner's Sons, New York.

Messel, M. (1918). Preface. In: L. Messel, *A garden flora trees and flowers grown in the gardens at Nymans … 1890–1915*: pp [vii]–ix,. Country Life & George Newnes, London; Charles Scribner's Sons, New York.

Metheny, D. (1970). *Erica multiflora* [sic]. *Bull. Heather Soc.* [1] no. 11 (Autumn): 5.

Metheny, D. (1977). *Erica australis* seedlings in the garden. *Yearb. Heather Soc.* **2** (6): 15–17.

Metheny, D. (1991). *Hardy heather species and some related plants*. North American Heather Society & Frontier Publishing, Seaside, Oregon.

Metheny, D. (2007). *Daboecia azorica*. *Heather News Quart.* **30** (2 Spring): 3.

Meusel, H., Jäger, E., Rauschert, S. & Weinert, E. (1978). *Vergleichende Chorologie der Zentraleuropaischen Flora*. Karten Bd **2**. Gustav Fischer, Jena.

Mikolajski, A. (1997). *Heathers*. (The new plant library.) Lorenz Books, London.

Miller, D. & Nelson, E. C. (2008). The Heather Society's Herbarium, and Ericaceae (*Bruckenthalia, Daboecia, Erica*) type specimens cited by D. C. McClintock (1913–2001). *Glasra* **4**: 109–117.

Milliken, W. & Bridgewater, S. (2004). *Flora celtica. Plants and people in Scotland*. Birlinn, Edinburgh.

Mitchell, A. (1982). *The trees of Britain and northern Europe*. London, Collins.

Mitchell, G. F., Scannell, M. J. P., Walsh, P. T. & Watts, W. A. (1980). Quaternary vegetable debris from a karstic cavity in Permian gypsum near Carrickmacross, Co. Monaghan. *Proc. Roy. Irish Acad.* **80** B: 335–342.

Mitchell, G. F. & Watts, W. A. (1979). The history of the Ericaceae in Ireland during the Quaternary Epoch. In: D. Walker & R. G. West (eds), *Studies in the vegetational history of the British Isles. Essays in honour of Harry Godwin*, pp. 13–39. University Press, Cambridge.

Mitchell, M. E. (1975). Irish botany in the seventeenth century. *Proc.Roy. Irish Acad.* **75** B (13): 275–284.

Moon, H. G. (1896). The Dorset heath (Erica ciliaris). *Garden (London 1871–1927)* **53** (30 April): 364–365, plate 1168.

Moore, D. (1852). On the distribution of *Erica mediterranea*, var. *hibernica*, and some other plants, in Ireland. *Phytol.* **4**: 597–599.

Moore, D. (1855). The white Mediterranean heath. *Gard. Chron.* (12 May): 317.

Moore, D. (1860). Observations on the prevailing and rare plants of Erris, and of some other portions of the county of Mayo. *Proc. Dublin Univ. Zool. Bot. Assoc.* **2**: 14–19 (and in *Nat. Hist. Rev.* **7**: 414–417).

Moore, D. & More, A. G. (1866). *Contributions towards a Cybele hibernica*. Hodges, Smith & Co., Dublin.

More, A. G. (1874). A new station for *Erica mackayana*. *J. Bot.* **12**: 306.

Morley, B. D. (1974). *Erica erigena*: origin of the specific epithet. *J. Roy. Hort. Soc.* **99** (10): 463.

Morley, B. D. & Scannell, M. J. P. (1972). *Erica erigena* R. Ross — the Mediterranean heath in Ireland. *Irish Naturalists' J.* **17** (6): 203–204.

Mouillot, F., Rambal, S. & Joffre, R. (2002). Simulating climate change impacts on fire frequency and vegetation dynames in a Mediterranean-type ecosystem. *Global change Biol.* **8**: 423–427.

Mouissie, A. M., Lengkeek, W. & van Diggelen, R. (2005). Estimating adhesive seed-dispersal distances: field experiments and correlated random walks. *Functional Ecol.* **19**: 478–486.

Mullaj, A. & Tan, K. (2010). *Erica multiflora* (Ericaceae), *Onosma pygmaeum* (Boraginaceae) and *Typha minima* (Typhaceae) in Albania. *Phytol. Balcan.* **16** (2): 267–269.

Mulligan, B. (1963). *Erica cinerea* 'Colligan Bridge'. *Gard. Chron. Gard. Ill.* 5 October: 247.

Nelson, E. C. (1971). An investigation of the colony of *Erica vagans*, L. at Carrickbrawn in the county of Fermanagh, Northern Ireland. Unpublished BSc thesis, Department of Botany, University College of Wales, Aberystwyth. [See Nelson & Coker (1974) for a published account.]

Nelson, E. C. (1978). Craiggamore — the earliest record. *Irish Naturalists' J.* **19**: 250.

Nelson, E. C. (1979a). Historical records of the Irish Ericaceae, with particular reference to the discovery and naming of *Erica mackaiana* Bab. *J. Soc. Bibliogr. Nat. Hist.* **9**: 289–299.

Nelson, E. C. (1979b). Ireland's flora: its origin and composition. In: E. C. Nelson & A. Brady (eds), *Irish gardening and horticulture*, pp. 17–35. Royal Horticultural Society of Ireland, Dublin.

Nelson, E. C. (1981a). Studies in *Erica mackaiana* Bab. I: distribution in Connemara, Ireland. *Irish Naturalists' J.* **20**: 198–202

Nelson, E. C. (1981b). The origin of *Calluna vulgaris* 'County Wicklow'. *Irish Naturalists' J.* **20**: 212.

Nelson, E. C. (1981c). William McCalla — a second 'panegyric' for an Irish phycologist. *Irish Naturalists' J.* **20**: 275–283.

Nelson, E. C. (1981d). Phytogeography of southern Australia. In: A. Keast (ed.), *Ecological biogeography of Australia*, pp. 735–759. Junk, The Hague.

Nelson, E. C. (1982a). Historical records of the Irish Ericaceae — additional notes on *Arbutus unedo, Daboecia cantabrica, Erica erigena, Erica mackaiana* and *Ledum palustre. Irish Naturalists' J.* **20**: 364–369.

Nelson, E. C. (1982b). *Erica ciliaris* in Connemara November 1981. *Yearb. Heather Soc.* **2** (11): 34–36.

Nelson, E. C. (1983). William McCalla, discoverer of *Erica mackaiana. Yearb. Heather Soc.* **3** (1): 28–32.

Nelson, E. C. (1984a). James Walker Porter of Carryduff. *Yearb. Heather Soc.* **3** (2): 24–34.

Nelson, E. C. (1984b). Dabeoc — a saint and his heather. *Yearb. Heather Soc.* **3** (2): 41–46

Nelson, E. C. (1986a). New cultivars of *Erica* and *Daboecia* and a new name in *Rhododendron. Moorea* **5**: 19–20.

Nelson, E. C. (1986b). Heathers from Carryduff for limestone pavements. *Home Gard.* **1** (5): 18–19.

Nelson, E. C. (1989a). Heathers in Ireland. *Bot. J. Linn. Soc.* **101**: 269–277.

Nelson, E. C. (1989b). Notes on Irish cultivars with some new names. *Moorea* **8**: 41–49.

Nelson, E. C. (1993a). *Bruckenthalia spiculifolia* f. *albiflora* — an early report from Bulgaria, 1911. *Yearb. Heather Soc.* **4** (1): 35–36.

Nelson, E. C. (1993b). Mapping plant distribution patterns: two pioneering examples from Ireland published in the 1860s. *Arch. Nat. Hist.* **20**: 391–403.

Nelson, E. C. (1994a). *Daboecia cantabrica* 'Celtic Star'. *Yearb. Heather Soc.* **1994**: 14.

Nelson, E. C. (1994b). *Daboecia cantabrica* 'Charles Nelson'. *Yearbook of the Heather Society* **1994**: 24.

Nelson, E. C. (1995a). *Erica × stuartii*: the authorship reconsidered. *Watsonia* **20**: 275–278.

Nelson, E. C. (1995b). *Erica mackaiana* forma *multiplicata*: a new name for the "multipetalled" form of Mackay's heath, with a history of Crawford's heath. *Yearb. Heather Soc.* **1995**: 33–40.

Nelson, E. C. (1996a). Casual observations (1982, 1994) on *Erica tetralix* L. in Picos de Europa, Asturias, Spain. *Yearb. Heather Soc.* **1996**: 47–48.

Nelson, E. C. (1996b). Hybrids of St. Dabeoc's heath, *Daboecia* (Ericaceae): some nomenclatural adjustments. *New Plantsman* **3**: 84–85.

Nelson, E. C. (1997). White-flowered *Bruckenthalia*; its name within *Erica. Yearb. Heather Soc.* 1997: 37.

Nelson, E. C. (1998). The heathers in John Gerard's *The Herball or generall Historie of Plantes* 1597–98, and Thomas Johnson's 'Very much Enlarged and Amended' edition 1633. *Yearb. Heather Soc.* **1998**: 39–54.

Nelson, E. C. (1999a). Farewell to *Bruckenthalia. Yearb. Heather Soc.* **1999**: ii.

Nelson, E. C. (1999b). *Erica umbellata* in Norfolk. *Yearb. Heather Soc.* **1999**: 10.

Nelson, E. C. (1999c). A table of hybrid heathers involving European species. *Yearb. Heather Soc.* **1999**: 22.

Nelson, E. C. (1999d). A Dabeocian miscellany. An editorial intervention. *Yearb. Heather Soc.* **1999**: 32–34.

Nelson, E. C. (2000a). A history, mainly nomenclatural, of St Dabeoc's heath. *Watsonia* **23**: 47–58.

Nelson, E. C. (2000b). Viking ale and the quest for the impossible: some marginalia leading, perhaps, to 'the most powerfullest drink ever known'. *Yearb. Heather Soc.* **2000**: 25–33.

Nelson, E. C. (2000c). Bell heather: *Erica cinerea*. *Yearb. Heather Soc.* **2000**: 34.

Nelson, E. C. (2000d). A mixed-up heather — what is *Erica mackaiana* 'Lawsoniana'. *Yearb. Heather Soc.* **2000**: 77–80.

Nelson, E. C. (2000e). *A heritage of beauty. The garden plants of Ireland. An illustrated encyclopaedia.* Irish Garden Plant Society, Dublin.

Nelson, E. C. (2001a). Dr Charles Stuart's heather rediscovered in Connemara, Ireland. *Yearb. Heather Soc.* **2001**: 35–37.

Nelson, E. C. (2001b). *Wild plants of Connemara and West Mayo. A souvenir and simple pocket guide.* Strawberry Tree, Dublin.

Nelson, E. C. (2002). [*Erica pallido-purpurea* L. ... nom. illeg.]. In: S. Cafferty & C. E. Jarvis (eds), Typification of Linnaean plant names in *Ericaceae*. *Taxon* **51**: 753.

Nelson, E. C. (ed.) (2003a). *William Robinson The wild garden a new illustrated edition.* Strawberry Tree, Dublin.

Nelson, E. C. (2003b). Who's that heather named after? 1 *Erica ciliaris* 'Mawiana'. *Bull. Heather Soc.* **6** (10): 5–6 (see also *Heather News Quart.* **26** (4): 15).

Nelson, E. C. (2003c). Robert Brown's manuscript descriptions of British and Irish plants (B.93) (1792–1800), with indexes to botanical names and places [in 3 parts]. *Scottish Naturalist* **115**: 115–179.

Nelson, E. C. (2003d). Cat heather. *Plant-lore notes & news* no. 76: 369–370.

Nelson, E. C. (2003e). Cat heather. *Bull. Heather Soc.* **6** (9): 7–8.

Nelson, E. C. (2004a). The enigma of the alien heathers of Britain, especially *Erica × darleyensis*. *BSBI news* no. **95**: 13–14 + illust. p. 2.

Nelson, E. C. (2004b). What is Irish heath? Or, the baffling matter of common/colloquial/vernacular names for heathers (with apologies to the editors of *Heather notes*)! *Bull. Heather Soc.* **6** (13): 13–15.

Nelson, E. C. (2004c). Publication dates of (early) "parts" of Thomas Martyn's edition of Philip Miller's *The gardener's and botanist's dictionary* (1795–1807). *Soc. Hist. Nat. Hist. Newslett.* no. **79**: 16–17.

Nelson, E. C. (translated by Schröder, J.) (2004d). Die Sortennamen für zwei bemerkenswerte Heidezüchtungen. *Der Heidegarten* **56**: 52–57. [See Nelson (2005c) below.]

Nelson, E. C. (2005a). *Erica mackaiana* Bab. and *Erica × stuartii* (MacFarl.) Mast. (Ericaceae): two heathers new to South Kerry (v. c. H1), Ireland. *Watsonia* **25**: 414–417.

Nelson, E. C. (2005b). For luck and love ... the origins of "lucky" white heather. *Scots Mag.*, n.s., **163**: 250–253.

Nelson, E. C. (2005c). Cultivar names for two remarkable heathers from Kurt Kramer, Edewecht, Germany. *Heathers* **2**: 19–22. [A revised version of the paper, translated into German by Jürgen Schröder, published in *Der Heidegarten* **56**: 52–57 (2004).]

Nelson, E. C. (2005d). The most rewarded heathers. *Bull. Heather Soc.* **6** (15): 5–9.

Nelson, E. C. (2005e). Clarification of a publication date for George Bentham's treatment of Ericeae (Ericaceae) published in A.-P. de Candolle's *Prodromus* volume 7 part 2 (December 1839). *Arch. Nat. Hist.* **32**: 106.

Nelson, E. C. (2006a). Notes on heather seeds II: dimensions of seeds. *Heathers* **3**: 53, 55.

Nelson, E. C. (2006b). Butterflies and heathers. *Bull. Heather Soc.* **6** (19): 18.

Nelson, E. C. (2006c). Lucky white heather: a sesquicentennial review of a Scottish Victorian conceit. *Heathers* **3**: 38–46.

Nelson, E. C. (2006d). Miss Irene Fawkes's *Heather*: the "cover story". *Heathers* **3**: ii, 19–21.

Nelson, E. C. (2007a). Richard Salisbury FLS and the discovery of elaiosomes in *Erica*. *Linnean* **23** (1): 26–30.

Nelson, E. C. (2007b). The Porters and their passion. *Irish Gard.* **16** (2, March): 62–65.

Nelson, E. C. (2007c). *Erica mackayana* — y? Oh! Why? *Bull. Heather Soc.* **6** (20): 6.

Nelson, E. C. (2007d). Williams's heath: the wild-collected clones. *Heathers* **4**: 45–56.

Nelson, E. C. (2007e). *Erica × arendsiana* (*E. terminalis × cinerea*): a hardy German hybrid re-created. *Heathers* **4**: 59–60.

Nelson, E. C. (2007f). *Erica × williamsii* Druce (*E. tetralix* L. × *vagans* L.): a note on typification. *Watsonia* **26**: 485–486.

Nelson, E. C. (2007g). The original material of two Turkish species of *Erica* (Ericaceae) described and named by Richard Anthony Salisbury (1761–1829). *Turkish J. Bot.* **31**: 463–466.

Nelson, E. C. (2008a). Identifying "*Erica hirsuta anglica*". *Linnean* **24** (2): 21–25.

Nelson, E. C. (2008b). Auctions, bankrupts and burglars, and the origin of "brier-wood pipes". *Heather News Quart.* **31** (1): 11–12, 17–20.

Nelson, E. C. (2008c). The formation of the North American Heather Society: The Pacific Northwest Group of The Heather Society, 1977–1978. *Heather News Quart.* **31** (2): 12, 17–19.

Nelson, E. C. (2008d). Typification of two horticultural hybrids in *Erica* (Ericaceae). *Glasra* **4**: 107–108.

Nelson, E. C. (2008e). [*Erica*] *australis* 'Trisha', p. 71, in Supplement VIII (2008) to *International register of heather names*. *Heathers* **5**: 65–74.

Nelson, E. C. (2009a). *Erica scoparia* and *Erica spiculifolia* (formerly *Bruckenthalia spiculifolia*) in interglacial floras in Ireland and Britain; confused nomenclature leading to misidentification of fossilized seeds. *Quatern. Sci. Rev.* **28**: 381–383.

Nelson, E. C. (2009b). Victorian royal wedding flowers: orange, myrtle, and the apotheosis of white heather. *Gard. Hist.* **37** (2): 232–237.

Nelson, E. C. (2009c). Some more about "rairb toor sepip"; Irish examples of brier-root pipes. *Heather News Quart.* **32** (3): 15–17.

Nelson, E. C. (2009d). *An Irishman's cuttings. Tales of Irish gardens, and gardeners, plants and plant hunters.* The Collins Press, Cork.

Nelson, E. C. (2010a). "The mythical *Calluna atlantica*": heather in Newfoundland. *Heathers* **7**: 27–35.

[Nelson, E. C.] (2010b). [*Erica*] *erigena* × *lusitanica* 'Lucy Gena', p. 69, in Supplement X (2010) to *International register of heather names*. *Heathers* **7**: 65–70.

Nelson, E. C. (2010d). Henry Andrews's *Dahliae mythologicae*. *Newslett. Soc. Hist. Nat. Hist.* **99**: 8–9.

Nelson, E. C. (2011a). *Erica manipuliflora* 'Bert Jones'. *Heathers* **8**: 41–42, 74.

Nelson, E. C. (2011b). *The gardener's and botanist's dictionary* (1795–1807). In: E. C. Nelson (ed.), *History and mystery. Notes and queries from the*

newsletter of *The Society for the History of Natural History,* pp. 146–147. SHNH, London.

Nelson, E. C. (in prep.). Sources of plants for, and distribution of plants from, the Royal Dublin Society's Botanic Gardens, Glasnevin, 1795–1879: an annotated checklist. [in preparation].

Nelson, E. C. & Cafferty, S. (2005). Proposal to reject the name *Erica viridipurpurea* (Ericaceae). *Taxon* **54** (1): 206.

Nelson, E. C. & Coker, P. D. (1974). Ecology and status of *Erica vagans* L. in County Fermanagh, Ireland. *Bot. J. Linn. Soc.* **69**: 153–195.

Nelson, E. C. & Grills, A. (1998). *Daisy Hill Nursery, Newry. A history of 'the most interesting nursery probably in the World'.* Northern Ireland Heritage Gardens Committee, Belfast.

Nelson E. C. & King, W. H. (2007). James Lothian (1817–1871) and his book *Practical hints on the culture and general management of alpine or rock plants.* *Huntia* **13** (2): 143–154.

Nelson, E. C. & McClintock, D. C. (1984). Two new white-flowered heathers (*Erica andevalensis* and *E. mackaiana*) from Spain. *Glasra* **7**: 35–40.

Nelson, E. C., McClintock, D. C. & Small, D. J. (1985). The natural habitat of *Erica andevalensis* in south-western Spain. *Kew Mag.* **2**: 324–330.

Nelson, E. C. & Oliver, E. G. H. (2005a). Cape heaths in European gardens: the early history of South African *Erica* species in cultivation, their deliberate hybridization and the orthographic bedlam. *Bothalia* **34** (2): 127–140.

Nelson, E. C. & Oliver, E. G. H. (2005b). Chromosome numbers in *Erica* — an updated checklist. *Heathers* **2**: 57–58.

Nelson, E. C. & Oswald, P. H. (2005). *Polifolia* revisited and explained. *Huntia* **12**: 5–11.

Nelson, E. C. & Parnell, J. (1992). *Flora hibernica* (1836): its publication, and aftermath as viewed by Dr Thomas Taylor. *Taxon* **41**: 35–42.

Nelson, E. C. & Rourke, J. P. (1993). James Niven (1776–1827), a Scottish botanical collector at the Cape of Good Hope: his *hortus siccus* at the National Botanic Gardens, Glasnevin, Dublin (**DBN**), and the Royal Botanic Gardens, Kew (**K**). *Kew Bull.* **48**: 663–682.

Nelson, E. C. & Small, D. J. (2000). *International register of heather names, Vol.* **1** *Hardy cultivars and European species* (in 4 parts). The Heather Society, Creeting St Mary.

Nelson, E. C. & Small, D. J. (2004–2005). *International register of heather names, Vol.* **2** *African species, hybrids and cultivars* (in 4 parts). The Heather Society, Creeting St Mary.

Nelson, E. C. & Walsh, W. F. (1984). *An Irish flower garden. The histories of some of our garden plants.* Boethius Press, Kilkenny.

Nelson, E. C. & Walsh, W. F. (1991). *The Burren: a companion to the wildflowers of an Irish limestone wilderness.* Boethius Press & The Conservancy of The Burren, Aberystwyth & Ennis. (Reprinted with amendments 1997.)

Nelson, E. C. & Walsh, W. F. (1995). *The flowers of Mayo. Dr Patrick Browne's Fasciculus plantarum Hiberniae (1788).* Éamonn de Búrca.

Nelson, E. C. & Walsh, W. F. (1997). *An Irish flower garden replanted. The histories of some of our garden plants.* Éamonn de Búrca, Castlebourke.

Nelson, E. C. & Wulff, E. M. T. (2007a). *Erica × gaudificans* (*E. spiculifolia* × *bergiana*): Kurt Kramer's second north-south hybrid. *Heathers* **4**: 57–58.

Nelson, E. C. & Wulff, E. M. T. (2007b). (1786) Proposal to conserve the name *Erica manipuliflora* against *E. forskalii* (Ericaceae). *Taxon* **56** (3): 959–961.

Nepi, C. (2006). Ericaceae. In: M. Thulin (ed.), *Flora of Somalia.* Vol. **3**: 5. Royal Botanic Gardens, Kew.

Nicholson, H. L. (1982). J. W. Sparkes. An appreciation. *Yearb. Heather Soc.* **2** (11): 10–12.

Niedermaier, K. (1985). *Bruckenthalia*-Untersuchunger I, Das aktuelle Areal von *Bruckenthalia spiculifolia* (Salisb.) Reichb. und seine Entstehungsgeschichte. In: H. Heltman & G. Wendelberge (eds), *Naturwissenschaftliche Forschungen über Siebenbürgens III Beiträge zur Pflanzengeographie des Südost-Karpatenraumes.* (*Siebenbürg. Arch.* Folge 3, Bd **20**: 217–255).

Niven, N. (1837). [Letter, dated 10 September 1836, to the Royal Dublin Society's Committee of Botany.] *Proc. Roy. Dublin Soc.* **73**: 6–8.

Obeso, J. R. & Vera M. L. (1996). Resprouting after experimental fire application and seed germination in *Erica vagans. Orsis* **11**: 155–163.

Oliver, E. G. H. (1986). The identity of *Erica vinacea* and notes on hybridization in *Erica. Bothalia* **16**: 35–38.

Oliver, E. G. H. (1989). The Ericoideae and the southern African heathers. *Bot. J. Linn. Soc.* **101**: 319–327.

Oliver, E. G. H. (1991). The Ericoideae (Ericaceae) — a review. *Contr. Bolus Herb.* **13**: 158–208.

Oliver, E. G. H. (1994a). Phytogeography and endemism in the Ericoideae (Ericaceae), vol. **2**: 941–951. In: J. H. Seyani & A. Chikuni (eds), *Proceedings of the XIIIth plenary meeting AETFAT, Malawi.* National Herbarium and Botanic Gardens of Malawi, Zomba.

Oliver, E. G. H. (1994b). Studies in the Ericoideae (Ericaceae). XIV. Notes on the genus *Erica. Bothalia* **24**: 25–30.

Oliver, E. G. H. (1995). *Erica* — update on species numbers. *Yearb. Heather Soc.* **1996**: 11–12.

Oliver, E. G. H. (1996). The position of *Bruckenthalia* versus *Erica. Yearb. Heather Soc.* **1996**: 6.

Oliver, E. G. H. (2000). Systematics of Ericeae (Ericaceae: Ericoideae) species with indehiscent and partially dehiscent fruits. *Contr. Bolus Herb.* **19**.

Oliver E. G. H. & Oliver, I. M. (1998). Three new species of *Erica* (Ericaceae) from South Africa. *Novon* **8**: 267–274.

O'Meara, J. J. (1988). *Eriugena.* Clarendon Press, Oxford.

O'Neill, J. & Nelson, E. C. (1995). Introduction of St Dabeoc's heath into English gardens, 1763. *Yearb. Heather Soc.* **1995**: 27–32.

Ostenfeld, C. E. H. (1912). Some remarks on the International Phytogeographic Excursion in the British Isles. *New Phytol.* **11** (4): 114–127.

Osti, G. (2009). Heathers and David McClintock. *Heathers* **6**: 9–12.

Oswald, P. H. & Nelson, E. C. (2009). Κουκουλόχορτο, koukoulóhorto, a Greek name for some heathers and other plants. *Heathers* **6**: 57–52.

Ottewill, D. (1989). *The Edwardian garden.* Yale University Press, New Haven & London.

Owen, E. (1922). *A catalogue of the manuscripts relating to Wales in the British Museum.* The Honourable Society of Cymmrodorion.

Oxbrow, A. & Moffat, J. (1979). Plant frequency and distribution on high-lead soil near Leadhills, Lanarkshire. *Pl. & Soil* **52**: 127–130.

Ozinga, W. A. & Schaminée, J. H. J. (eds) (2005). *Target species — species of European concern. A database driven selection of plant and animal species for the implementation of the Pan European Ecological Network.* (Alterra-report 1119.) Alterra, Wageningen.

Padley, D. (2006). My ancestor, George Sinclair, and *Hortus ericæus woburnensis*. *Heathers* **3**: 30–37.

Page, C. N. (1979). Macaronesian heathlands. In: R. L. Specht (ed.), *Ecosystems of the World.* **9A** *Heathlands and related shrublands descriptive studied,* chapter 5 (pp. 117–123). Elsevier, Amsterdam.

Pakeman, R., Stolte, A., Malcolm, A. & Marrs, R. (2002). *Heather beetle outbreaks in Scotland.* The Scottish Executive Environment Group, Edinburgh.

Pardo de Santayana, Ramón-Laca, L. & Morales, R. (2002). Traditional uses of heathers in Spain. *Yearb. Heather Soc.* **2002**: 47–56.

Parris, A. (1976). Preliminary notes on a cross between *Erica erigena* and *E. carnea*. *Yearb. Heather Soc.* **2** (5): 48–49.

Parris, A. (1980). Further notes on *E. × darleyensis* induced hybrids. *Yearb. Heather Soc.* **2** (9): 67–70.

Parry, J. (2003). *Living landscapes. Heathland.* The National Trust, London.

[Patrick, P. S.] (1963). Editorial. *Yearb. Heather Soc.* [**1** (1)]: 5–6.

Patrick, P. S. (1964). Where have they come from? *Yearb. Heather Soc.* [**1** (2)]: 23–31.

Paula, S. & Ojeda, F. (2006). Resistance of three co-occurring resprouter *Erica* species to highly frequent disturbance. *Pl. Ecol.* **183**: 329–336.

Pausas, J. G. & Paula, S. (2005). *EUFIRELAB: Euro-Mediterranean Wildland Fire Laboratory, a "wall-less" laboratory for wildland fire sciences and technologies in the Euro-Mediterranean Region … Plant functional traits database for Euro-Mediterranean ecosystems.* (URL http://www.eufirelab.org/prive/directory/units_section_4/D-04-06/D-04-06.pdf accessed 25 February 2007).

Pérez Latorre, A. V., Navas Fernández, P., Navas, D., Gil, Y. & Cabezudo, B. (2002). Datos sobre la flora y vegetación de la cuenca del río Guadiamar (Sevilla-Huelva, España). *Acta Bot. Malacit.* **27**: 189–228.

Persano Oddo, L., Piana, L., Bogdanov, S., Bentabol, A., Gotsiou, P., Kerkvliet, J., Martin, P., Morlot, M., Ortiz Valbuena, A., Ruoff, K. & von der Ohe, K. (2004). Botanical species giving unifloral honey in Europe. *Apidologie* **35**: S82–S93.

Peşmen, H. (1968). *Pentapera bocquetii, specie nova*, in Apportationes ad Floram turcicam 1–2. *Candollea* **23**: 271–273.

Petiver, J. (1704). *Gazophylacii naturæ et artis decas tertia*: … . Christopher Bateman, London.

Petterssen, E. (2004). My garden. *Heather news* **27** (2 Spring): 10–13.

Petterssen, E. (2006). [Letter to the editor.] *Bull. Heather Soc.* **6** (17, Spring): 11.

Phillips, L. & Sparks, B. W. (1976). Pleistocene vegetational history and geology in Norfolk … with an appendix on the non-marine Mollusca from Swanton Morley. *Philos. Trans.,* Ser. B, **275**: 215–286.

Phillips, V. R., Scotford, I. M., White, R. P. & Hartshorn, R. L. (1995). Minimum-cost biofilters for reducing odours and other aerial emissions from livestock buildings: 1. Basic airflow aspects. *J. Agric. Engin. Res.* **62**: 203–214. [Abstract only seen.]

Pichi-Sermolli, R. & Heiniger, H. ("1952" [1953]). Adumbratio floræ aethiopicae. II. Ericaceae. *Webbia* **9** (1): 9–48.

Pignatti, S. (1982). *Flora d'Italia.* Vol. **2**. Edagricole, Bologna.

Pinto de Silva & Teres (1971). Catálogo das plantas herborizadas. *Mem. Soc. Brot.* **21**: 51–323.

Pirie, M. D., Oliver, E. G. H. & Bellstedt, D. U. (2011). A densely sampled ITS phylogeny of the Cape flagship genus *Erica* L. suggests numerous shifts in floral macro-morphology. *Molec. Phylogenet. Evol.* **61**: 593–601.

Pizarro Domínguez, J. M. (2007). About *Erica × lazaroana* Rivas Goday & Bellot (*E. arborea* Linnaeus × *E. umbellata* Linnaeus). *Fontqueria* **55** (53): 439–442.

Platt, J. (1982). New acquisitions. *Yearb. Heather Soc.* **2** (11): 47–49.

Platt, J. (1991). New acquisitions. *Yearb. Heather Soc.* **3** (9): 34–37.

Platt, J. (1992). New acquisitions. *Yearb. Heather Soc.* **3** (10): 61–65.

Polunin, O. & Smythies, B. E. (1973). *Flowers of south-west Europe a field guide.* Oxford University Press, London.

Post, G. E. & Dinsmore, J. E. (1923–1933). *Flora of Syria, Palestine and Sinai.* Second edition. American Press, Beirut.

Praeger, R. Ll. (1925). St. Dabeoc's heath. *Garden (London 1871–1927)* **89**: 180.

Praeger, R. Ll. (1934). *The botanist in Ireland.* Hodges, Figgis & Co., Dublin.

Praeger, R. Ll. (1938). Cornish heather in Ireland. *Irish Naturalists' J.* **7**: 3–5.

Pratt, A. [1855]. *The flowering plants of Great Britain.* London. 3 vols.

Preston, C. P & Hill, M. O. (1997). The geographical relationships of British and Irish vascular plants. *Bot. J. Linn. Soc.* **124**: 1–120.

Preston, C. D., Pearman, D. A. & Dines, T. D. (eds) (2002). *New atlas of the British and Irish flora: an atlas of the vascular plants of Britain, Ireland, the Isle of Man and the Channel Islands.* Oxford University Press, Oxford.

Proctor, M., Yeo, P. & Lack, A. (1966). *The natural history of pollination.* (New naturalist.) HarperCollins, London.

Proudley, B. & Proudley, V. (1974). *Heathers in colour.* Blandford Press & Hippocrene Books, London & New York.

Q. [Quiller-Couch, A.] (1895). *Wandering heath; stories, studies and sketches.* Cassell, London. (The Duchy edition (1928). J. M. Dent & Sons, London)

Quézel, P. & Santa, S. (1963). *Nouvelle flore de l'Algérie et des régionas désertiques méridionales.* Vol. **2**. Centre national de la recherche scientifique, Paris.

Quintana, R., Cruz, A., Fernández-González, F. & Moreno, J. M. (2004). Time of germination and establishment success after fire of three obligate seeders in a Mediterranean shrubland of central Spain. *J. Biogeogr.* **31**: 241–249.

Rackham, O. (2003). *The illustrated history of the countryside.* Weidenfeld & Nicolson, London.

Ramón-Laca, L., de Luarca, M. & Morales Valverde, R. (eds) (2005). *Charles de l'Écluse de Arras. Descripción de algunas plantas raras encontradas en España y Portugal.* Junta de Castilla y León, Valladolid.

Ramón-Laca, L. & Morales, R. (2000). Heathers from Spain and Portugal in Charles de l'Écluse's works. *Yearb. Heather Soc.* **2000**: 81–88.

Ramsay, A. (1808). *The gentle shepherd, a pastoral comedy; with illustrations of the scenary: … .* Vol. **2**. W. Martin, Edinburgh.

Raven, J. E. (2000). *Plants and plant lore in Ancient Greece.* Leopard's Head Press Ltd, Oxford.

Ray, J. (1670). *Catalogus plantarum angliae, et insularum adjacentium.* J. Martyn, London.

Ray, J. (1704). *Historia plantarum species hactenus editas.* Vol. **3**. Samuel Smith & Benjamin Walford, London.

Rebelo, A. G., Siegfried, W. R. & Oliver, E. G. H. (1985). Pollination symdromes of *Erica* species in the south-western Cape. *S. African J. Bot.* **51**: 270–280.

Registrar [McClintock, D.] (1990). Cultivars registered during 1989. *Yearb. Heather Soc.* **3** (8): 62–63.

Registrar [Nelson, E. C.] (2004). Denomination classes. In: Supplement to *International register of heather names* — IV (2004): p. 65. *Heathers* **1**: 64–74.

Registrar [Nelson, E. C.] (2007). Supplement to *International register of heather names* — VII (2007). *Heathers* **4**: 67–74.

Reichenbach, H. G. L. (1831). *Flora germanica excursoria.* Vol. **1**. C. Cnobloch, Leipzig.

Rendall, V. (1934). *Wild flowers in literature.* George Allen & Unwin Ltd, London.

Rendell, S. & Ennos. R. A. (2002). Chloroplast DNA diversity in *Calluna vulgaris* (heather) populations in Europe. *Molec. Ecol.* **11**: 69–78.

Reynolds, S. (2002). *A catalogue of alien plants in Ireland.* National Botanic Gardens, Glasnevin.

Ribeiro, S. P., Borges, P. A. V. & Gaspar, C. S. (2003). Ecology and evolution of the arborescent *Erica azorica* Hochst (Ericaceae). *Arquipélago. Agric. & Environm. Sci.* **1** (1): 41–49 [+ 3pp].

Ribeiro, S. P., Borges, P. A. V., Gaspar, C., Melo, C., Serrano, A. R. M., Amaral, J., Aguiar, C., André, G. & Quartau, J. A. (2005). Canopy insect herbivores in the Azorean Laurisilva forests: key host plant species in a highly generalist insect community. *Ecography* **28** (3): 315–330.

Richards, D. A. (1976). Mostly *Erica maderensis* and *Daboecia azorica. Yearb. Heather Soc.* **2** (5): 15–19.

Richards, J. (1990). 'Lucky white heather?'. *BSBI news* no. **54**: 32.

Richter, H. E. (1840). *Codex botanicus linnaeanus.* O. Wigand, Leipzig. (Facsimile reprint, 2003. *Regnum Veg.* **140**).

Riddell, J. & Stearn, W. T. (1994). *By Underground to Kew: London Transport posters: 1908 to the present.* London.

Rivas Goday, S. & Bellot Rodriguez, F. (1946). [X *Erica lazaroana* hibr. nov. = *Erica arborea* L. × *umbellata* L.], in Estudios sobre la Vegetacion y Flore de la comarca Despeñaperros-Santa Elena. *Anales Jard. Bot. Madrid* **6** (2): 152–153.

Rivas-Martínez, S., Díaz González, T. E., Fernández-González, F., Izco, J., Loidi Arregui, J, Lousã, M. & Penas Merino, A. (2002). Vascular plant communities of Spain and Portugal. *Itin. Geobot.* **15**: 5–922.

Rivas-Martínez, S., Wilpret de la Torre, W., Arco Aguilar, M., Rodríguez Delgado, P., Pérez de Paz, P. L., García-Gallo, A., Acebes Ginovés, J. R., Díaz González, T. E. & Fernández-González, F. (1993). Las comunidades vegetales de la Isla de Tenerife (Islas Canarias, España). *Itin. Geobot.* **7**: 169–374.

Robinson, T. (1990). *Connemara Part 1: Introduction and gazetteer.* Folding Landscapes, Roundstone.

Robinson, T. (2006). *Connemara. Listening to the wind.* Penguin, Dublin.

Robinson, W. (1870). *The wild garden.* J. Murray, London. (For an illustrated, modern edition of Robinson's first edition, see Nelson 2003a.)

Robinson, W. (1883). *The English flower garden.* First edition. J. Murray, London.

Robinson, W. (1897). *The English flower garden.* Fifth edition. J. Murray, London.

Robinson, W. (1898). *The English flower garden.* Sixth edition. J. Murray, London.

Robinson, W. (1911). *Gravetye Manor or twenty years' work round an old manor house.* J. Murray, London. (Facsimile edition (1984). Introduction by Graham Stuart Thomas with a contribution by Will Ingwersen. Afterword by Peter Herbert. Sagapress, New York.)

Robinson, W. (1918). Foreword. In: L. Messel, *A garden flora trees and flowers grown in the gardens at Nymans ... 1890–191,* pp [v]–vi, 5. Country Life & George Newnes, London; Charles Scribner's Sons, New York.

Rodríguez, N., Amils, R., Jiménez-Ballesta, R. & Rufo, L. (2007). Heavy metal content in *Erica andevalensis*: an endemic plant from the extreme acidic environment of Tinto River and its soils. *Arid Land Res. & Managem.* **21**: 51–65.

Rogers, C. (2006). Sunburn or frostbite? Observations on the changing colour of some heathers. *Heathers* **3**: 9–13.

Roles, S. J. (1960). *Flora of the British Isles: illustrations.* Part 2. Cambridge University Press, Cambridge.

Rollisson, G. (1843). Hybrid plants. *Gard. Chron.* 8 July: 461.

Rose, R. J., Bannister, P. & Chapman, S. B. (1996). Biological flora of the British Isles. *Erica ciliaris* L. *J. Ecol.* **84**: 617–628.

Ross, R. (1967). On some Linnaean species of *Erica. J. Linn. Soc., Bot.* **60**: 61–73.

Ross, R. (1969). Nomenclatural changes in British plants 357/7 *Erica erigena* R. Ross, nom. nov. *Watsonia* **7**: 164.

Ross-Craig, S. (1963). *Drawings of British plants.* Part 19 (Lobeliaceae, Campanulaceae, Ericaceae, Pyrolaceae, Monotropaceae, Diapensiaceae). G. Bell, London.

Rochefoucauld, B. de la (1979). *La bruyère.* Dargaud Editeur, Neuilly-sur-Seine.

Rochefoucauld, B. de la (1997). *La bruyère.* Éditions Rustica, Paris.

Rouleau, E. & Lamoureux, G. (1992). *Atlas of the vascular plants of the island of Newfoundland and of the islands of St Pierre and Miquelon.* Fleurbec, Quebec.

Rozefelds, A. C. F., Cave, L., Morris, D. I. & Buchanan, A. M. (1999). The weed invasion in Tasmania since 1970. *Austral. J. Bot.* **47**: 23–48.

Rudolphi, I. K. A. (1799). "14. Erica *scoparia* Linn. und Thunb. ...". *J. Bot. (Schrader)* Bd. **2** (pt 4): 283–286.

Ruiz, M. R., Martín-Cordero, C., Ayuso González, M. J., Toro Sainz, M. V. & Alarcón de la Lastra, C. (1996). Antiulcer activity in rats by flavonoids of *Erica andevalensis* Cabezudo & Rivera. *Phytotherapy Res.* **10**: 300–303.

Ruppius, H. B. (1726). *Flora ienensis sive enumeratio plantarum, ...: in usum botanophilorum ienensium edita multisque in locis correcta et aucta.* E. C. Bailliar, Frankfurt & Leipzig.

Ruskin, J. (1875). *Proserpina. Studies of wayside flowers while the air was yet pure upon the Alps and in the Scotland and England which my father knew.* London

Salisbury, R. A. (1796). *Prodromus stirpium in horto de Chapel Allerton.* London.

Salisbury, R. A. (ed.) (1800). *Dissertatio botanica de Erica quam præside C. P. Thunberg, subjicit J. B. Struve, &c. Edito altera curante R. A. Salisbury.* Featherstone,

Salisbury, R. A. (1802). Species of Erica. *Trans. Linn. Soc. London* **6**: 316–388.

Salisbury, R. A. (1806). *The generic characters in The English botany collated with those of Linné.* J. Hatchard, London.

Salisbury, R. A. (1866). *The genera of plants ... a fragment containing part of Liriogamæ.* J. Van Voorst, London.

Sallinen, V. (2007). *Erica tetralix* (kellokanerva, cross-leaved heath) in Finland. *Heathers* **4**: 24–26.

Sandeman, P. W. (1957). Snow-bunting feeding on heather seeds. *Scott. Naturalist* **69**: 192.

Sands, M. J. S. (1991). *Paxos walking and wild flowers.* Second edition. The Greek Islands Club, Walton-on-Thames.

Savory, C. J. (1978). Food consumption of red grouse in relation to the age and productivity of heather. *J. Anim. Ecol.* **47**: 269–282

Sawyer, F. C. (1971). A short history of the libraries and list of manuscripts and original drawings in the British Museum (Natural History). *Bull. Brit. Mus. (Nat. Hist.)*, Hist. Ser., **4** (2): 77–204.

Scannell, M. J. P. (1975). Craiggamore and Craiggabeg in West Galway. *Irish Naturalists' J.* **18** (7): 224–22.

Scannell, M. J. P. (1976a). *Erica erigena* R. Ross: a note on the derivation of the specific epithet. *Garden* **101** (9): 489–490.

Scannell, M. J. P. (1976b). *Erica × watsonii* Benth. on the flora of Ireland. [*Glasra*] *Contr. Natl. Bot. Gard., Glasnevin* **1**: 40–47.

Scannell, M. J. P. (1983). *Erica erigena* R. Ross and *Daboecia cantabrica* (Huds.) C. Koch — some further notes on the historical records. *Irish Naturalists' J.* **21** (1): 29–30.

Scannell, M. J. P. (1985). '*Erica mackaii*' — unpublished notes by M. C. Knowles written about 1910. *Irish Naturalists' J.* **21** (10): 425–430.

Scannell, M. J. P. & McClintock, D. C. (1974). *Erica mackaiana* Bab. in Irish localities and other plants of interest. *Irish Naturalists' J.* **18** (3): 81–85.

Scannell, M. J. P. & Synnott, D. M. (1987). *Census catalogue of the Irish flora.* Second edition. Stationery Office, Dublin.

Scarth, A. & Tanguy, J.-C. (2001). *Volcanoes of Europe.* Terra Publishing, Harpenden.

Schacht, H. (1859). *Madeira und Tenerife mit ihrer Vegetation.* G. W. F. Müller, Berlin.

Schäfer, H. (2002). *Flora of the Azores a field guide.* Margraf Verlag, Weickersheim.

Schouw, J. F. (1823). *Pflanzengeographischer Atlas.* Reimer, Berlin.

Schröder, J. (2005). The explosion of bud-flowerers. *Heathers* **2**: 17–18.

Schumann, D. & Kirsten, G. (1992). *Ericas of South Africa.* Fernwood Press, Vlaeberg.

Scott, M. H. Baillie (1906). *Houses and gardens.* G. Newnes, London. (Reprint 1995. Antique Collectors' Club, Woodbridge.)

Scully, R. W. (1916). *Flora of County Kerry.* Hodges, Figgis & Co., Dublin.

Sealy, J. R. (1949). *Daboecia azorica. Curtis's Bot. Mag.,* n.s., **166**: tab. 46.

Seemann, B. (1866). On the Newfoundland heather. *J. Bot.* **4**: 305–306, plate 53.

Sell, P. (1990). The typification and nomenclature of *Erica mackaiana* Bab. *Watsonia* **18**: 92–93.

Sellers, B. (1994). Spring 1994 visit to the Netherlands and Germany. *Bull. Heather Soc.* **5** (2): 3–5.

Sellers, B. (1999). Propagation of heathers from seed. *Yearb. Heather Soc.* **1999**: 23–29.

Serra, L. (1966). Ricerche geobotaniche su «*Erica cinerea*» in Italia. *Webbia* **21**: 801–837.

Seubert, M. A. (1844). *Flora azorica: quam ex collectionibus schedisque Hochstetteri patris et filii.* A. Marc, Bonn.

Shanly, C. D. (1861). The brier-wood pipe. *Vanity fair* **4** (no. 80, 6 July): 5.

Sharpe, J. (1992). This was "our Mr Richard Potter". *Yearb. Heather Soc.* **3** (10): 32–36.

Shaw, M. F. (1955). *Folksongs and folklore of South Uist.* Routledge & Kegan Paul, London.

Shaw, M. W. & Sutherland, J. P. (1966). Damage by the gall midge *Wachtliella ericina* (F. Loew) to *Erica carnea* in Scotland. *Pl. Pathol.* **15** (4): 192.

Sheppard, A., Ainsworth, N., Peterson, P., Iaconis, L. & Thorp, J. (2003). Alert list for environmental weeds. Weed management guide. Heather — *Calluna vulgaris.* CRC for Australian Weed Management (available on-line).

Sibthorp, J. & Smith, J. E. (1806). *Floræ græcæ prodromus: sive plantarum omnium enumeratio quas in provinciis aut insulis Graæciæ invenit Johannes Sibthorp. Characteres et synonyma, omnium cum annotationibus elaboravit Jacobus Edvardus Smith.* Vol. **1**. R Taylor, London.

Sibthorp, J. & Smith, J. E. (1823). *Flora graeca: sive plantarum rariorum historia, quas in provinciis aut insulis Graeciae legit, investigavit et depingi curavit Johannes Sibthorp … Characteres omnium, descriptiones et synonyma elaboravit Jacobus Edvardus Smith.* Cent. 4. R Taylor, London.

Siddiqi, M. A. (1978). Ericaceae. In: S. M. H. Jafri & A. El-Gadi (eds). *Flora of Libya* **54**: 1–8. Al Faateh University, Tripoli.

Siebert, S. & Ramdhani, S. (2005). Heathers on the roof of Africa: The Bale Mountains of Ethiopia. *Heathers* **2**: 49–54 (see also *Veld & flora* **90** (2): 54–59. 2004).

Sigerson, G. (1871). Anomalous form of corolla in *Erica tetralix*. *Proc. Roy. Irish Acad.* **13**: 191–192.

Silva, J. S., Rego, F. C. & Martins-Loução, M. A. (2002). Below-ground traits of Mediterranean woody plants in a Portuguese shrubland. *Ecol. Medit.* **28** (2): 5–13.

Silva, L., Pinto, N., Press, J. R., Rumsey, F., Carine, M., Henderson, S. & Sjögren, E. (2005). Lista das plantas vasculares (Pteridophyta e Spermatophyta). In: P. A. V. Borges, R. Cunha, R. Gabriel, A. F. Martins, L. Silva & V. Vieira (eds), *Listagem de fauna e flora (Mollusca e Arthropoda) (Bryophyta, Pteridophyta e Spermatophyta) terrestres dos Açores*. Direcção Regional do Ambiente & Universidade dos Açores, Horta, Angra do Heroísmo & Ponta Delgada.

Simpson, N. B. (1960). *Bibliographic index to the British flora*. Bournemouth.

[Sinclair, G.] (1825). *Hortus ericaeus woburnensis: or, a catalogue of heaths, in the collection of the Duke of Bedford at Woburn Abbey*. [London.]

Sjögren, E. A. (1972). Vascular plant communities of Madeira. *Bol. Mus. Munic. Funchal* **26** (114): 45–125.

Sjögren, E. A. (1973). Conservation of natural plant communities on Madeira and the Azores. *Monogr. Biol. canarienses* no. **4** (Proceedings of the first international congress pro Flora Macaronesica): 148–153.

Sjögren, E. A. (1974). Local climatic conditions and zonation of vegetation on Madeira. *Agron. Lusit.* **36**: 95–139.

Sjögren, E. A. (2001). *Plants and flowers of the Azores*. Espaço Talassa Azorean Whale Watching Base, Lajes do Pico.

Skan, S. A. (1906). *Erica terminalis*. *Curtis's Bot. Mag.* **132**: tab. 8063.

Skuhravá, M. & Skuhravy, V. (2003). Die Gallmückenfauna (Cecidomyiidae, Diptera) Südtirols: 3. Die Gallmücken der Sextener Dolomiten (Gall midge fauna (Cecidomyiidae, Diptera) of South Tyrol 3. Gall midges of Sexten Dolomites). *Gredleriana* **3**. (Abstract only).

Slinger, L. (2007). A life among shrubs: heathers. *Bull. Heather Soc.* **7** (2): 10–14.

Small, D. J. (1973). Propagation experiments to determine rooting responses of heather cuttings. *Heather Soc. Techn. Rep.* 1/73.

Small, D. J. (1974). Observations on rooting cuttings. *Yearb. Heather Soc.* **2** (3): 28–31.

Small, D. J. (1992). *Erica × stuartii* 'Pat Turpin'. *Yearb. Heather Soc.* **3** (10): 37–38.

Small, D. J. (1995). Propagation of hardy heathers in the garden. *Yearb. Heather Soc.* **1995**: 23–26.

Small, D. J. (1996a). *Erica arborea* 'Estrella Gold'. *Yearb. Heather Soc.* **1996**: 36.

Small, D. J. (1996b). *Calluna vulgaris* 'Alexandra'. *Yearb. Heather Soc.* **1996**: 52.

Small, D. J. (2001). Award of honour. *Yearb. Heather Soc.* **2001**: 63–64.

Small, D. J. (2004a). The Old 804: European heathers naturalized in Oregon. *Heathers* **1**: 27–29.

Small, D. J. (2004b). Herman M. J. Blum (1934–2003). *Heathers* **1**: 54.

Small, D. J. (2008). [Pruning.] *Bull. Heather Soc.* **7** (4): 00.

Small, D. J. & Cleevely, R. C. (1999). *The Heather Society's guide to recommended heathers. The 100 best heathers chosen by experts*. The world of heather booklet series, no. 2. The Heather Society, Creeting St Mary, Suffolk.

Small, D. J. & Small, A. (1998). *The Heather Society's handy guide to heathers*. Second edition. The Heather Society, Creeting St Mary, Ipswich.

Small, D. J. & Small, A. (2001). *The Heather Society's handy guide to heathers*. Third edition. The Heather Society, Creeting St Mary, Ipswich.

Small, D. J. & Wulff, E. M. T. (2008). *Gardening with hardy heathers*. Timber Press, Portland.

Small, I. & Alanine, H. (1994). Some speculation on colourful foliage in heathers. *Yearb. Heather Soc.* **1994**: 27–34.

Small, I. & Kron, K. (2001). Placing heathers on the "Tree of Life". *Yearb. Heather Soc.* **2001**: 15–21.

Smith, J. E. (1791). *Erica dabeoci* Irish heath. In: J. Sowerby & J. E. Smith, *Engl. Bot.* **1**: tab. 35.

S[mith, J. E.] (1809). *Erica*. In: A. Rees, *Cyclopaedia*. Longman, Hurst, Rees, Orme & Brown, London.

Smith, J. E. (1821). *A selection of the correspondence of Linnaeus and other naturalists, from the original manuscripts*. Longman *et al.*, London. 2 vols.

Smith, J. E. (1824). *The English flora*. Vol. **2**. Second edition. Longman, Rees, Orme, Brown & Green, London.

Smith, M. D. (2004). Couch, Sir Arthur Thomas Quiller- (1863–1944). In: H. C. G. Matthew & B. Harrison (eds), *Oxford dictionary of national biography*. OUP, Oxford. [http://www.oxforddnb.com/view/article/35640 (accessed February 2, 2007)].

Smith, M. H. (1930). Leaf anatomy of the British heaths. *Trans. Bot. Soc. Edinburgh* **30**: 198–205.

Smith, P. (ed.) (1832). *Memoir and correspondence of the late Sir James Edward Smith ... .* Vol. **2**. Longman, Rees, Orme, Brown, Green, & Longman, London.

Snow, J. L. (1847). Heather-edgings. *Gard. Chron.* 2 January: 5.

Sowerby, J. (1790). *Erica vagans* Cornish heath. In: J. Sowerby & J. E. Smith, *Engl. Bot.* **1**: tab. 3.

Sowerby, J. (1791). *Erica dabeoci* Irish heath. In: J. Sowerby & J. E. Smith, *Engl. Bot.* **1**: tab. 35.

Stace, C. A. (1991). *New flora of the British Isles.* Cambridge University Press, Cambridge.

Stace, C. A. (2005). Plants found in Ireland but not in Britain. *Watsonia* **25** (3): 296–298.

Stafleu, F. A. (1967). The great Prodromus. In: *Adanson, Labillardière, de Candolle. Introductions to four of their books reprinted in the series Historiae naturalis classica,* pp. 64–100. J. Cramer, Lehre.

Stafleu, F. A. (1971). *Linnaeus and the Linnaeans. The spreading of their ideas in systematic botany, 1735–1789.* IAPT, Utrecht.

Starling, B. (1998). Souvenirs of the Azores. *The Garden* **123** (2): 121.

Stearn, W. T. (2007). Society for the Bibliography of Natural History, later the Society for the History of Natural History, 1936–1985. A quinquagenary record. *Arch. Nat. Hist.* **35**: supplement.

Stefanesco, E. & Vilhena, H. (1966). *Erica mediterranea* L. et *Vicia bithynica* L. *Al Awamia* **21**: 127

Stephens, P. & Browne, W. (1658). *Catalogus horti botanici Oxoniensis ... .* Second edition. W. Hall, Oxford.

Stevens, P. F. (1971). A classification of the Ericaceae: subfamilies and tribes. *Bot. J. Linn. Soc.* **64**: 1–53.

Stevens, P. F. (1978). *Erica* L.; *Bruckenthalia* Reichenb.; *Calluna* Salisb. In: P. H. Davis, J. Cullen & M. Coode (eds), *Flora of Turkey and the East Aegean islands,* vol. **6**, pp 95–98. Edinburgh University Press, Edinburgh.

Stevens, P. F. (2003). George Bentham (1800–1884): the life of a botanist's botanist. *Arch. Nat. Hist.* **30**: 189–202.

Stevens, P. F., Luteyn, J., Oliver, E. G. H., Bell, T. L., Brown, E. A., Crowden, R. K., George, A. S., Jordan, G. J., Ladd, P., Lemson, K., McLean, C. B., Menadue, Y., Pate, J. S., Stace, H. M. & Weiller, C. M. (2004). Ericaceae. In: K. Kubitzki (ed.), *Families and genera of vascular plants.* Vol. **6**: 145–194 (Flowering plants: Dicotyledons: Celastrales, Oxalidales, Rosales, Cornales, Ericales). Springer, Berlin.

Stewart, R. & Anderson, C. W. N. (2005). *Eric andevalensis,* a unique "copper flower" from Spain. *Heathers* **2**: 23–26.

Stokes, J. (1812). *A botanical materia medica, consisting of the generic and specific characters of the plants used in medicine and diet.* 4 vols. Johnson, London.

Story, G. M. (2000). Cormack, William Eppes. In: *Dictionary of Canadian biography online* [accessed 7 January 2007].

Story, W. H. (1848). Seedling heaths. *Florist* 1848: 314–316.

Stow, A. J. (2004). Forty years on. *Heathers* **1**: 13–17

Strid, A. (ed.) (1986). *Mountain flora of Greece* Vol. **1**. University Press, Cambridge.

Sykes, W. R. (1981). Checklist of dicotyledons naturalised in New Zealand 11. Apocynales, Campanulales, Ericales, Gentianles, Loganiales, Plantaginales, Primulales, and Rubiales. *New Zealand J. Bot.* **19**: 319–325.

Syme, J. T. Boswell (ed.) (1866). *Erica hibernica. English botany; or, coloured figures of British plants.* Vol. **6**: 42, tab. 893. Third edition. Hardwicke, London.

Synge, P. M. (ed.) (1956). *The Royal Horticultural Society dictionary of gardening.* Vol. **2**: Co–Ja. Second edition. Clarendon Press, Oxford.

Syrett, P., Smith L. A., Bourner, T. C., Fowler, S. V. & Wilcox, A. (2000). A European pest to control a New Zealand weed: investigating the safety of heather beetle, *Lochmaea suturalis* (Coleoptera: Chrysomelidae) for biological control of heather, *Calluna vulgaris. Bull. Entomol. Res.* **90**: 169–178.

Szkudlarz, P. (2001). Morphological and anatomical structure of seeds in the family Ericaceae. *Biol. Bull. Poznan* **38** : 113–132.

Tabernaemontanus, J. T. (1588). *Neuw Kreuterbuch.* N. Bassaeus. Frankfurt.

Tabernaemontanus, J. T. (1590). *Eicones plantarum seu stirpium, arborum nempe, fructicum, herbarum, fructuum, lignorum, radicum, omnis generis.* N. Bassaeus. Frankfurt.

Tan, K., Bonetti, A. & Lafranchis, T. (2010). Report 76 *Erica multiflora* L. In: V. Vladimirov, F. Dane & K. Tan (compliers), New floristic records in the Balkans: 13: p. 158. *Phytol. Balcan.* **16** (1): 143–165.

Tausch, J. F. (1834). Bemerkungen über *Erica*. *Flora* **17** (38): 593–606.

Thomas, G. S. (1957). *Colour in the winter garden.* (Third edition 1984). J. M. Dent, London.

Thomas, G. S. (1970). *Plants for ground-cover.* (Revised edition 1984.) J. M. Dent, London.

Thomas, G. S. (1979). *Gardens of the National Trust.* The National Trust / Weidenfeld & Nicolson, London.

Thomas, G. S. (1984). *The art of planting or the planter's handbook.* J. M. Dent, London.

Thomas, G. S. (1997). Heathers in the garden. In: *Cuttings from my garden notebooks,* chapter 61. Sagapress, Sagaponack.

Thomas, G. S. (2003). *Recollections of great gardeners.* Frances Lincoln, London.

Thompson, D. B. & Griswold, R. E. (1963). *Garden lore of Ancient Athens.* (Excavations of the Athenian Agora picture book no. 8.) American School of Classical Studies at Athens, Princeton, New Jersey.

Thompson, D. B. A., Hester, A. & Usher, M. B. (eds) (1995). *Heaths and moorland: cultural landscapes.* HMSO, Edinburgh.

Threlkeld, C. (1726). *Synopsis stirpium Hibernicarum.* The author, Dublin. (Facsimile edition, 1984. Boethius Press, Kilkenny).

Thunberg, C. P. (1785). *Dissertatio botanica de Erica.* J. Erdman, Uppsala.

Thurston, E. (1930). *British and foreign trees and shrubs in Cornwall.* Cambridge University Press, Cambridge.

Toro Sainz, M. V., García Gimenez, M. D. & Pascual Matinez, M. (1987). Isolement et identification des acides phénols chez *Erica andevalensis* Cabezudo & Rivera: leur contribution à l'activité antimicrobienne de l'espèce. *Ann. Pharm. Franç.* **45**: 401–407.

Toro Sainz, M. V., García Gimenez, M. D. & Aumente, M. D. (1988). Étude de l'activité diuretique d'*Erica andevalensis* Cabezudo & Rivera. *Pl. Méd. Phytothérap.* **22**: 171–174.

Tournefort, J. P. de (1694). *Elemens de botanique, ou methode pour connoître les plantes.* L'Imprimerie Royale, Paris.

Treleease, W. (1897). Botanical observations on the Azores. *Eighth annual report of the Missouri Botanical Garden.*

Tucker, A. O., Maciarello, M. J. & Tucker, S. S. (1991). A survey of color charts for biological descriptions. *Taxon* **40**: 201–214.

Turland, N. J., Chilton, L. & Press, J. R. (1993). *Flora of the Cretan area.* HMSO, London.

Turner, A. (1973). Heather from seed ... an interesting experiment. *Yearb. Heather Soc.* **2** (2): 37–39.

Turner, W. (1548) *The names of herbes.* London. (Facsimile (1965) in *William Turner ... Facsimiles with introductory matter by James Britten, B. Draydon Jackson & W. T. Stearn.* The Ray Society, London.)

Turner, W. (1551). *A new herball.* London. (Facsimile edition, 1989, edited by G. Chapman & M. Tweddle. The Mid Northumberland Arts Group & Carcanet Press.)

Turpin, P. G. (1979a). The tree heaths 'Gold Tips' and 'Pink Joy'. *Yearb. Heather Soc.* **2** (8): 34–38.

Turpin, P. G. (1979b). The bud-flowering forms of *Calluna. Yearb. Heather Soc.* **2** (8): 41–43.

Turpin, P. G. (1981). *Erica arborea* 'Alpina'. *Yearb. Heather Soc.* **2** (10): 43–47.

Turpin, P. G. (1982a). The heathers of the Lizard district of Cornwall. *University of Bristol Lizard project report* no. 2. The University, Bristol.

Turpin, P. G. (1982b). A new and curious form of *Erica vagans* L. *Watsonia* **14**: 184–185.

Turpin, P. G. (1984). The heather species, hybrids and varieties of the Lizard district of Cornwall. *Cornish Stud., J. Inst. Cornish Stud.* **10** (for 1982): 5–17, plates 1–4.

Turpin, P. G. (1985). Two new plants of *E. × williamsii* found at the Lizard. *Yearb. Heather Soc.* **3** (3): 66.

Turpin, P. G. (1991). Another plant of *Erica × williamsii. Yearb. Heather Soc.* **3** (9): 27.

T[urrill], W. B. (1911). A hybrid heath. *Bull. Misc. Inform., Kew* **1911**: 378–379.

Turrill, W. B. (1922). *Erica vagans,* L., var. *kevernensis,* Turrill. *Bull. Misc. Inform., Kew* **1922**: 175–176.

Tutin, T. G. (1953). The vegetation of the Azores. *J. Ecol.* **41**: 53–61.

Tutin, T. G. & Warburg, E. F. (1932). Contributions from the University Herbarium, Cambridge.— Notes on the flora of the Azores. *J. Bot.* **70**: 7–12.

Tyas, R. (1875). *Speaking flowers: or, flowers to which a sentiment has been assigned: with illustrative poetry.* London.

Tyler, J. (2001). *Erica umbellata* 'David Small' [painting]. *Yearb. Heather Soc.* **2001**: iv.

Underhill, T. (1971). *Heaths and heathers Calluna, Daboecia and Erica.* David & Charles, Newton Abbot.

Underhill, T. (1990). *Heaths and heathers the grower's encyclopaedia*. David & Charles, Newton Abbot.

Utinet, –. (1839). Bruyère d'hiver. *Erica hibernica*. Hort. *Ann. Fl. Pomone* **1839–1840**: 94.

Valbuena, L., Nuñez, R. & Calvo, L. (2001). The seed bank in *Pinus* stand regeneration in NW Spain after wildfire. *Web Ecol.* **2**: 22–31.

Valbuena, L., Tárrega, R. & Luis-Calabuig, E. (2000). Seed banks of *Erica australis* and *Calluna vulgaris* in a heathland subjected to experimental fire. *J. Veg. Sci.* **11**: 161–166.

Valbuena, L. & Vera, M. L. (2002). The effects of thermal scarification and seed storage on germination of four heathland species. *Pl. Ecol.* **161**: 137–144.

van de Laar, H. (1974). *Het heidetuinboek*. Zomer & Keuning Boeken B. V., Wageningen.

van de Laar, H. (1978). *The heather garden … Translated by P. Rowe-Dutton. Adapted … by D. McClintock*. Collins, London.

van de Laar, H. (1998). *Erica cinerea* — a summary of the Royal Boskoop Horticultural Society trials 1993–1995. *Yearb. Heather Soc.* **1998**: 3–10.

van Doorslaer, L. (1990). *Erica mackaiana* in Mayo. *Irish Naturalists J.* **23**: 268–270.

van Oostrom, S. J. & Reichgelt, T. J. (1956). Floristische notities no. 11. *Acta Bot. Néerl.* **5** (1): 109.

Vickers, G. P. (1976). *A report on a trial to determine the characteristics of cultivated heathers … at Harlow Car Harrogate*. The Heather Society. (See also Anonymous 1983.)

Victoria, Queen (1868). *Leaves from the journal of our life in the Highlands from 1848 to 1861*. (Edited by Arthur Helps.) Smith, Elder & Co., London.

Victoria, Queen (1884). *More leaves from the journal of a life in the Highlands from 1862 to 1882*. Smith, Elder & Co., London.

Vila, M. & Terradas, J. (1995). Effects of competition and disturbance on the resprouting performance of the Mediterranean shrub *Erica multiflora* L. (Ericaceae). *Amer. J. Bot.* **82**: 1241– 1248.

Villar, I. (1993). *Calluna*. In: *Flora iberica*. Vol. **4**: 506. CSIC, Madrid.

Vines, S. H. & Druce, G. C. (1914). *An account of the Morisonian herbarium*. Claredon Press, Oxford.

Viney, D. E. (1994a). *Erica* spp. in North Cyprus. *Yearb. Heather Soc.* **1994**: 25–26.

Viney, D. E. (1994b). *An illustrated flora of north Cyprus*. Koeltz Scientific Books, Koenigstein.

Visset, L. (1975). Etude au microscope électronique à balayage des pollens des espèces européennes du genre *Erica* L. *Bull. Soc. Bot. France* **122**: 203–216.

Vitman, F. (1789). *Summa plantarum*. Tomus **II**. Milan.

Vogel, H. (1982). *Azaleen, Eriken, Kamalien*. Second edition. Paul Parey, Berlin & Hamburg.

Volk, F., Forshaw, N., Oliver, T. & Oliver, I. (2005). Genus *Erica*: interactive identification key. CD-ROM (v. 2.01).

Wace, N. M. & Holdgate, M. W. (1958). The vegetation of Tristan da Cunha. *J. Ecol.* **46**: 593–620.

Waitz, C. F. (1805). *Beschreibung der Gattung und Arten der Heiden, nebst einer Anweisung zur zweckmäszigen Kultur derselben*.

W[alker], A. (1934). A lucky find. *My garden* **2**: 105–106. (Reprinted in *Heathers* **4**: 33. 2007.)

Walker, A. (1934). "A lucky find": the finding of *Erica carnea* 'Springwood White'. *Heathers* **4**: 33, 34. 2007. (Reprinted, with editorial addition, from *My garden* **2**: 105–106. 1934).

Wallace, A. (1903). *The heather in lore lyric and lay*. A. T. de la Mare, New York.

Walsh W. F. & Nelson, E. C. (1987). *An Irish florilegium* **II**. Thames & Hudson, London.

Walsh W. F., Ross, R. I. & Nelson, E. C. (1983). *An Irish florilegium*. Thames & Hudson, London.

Waterston, A. R. (1936). Partridge *versus* heather beetle. *Scott. Naturalist* **1936**: 30.

Watson, H. C. (1843). Notes on a botanical tour in the western Azores. (In a letter ... to the editor [Sir W. J. Hooker] dated, November, 1842.) *London J. Bot.* **2**: 1–9, 125–131, 394–408.

Watson, H. C. (1844). Notes on the botany of the Azores. *London J. Bot.* **3**: 582–617.

Watson, H. C. (1865). *Calluna vulgaris* in Cape Breton, in North America. *Nat. Hist. Rev.* **5**: 149–150.

Watts, W. A. (1959). Interglacial deposits at Kilbeg and Newtown, Co. Waterford. *Proc. Roy. Irish Acad.* **60 B** (2): 79–134.

Watts, W. A. (1964). Interglacial deposits at Baggotstown near Bruff, Co. Limerick. *Proc. Roy. Irish Acad.* **63 B**: 167–189.

Weathers, J. (1901). *A practical guide to garden plants*. London.

Webb, D. A. (1948). Some observations on the *Arbutus* in Ireland. *Irish Naturalists' J.* **9**: 198–203.

Webb, D A. (1954). Notes on four Irish heaths. *Irish Naturalists' J.* **11**: 187–192, 215–219.

Webb, D. A. (1955). Biological flora of the British Isles. *Erica mackaiana* Bab. *J. Ecol.* **43**: 319–330.

Webb, D. A. (1966a). *Erica ciliaris* L. in Ireland. *Proc. Bot. Soc. Brit. Isles* **6**: 221–225.

Webb, D. A. (1966b). The flora of European Turkey. *Proc. Roy. Irish Acad.* **65** B (1): 1–100.

Webb, D. A. (1967). *Erica praegeri* Ostenfeld and *E. stuartii* E. F. Linton. *Watsonia* **6** (5): 296–297.

Webb, D. A. (1970). White-flowered *Daboecia cantabrica*. *Irish Naturalists J.* **16** (10): 319.

Webb, D. A. (1972a). *Bruckenthalia* Reichenb., *Calluna* Salisb. & *Daboecia* D. Don. In: *Flora europaea*. Vol. **3**. Cambridge University Press, Cambridge.

Webb, D. A. (1972b). *Erica* L., in *Flora europaea*. Notulae systematicae no. 12. *Bot. J. Linn. Soc.* **65**: 259.

Webb, D. A. (1975). *Erica* L. In: C. A. Stace (ed.), *Hybridization and the flora of the British Isles*, pp. 339–341. Botanical Society of the British Isles & Academic Press, London.

Webb, D. A. (1983). The flora of Ireland in its European context. *J. Life Sci. Roy. Dublin Soc.* **4**: 143–160.

Webb, D. A. & Rix, E. M. (1972). *Erica* L. In: *Flora europaea*. Vol. **3**: 5–8. Cambridge University Press, Cambridge.

Webb, D. A. & Scannell, M. J. P. (1983). *Flora of Connemara and The Burren*. Royal Dublin Society & Cambridge University Press, Cambridge

Webb, P. B. & Berthelot, S. (1844). *Histoire naturelle des îles Canaries tom. 3 pt 2. Phytographia canariense*. Béthune, Paris.

Welch, D. (1985). Studies in the grazing of heather moorland in north-east Scotland. IV. Seed dispersal and plants establishment in dung. *J. Appl. Ecol.* **22**: 461–472.

Weston, R. (1770). *Botanicus universalis et hortulanus*. Vol. **1**. J. Bell, London.

Weston, R. (1775). *Flora anglicana: seu, arborum, fruticum, plantarum, et fructum, tam indigenorum quam exoticorum … catalogus*. The author, London.

Whittington, G. (1994). *Bruckenthalia spiculifolia* (Salisb.) Reichenb. (Ericaceae) in the Late Quaternary of Western Europe. *Quatern. Sci. Rev.* **13**: 761–768.

Whittington, G., Hall, A. M. & Jarvis, J. (1993). A pre-Late Devensian pollen site from Camp Fauld, Buchan, north-east Scotland. *New Phytol.* **125**: 867–874.

Wiksten, J. (2000). Heather follies of 1861. *Yearb. Heather Soc.* **2000**: 35–42.

Williams, F. N. (1910). Ericales … Ericaceae [*Erica, Calluna*]. *Prodromus florae britannicae* vol. **1**: 449–455. C. Stutter, Brentford.

Wilson, D. (2000). New Canadian heathers. *Yearb. Heather Soc.* **2000**: 18–21.

Wilson, D. G., Marchant, R. & Oldfield, F. (1973). Fossil seeds of *Erica* from the Cromer Forest Bed series. *New Phytol.* **72**: 1235–1237.

Wilson, K. (1998). Heathers in a plantsman's life. *Yearb. Heather Soc.* **1998**: 23–25 (Revised & republished in *Heather News Quart.* **29** (4): 3–5.)

Woodell, S. R. J. (1958). *Daboecia cantabrica* K. Koch (*D. polifolia* D. Don; *Menziesia polifolia* Juss.). *J. Ecol.* **46**: 205–216.

Woolley, R. V. G. (1939). The possibilities of Daboecias. *Gard. Chron.* **106**: 82.

Woolner, L. R. (1974). *Erica scoparia*. *Yearb. Heather Soc.* **2** (3): 13–15.

Wulff, E. M. T. (2007). In praise of 'Anette'. *Heather News Quart.* **30** (1): 8.

X. (1884). Winter bedding. *Gard. Chron.*, n.s., **21**: 371–372.

Yaltirik, F. (1967). Contributions to the taxonomy of woody plants in Turkey. *Notes Roy. Bot. Gard. Edinburgh* **28**: 9–16.

Yates, G. (1978). *Pocket guide to heather gardening*. Tabramhill Gardens, Ambleside.

Yates, G. (1985). *The gardener's book of heathers*. Frederick Warne, London.

ACKNOWLEDGEMENTS

This monograph could not have been undertaken and published without grants from The Appleyard Fund of The Linnean Society of London, from The Authors' Foundation of The Society of Authors, and from The Heather Society. Research was also enabled less directly by The Heather Society and the Society for the History of Natural History. To all four societies I am most grateful for the generous support.

I am also grateful to the following individuals, noted in no particular order, for their contributions, comments and criticisms: Prof. Dr Salvador Rivas-Martínez (CIF-Phytosociological Research Center, Madrid, Spain), Dr Luis Ramon-Laca (Madrid, Spain), Dr Torbjørn Alm (University of Tromsø, Norway), Terhi Ryttäri (Helsinki, Finland), Dr Gerhard Cadée (NIOZ, Netherlands), Piet Zumkier (Terschelling, Netherlands), Prof. Graeme Whittington (University of St Andrews, Scotland), Dr Paul Coxon (Department of Geography, Trinity College, Dublin, Ireland), Prof. John Parnell (School of Botany, Trinity College, Dublin, Ireland), Dr Naomi Kingston (OPW, Dublin, Ireland), Dr Tony Orchard (Centre for Plant Biodiversity Research, Canberra, Australia), Dr Yves Frenot (Institut Polaire Français Paul-Emile Victor, Plouzane, France), Dr Andreas Kaffke (Ernst Moritz Arndt University, Greifswald, Germany), Dr Charlie Jarvis, Vicki Papworth and Roy Vickery (Botany Department, The Natural History Museum, London), Anja Maubach (Wuppertal, Germany), Dr Jacqueline van Leeuwen (University of Berne, Switzerland), Dr Mats Hjertson (Botany Section, Museum of Evolution, Uppsala), Dr Keith Harris, Miss Rose Murphy, Ian Bonner, Nigel Brown (Treborth Botanic Garden, Bangor, North Wales), Dr Erika Schneider (University of Karlsruhe, Rastatt, Germany), Dr Donald Mackay (Pleasantvile, New York, USA), Prof. John McNeill (Royal Botanic Garden, Edinburgh, Scotland), Dr Carlo Violani (University of Pavia, Italy), Dr Stephen Jury (University of Reading, England), Mag. Christa Riedl-Dorn (Archiv Naturhistorischen Museum, Vienna, Austria), Maria Lomonosova (Herbarium, Central Siberian Botanic Garden, Novosibirsk, Russia), Magda Sagarzazu (Archivist, Canna House, The National Trust for Scotland, Isle of Canna), Dr Mike Nesbitt (Royal Botanic Gardens, Kew), Prof. Dr Elmar Robbrecht and Dr Piet Stoffelen (National Botanic Garden of Belgium, Meise), Dr W. R. P. Bourne (Dufftown, Scotland), Laura Wiggins (Entomology Library, Natural History Museum, London), Dr Peter Barnard (Leigh-on-Sea, England), Clara Gaspar (Univeristy of Sheffield, England), Prof. Livio Ruggiero (Museo dell'Ambiente dell'Università, Lecce, Italy), Dr Jaimé Faguñdez (Santiago de Campostela, Spain), Matthew Baker and Andrew Crane (Tasmania), Robyn Barker (South Australia), and, most recently, Dr Michael D. Pirie (Department of Biochemistry, University of Stellenbosch, South Africa) for discussion about the phylogeny of *Erica* based on his continuing DNA studies. My thanks are here reiterated to Ella May Wulff and Dr Ted Oliver for sharing their knowledge of heathers and reading the early drafts of this work.

I must also thank my past and present colleagues including the curators, foremen and gardeners at the National Botanic Gardens, Glasnevin, Dublin, both for growing heathers (especially those from Spain collected in 1982) and for providing invaluable assistance with herbarium and bibliographic sources: in particular, I am grateful to Colette Edwards, Alexandra Caccamo, the late Grace Pasley, Brendan Sayers, Matthew Jebb and Donal Synnott.

Regarding hybrids and cultivars both extant and extinct, and their creation and cultivation, I am indebted to Prof. John Griffiths (Leeds, Yorkshire), David Wilson (Chilliwack, British Columbia, Canada), Barry Sellers (London), Kurt Kramer (Edewecht, Germany), Karl Lortz (Shelton, Washington, USA), Mrs Brita Johansson (Vargön, Sweden), Mrs Heather Dobbin and her mother, the late Mrs Eileen Porter (Carryduff, Northern Ireland), Richard Harnett (Kernock Park Plants, Cornwall), Jean McCrindle (Dunblane, Perthshire), Dr Colin Rogers (Tintwistle, Derbyshire), John Hall jun. and his father (Bordon, Hampshire), Allen Hall (Loughborough), Phil Joyner (Totton, Hampshire), Susie Kay (Connemara, Ireland), Richard Canovan (Swindon, Wiltshire), Peter Bingham (Spalding, Lincolnshire), David Edge (Wimborne, Dorset), Daphne Everett (Bringsty, Herefordshire), Jean Julian (now Preston; Askham Richard, Yorkshire), and, last but not least, Anne and David Small (Creeting St Mary, Suffolk).

To Philip Oswald for invaluable help with Latin texts and etymology, and with both ancient and modern Greek; Ian Johnston (Honorary Research Associate, Malaspina University-College, Nanaimo, British Columbia, Canada) for permission to quote his translation of Aeschylus's *The Oresteia*. For permission to publish text or images, I also acknowledge Dr Patricia Macdonald (Aerographica Aerial Photography) (Fig. 1); John Herriott Photography & Design (Bandon, County Cork) (Fig. 24); The National Trust for Scotland (Fig. 33); The Ashmolean Museum, Oxford (Fig. 28); The British Library (Fig. 116); The Linnean Society of London (Plate 3); Transport for London (London's Transport Museum) (Fig. 25); Royal Irish Academy, Dublin (Fig. 104); Oxford University Press (for the quotation from *Oxford English dictionary*, p. 12 (note 1)).

While I have very little input into the colour plates by Christabel King, I express my gratitude to her, and also to Joanna Langhorne for their splendid illustrations for this monograph, and to the Kew Publications staff who have seen this book into press, especially Ruth Linklater, Lloyd Kirton, and Chris Beard for her excellent layout and design. Other friends, either artists or photographers, have contributed pictures to enhance this book: Wendy Walsh, with whom I have collaborated with the greatest pleasure for more than three decades, Deborah Lambkin (for the new plate of *Erica erigena*), Brita Johansson (for her studies of the florets and leaves of *Erica carnea*; Fig. 217), Allen Hall, Christopher Massy-Beresford, Dr Helmut Pirc, Michael Kesl, Dr Martyn Rix, Kostas Konstantinidis and Martyn Turner for the bookplate tailpiece.

Cultivar descriptions are derived mainly from The Heather Society's database of cultivar names, which in turn is based on Vickers (1976). Descriptions of some cultivars comes from the several editions of the *Handy guide to heathers* (Small & Small 1998, 2001).

I owe immense gratitude to three people whose deaths during the last decade have deprived the community of heather enthusiasts of much wisdom and knowledge: Dr Paddy Coker, David McClintock and David Small.

Ultimately my wife Sue deserves the greatest thanks, for everything: without her support this work would never have been possible.

INDEX OF SCIENTIFIC NAMES

(page nos for principal references to taxa and those including illustrations are in **bold**)

GENERAL INDEX
Common names, people, places and publications

(book and journal titles are italicised; page nos of principal references to species and those including illustrations are in **bold**)